Developing drought-resistant cereals

It is widely recognised that agriculture is a significant contributor to global warming and climate change. Agriculture needs to reduce its environmental impact and adapt to current climate change whilst still feeding a growing population, i.e. become more 'climate-smart'. Burleigh Dodds Science Publishing is playing its part in achieving this by bringing together key research on making the production of the world's most important crops and livestock products more sustainable.

Based on extensive research, our publications specifically target the challenge of climate-smart agriculture. In this way we are using 'smart publishing' to help achieve climate-smart agriculture.

Burleigh Dodds Science Publishing is an independent and innovative publisher delivering high quality customer-focused agricultural science content in both print and online formats for the academic and research communities. Our aim is to build a foundation of knowledge on which researchers can build to meet the challenge of climate-smart agriculture.

For more information about Burleigh Dodds Science Publishing simply call us on +44 (0) 1223 839365, email info@bdspublishing.com or alternatively please visit our website at www.bdspublishing.com.

Related titles:

Achieving sustainable cultivation of maize Volume 1: From improved varieties to local applications
Print (ISBN 978-1-78676-008-1); Online (ISBN 978-1-78676-011-1, 978-1-78676-010-4)

Achieving sustainable cultivation of wheat Volume 1: Breeding, quality traits, pests and diseases
Print (ISBN 978-1-78676-016-6); Online (ISBN 978-1-78676-018-0, 978-1-78676-019-7)

Achieving sustainable cultivation of sorghum Volume 1: Genetics, breeding and production techniques
Print (ISBN 978-1-78676-120-0); Online (ISBN 978-1-78676-122-4, 978-1-78676-123-1)

Achieving sustainable cultivation of rice Volume 1: Breeding for higher yield and quality
Print (ISBN 978-1-78676-024-1); Online (ISBN 978-1-78676-027-2, 978-1-78676-026-5)

Advances in breeding techniques for cereal crops
Print (ISBN 978-1-78676-244-3); Online (ISBN 978-1-78676-246-7, 978-1-78676-247-4)

Achieving durable disease resistance in cereals
Print (ISBN 978-1-78676-601-4); Online (ISBN 978-1-78676-603-8, 978-1-78676-604-5)

Achieving sustainable cultivation of barley
Print (ISBN 978-1-78676-308-2); Online (ISBN 978-1-78676-310-5, 978-1-78676-311-2)

Chapters are available individually from our online bookshop:
https://shop.bdspublishing.com

climate•SMART•publishing
IN AGRICULTURAL SCIENCE

BURLEIGH DODDS SERIES IN AGRICULTURAL SCIENCE
NUMBER 124

Developing drought-resistant cereals

Edited by Professor Roberto Tuberosa, University of Bologna, Italy

burleigh dodds
SCIENCE PUBLISHING

Published by Burleigh Dodds Science Publishing Limited
82 High Street, Sawston, Cambridge CB22 3HJ, UK
www.bdspublishing.com

Burleigh Dodds Science Publishing, 1518 Walnut Street, Suite 900, Philadelphia, PA 19102-3406, USA

First published 2023 by Burleigh Dodds Science Publishing Limited

Library of Congress Control Number: 2023940854

British Library Cataloguing in Publication Data
A catalogue record for this book is available from the British Library

ISBN 978-1-78676-985-5 (print)
ISBN 978-1-78676-988-6 (PDF)
ISBN 978-1-78676-987-9 (ePub)
ISSN 2059-6936 (print)
ISSN 2059-6944 (online)

DOI: 10.19103/AS.2022.0109

Typeset by Deanta Global Publishing Services, Dublin, Ireland

Contents

Part 2 Techniques for improving resistance

Series list

Title	Series number
Achieving sustainable cultivation of maize - Vol 1 From improved varieties to local applications Edited by: Dr Dave Watson, CGIAR Maize Research Program Manager, CIMMYT, Mexico	001
Achieving sustainable cultivation of maize - Vol 2 Cultivation techniques, pest and disease control Edited by: Dr Dave Watson, CGIAR Maize Research Program Manager, CIMMYT, Mexico	002
Achieving sustainable cultivation of rice - Vol 1 Breeding for higher yield and quality Edited by: Prof. Takuji Sasaki, Tokyo University of Agriculture, Japan	003
Achieving sustainable cultivation of rice - Vol 2 Cultivation, pest and disease management Edited by: Prof. Takuji Sasaki, Tokyo University of Agriculture, Japan	004
Achieving sustainable cultivation of wheat - Vol 1 Breeding, quality traits, pests and diseases Edited by: Prof. Peter Langridge, The University of Adelaide, Australia	005
Achieving sustainable cultivation of wheat - Vol 2 Cultivation techniques Edited by: Prof. Peter Langridge, The University of Adelaide, Australia	006
Achieving sustainable cultivation of tomatoes Edited by: Dr Autar Mattoo, USDA-ARS, USA and Prof. Avtar Handa, Purdue University, USA	007
Achieving sustainable production of milk - Vol 1 Milk composition, genetics and breeding Edited by: Dr Nico van Belzen, International Dairy Federation (IDF), Belgium	008
Achieving sustainable production of milk - Vol 2 Safety, quality and sustainability Edited by: Dr Nico van Belzen, International Dairy Federation (IDF), Belgium	009
Achieving sustainable production of milk - Vol 3 Dairy herd management and welfare Edited by: Prof. John Webster, University of Bristol, UK	010
Ensuring safety and quality in the production of beef - Vol 1 Safety Edited by: Prof. Gary Acuff, Texas A&M University, USA and Prof. James Dickson, Iowa State University, USA	011
Ensuring safety and quality in the production of beef - Vol 2 Quality Edited by: Prof. Michael Dikeman, Kansas State University, USA	012
Achieving sustainable production of poultry meat - Vol 1 Safety, quality and sustainability Edited by: Prof. Steven C. Ricke, University of Arkansas, USA	013
Achieving sustainable production of poultry meat - Vol 2 Breeding and nutrition Edited by: Prof. Todd Applegate, University of Georgia, USA	014

Introduction

Drought remains the biggest single threat from climate change to the production of the key cereal crops that feed mankind. Cereals also respond in complex ways to drought stress, making improved drought tolerance a challenging goal. With many cereals recognised as staple food crops due to their nutritional value, more research is required into improving drought tolerance as a means of ensuring the future food security of mankind.

This volume reviews the wealth of research which addresses how to overcome the challenge of achieving drought tolerance in order to mitigate climate change effects in cereal production. The chapters are split into two parts: Part 1 chapters focus on understanding different mechanisms of drought tolerance, such as options to increase crop yield, the role of plant growth regulators, drought-induced proteins and the genetics/genomics of drought tolerance. Chapters in Part 2 review techniques for improving resistance, focusing specifically on advances in phenotyping and identifying genes that control root system architecture - the hidden and often neglected half - and photosynthetic response. Chapters also report and update on genomic selection, gene editing, genetic engineering, identifying genes for yield-related traits under drought stress conditions and also developing hybrids with improved performance under water deficits.

Part 1 Understanding mechanisms of drought tolerance

The first chapter of the book poses the question what is 'drought stress' and the current options available to increase crop yield. Chapter 1 provides a series of axioms and propositions that help to establish a framework in order to better understand how crop yield in the field can be improved through improving the crop's response to certain stresses such as drought. After this framework is established, the chapter then draws to a close by providing a renewed perspective for increasing crop drought resilience through developing research on the water-deficit limitation in crop yield. The chapter also highlights how identifying traits relating to potential drought-resilience is important in improving future crop yield.

The next chapter focuses on the role of plant growth regulators in adaptation to drought stress in cereals. Chapter 2 begins by describing how stress-tolerant genotypes can be defined and how researchers can choose the right genotype depending on their intended outcome. The chapter then reviews plant growth regulators and yield development of drought-stressed crops followed by a discussion on hormonal regulation of crop productivity and resilience. The chapter presents wheat as a case study to show how a hormonal signature (ABA and ethylene) can be used as a physiological proxy trait for crop improvement. It concludes by providing an overview of potential opportunities for future work in the research field.

Chapter 3 examines the genetics of drought tolerance in cereals. The chapter provides some of the latest findings related to the genetics of drought resistance of the world's most widely grown cereals. It begins by first reviewing the impact drought stress has on cereal crops, then goes on to provide separate sections on understanding and optimising drought tolerance in particular cereals. The chapter focuses on rice first due to its key role as an ancestral founder species and also its phylogenetically pivotal role for cereal evolution. It then goes on to review barley, durum wheat, bread wheat, sorghum and maize. The chapter concludes by emphasising how integrating advanced genomics and bioinformatics with sound phenotyping will lead to more effective haplotype selection in terms of drought-resistance.

The final chapter of Part 1 reviews the role of drought-induced proteins in regulating drought tolerance in cereals. Chapter 4 first highlights current drought tolerance mechanisms, focusing specifically on the various biochemical, physiological, morphological and anatomical mechanisms that can influence the potential adaptation of drought resistance in crops. The chapter then goes on to describe drought-related proteins and proteomic studies. This is followed by a section that summarises recent advances in drought-adaptive proteins using transgenesis using major cereal crops such as maize, rice and wheat. The chapter also provides an overview of recent advances in drought-adaptive proteins using genome editing.

Part 2 Techniques for improving resistance

The first chapter of Part 2 discusses advances in phenotyping to identify drought-resistance traits in cereal roots. Chapter 5 first describes current root phenotyping strategies and challenges, which is then followed by an overview of important root traits to improve drought adaptation, such as root system architecture, root anatomy and changes in roots in response to drought. A section on phenotyping in controlled ex-situ conditions is also provided, which is then followed by an overview of field phenotyping, three-dimensional imaging and root interactions with the soil as well as imaging and trait data extraction. The chapter also discusses above ground correlated traits in the field, trait by environment interactions and phenotyping methods for genomic selection and molecular breeding. A case study on late, deep root development in sub-tropical wheat is also included.

Chapter 6 explores identifying and exploiting genes controlling root system architecture for improving drought tolerance in cereals. It starts by highlighting root system ideotypes for water-limited environments, such as genes controlling root growth angle and roots in deep soil layers, root hairs, allocation of biomass to roots and its spatial distribution and the ability of roots to penetrate soils. This is then followed by an overview of root-derived hydraulic and hormonal signals, focusing specifically on the regulation of plant hydraulic conductivity and hydraulic lift in mixed cropping and agroforestry. A section on the plasticity of root growth is also included, where biomass allocation patterns are reviewed, followed by hydrotropism and hydropatterning, as well as developmental instability, cryptic variation and non in-heritable variation. The chapter also discusses root-microbe interactions, the role of roots in supporting yield stability across water regimes and ways to improve drought tolerance via breeding for specific root traits.

The subject of Chapter 7 is identifying and exploiting photosynthetic genes in improving drought resistance in cereals. It first describes the indicators for photosynthetic traits, then goes on to discuss the different factors that can affect photosynthetic efficiency. The chapter also considers how drought stress can affect photosynthetic assimilation efficiency and examines the photosynthetic metabolic pathways of photosynthesis associated with drought tolerance in crops. The chapter also looks at how maintaining photosynthetic rate under drought stress can contribute to crop drought tolerance. A section on how the engineering of C_4 traits into C_3 plants could improve crop drought tolerance is also provided, followed by an overview of drought tolerance genes related to photosynthesis. The chapter examines the use of quantitative trait loci (QTL) mapping for photosynthetic traits, considers how the stay-green phenotype can lead to increased crop yield under drought conditions and reviews how introducing this stay-green trait can enhance photosynthesis under drought stress. A case study on the enhancement of sorghum tolerance to post-flowering drought stress using stay-green QTL is also included.

Chapter 8 provides the state of the art on genomic prediction of estimated breeding values (GEBV) for selection (or genomic selection) as well as on the use of genetic engineering for both gene editing and transgenic breeding in main cereals crops. Genomic selection has been successfully used for breeding maize in drought-prone sites of Africa and Asia, while progress was noted for using GEBV for improving pearl millet, rice, sorghum and wheat germplasm. Transgenic maize-derived hybrids were released about one decade ago for growing in the US Corn Belt, and there are available transgenic rice and wheat germplasm though not yet released as new cultivars. Gene editing shows promising results for maize while research advances are noted in both rice and wheat. Combining genomic selection and gene editing may further elevate genetic gains while improving cereals crops for drought-prone environments elsewhere.

The next chapter focuses on identifying genes for yield-related traits under drought stress conditions, specifically drawing attention to durum wheat.

Chapter 9 begins by providing an overview of QTL related to yield in durum wheat, which is then followed by a review of how meta-QTL analysis can be used to improve the identification of QTL in durum wheat. A section on identifying QTL related to yield and drought-stress response is also provided. The chapter also highlights candidate genes for yield-related traits under drought stress conditions.

The final chapter of the book focuses on developing corn hybrids with improved performance under water deficits. Chapter 10 reviews the key stages in corn plant development when water deficits can most impact on productivity. It reviews progress made in genotyping and phenotyping that enable selection for improved productivity in water-limited growing conditions. Technical approaches to experimental design and data analysis that can facilitate discovery, validation and implementation of research outputs at a commercial level are also described.

Acknowledgements

We wish to acknowledge the following for their help in reviewing particular chapters:

- Chapter 3: Prof. Rajeev Varshney, Murdoch University, Australia
- Chapter 5: Dr Hannah Schneider, Wageningen University and Research, The Netherlands and Dr Fabio Fiorani, Forschungszentrum Jülich, Germany

Part 1

Understanding mechanisms of drought tolerance

Chapter 1

What is 'drought stress' and what are options to increase crop yield?

Thomas R. Sinclair, North Carolina State University, USA; and Michel E. Ghanem, Centre de coopération internationale en recherche agronomique pour le développement (CIRAD), France and Mohammed VI Polytechnic University, Morocco

1 Introduction

Drought is commonly at the top of the list of abiotic stresses for which crop 'resistance' needs to be developed to increase crop yields. Indeed, the focus of this book is explicitly on 'drought resistance,' which reflects the common

http://dx.doi.org/10.19103/AS.2022.0109.01

hope since humans first sowed seeds that the resulting plants would be able to cope with the possibility of inadequate water. While management techniques including irrigation resulted in progress toward this hope, until very recently, there has been little evidence of plant physiological modifications for improving crop performance under drought (Sadras and Richards, 2014; Hammer et al., 2021).

The challenge in trait modification to improve drought resistance results from the ambiguous nature of 'drought' and 'drought resistance.' Implicitly, it seems that drought is assumed to be a well-defined stress that can be resolved by specific trait modification. That is, drought resistance can be genetically improved in a straightforward manner by gene identification and plant improvement in the same way that plants are selected for specific disease or insect tolerance. However, crop response to water deficit is very complex since it involves multifaceted processes in plants, and drought itself is highly variable among growing seasons and in the temporal dynamics of water deficit within each season. The simple approach of exposing a population of plants to water-deficit stresses and subjecting the population to a G×E×M analysis is insufficient to give field-relevant insight and predictions about crop improvement. While drought resistance can be characterized by some very specific water-deficit criteria under controlled-environment conditions, the results of these experiments have not translated to increased crop yield under field water deficits (Tardieu et al., 2018).

The failure in achieving yield increase under water deficit in the field from trait selection in a controlled environment in nearly all cases exposes an under appreciation of the nature of water deficit in the field. The water conditions to which a crop is subjected in the field are ever-changing through a growing season as a result of the temporal dynamics of soil water status, plant development and growth, and environmental conditions. Due to the temporal dynamics influencing the development of water deficit in the field, it is essentially impossible to define a static experimental condition to test for drought resistance that is applicable to the range of conditions that may be experienced in the field. Nevertheless, this situation has not deterred a steady stream of drought studies based on static plant responses observed under stable environmental conditions (Ghanem et al., 2015; Poorter et al., 2016; Chiang et al., 2020; Langstroff et al., 2022).

In an attempt to gain better insight about improving crop yield in the field under the almost inevitable condition of soil water deficit, we propose in this review a series of axioms and propositions that help to establish a framework for understanding under field conditions water deficit, crop response and realistic expectations for yield increase. These axioms and propositions are presented as a reminder of the critical importance of the temporal dynamics of plant development and water status under field conditions. These axioms and propositions lead to an enhanced perspective in evaluating the relative advantage of specific trait modifications for altering crop yield.

2 Axiom I: The term 'drought' is essentially meaningless in describing the temporal dynamics of ever-changing water deficit in the field and crop responses to the infinity of possible water-deficit scenarios

'Drought' and 'drought resistance' are terms that only offer a vague, generic impression of water status in the field. These terms are useful only in the analogous sense of defining a 'vehicle.' While a vehicle is a machine having wheels and a motor, such a definition does little to narrow the nature of the machine needed for your desired purpose. Can a Ferrari be used to work in a field to sow a crop? In reality, the terms drought and drought resistance do little to define under field conditions the nature of the water-deficit stress and the crop response (Passioura, 2020). The nature of drought is essentially infinitely variable due to the variability in the temporal dynamics of the weather and the seasonal progression of crop development and growth.

There is no such thing as a unique, repetitious drought scenario for a location; but rather a suite of water-deficit scenarios that vary in intensity and timing and are modified by soil characteristics, weather and management practices. To complicate matters further, new weather extremes due to climate change are expanding the range of drought scenarios to which a crop may be subjected through a growing season. To improve crop response to water deficit, there is a need to consider first the stochastic and contextual nature of the development of soil water deficit in the specific climate and soil in which the crop is to be grown.

One approach advocated to deal with the impact of environmental complexity on breeding for yield increase has been labelled as 'envirotyping.' Envirotyping calls for a measurement of "all environmental factors that affect plant growth and production" (Xu, 2016) and accounting for these in defining phenotypic response. Aside from the ambiguity of such a task, there is no accounting for the essentially infinite possibilities of temporal dynamic interactions through the growing season among all the abiotic and biotic environmental factors that influence crop development and growth. Envirotyping is a statistically challenging attempt to represent crop behavior without grappling with the temporally dynamic and unique nature of the environment and plant processes through each individual growing season. While expanded consideration of environmental variability on crop growth has been applied to multienvironment trials (Cooper et al., 2016; Resende et al., 2021), there remains uncertainty in the application of the approach in what is an inherently empirical analysis.

3 Axiom II: There is no universal 'drought resistance' trait that always results in yield increase

The variability in drought scenarios and the complexity of the interaction between the environment and plants mean that the goal of identifying a specific plant characteristic, sometimes referred to as an 'ideotype,' that is expected to always result in yield increase under water deficit is futile. As stated by Tardieu (2012), the 'most relevant question on drought tolerance is probably not "Does a given allele or trait confer drought tolerance?" but rather, "Does a given allele confer a positive effect on yield in an appreciable proportion of years/scenarios in a given area?"' The year-to-year and site-to-site weather variability in genetic analyses of plant performance in both current and future climatic scenarios need to be taken into account (Millet et al., 2016, for example).

While in one growing season a trait may result in a positive yield response, the water-deficit scenario and plant status will differ in the next growing season and the crop response will inevitably be different in some way. In fact, a positive trait in one growing season may well be negative in the next growing season (Tardieu, 2012; Tardieu et al., 2018). As an aside, such variation completely undermines the requirement imposed by some scientific journals that somehow 2 years of experimental data improve insight about drought response. To resolve the issue of drought resistance requires an understanding of the physiological bases of yield response over the full range of weather conditions that are experienced at any given location. The criterion for crop improvement will almost certainly be better indicated by the probability of yield increase in a given location over a number of growing season rather than average yield across seasons.

The year-to-year variation in response to drought was well illustrated in a test by Morgan (2000) of the yield difference associated with osmotic accumulation in a number of wheat (*Triticum aestivum* L.) lines grown at a number of sites in New South Wales, Australia. The ratio of yield among genotypes expressing higher osmotic accumulation and lower osmotic accumulation varied across environments between positive and negative yield responses. Based on the data of Morgan (2000), it was not possible to claim a benefit from osmotic adjustment unless the temporal pattern of water deficit was known for each individual growing season.

The variability in response to specific traits was further illustrated in a geospatial simulation study of the impact of a number of physiological traits in soybean (*Glycine max* L. Merr.) yield in the United States (Sinclair et al., 2010). No plant trait was identified that universally resulted in an increase or decrease in yield across the 50 years of simulated growing season at any of the 2655 locations across the United States. For example, increasing rate of daily extension of rooting depth from 20 to 25 mm tended to result in yield increase

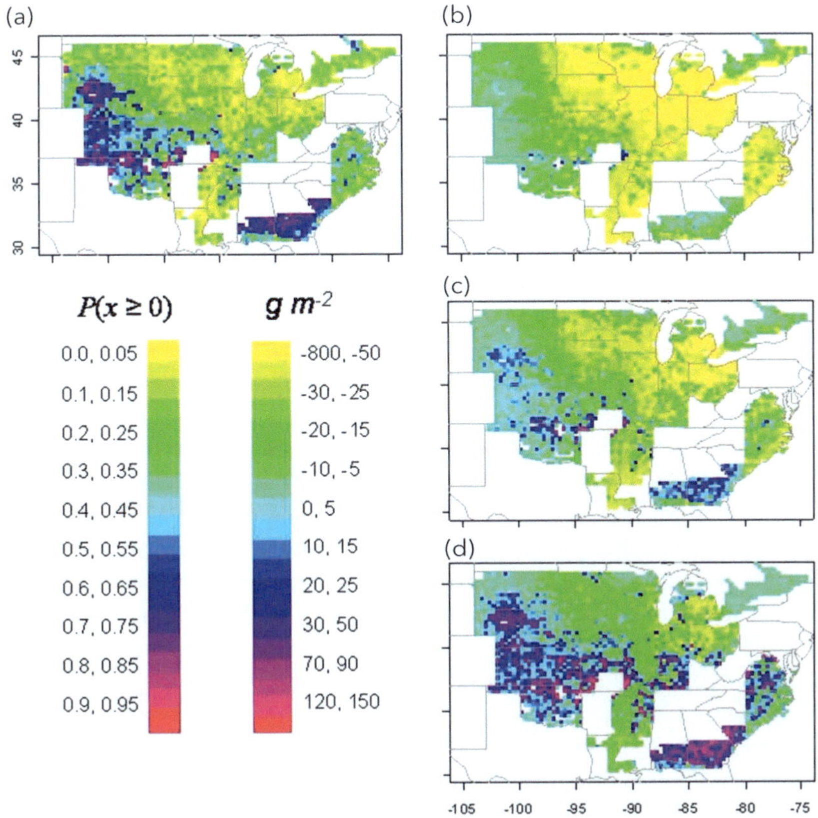

Figure 1 Simulation results for soybean yield with daily root extension increased from 20 mm d^{-1} to 25 mm d^{-1} (Sinclair et al., 2010). Panel (a) is the probability of yield increase with the legend color bar on the left of the two color bars. The absolute yield change (g m^{-2}) for three percentile rankings of 75% (wet, Panel b), 50% (median, Panel c) and 25% (dry, Panel d) with the legend color bar on the right of the two color bars.

at drier locations with a probability of yield increase greater than 50% (Fig. 1). However, in most locations in the United States, yield tended to be *decreased* by a greater rate of root extension because a more rapid increase in the depth of rooting allowed greater soil water extraction early in the growing season resulting in the crop being more vulnerable to late-season water deficits.

4 Axiom III: Be wary of 'drought trait' suggestions that have not been subjected to a probabilistic analysis using realistic varying weather scenarios

Despite the temporal variability of water deficit in the field, claims are still made about the possibility of various 'physiological' trait alterations leading to universal crop yield improvement under water-deficit conditions (Gupta et al., 2020). One of the more extensive discussions of individual drought traits was presented by Ludlow and Muchow (1990), who considered more than 16 traits to improve crop yield under drought. Their presentation was a valuable review of plant trait modifications that might be associated with yield alteration under water-deficit conditions. However, their intuitive approach in examining drought traits tended to rely on a rather 'static' view of drought stress. That is, their views may have applied to a specific drought condition they had in mind, but the variation and complexity of the impact of changing water deficit through each growing season was not a part of their consideration.

The assessment of the variability in crop response to water deficit is likely not well resolved by measuring average yield across several environmental scenarios since the nature of the variable response is disguised in the average value. A more appropriate metric for assessing the resilience of crops subjected to water deficit is to document the probability of yield increase over the range of environmental conditions at a location, as noted by Tardieu (2012). Probabilistic assessments either based on simulations as shown in Fig. 1 or experimental observations allow an assessment of an alteration in a plant trait or management regime across seasons at a given location. Those alterations resulting in a high probability of yield increase across growing seasons provide information for the grower to make a risk assessment in deciding whether to adopt a new cultivar or management, even if the average yield increase is not large. Specifically, what is the probability across growing seasons of yield increase in lower yield seasons when economic risk is greatest vs. the probability of possible yield decrease in higher yield seasons?

5 Axiom IV: Be wary of simplistic hypotheses for improving drought resilience such as osmotic adjustment

Failure to consider the temporal dynamics of soil water content and the fundamental physiological impact on crop response in the field can result in confusion about the contribution of a trait alteration on yield. One such example of such confusion is illustrated in conclusions about hypothesized benefits resulting from plant osmotic accumulation (Serraj and Sinclair, 2002; Turner, 2018). The confounding interactions in the seasonal dynamics of soil water availability and plant growth hinder an unqualified conclusion about the benefit of osmotic accumulation in plants. One of the best documentations of this variation in response is illustrated in the study with wheat by Morgan (2000), which was previously referenced.

One common experimental approach to assess the benefit of osmotic accumulation is to subject plants to extreme water deprivation and identify those genotypes that survive. While discriminatory results among genotypes can be identified, the survival criterion has virtually no relevance in the field. In the field, such survival conditions develop only after a prolonged time of lack of rainfall during which there have been major losses in crop physiological activity. Survival of such severe, extended drought allows little time or capability for crop recovery and grain growth so that extremely low yields will necessarily be the outcome of such severe drought even if the crop survives. Achieving crop survival of severe water deficit as a result of osmotic accumulation, or any other trait, is likely to have virtually no commercial relevance.

It is also often suggested that osmotic accumulation allows plants to remain turgid when subjected to water deficit so that leaf gas exchange and tissue expansion can be sustained during drought. In fact, the retention of turgidity with the onset of water deficit in many cases may be exactly the opposite of what will benefit a crop. If water deficits are developing, the more appropriate response by the crop may be initiation of water conservation of the remaining soil water by decreasing crop water loss, i.e. induce partial stomatal closure and limit leaf area expansion. A water conservation approach allows sustained physiological activity as water deficit continues to become more severe. In the simulations of Sinclair et al. (2010) for soybean, delayed stomata closure with soil drying as a suggested benefit of osmotic accumulation resulted in a preponderance of *decreased* soybean yield at most locations in the United States. Generally, drought response was benefited by early stomata closure in a soil drying cycle; just the opposite response of what is suggested as a benefit of osmotic accumulation. But as with all traits, the results are variable, and simulations for wheat in Tunisia showed at some locations that there was a probability of yield increase resulting from a shift in stomata closure to a lower soil water content (Sadok et al., 2019).

A more focused suggestion for increased cell turgor as a result of osmotic accumulation has pointed to the meristem of root tips. The hypothesis is that increased turgor in the root meristem will allow greater root extension rate and give greater access to soil water. While differences in root tip osmotic potential have been measured in germinating seedlings, little evidence exists for genotypic differences in more mature plants. Beseli et al. (2019) failed to find differences in root tip osmotic potential between two maize lines, which differed in leaf osmotic potential, at several stages during vegetative development. Even more concerning is that simulations of differences in root extension rate by soybean as discussed previously showed no clear advantage, commonly a disadvantage, for greater root extension rate except in the most arid region of the United States (Fig. 1).

Another suggested benefit of osmotic accumulation in the plant is an increased hydraulic potential gradient between the soil and the plant as a result of osmotic accumulation in the plant. Certainly, an increased hydraulic potential gradient will allow greater rates of soil water extraction at equivalent soil water contents. A benefit of this greater soil water extraction rate is expected to be greater leaf gas exchange and crop growth. However, within the temporal dynamic context of the growing season and increasing severity of soil water deficit, the greater rate of soil water extraction as a result of osmotic accumulation will potentially cause the crop to reach more severe soil water deficits earlier in the growing season and ultimately result in a disadvantage in terms of sustained crop physiological activity.

Beseli et al. (2020) observed such a differential response to drying soil by a pair of maize (*Zea mays* L.) near-isolines that differed in leaf osmotic accumulation. In their controlled-environment studies, the near-isoline with high osmotic accumulation initially had greater transpiration rates than the low osmotic-accumulating near-isoline (Fig. 2), as expected. However, since the soil dried faster to a lower soil water content for the high osmotic-accumulating near-isoline as compared to the low osmotic-accumulating near-isoline, the relative transpiration rates of the two near-isolines was eventually reversed. Full stomata closure of the high osmotic-accumulating near-isoline was 3 to 6 d 'earlier' than the low osmotic-accumulating near-isoline. While the high osmotic-accumulating near-isoline had an advantage early in the soil drying cycle, the low osmotic-accumulating near-isoline had a clear advantage late in the soil drying cycle when the development of water deficit was prolonged. These observed differences for these maize near-isolines grown in controlled environments were fully consistent with the observations of the behavior of the same near-isolines grown under field conditions (Chimenti et al., 2006).

Given the importance of the temporal dynamics of water use as described earlier, the framework for understanding and improving crop resilience in response to developing water deficit needs to be reviewed. In the following

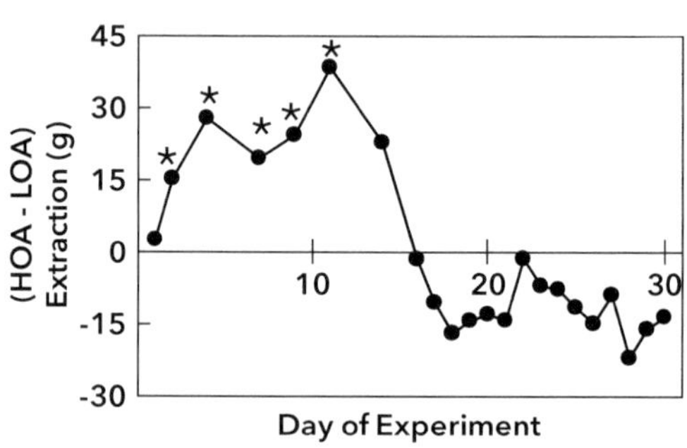

Figure 2 Difference in daily transpiration between the high osmotic accumulator line (HOA) and low osmotic accumulation line (LOA) measured over a 30-day soil drying cycle (Beseli et al., 2020). Difference values above 0 are cases where the HOA line had higher transpiration rate than the LOA line. The asterisks indicate differences with *P*-values ≤ 0.05.

sections, we offer a discussion of several 'propositions' for evaluating and understanding crop yield response to the temporal dynamics of water deficit under realistic field conditions.

6 Proposition A: Be wary of phenomenological equations for insight about the mechanistic basis for overcoming water deficit

The critical importance of temporal dynamics through the growing season of changing soil water content and of plant development and growth are difficult to capture in a single equation, especially one that does not explicitly consider temporal changes. The challenge is even greater, if not impossible, when the model is a phenomenological description of the static response to crop growth under limited water availability. One model that has been widely used that highlights the challenges in increasing crop yield with limited water availability is the phenomenological equation presented by John Passioura (1977). While this equation offers oversight about features of the cropping system impacting yield, the mechanistic impact of changing conditions through the growing season is not considered.

In the phenomenological equation given by Passioura (1977), yield (Y) is described as a function of the amount of soil water extracted for transpiration (T), harvest index (HI) and water use efficiency (WUE).

$$Y = T \bullet HI \bullet WUE \tag{1}$$

An important feature of this phenomenological equation is that it describes the fact that yield cannot exceed the total amount of water available for transpiration. This is an important point to remember when considering maximum yield increases that might be realistically achieved under water-deficit conditions. Ultimately for many field situations, yield will be rigidly limited by the amount of available water.

Another important aspect of Eqn (1) is the dependence of achieving a high yield on having a high HI. Interestingly, the HI of modern crops remains high and nearly stable even when fairly severe water deficits are experienced (Spaeth et al., 1984; Sinclair et al., 1990). This HI stability results from the fact that yield is often well correlated with total crop mass so their ratio remains nearly constant. However, in some cases where the intercept of the ordinate of the graph of seed mass versus total crop mass is negative, the HI does decrease with decreasing crop mass, although not in a linear fashion. For instance, in sweet corn, Jafarikouhini et al. (2021) found over a wide range of growth conditions that $Y = 0.458 \bullet W - 112$, where W equals total plant dry weight (g m^{-2}). Rearranging this equation, $HI = 0.458 - 112 / W$ shows that large decreases in HI occur only when W is small.

The above suggestion that HI may be fairly stable generally applies no matter the timing of the imposition of water deficit. That is, water deficit during what is often referred to as the 'critical period' around anthesis also commonly does not greatly alter the stability of HI (Sinclair et al., 1990). The conclusion that anthesis is a critical period for water deficit is a result of the fact that this period is usually associated with maximum leaf area and growth rate so that water deficit at this time develops most rapidly and can have the largest impact on crop mass accumulation. Therefore, due to decreased mass accumulation during the period around anthesis, water deficit has a large negative impact on yield even when HI remains stable. Under field conditions, it is necessary to document possible variation of HI in response to water deficits to resolve whether sensitivity of the processes involved in the transition to reproductive growth is actually a main contributor to crop yield loss.

The remaining term defining yield in Eqn (1) is WUE. Due to the static nature of the phenomenological equation, it is probably best to consider this term as a 'placeholder.' Nevertheless, many studies have undertaken extensive searches for genotype variation in WUE based on static measures of leaf gas exchange. However, the crucial dependence of transpiration rate on atmospheric vapor pressure deficit (VPD) (de Wit, 1958), which is not explicitly included in the phenomenological equation, has to be considered a major component, if not the dominating component, in determining crop water use and WUE.

The impact of VPD in Eqn (1) has been overlooked by some even though VPD is a key variable in determining plant water use. This oversight may occur because VPD is considered by some as an 'environmental' variable. However, as discussed later, limiting crop water loss rate by partial stomata closure in response to elevated VPD, results in a lowered value for the 'effective VPD' driving transpiration rate. Depending on the drought scenario, the conservation of soil water early in the growing season as a result of partial stomata closure and decreased effective VPD allows for increased water availability later in the growing season to sustain physiological activity (Sinclair, 2017). The possible variation of effective VPD as a result of plant regulation is an especially important physiological variable for improving crop yields under water deficit and confounds attempts to define genotype WUE.

7 Proposition B: A more mechanistic description of yield under water deficit

A mechanistically based description of water-deficit limitation on yield based on leaf gas exchange can be obtained from the derivation of Tanner and Sinclair (1983).

$$\int Y\,dt = HI \bullet k \bullet \int (T / VPD)\,dt, \tag{2}$$

where k = mechanistic coefficient explicitly defined by crop physical and physiological characteristics (Pa). The value of the variable k includes dependence on two key variables, which are the energy content of the synthesized plant products and the CO_2 concentration ratio between the leaf interior and leaf-surrounding atmosphere. In seed crops, seed composition is generally narrowly restricted by the commercial requirements in the uses of the grain. Consequently, changes in seed composition are not generally viable options for alteration in many crop species to allow major increases in k.

The CO_2 concentration ratio in defining k is determined to a large extent by the C-assimilation pathway, i.e. C3 vs. C4. Only in cases where the CO_2 concentration of the leaf interior is unusually high relative to the expected ratios for C3 and C4 species is there an opportunity to increase k. Since genotypes with high CO_2 ratios would almost necessarily have low growth rates, and therefore low yields, a focus on lowering this ratio is likely to have a primary benefit in simply improving the deficient carbon assimilation rate of these genotypes.

The mechanistic expression of Eqn (2) indicates explicitly that a major variable influencing yield is VPD. Conditions where VPD is high cause higher transpiration rates and a decrease in allowable yield for a given amount of water, i.e. T. As discussed previously, however, VPD is not necessarily simply determined by atmospheric conditions. Partial stomata closure under elevated VPD will result in a decrease in the 'effective VPD' gradient defining transpiration rate. Such a decrease in effective VPD will shift the seasonal pattern of water use to greater water availability later in the growing season and thereby allow sustained physiological activity during reproductive development and avoid premature decrease in seed growth (Sinclair, 2017). This approach has the possibility of resulting in more effective use of the season-long availability of limited soil water, which Blum (2009) and Sinclair (2018) emphasized as a critical perspective for increasing crop yield.

8 Proposition C: Increase effectiveness of water use as a result of plant traits resulting in early-season water conservation

There are several mechanisms that result in early-season soil water conservation. Two traits that decrease early-season water loss are slow development of leaf area and low plant growth rate. These two traits were examined by Muchow (1989) by measuring radiation interception and crop radiation use efficiency. He found that the values for these parameters were lowest in pearl millet (*Pennisetum americanum* L. Leeke) followed by sorghum (*Sorghum bicolor* L. Moench), and greatest in maize. This ranking matches with the general view of crop species adaptability to water-deficit environments with the species with the greatest early-season water conservation resulting in the greatest drought resilience.

Decreased development of leaf area in stay-green sorghum near-isolines was found by Borrell et al. (2014) to result in greater yield as compared to the normal parent under terminal drought. Lower leaf area of the stay-green genotypes resulted from fewer culms m^{-2} and smaller leaves at the top of the canopy. The key consequence of lower leaf area was decreased transpiration rate and lower water use during the pre-anthesis period. The advantage of the temporal shift in water use from the pre-anthesis period to grain fill was found especially at yields less than 5.5 t ha^{-1}. The lack of advantage at higher yield levels likely resulted from the fact that the slow leaf area development and conserved water use at higher water availability can result in an offsetting lower crop growth rate.

Slow root extension is another trait that results in water conservation when low water extraction from the soil limits the transpiration rate. As discussed previously, slow root extension conserves deeper soil water reserves for use late in the season during reproductive growth. A slow root extension trait, however, has the negative feature of an inability to have greater rates of water extraction if high rainfall in a given season results in high levels of water availability deep in the soil.

A constitutive water trait that results in water conservation only when soil water deficits develop is an alteration in the response mechanism for the initiation of partial stomatal closure with soil drying. Commonly, most crop plants initiate partial stomata closure when the fraction of extractable soil water decreases to the range of 0.3 to 0.4 (Sadras and Milroy, 1996). This response appears to be coupled with a decrease in soil hydraulic conductance as the soil dries (Sinclair, 2005). However, some genotypes initiate partial stomata closure at high soil water contents so that their rate of soil water consumption is decreased earlier in the soil drying cycle, thereby resulting in soil water conservation. As a result of soil water conservation, plants being subjected to water deficit can have prolonged physiological activity if water deficit continues to increase. In all crop species studied except soybean, at least a few genotypes have been identified that initiate partial stomata closure at higher soil water contents (Sinclair, 2017).

Figure 3 shows contrasting genotypes of peanut (*Arachis hypogaea* L.) in their initiation of the decrease in transpiration rate with soil drying (Devi et al., 2009). In this experiment, soil water content was defined by the fraction of transpirable soil water (FTSW), which reflects directly measured volumetric water availability in the soil for plant extraction to support transpiration.

In peanuts, genotype N12006ol with early initiation of stomata closure with soil drying was identified as having delayed wilting under developing water deficit and a yield advantage under dryland conditions (Sinclair et al., 2018). In

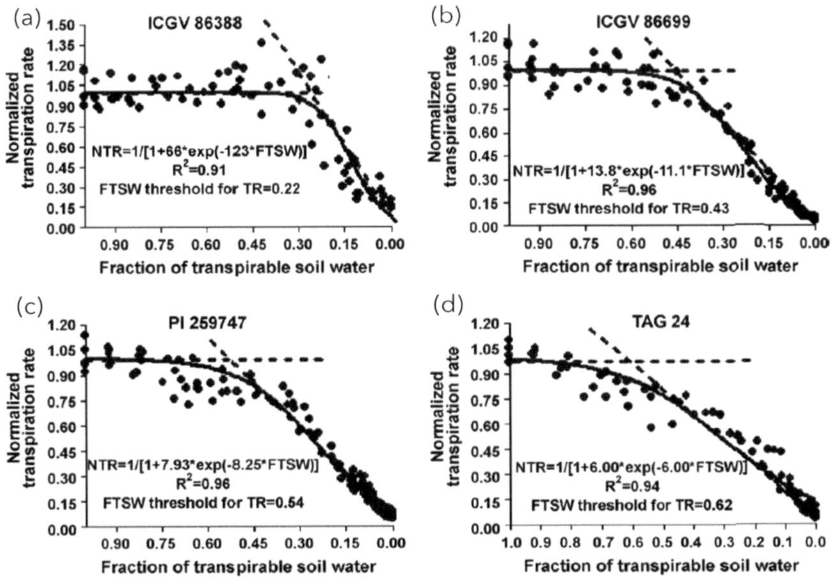

Figure 3 Graphs of four peanut genotypes of normalized transpiration rate vs. fraction of transpirable soil water (FTSW) in an experiment of continuous soil drying over about 2 weeks (Devi et al., 2009). The data for each genotype were analyzed using a two linear-segment regression (dashed lines) to obtain an estimate for the FTSW threshold for the decrease in normalized transpiration rate.

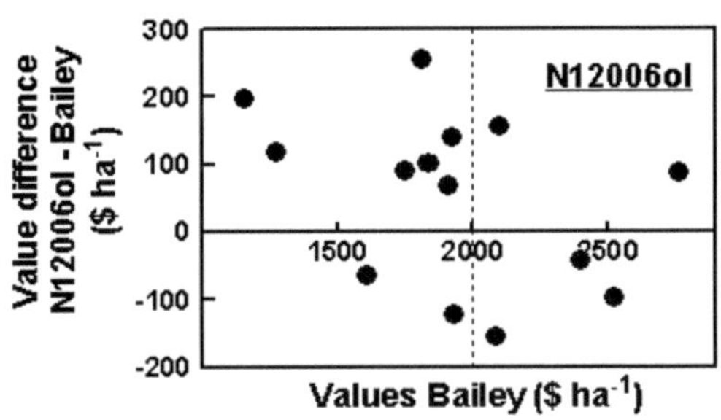

Figure 4 Difference in monetary value of yield of genotype N12006ol minus value of cultivar Bailey graphed against the monetary value in each environment of Bailey (Sinclair et al., 2018). In these environments, the variation in yield of Bailey was mainly a result of differences in water availability. Data above the solid horizontal line show a yield advantage for N12006ol. The vertical dashed line is a reference below which N12006ol has superior yield in seven of nine environments.

Fig. 4, the yield difference between genotype N12006ol and the commercial genotype 'Bailey' in individual environments is graphed against the yield in the same environments of Bailey, which indexes the water availability in each environment. In the drier environments with lower yields, N12006ol had a greater yield than Bailey. Due to the yield advantage of N12006ol, this genotype is being prepared for commercial release for peanut production in the eastern United States for dryland conditions.

9 Proposition D: Responsiveness to vapor pressure deficit to achieve early-season water conservation

Previously, the possibility of plants regulating an 'effective VPD' that influences transpiration rate was introduced. One approach to achieve lower effective VPD is partial stomata closure at elevated VPD to minimize integrated transpiration rate (Eqn (2)). Figure 5 illustrates the genotypic difference in soybean in the expression of limited-transpiration rate at elevated VPD. In most soybean genotypes, the transpiration rate increases linearly as VPD increases, but in genotype PI 416937, partial stomata closure is initiated when VPD exceeded 2.13 kPa so that in this genotype the transpiration rate remained constant at elevated VPD (Fletcher et al., 2007). The water conservation of soil water under elevated VPD early in the season provides more water for use during water-deficit conditions that may develop later in the growing season. Genotype PI 416937 is in the pedigree of a soybean cultivar for high yield under water-deficit conditions (Carter et al., 2016).

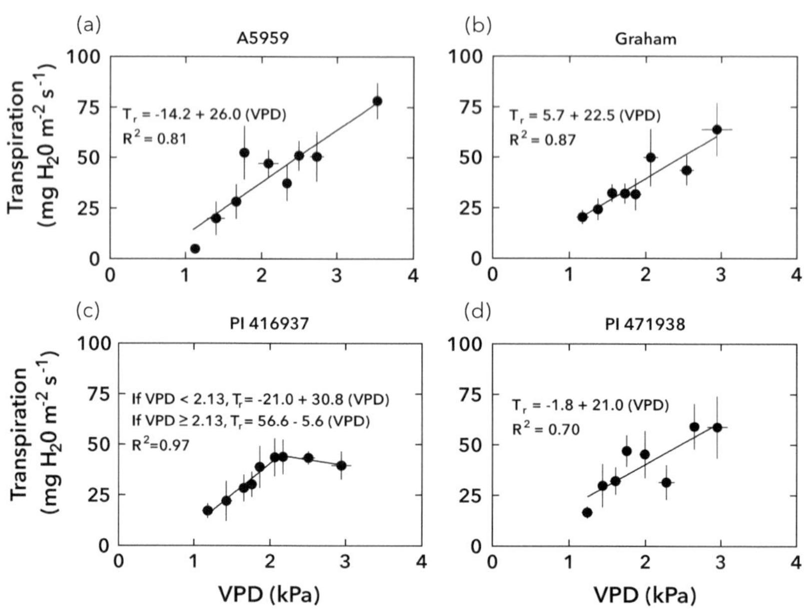

Figure 5 Graph for four soybean cultivars of plant transpiration rate vs. vapor pressure deficit (VPD) (Fletcher et al., 2007). Unlike the other three genotypes, genotype PI 416937 was found to have initiation of partial stomata closure at VPD of 2.13 kPa.

An advantage of the limited-transpiration trait at elevated VPD is that it is expressed only under hot, arid conditions usually associated with drought. That is, the expression of this trait is specifically focused on the environmental conditions commonly associated with developing soil water deficit. If there are no hot, dry conditions resulting in elevated VPD, this trait is not expressed and there is no limitation on leaf gas exchange or crop growth.

However, the limited-transpiration response under elevated VPD, as with all drought-resilience traits, can result in either yield increase or yield decrease depending on the unique weather scenarios of each growing season and the crop development and growth patterns. As discussed earlier, yield increase is likely when late-season water deficits develop. On the other hand, partial stomata closure under elevated VPD will restrict carbon assimilation, and hence, potentially limit crop growth. The tradeoff between immediate loss of growth capacity and sustained growth during late-season water-deficit events was found in geospatial analyses to vary with each growing season in soybean in the United States and Africa (Sinclair et al., 2010, 2014), maize in the United States (Messina et al., 2015), and wheat in Tunisia (Sadok et al., 2019) and in Australia (Collins et al., 2021). As illustrated in Fig. 6 in the results for wheat in Australia (Collins et al., 2021), a few locations had decreased yield as a result of the limited-transpiration trait but more locations had yield increase, although many locations had no yield change. In the Australian wheat simulations, greater expression of the limited-transpiration trait (LTR++ vs. LTR+) tended to result in greater yield increases.

Commercial yield advantages of the VPD-sensitive trait have been documented for cultivars of maize (Gaffney et al., 2015) and soybean (Carter et al., 2016) that were developed for dryland conditions. The trait has also been found in other species. Gholipoor et al. (2010) identified 17 out of 26 sorghum cultivars that expressed the VPD-sensitive trait. Devi et al. (2010) found in a test of nine peanut cultivars grown commercially in the southern United States

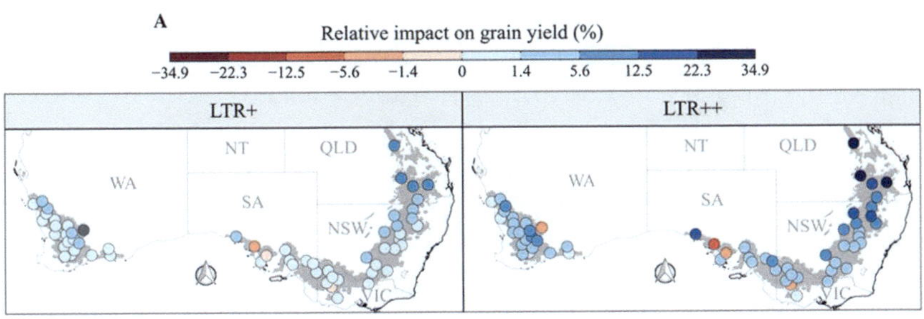

Figure 6 Projected average yield change for 60 locations across the Australian wheat belt in response to expression of the limited-transpiration trait when VPD exceeded 1.3 kPa (Collins et al. 2021). The degree of expression of the trait was simulated for two cases in which transpiration was decreased by 30% (LTR+) or 100% (LTR++).

that eight showed partial stomata closure at elevated VPD in the range of 1.8 to 2.2 kPa. Sheekoofa et al. (2017) found that about 30% of the genotypes in two populations of peanuts grown in the mid-Atlantic region of the United States expressed the VPD-sensitive trait. In a study of 23 wheat lines released in Australia over 120 years, Schoppach et al. (2017) found that all cultivars expressed the VPD-sensitive trait.

Our experience in the development of the commercial cultivar USDA-N8002 in soybean, which has PI 416937 in its pedigree, illustrates several important steps in developing the limited-transpiration trait for commercial cultivars. Carter et al. (2016) undertook a field screen for tolerance to water deficit of a broad genetic base by documenting the initial appearance of leaf wilting following the termination of irrigation. Those lines with delayed wilting were hypothesized to potentially express a water conservation trait. Fletcher et al. (2007) tested several of the genotypes expressing delayed wilting and found that PI 416937 expressed transpiration sensitivity to VPD (Fig. 5). Yield tests of progeny lines developed from PI 416937 identified 'USDA-N8002' as a high-yielding line for dryland conditions (Carter et al., 2016). Devi et al. (2014) confirmed that USDA-N8002 expressed the VPD-sensitive trait.

10 Conclusion: A renewed perspective for increasing crop drought resilience

Research of water-deficit limitation on crop yield using the phenomenological model (Eqn (1)) and the huge amount of experimentation from the 'static' perspective of water deficit have enhanced our understanding of the interaction of crop growth and water deficit. However, there has been limited progress in altering specific plant traits to increase commercial crop yield other than the research exploiting the VPD-sensitive trait as discussed earlier. However, even in the case of the VPD-sensitive trait, much of the progress with this trait has been a result of 'cryptic' selection in breeding programs rather than direct trait selection.

We suggest that the limited progress in overcoming the water-deficit limitation in crop yield may be to a large extent a failure to look beyond the static view of water deficit. Unique temporal dynamics in each growing season result from a wide range of weather scenarios and crop conditions. The yield response to alteration of a drought-resilience trait modification is very much dependent on the nature of these temporal dynamics. There appears to be no universal drought-resilience trait. Instead, it is concluded that the goal of water-deficit research needs to focus on identifying specific traits and the environments in which there is evidence indicating a high probability among seasons for yield increase. This conclusion requires a new perspective in developing cultivars for water-deficit conditions. Rather than base breeding decisions on average yields across environments, a full understanding of the range of yield response over locations and growing seasons is required to resolve the probability of yield increase. Recommendations for the suitability of a specific cultivar will likely require geospatial analysis based on crop simulation studies to extrapolate an understanding of trait impact on the 'probability geography' for positive yield response as well as possible information on average and range of yield increase. A key advantage of such geospatial analysis is insight about possible major yield gains specifically in 'drier' growing seasons even though the average increase in yield over all seasons may not be particularly large. A temporal dynamic perspective of water deficit focused on a more effective use of available water through growing seasons offers a renewed perspective on options to improve crop yields under water-deficit conditions.

11 References

Beseli, A. L., Hall, A. J., Manandhar, A. and Sinclair, T. R. (2019). Root osmotic potential and length for two maize lines differing in leaf osmotic potential. *J. Crop Improv.* 33(4), 429–444.

Beseli, A. L., Shekoofa, A., Ali, M. and Sinclair, T. R. (2020). Temporal water use by two maize lines differing in leaf osmotic potential. *Crop Sci.* 60(2), 945-953.

Blum, A. (2009). Effective use of water (EUW) and not water-use efficiency (WUE) is the target of crop yield improvement under drought stress. *Field Crops Res.* 112(2-3), 119–123.

Borrell, A. K., Mullet, J. E., George-Jaeggli, B., van Oosterom, E. J., Hammer, G. L., Klein, P. E. and Jordan, D. R. (2014). Drought adaptation of stay-green sorghum is associated with canopy development, leaf anatomy, root growth, and water uptake. *J. Exp. Bot.* 65(21), 6251-6263.

Carter, T. E. Jr., Todd, S. M. and Gillen, A. M. (2016). Registration of 'USDA-N8002' soybean cultivar with high yield and abiotic stress resistance traits. *J. Plant Reg.* 10(3), 238–245.

Chiang, C., Bånkestad, D. and Hoch, G. (2020). Reaching natural growth: The significance of light and temperature fluctuations in plant performance in indoor growth facilities. *Plants (Basel)* 9(10), 1312.

Chimenti, C. A., Marcantonio, M. and Hall, A. J. (2006). Divergent selection for osmotic adjustment results in improved drought tolerance in maize (*Zea mays* L.) in both early growth and flowering phases. *Field Crops Res.* 95(2-3), 305–315.

Collins, B., Chapman, S., Hammer, G. and Chenu, K. (2021). Limiting transpiration rate in high evaporative demand conditions to improve Australian wheat productivity. *In Silico Plants* 3(1), 1-16.

Cooper, M., Technow, F., Messina, C., Gho, C. and Totir, L. R. (2016). Use of crop growth models with whole-genome prediction: Application to a maize multienvironment trial. *Crop Sci.* 56(5), 2141-2156.

de Wit, C. T. (1958). *Transpiration and Crop Yields* (Vol. 646). Wageningen, The Netherlands: Institute of Biological Chemistry Research, Field Crops and Herbage.

Devi, J. M., Sinclair, T. R., Chen, P. and Carter, T. E. (2014). Evaluation of elite southern maturity soybean breeding lines for drought-tolerant traits. *Agron. J.* 106(6), 1947–1954.

Devi, M. J., Sinclair, T. R. and Vadez, V. (2010). Genotypic variation in peanut transpiration response to vapor pressure deficit. *Crop Sci.* 50(1), 191–196.

Devi, M. J., Sinclair, T. R., Vadez, V. and Krishnarmurthy, L. (2009). Peanut genotypic variation in transpiration efficiency and decreased transpiration during progressive soil drying. *Field Crops Res.* 114(2), 280–285.

Fletcher, A. L., Sinclair, T. R. and Allen, L. H., Jr. (2007). Transpiration responses to vapor pressure deficit in well watered 'slow wilting' and commercial soybean. *Environ. Exp. Bot.* 61(2), 145–151.

Gaffney, J., Schussler, J., Loffler, C., Cai, W., Paszkiewicz, S., Messina, C., Groeteke, J., Keaschall, J. and Cooper, M. (2015). Industry-scale evaluation of maize hybrids selected for increased yield in drought-stress conditions of the US corn belt. *Crop Sci.* 55(4), 1608–1618.

Ghanem, M. E., Marrou, H. and Sinclair, T. R. (2015). Physiological phenotyping of plants for crop improvement. *Trends Plant Sci.* 20(3), 139–144.

Gholipoor, M., Prasad, P. V. V., Mutava, R. N. and Sinclair, T. R. (2010). Genetic variability of transpiration response to vapor pressure deficit among sorghum genotypes. *Field Crops Res.* 119(1), 85–90.

Gupta, A., Rico-Medina, A. and Cano-Delgado, A. I. (2020). The physiology of plant responses to drought. *Science* 368(6488), 266–269.

Hammer, G. L., Cooper, M. and Reynolds, M. P. (2021). Plant production in water-limited environments. *J. Exp. Bot.* 72(14), 5097–5101.

Jafarikouhini, N., Kazemeini, S. A. and Sinclair, T. R. (2021). Fresh sweet corn yield sensitivity to deficit nitrogen and water conditions. *J. Crop Improv.* 36(4), 593–603.

Langstroff, A., Heuermann, M. C., Stahl, A. and Junker, A. (2022). Opportunities and limits of controlled-environment plant phenotyping for climate response traits. *Theor. Appl. Genet.* 135(1), 1–16.

Ludlow, M. M. and Muchow, R. C. (1990). A critical evaluation of traits for improving crop yield in water-limited environments. *Adv. Agon.* 43, 107–153.

Messina, C. D., Sinclair, T. R., Hammer, G. L., Duran, D., Thompson, J., Oler, Z., Gho, C. and Cooper, M. (2015). Limited-transpiration trait may increase maize drought tolerance in the US Corn Belt. *Agron. J.* 107(6), 1978–1986.

Millet, E. J., Welcker, C., Kruijer, W., Negro, S., Coupel-Ledru, A., Nicolas, S. D., Laborde, J., Bauland, C., Praud, S., Ranc, N., Presterl, T., Tuberosa, R., Bedo, Z., Draye, X., Usadel, B., Charcosset, A., Van Eeuwijk, F. and Tardieu, F. (2016). Genome-wide analysis of yield in Europe: Allelic effects vary with drought and heat scenarios. *Plant Physiol.* 172(2), 749–764.

Morgan, J. M. (2000). Increases in grain yield of wheat by breeding for an osmoregulation gene: Relationship to water supply and evaporative demand. *Aust. J. Agric. Res.* 51(8), 971–978.

Muchow, R. C. (1989). Comparative productivity of maize, sorghum and pearl millet in a semi-arid tropical environment I. Yield potential. *Field Crops Res.* 20(3), 191–205.

Passioura, J. B. (1977). Grain yield, harvest index and water use of wheat. *J. Aust. Inst. Agric. Sci.* 45, 117–121.

Passioura, J. B. (2020). Translational research in agriculture: Can we do it better? *Crop Pasture Sci.* 71(6), 517–528.

Poorter, H., Fiorani, F., Pieruschka, R., Wojciechowski, T., van der Putten, W. H., Kleyer, M., Schurr, U. and Postma, J. (2016). Pampered inside, pestered outside? Differences and similarities between plants growing in controlled conditions and in the field. *New Phytol.* 212(4), 838-855.

Resende, R. T., Piepho, H. P., Rosa, G. J. M., Silva-Junior, O. B., E Silva, F. F., de Resende, M. D. V. and Grattapaglia, D. (2021). Enviromics in breeding: Applications and perspectives on envirotypic-assisted selection. *Theor. Appl. Genet.* 134(1), 95-112.

Sadok, W., Schoppach, R., Ghanem, M. E., Zucca, C. and Sinclair, T. R. (2019). Wheat drought-tolerance to enhance food security in Tunisia, birthplace of the Arab Spring. *Eur. J. Agron.* 107, 1-9.

Sadras, V. O. and Milroy, S. P. (1996). Soil-water threshold for the responses of leaf expansion and gas exchange: A review. *Field Crops Res.* 47(2-3), 253-266.

Sadras, V. O. and Richards, R. A. (2014). Improvement of crop yield in dry environments: Benchmarks, levels of organization and the role of nitrogen. *J. Exp. Bot.* 65(8), 1981-1995.

Schoppach, R., Fluery, D., Sinclair, T. R. and Sadok, W. (2017). Transpiration sensitivity to evaporative demand across 120 years of breeding of Australian wheat cultivars. *J. Agron. Crop Sci.* 203(3), 219-226.

Serraj, R. and Sinclair, T. R. (2002). Osmotic accumulation: Can it really help to increase crop yield under drought conditions? *Plant Cell Environ.* 25(2), 333-341.

Sheekoofa, A., Sinclair, T. R., Aninbon, C., Holbrook, C. C., Isleib, T. G., Ozias-Akins, P. and Chu, Y. (2017). Expression of the limited-transpiration trait under high vapour pressure deficit in peanut populations: Runner and Virginia types. *J. Agro Crop Sci.* 203(4), 295-300.

Sinclair, T. R. (2005). Theoretical analysis of soil and plant traits influencing daily plant water flux on drying soils. *Agron. J.* 97(4), 1148-1152.

Sinclair, T. R. (Ed.). (2017). *Water-Conservation Traits to Increase Crop Yields in Water-Deficit Environments*. Switzerland: Springer Briefs in Environmental Science.

Sinclair, T. R. (2018). Effective water use required for improving crop growth rather than transpiration efficiency. *Front. Plant Sci.* 9, 1442.

Sinclair, T. R., Bennett, J. M. and Muchow, R. C. (1990). Relative insensitivity of grain yield and biomass accumulation to drought in field-grown maize. *Crop Sci.* 30(3), 690-693.

Sinclair, T. R., Marrou, H., Soltani, A., Vadez, V. and Chandolu, K. C. (2014). Soybean production potential in Africa. *Glob. Food Sec.* 3(1), 31-40.

Sinclair, T. R., Messina, C. D., Beatty, A. and Samples, M. (2010). Assessment across the United States of the benefits of altered soybean drought traits. *Agron. J.* 102(2), 475-482.

Sinclair, T. R., Shekoofa, A., Isleib, T. G., Balota, M. and Zhang, H. (2018). Identification of Virginia-type peanut genotypes for water-deficit conditions based on early decrease in transpiration rate with soil drying. *Crop Sci.* 58(6), 2607-2612.

Spaeth, S. C., Randall, H. C., Sinclair, T. R. and Vendeland, J. S. (1984). Stability of soybean harvest index 1. *Agron. J.* 76(3), 482-486.

Tanner, C. B. and Sinclair, T. R. (1983). Efficient water use in crop production: Research or re-search? In: Taylor, H. R., Jordan, W. R. and Sinclair, T. R. (Eds.), *Limitations to Efficient Water Use in Crop Production*. Madison, WI: American Society of Agronomy, pp. 1-27.

Tardieu, F. (2012). Any trait or trait-related allele can confer drought tolerance: Just design the right drought scenario. *J. Exp. Bot.* 63(1), 25-31.

Tardieu, F., Simonneau, T. and Muller, B. (2018). The physiological basis of drought tolerance in crop plants: A scenario-dependent probabilistic approach. *Annu. Rev. Plant Biol.* 69, 733–759.

Turner, N. C. (2018). Turgor maintenance by osmotic adjustment: 40 years of progress. *J. Exp. Bot.* 69(13), 3223-3233.

Xu, Y. (2016). Envirotyping for deciphering environmental impacts on crop plants. *Theor. Appl. Genet.* 129(4), 653-673.

Chapter 2

The role of plant hormones in adaptation to drought stress in cereals

Arnauld A. Thiry, Lancaster Environment Centre, UK; Matthew P. Reynolds, International Maize and Wheat Improvement Center (CIMMYT), Mexico; and William J. Davies and Ian C. Dodd, Lancaster Environment Centre, UK

1 Introduction

Drought is by far the most important environmental stress that constrains crop growth, development and yield (Blum, 2011; Zenda et al., 2021). Water availability for agriculture is already negatively impacting agriculture on most continents (IPCC, 2021) and is predicted to soon become increasingly limiting in many food-producing regions of the world (Israeli and Lahav, 2017; Zenda et al., 2021). Furthermore, future increases in temperature and atmospheric vapour pressure deficit (VPD) will negatively affect yield (López et al., 2021; Ort and Long, 2014; Wardlaw et al., 1989). Agriculture must respond to the growing demand for food, but do this without causing further damage to the planet (Chávez-Dulanto et al., 2021), for example, by overusing ground water (e.g. Willett et al., 2019). If breeders are to develop genotypes that will continue to yield significantly under these constraints (Sareen et al., 2012), a greater understanding of mechanisms involved in plant tolerance/resilience to high temperature and drought stress will be a priority for improved crop production. Plant growth regulators (PGRs) are involved in regulating many plant responses to drought stress. Exploiting this new knowledge for crops grown in target environments can lead to new genotypes to combat growing

http://dx.doi.org/10.19103/AS.2022.0109.03

climate challenges, helping to meet the demand for more good quality and more sustainable food. However, defining/identifying the PGR ideotype is not easy, as endogenous PGR accumulation and the plant's sensitivity to it (Abley et al., 2021) can differentially affect plant resilience, development and productivity according to the developmental stage of the crop (Blum, 2015).

Furthermore, key agronomical developments such as altering sowing dates, using soil mulches and water-saving irrigation techniques can increase the efficiency of water use in agriculture. Using these techniques in specific environments can optimise the performance of novel genotypes, by exploiting our knowledge of the impact of PGRs on plant growth, development and functioning (e.g. Dodd and Kudoyarova, 2021; Yang and Zhang, 2006).

1.1 How to define a stress-tolerant genotype

More than 80 years of plant breeding activities have focused on increasing water-limited crop yields in dry environments, but the drought-resilient ideotype is not always well-defined (Blum, 2005). Fundamental research has significantly enhanced our understanding of the physiological, genomic and molecular responses of plants to water deficits (Cattivelli et al., 2008; Collins et al., 2008; Langridge and Reynolds, 2015), but it is not always clear which traits might identify genotypes with higher productivity under drought stress (Blum, 2005; Langridge and Reynolds, 2021; Reynolds and Tuberosa, 2008). Exploiting our understanding of plant hormone biology for crop improvement requires greater precision in our understanding of the physiology contributing to key drought resistance traits. For crop species, drought tolerance is defined as the ability of plants to grow and reproduce satisfactorily while producing harvestable yields with limited water supply or under periodic water deficit (Fleury et al., 2010; Turner, 1979). However, as long as yield stability is high, it may better indicate genotypic drought resistance than grain yield (Blum et al., 1989). Drought resistance is often considered a compromise between 'dehydration avoidance' and 'dehydration tolerance,' both of which can have variable impacts on yield (Blum and Tuberosa, 2018; Fischer and Maurer, 1978; Levitt, 1980; Ludlow and Muchow, 1990; Turner, 1979, 1986). Some genotypes accelerate their phenological development to avoid intense stress during critical phenological stages, allowing better yield performance. Genotypes with such an 'escape strategy' should not be considered drought-resilient *per se* (Fischer and Maurer, 1978; Turner, 1979, 1986; Zhou et al., 2017) but can yield well if farmers adjust sowing dates or other agronomical practices to ensure higher yields (Hunt et al., 2019; Zhou et al., 2017).

Around half of the recent increases in global crop production have been attributed to breeding and genetic gains, with improved field management and agronomic practices (Araus et al., 2004; Richards, 2006) and changes in global food policies (Robinson, 2018) accounting for the remainder. Fully exploiting gains made by genetic interventions often requires concurrent, targeted improvements in agronomic practice (Fernández et al., 2019; Hunt et al., 2019; Kirkegaard and Lilley, 2021; Lilley and Kirkegaard, 2016; Passioura and Angus, 2010; Wang et al., 2020). Understanding the nature and frequency of drought development in a target environment (Chenu et al., 2013) with recommended agronomical practices specific to each released genotype and region will help farmers and breeders identify their 'best' options for particular environmental conditions (Zhao et al., 2022).

1.1.1 *Current breeding selection methods and the need for new approaches*

Identifying and/or creating new varieties adapted to a changing climate is a major challenge for cereal crop breeders, but selecting improved genotypes under stress conditions is not common (Monneveux et al., 2012). Many breeding and pre-breeding programmes focusing on abiotic stress adaptation aim to increase yield potential (yield without any biotic and/or abiotic stresses), an approach that has contributed much to historical improvements in crop yield (Fischer and Edmeades, 2010; Reynolds et al., 2009; Slafer and Araus, 2007). High-yielding genotypes under optimal conditions usually also perform well in a wide range of environments including moderate drought but not under severe stress conditions such as prolonged drought periods, high temperatures and salinity (Ceccarelli and Grando, 1991; Fischer and Edmeades, 2010; Reynolds et al., 2009; Slafer and Araus, 2007). Furthermore, selection under optimal conditions is considered more effective, allowing higher heritability (Richards, 1996; Tuberosa, 2012).

A high-yielding genotype under both stressed and non-stressed environments does not necessarily indicate the existence of a resilience/tolerance trait (Ceccarelli and Grando, 1991). To better identify such traits, the importance of selecting genotypes under a target stress environment has been emphasised (Ceccarelli and Grando, 1991; Chenu et al., 2013; Rebetzke et al., 2013). While high-yielding genotypes under non-stress conditions often show limited drought resilience capacity when the stress is sufficiently severe, some rare exceptions to this generality do exist (Blum, 1996). Currently, there are no really effective means to rapidly screen these exceptional genotypes which show combined high yield and stress tolerance, and quicker and more reliable methods are needed to address this challenge (Thiry et al., 2016). Nevertheless, high-throughput phenotyping for cooler canopies and vegetative indices like Normalized difference vegetation index (NDVI) show considerable promise for selecting drought and heat adaptative traits at breeding scale (e.g. Tattaris et al., 2016).

1.1.2 Performance depends on stress environment, timing of the stress and genotype

Increasingly, the importance of screening plant material under realistic drought stress environments is recognised. Despite the increasing sophistication of phenotyping facilities, effective environmental control during such exercises is not easy to deliver. Soil characteristics, atmospheric properties and rainfall patterns all influence the intensity and timing of environmental stresses, as do the sensitivity of different developmental and physiological processes (e.g. transpiration and growth) contributing to crop yield (Rebetzke et al., 2013). PGRs affect water loss, growth and development of drought-stressed vegetative plants, but drought stress also adversely affects pollen development in wheat and rice (Dorion et al., 1996; Dong et al., 2017; Ji et al., 2010; Sheoran and Saini, 1996) and heat stress affects both the release of mature pollen reducing anther dehiscence and grain filling in cereals (Wardlaw and Wrigley, 1994; Jagadish et al., 2007; Jagadish, 2020). In wheat, the potential grain number is determined during the early stages of plant development, a critical period which is highly sensitive to abiotic stress. Drought occurring during the young microspore stage of pollen development, which starts at booting stage, is a particular problem (Acevedo et al., 2002; Dolferus et al., 2011). Drought stress decreased yields when applied at multiple growth stages (tillering, booting, heading, anthesis and maturity), but booting and grain filling were the most sensitive growth stages observed by Ihsan et al. (2016).

Better understanding the physiological and molecular processes leading to stress-induced pollen abortion may help to design approaches to allow substantial grain set and development of a high grain number under stress conditions (Dolferus et al., 2011; Dong et al., 2017). Heat stress-induced ethylene accumulation can induce pollen sterility and kernel abortion in wheat (Hays et al., 2007) with PGR applications such as abscisic acid (ABA) in rice (Rezaul et al., 2019), auxin in barley (Zhang et al., 2018) and salicylic acid in rice (Feng et al., 2018) under heat stress preventing this negative effect on pollen viability. Indeed, pollen sterility may be the main cause of grain loss in wheat during drought conditions, especially when the abiotic stress coincides with meiosis (Ji et al., 2010). During the pre-anthesis period, extending the duration of stem elongation (Jointing stage) and the spike growth periods (booting stage) can improve floret survival, thereby increasing grain set (i.e. grain number) (González et al., 2011; Guo et al., 2018; Miralles et al., 2000). All these growth and development stages are sensitive to PGRs, with drought timing and intensity in the target environment determining which developmental stages should receive more attention. One conclusion is that drought-resilient genotypes should be associated with a particular environmental region (Rebetzke et al., 2013) or a specific mega-environment (Braun and Payne, 2012).

Numerous traits, genes and/or quantitative trait loci (QTLs) can improve crop performance, but relatively few are used by commercial breeding programmes, mainly because their value has not been clearly demonstrated and especially because QTL effects are context-dependent (Collins et al., 2008; Monneveux et al., 2012; Rebetzke et al., 2013; Richards, 1996; Tardieu et al., 2010b). Rapidly and accurately measuring the phenotype remains the major limitation to improved selection methods under drought environments (Richards et al, 2010). Improving the accuracy and speed of trait phenotyping, including yield, should enhance crop improvement efforts and boost the value of new molecular technologies. New techniques to precisely and rapidly quantify PGRs in different plant compartments will also be important (e.g. Albacete et al., 2008; Huang et al., 2014; Pu et al., 2018).

1.2 Identifying highly productive and resilient genotypes – which traits to choose?

Identifying physiological traits that simultaneously or jointly indicate high yield potential and stress tolerance can be critical to crop yield improvement (Gleason et al., 2022; Slafer and Araus, 2007). A conceptual model of a breeding programme for drought adaptation usually considers yield under drought to be a function of (1) yield potential, (2) flowering date (which indicates whether the crop will avoid drought stress) and (3) secondary traits that provide drought resistance (Monneveux et al., 2012). Significant progress in understanding wheat plant physiology indicates that yield potential may be boosted by enhancing (1) the efficiency of carbon gain (e.g. South et al., 2019), (2) spike fertility (SF) and the partitioning of assimilates to the growing spike and grains and (3) lodging resistance of these genotypes with bigger spikes (Reynolds et al., 2009, 2021; Parry et al., 2011; Foulkes et al., 2011).

Reliable screening methods (e.g. field phenotyping) are a major bottleneck for improving stress tolerance during the reproductive stage (Araus and Cairns, 2014; Dolferus et al., 2011). Despite extensive research identifying drought-adaptive traits such as leaf waxiness and/or leaf rolling, tiller survival and large seed size, few breeders have incorporated these traits into their selection programmes (Monneveux et al., 2012; Richards, 1996). Many breeders are not convinced by the proposed criteria, and their evaluation is usually time-consuming and/or expensive and therefore not readily applied to trials of thousands of genotypes (Monneveux et al., 2012; Richards, 1996). Advances in high-throughput field phenotyping methods (Reynolds et al., 2020) should reduce costs (economic and time), allowing physiological traits to be incorporated into the selection criteria. Ideal secondary physiological traits should be (1) genetically associated with grain yield under drought; (2) genetically variable; (3) highly heritable; (4) easily, inexpensively and rapidly measured or observed; (5) non-destructive; (6) stable over the measurement period and (7) not associated with yield penalties under favourable conditions (Monneveux et al., 2012; Richards, 1996). This last condition conflicts with many studies reporting opposing effects of traits when stress intensity varies (see e.g. Tardieu, 2012). These criteria may be too strict (yield itself does not meet criteria 2–6), particularly when considering elite materials. A simple-to-measure trait like canopy temperature, while of similar heritability to yield, can explain a large degree of variation in yield under drought and heat stress (Pinto et al., 2010) and is strongly associated with root mass at depth (Lopes and Reynolds, 2010).

Yield is a complex trait as it is formed continuously from sowing to maturity, and almost all plant genes are involved either directly or indirectly in producing grain yield by affecting plant growth, development and physiology. Different environmental factors will interact with these component processes, causing genotype × environment interactions (G×E) (Slafer, 2005). Rather than focusing on wheat yield itself, Slafer (2003) recommends measuring simpler traits that compose yield. The two major yield components are the weight of an individual grain (sometimes expressed as thousand grain weight (TGW)) and the grain number per square metre (G#/m^2) which has more subdivisions to further simplify the trait (Slafer, 2003, 2005).

Under yield potential conditions, the yield of modern wheat cultivars is typically sink – rather than source-limited during the grain filling period (e.g. Borrás et al., 2004; Reynolds et al., 2021); thus, grain yield improvement has been highly associated with increased grain number per unit area (Peltonen-Sainio et al., 2007; Sayre et al., 1997). To improve yield, enhancing the number of grains per unit area requires improved SF, defined as the number of grains per spike chaff dry weight (Abbate et al., 2013). Variation in the production and/or interaction of plant hormones may potentially explain the environmental and/or genetic regulation of SF (Davies et al., 2013; Feng et al., 2018; Hays et al., 2007; Rezaul et al., 2019; Sakata et al., 2010; Wilkinson et al., 2012; Yang et al., 2006a,b; Zeng and King, 1986). Contrasting germplasm, with tolerant and resilient reproductive stage stress responses, should help identify the molecular and physiological basis of the resilience mechanism (Dolferus et al., 2011).

Different methods such as the use of stress indices have been used to quantify variation in genotype responses in contrasting environments (Tsenov et al., 2017), and these approaches can help focus our mechanistic studies. More than 20 stress indices (principally for drought evaluation) exist with 5 of these commonly used. These are (1) tolerance indices – such as the mean productivity (MP) (Rosielle and Hamblin, 1981), geometric mean productivity (GMP) (Fernandez, 1992) and stress tolerance index (STI) (Fernandez, 1992) and (2) susceptibility indices – such as the stress susceptibility index (SSI)

(Fischer and Maurer, 1978) and tolerance index (TOL) (Rosielle and Hamblin, 1981). Currently, STI, GMP and MP are the most commonly recommended indices used to identify genotypes with high yield under both non-stress and stress environments (heat and drought) as they focus mainly on productivity (Khodarahmpour et al., 2011; Mohammadi et al., 2011; Sareen et al., 2012; Thiry et al., 2016). Fernandez (1992) classified genotypes into four categories (Table 1, Fig. 1a):

1 A genotypes are highly productive under both environments (stress and no-stress).
2 B genotypes perform well only under yield potential but not under stress conditions.
3 C genotypes show better yield performance only under stress conditions.
4 D genotypes show poor yield performance under both environments.

Over many years, none of these five indices has clearly distinguished Group A cultivars from Groups B and C (Fernandez, 1992; Khayatnezhad et al., 2010) (Table 1). Tolerance indices are more related to yield production but fail under severe stress (Class 2 in Table 1), while susceptibility indices are more related to yield stability under severe stress conditions (Class 1 in Table 1) (Bennani et al., 2017; Fernandez, 1992; Nouri et al., 2011; Talebi et al., 2009; Thiry et al., 2016). Therefore, discriminating genotypes with higher performance

Table 1 Summary of the classification of genotypes using the five most commonly-used stress indices (Stress Susceptibility Index (SSI), Tolerance Index (TOL), Mean Productivity Index (MP), Geometric Mean Productivity Index (GMP) and Stress Tolerance Index (STI)) and showing their failures to distinguish the different response groups of plants defined by Fernandez (1992) where group A genotypes express uniform superiority in both stress and no-stress condition; group B genotypes express good performance only in yield potential but not under stress conditions; group C genotype shows a relatively higher yield only under stress environment and group D poor yield performance under both environments. Classes 1 and 2 correspond to susceptibility indices and tolerance indices, respectively, where Class 1 tends to distinguish between the stress-tolerant and the stress-susceptible genotypes and Class 2 tends to identify genotypes with stress-tolerance and high average yield (Thiry et al., 2016)

Index	Index value	Tolerance	Fails	Class
SSI	High	Low	Fails to distinguish A and C	1
TOL	High	Low	Fails to distinguish A and C	1
MP	High	High	Fails to distinguish A and B	2
GMP	High	High	Same failure as MP but better at distinguishing A than MP	2
STI	High	High	Same failure as MP but better at distinguishing A than MP and GMP	2

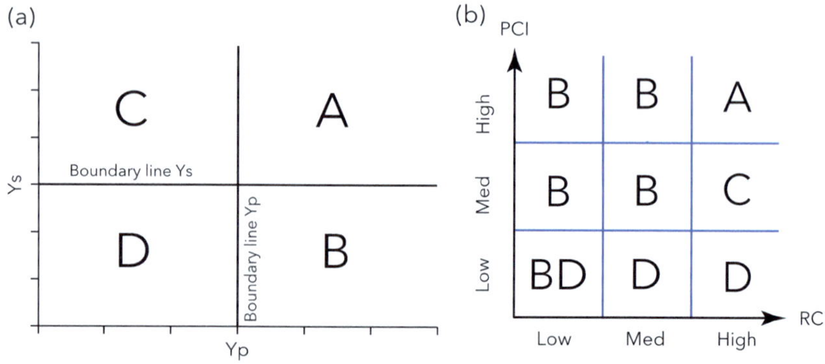

Figure 1 Diagram of genotype distribution into four genotype classes (A, B, C and D) (Fernandez, 1992); (a) as a function of yield under stress (YS) and non-stress (YP) and (b) as a function of the resilient capacity index (RCI) and the productivity capacity index (PCI) (Thiry et al., 2016). Fernandez (1992) grouped A genotypes as expressing uniform superiority in both stress and no-stress conditions; group B genotypes expressing good performance only in yield potential but not under stress conditions; group C genotypes showing a relatively higher yield only under stress environments and group D genotypes with poor yield performance under both environments.

under both environments from those with good performance only in one environment (Farshadfar and Elyasi, 2012) requires a more generally applicable stress index.

A more effective index to select better grain yield under drought should combine both productivity and resilience. Since stress indices all have their own scaling range, Thiry et al. (2016) used a scoring scale (from 1 to 10) of five previously developed stress indices to evaluate abiotic stress adaptation (SSI, TOL, MP, GMP and STI). The scores are attributed in function to the population range response within each index. Once obtained, score indices were classified into two new categories called Resilience Capacity Index (RCI) and Productivity Capacity Index (PCI). Combining RCI and PCI score indices develops three new indices highly correlated with yield under non-stress (Yield Potential Score index - YPSI) and stress (Yield Stress Score Index, YSSI and Mean Stress Index, MSI) conditions. Wheat yield under drought stress was tightly related to YSSI across different germplasm collections tested in different environments over multiple years (Table 2 from Thiry et al., 2016). Based on Thiry et al. (2016), the distributions of genotypes classified by Fernandez (1992) (A, B, C and D; Fig. 1a) can be expressed on the basis of RCI and PCI values (Fig. 1b) where:

- Group A genotypes show high scores for both RCI and PCI.
- Group B genotypes show moderate to high PCI and low RCI.
- Group C genotypes show moderate PCI and high RCI.
- Group D-genotypes show low score values for both.

Table 2 Summary table showing the Pearson correlation coefficient and coefficient of determination (R^2) of Yield Score Stress Index (YSSI) vs yield under stress (Ys) from multi-year populations (WAMII, Seri/Babax, CIMCOG-ROOT) previously studied in CIMMYT under different abiotic stresses (heat, drought, drought under drip irrigation; semi-drought: drought applied at booting stage). The consistency of correlations demonstrates the reliability of the index (from Thiry et al., 2016)

Trials	Entries	Environments	Years	Correlations	R^2
WAMII	294	Heat	2009-2010	0.988	0.976
WAMII	294	Heat	2010-2011	0.983	0.967
WAMII	294	Heat	2011-2012	0.987	0.975
WAMII	294	Drought	2009-2010	0.987	0.973
Seri/Babax	169	Heat	2004-2005	0.993	0.986
Seri/Babax	169	Heat	2005-2006	0.992	0.984
Seri/Babax	169	Heat	2009-2010	0.955	0.912
Seri/Babax	169	Drought	2005-2006	0.994	0.988
Seri/Babax	169	Drought (Drip)	2007-2008	0.989	0.978
Seri/Babax	169	Drought	2008-2009	0.991	0.982
Seri/Babax	169	Drought (Drip)	2009-2010	0.974	0.948
CIMCOG ROOT	10	Semi-drought	2012-2013	0.996	0.992
CIMCOG ROOT	10	Heat	2012-2013	0.975	0.950
CIMCOG ROOT	10	Drought	2013-2014	0.981	0.962

While C genotypes are adapted to stress target environment, pure A genotypes are rare (Blum, 1996). The RCI and PCI score indices (Thiry et al., 2016) are used below to focus on new traits to identify adaptive and/or productive genotypes. This method offers a quantitative evaluation of resilience and productivity of yield components under stress. We recommend the use of this method to help the understanding of the physiological/molecular basis of resilience mechanisms such as PGR signalling associated with one of the two categories PCI and RCI of agronomical traits such as yield and yield components.

2 Plant hormones and yield development of crops experiencing water scarcity

2.1 Biological targets for stress tolerance and high productivity

We now assess the case that different PGRs are involved in regulating biology that will support high productivity and/or drought stress resilience. Passioura (1977) first used the following identity to illustrate the importance of water availability to the accumulation of crop dry matter and yield:

Yield = Water Used × Water Use Efficiency (WUE) × Harvest Index (HI)

Both the availability (and use) of water and the efficiency with which the plant trades water for carbon (WUE) are key in delivering high plant productivity. To achieve high crop productivity when the soil is well supplied with water, the plant must access this water before it is lost by evaporation from the soil surface or lost by run-off or deep drainage. As the soil dries, continued root growth and development is required for effective water uptake by the plant. Root morphology (Kholova et al., 2014), depth of rooting (Kirkegaard et al., 2007) and root hydraulic properties (Draye et al., 2010; Tardieu et al., 2017) all determine how much water is accessed from the soil and when during plant development this 'extra' water is accessed. Traits influencing both plant water loss and CO_2 uptake determine WUE, including stomatal behaviour, stomatal density (Hughes et al., 2017) and the biochemistry of carbon fixation (López-Calcagno et al., 2020). Early-season leaf development helps to maximise shoot carbon gain (Dohlman and Long, 2009) and reduce wasteful evaporative losses from soil surfaces (Zhang et al., 2015). Nevertheless, early vigour in wheat could lead to more severe terminal drought stress under elevated atmospheric [CO_2] and semi-arid conditions (Bourgault et al., 2020). Drought stress (water scarcity, high temperatures, high VPD and high radiation load) affects vegetative growth and crop reproductive development, thereby determining crop stress resilience. These effects will manifest as lower HI in the Passioura (1977) identity shown earlier. We now consider the involvement of different plant hormones in regulating plant physiology at key developmental stages.

2.2 Root growth and functioning as soil dries

2.2.1 Growth of the primary root

Effective crop establishment requires vigorous primary root growth, with plant hormones controlling the growth of primary roots at low substrate water potentials (ψ) (Sharp et al., 1994). Under these conditions, shoot growth is usually limited to a greater extent than the growth of roots, thereby increasing root/shoot ratio (Spollen et al., 1993). Maintaining cell growth rates in maize primary roots at low ψ requires regulation of solute accumulation to maintain some turgor in the cell expansion zone of the root. ABA accumulation in growing roots promotes solute accumulation and turgor maintenance and is important in sustaining maize root growth in substrates of low ψ (Saab et al., 1990). In contrast, similar endogenous ABA concentrations in roots well supplied with water limits primary root elongation (Sharp et al., 1994). ABA not only promotes turgor maintenance (Ober and Sharp, 1994) but also alleviates the effects of low ψ by restricting accumulation of the growth inhibitor ethylene (Li et al., 2017; Luo et al., 2014; Spollen et al., 2000; Steffens et al., 2006). Treating ABA-deficient maize roots with ethylene biosynthesis inhibitors or antagonists sustained comparable rates of root elongation to wild-type (ABA-replete) root systems. Nevertheless, within 12 inbred maize lines, there was no correlation between root tip ABA concentration and the rate of root elongation at low ψ (Leach et al., 2011), which should not be surprising as water stress also affects root synthesis and accumulation of ethylene, cytokinins and auxins that interact with ABA to modulate root growth. Thus, supplying auxin to an ABA-deficient tomato mutant partially recovered root growth in drying soil (Zhang et al., 2022).

2.2.2 Growth of the secondary root system

Once established in the soil, an effective root system is required for productive growth and crop development. This depends on the hydraulic properties of individual roots as well as the morphology of the root system. Root System Architecture (RSA) can be defined by considering the length of the main roots, the density and elongation of lower order roots, their gravitropic set-point angle (Uga et al. 2013) and the possible presence of crown roots or adventitious roots (Maurel and Nacry, 2020). Root hormone accumulation can significantly affect each of these traits. For example, root-specific overexpression of cytokinin oxidase-dehydrogenase (CKX) genes increased root growth and thereby drought resistance of barley, without reducing grain yield (Ramireddy et al., 2018).

The adaptive significance of different root morphologies of plants in drying soil can be summarised by the paradigm of 'steep, cheap and deep' (e.g. Lynch, 2019). Root systems which require only limited resource commitment to extend into the soil profile can be classified as 'cheap,' those that commit fewer resources to shallow rooting are defined as 'steep' (often reflecting the growth angle of the seminal roots) and those that enable access to water deeper in the profile are 'deep.' Such properties will generally benefit whole plant performance in drying soil and even at higher air and soil temperatures (Zhao et al, 2022). However, this is not always the case. While deeper rooting of wheat in many cropping regions of Australia can generally result in enhanced water extraction from the soil, crop yields often benefit only if this enhanced root growth occurs early in the season as earlier sown crops have better access to subsoil water (Flohr et al., 2020; Kirkegaard and Lilley, 2021). More generally, simulating root system traits indicated that breeding success remained highly probabilistic and that maximum resilience and economic benefits depend on drought frequency (Kholova et al., 2014).

Beyond focusing on the importance of root morphology, individual root hydraulics affect plant water uptake capacity. Water moves radially from the soil to the stele and to the xylem vessels and must cross a membrane at some point. Water channel proteins or aquaporins in the cell plasma membranes (e.g. Maurel et al., 2015) facilitate this transport. Soil water availability and nutrient status can both up- and downregulate aquaporin activity. Molecular genetics has started to provide more direct approaches for isolating genes controlling radial root hydraulic conductivity (e.g. Tang et al., 2018). Root hydraulics and morphology, together described as hydraulic architecture, are a crucial integrative trait for root performance under drought, but plant breeders have largely neglected this variable as it is difficult to make a reasonable, rapid estimate. Many laboratories undertake RSA phenotyping, but detailed evaluations of root water uptake capacity would help classify an important set of properties for plant improvement. Perhaps even more importantly, Maurel and Nacry (2020) identify a key enigma in plant research as being the issue of how plant roots integrate signals from a heterogeneous below-ground environment to regulate developmental strategies for plant resource acquisition. Perhaps plant hormones play a key role in this regard.

2.2.3 Root growth and functioning

ABA is suggested to play a controlling role in plant functioning and development under stress (e.g. Li et al., 2017; Maurel and Nacry, 2020; Rowe et al., 2016), based on consistent responses of mutants in ABA-signalling and biosynthesis to drought, as well as studies applying ABA to wild types (Rosales et al., 2019). Alternatively, others emphasise that ABA interacts within a hormonal network involving at least auxin, ethylene and cytokinins (Dodd and Kudoyarova, 2021; Reynolds et al., 2021; Rowe et al., 2016).

Reynolds et al. (2021) argue that ethylene–auxin crosstalk is the best-established hormone interaction controlling primary root growth by modulating both cell production in apical meristems and cell elongation in the elongation zone. Manipulating auxin transport can regulate root architecture. For example, decreased expression of the auxin transporter OsPIN5b enhanced root vigour and rice yield (Lu et al., 2015). Also, auxin efflux carriers can regulate root growth orientation via dynamic modulation of auxin transport (Ogura et al., 2019). Nevertheless, ABA regulates root elongation at low substrate ψ by modulating ethylene sensitivity and also by impacting auxin transport (Xu et al., 2013).

Other hormones and gene networks also regulate root morphology. Positive root hydrotropism in response to gradients in soil water status is believed to enhance the effectiveness of plant water acquisition, with the genes MIZ1 and MIZ2 essential for hydrotropism in roots (Moriwaki et al., 2013). Light and ABA independently regulate MIZ1 expression, and overexpression of this gene enhances hydrotropism and inhibits lateral root formation. Gravitropism is required for MIZ1-regulated hydrotropism in roots in both the oblique orientation and the vertical orientation (Li et al., 2020). Both these groups suggest that MIZ1 probably plays a crucial role in hydrotropic response by regulating the endogenous auxin levels of Arabidopsis roots.

Another important root response to soil water status which affects root morphology is the sensing of differences in the availability of water across the circumferential axis of the root. These differences create spatial cues that determine the position of lateral root branches (Bao et al., 2014). Roots of several plant species have been shown to distinguish between a wet surface and an air-filled pore, thereby determining root hair patterning. This hydropatterning response is independent of ABA signalling (Bao et al., 2014), thus distinguishing it from a classic water-stress response. Higher water availability induces the biosynthesis and transport of auxin, the lateral root-inductive signal.

In compacted soil, accumulation of the volatile hormone ethylene actively suppresses rice primary root growth (Pandey et al., 2021). Soil compaction lowers gas diffusion by decreasing soil air-filled porosity, thereby causing ethylene to accumulate in root tissues and trigger hormone responses that restrict growth. This result could be relevant to the breeding of crops resilient to soil compaction, a soil property that adversely affects farming across the globe.

2.3 Shoot properties that contribute to stress resilience and productivity

2.3.1 Shoot growth

In 2012, Australia's Grains Research and Development Corporation started a research initiative across the southern and western regions of the country, aiming to improve WUE of grain-based farming systems by 10%. While farm management was one focus, desirable crop characteristics that can be enhanced by plant breeding included longer coleoptiles, early vigour, canopy architecture, tiller number, leaf glaucousness and high transpiration efficiency. Early seedling vigour is key for crop establishment by increasing ground cover and radiation interception. More rapid ground coverage reduced evaporative soil water losses while enabling greater plant water uptake, leading to efficient water use and higher biomass production at early growth stages (Zhao et al., 2019). This translated to greater biomass production at anthesis as well as at maturity to increase grain yield in years with less terminal drought.

While many plant hormones affect leaf growth, ABA attracts attention since ABA concentrations *in planta* substantially increase as the soil dries, with relevant effects on shoot growth. By limiting stomatal opening, ABA accumulation buffers the day-night alternations of leaf growth rate and the negative effect of high evaporative demand. Furthermore, ABA can improve leaf growth by increasing root conductance to water transfer, while there is also a modest negative, non-hydraulic effect of ABA on leaf growth of plants subjected to water deficit (Tardieu et al., 2010a). Indeed, overexpression of *OsNCED3* (a key rate-limiting enzyme in the ABA biosynthetic pathway) in transgenic rice plants grown under well-watered conditions restricted leaf development compared to the wild type and increased ABA accumulation by 15–20% (Huang et al, 2018). While additional ABA accumulation in such transgenic plants may restrict water use (by limiting canopy development and causing stomatal closure) early in the season, this may ensure water is available later in the season when grain filling occurs.

2.3.2 Stomatal behaviour

Increasing atmospheric VPDs are predicted to cause substantial problems for crops (Lopez et al., 2021), and many authors have shown that restricting stomatal conductance (gs) in response to increased atmospheric VPD increases crop WUE (e.g. Ryan et al., 2016). However, modelling has demonstrated that an enhanced stomatal VPD response enhances maize yield only in regions where drought frequently occurs during the growing season (Messina et al., 2015). Enhanced stomatal sensitivity to VPD decreased yield in regions of the country with high soil water availability, illustrating that introducing specific drought resistance traits often succeeds only in particular 'target environments' (e.g. Condon et al., 2004).

As with most crop species, maize shows pronounced genotypic variation of whole plant WUE (Ryan et al., 2016), which is related to transpirational responses to VPD. While transpiration of some genotypes increased linearly with VPD (with the slope of this relationship determining WUE), other genotypes showed a non-linear response with a pronounced change point at high VPD (low WUE) or low VPD (high WUE). In the latter group, high WUE genotypes required a significantly lower VPD before transpiration was restricted and had a significantly lower rate of transpiration in response to VPD after this point when compared to low WUE genotypes. Change point values were significantly positively correlated with stomatal sensitivity to VPD. Furthermore, foliar ABA concentration was weakly but significantly positively related to increasing WUE. The development of high-throughput phenotyping platforms (Halperin et al., 2017) allows the significance of these relationships to be tested in a wider range of crop species.

Soil water deficit can increase shoot ABA concentrations by two orders of magnitude or more, according to the degree of plant water deficit (e.g. Wright, 1977). Very low ABA concentrations (eg. 10^{-10}M – Ogunkanmi et al., 1973) cause partial stomatal closure. These two observations led to the general acceptance that ABA is an effective and sensitive chemical regulator of gas exchange. ABA promotes stomatal closure via a direct biochemical effect on guard cells and via an indirect hydraulic effect by decreasing water permeability within leaf vascular tissues (Pantin et al., 2012). Variability in sensitivity of leaf hydraulic conductance to ABA among species could provide a physiological basis for isohydric or anisohydric behaviour (Pantin et al., 2012). Moderate water deficit also causes comparatively limited changes in tissue auxin and cytokinin concentrations (e.g. Masia et al., 1994). Although these hormones have often been considered as less sensitive regulators of shoot properties in response to varying plant water status, both interact with ABA to alter stomatal opening (Cox et al., 1985; Blackman and Davies, 1983). Thus, considerations of chemical signalling should include many plant hormones.

There is some uncertainty about the effects of plant water deficits on ethylene signalling. While drought and heat stress can increase ethylene synthesis (e.g. Hays et al., 2007; Wright, 1977), soil water deficit can also suppress ethylene synthesis (Morgan and Drew, 1997). This discrepancy may relate to variation in the rapidity and intensity of soil drying. Different experiments have measured ethylene in plants at different phenological stages and at different times of day. Re-watering drought-stressed plants increased the export of the ethylene precursor (ACC – 1-amino-cyclopropane-carboxylic acid) from roots to shoots (Gomez-Cadenas et al., 1996; Pérez-Pérez et al., 2020). Since ethylene accumulation limits ABA-induced stomatal closure (Desikan et al, 2006; Wilkinson and Davies, 2009), this extra ethylene may help promote rapid re-opening of stomata of re-watered stressed plants and therefore enhance carbon gain. Nevertheless, many effects of ethylene accumulation (perhaps in combination with other hormones) on growth, abscission and senescence (Abeles et al., 1992) and physiology may limit crop yields (e.g. Pierik et al., 2006). Indeed, downregulating expression of ACC synthase (the rate-limiting step in ethylene synthesis) delayed leaf senescence of maize under optimal and droughted conditions, thereby extending the duration of net carbon gain (Young et al., 2004) and enhancing grain yields (Habben et al., 2014).

2.4 Growth and development of reproductive plant parts

The timing of drought stress relative to plant developmental stage determines its effects on grain yield, resulting in fewer and/or smaller grains (e.g. Griffiths et al., 2020). Drought impacts on growth and physiology of the vegetative plant (reviewed earlier), in combination with variation of transport of assimilates to the developing grains, will commonly reduce grain size and number and affect crop resilience and productivity. Severe soil drying during the grain-filling period usually induces early senescence, reduces photosynthesis and shortens the grain-filling period, restricting kernel size. Monocarpic plants such as rice mobilise pre-stored carbon reserves in the stem and sheath during grain filling, and these reserves can make up 20–40% of the grain weight (Yang et al., 2001). Delayed senescence can cause a substantial yield penalty. Controlled, mild soil drying at the grain-filling stage saved water and promotes whole plant senescence and therefore carbon remobilization (Yang and Zhang, 2006). Although soil drying may decrease concurrent photosynthesis, the benefit of increased utilization of pre-stored reserve outweighs the loss caused by reduced gas exchange. Exploiting this drought-stress physiology within deficit irrigation techniques such as partial root-zone drying and alternate wetting and drying demonstrates the benefits of modifying PGR balance (see e.g. Dodd and Kudoyarova, 2021) to both save water and increase WUE.

ABA and cytokinins seem important in regulating plant senescence (Haberer and Kieber, 2002; Zeevaart and Creelman, 1988) thereby affecting grain yield. While ABA can promote foliar senescence (e.g. Huang et al., 2018), it may also regulate assimilate transport to developing seeds or fruits (Brenner and Cheikh, 1995; Dewdney and McWha, 1979). Regulated expression of a cytokinin biosynthesis gene IPT delayed leaf senescence and improved canola yields under rainfed and irrigated conditions (Kant et al, 2015). Moreover, downregulating a gene coding for cytokinin oxidase in rice caused cytokinin accumulation in inflorescence meristems that increased the number of reproductive organs (Ashikari et al., 2005). Similarly, downregulating a gene coding for cytokinin oxidase in barley increased yields by increasing spike and grain number, although at the expense of TGW (Holubova et al., 2018). The advent of genome editing techniques allows more targeted regulation of specific isoforms of this multi-gene family (Gasparis et al., 2018).

Rice grains produce more ethylene during the early grain-filling stage and less during the linear period of grain growth (Yang et al., 2004). In this study, moderate soil drying reduced ethylene evolution, which was significantly and negatively correlated with the grain-filling rate. Increased ethylene concentrations can also accelerate leaf senescence and reduce sink size by accelerating embryo and grain abortion, thereby reducing grain yields (Hays et al., 2007; Shi et al., 2015). A population of wheat with wide genetic diversity exhibited substantial genotypic variation for spike ethylene accumulation (SET) and spike dry weight (SDW) at anthesis, with these variables strongly negatively correlated (Valluru et al., 2017). Applying silver nitrate, an ethylene response inhibitor, significantly decreased SET and subsequently contributed to increased SDW and grain yield under heat stress. Numerous putative SNPs explaining genotypic variation of spike ethylene and dry weight were detected, suggesting that we can select for both variables. QTL studies are starting to reveal apparent ethylene-mediated effects on cereal yields. For example, Kuanar et al. (2019) have shown that introgression of a Sub1 (SUB1) QTL in mega rice cultivars increased ethylene production to the detriment of grain filling under flooding.

2.5 Hormonal signalling regulating crop yield under drought stress

Reynolds et al. (2021) identified several under-researched and under-funded areas where more research efforts should be able to enhance crop yield in water-scarce cropping regions. One of these areas is 'the role of hormone crosstalk' since a multi-hormone stress signal might integrate the effects of drought and/or heat on crop yield. To conceptualise the idea that the 'stress hormone' ABA acts within a hormonal interaction network involving auxin, ethylene and cytokinins, we examine the possibility that ABA accumulation under water deficit integrates root responses to different stresses. In a field-grown maize crop, gs was not related to either current leaf water status or bulk leaf ABA concentration but was tightly related to xylem ABA concentration (Tardieu et al., 1992). This relationship was common for variations in xylem [ABA] linked to gradual decreases in soil water reserve, simultaneous differences between plants grown on compacted, non-compacted and irrigated soil, and plant-to-plant variability. However, bulk leaf ABA concentration was not related to gs, likely because much of the ABA synthesised in the leaves can be sequestered into leaf mesophyll cells that act as an anion trap for the charged ABA molecule (away from guard cell ABA binding sites) (Hartung et al., 1988). Furthermore, daily uptake of ABA from the xylem into leaves of grapevines accounted for diurnal changes in leaf ABA concentration (Loveys, 1984). These relationships have led to suggestions that stomata can respond to chemical 'root signals' (e.g. Tardieu et al., 1992), with ABA accumulating under water deficit integrating root responses to the associated stress (Maurel and Nacry, 2020).

While the xylem transports many compounds from roots to shoots, Jackson (1993) suggested that xylem-mobile compounds should only be considered 'root signals' if they are synthesised in roots. Much recent debate has discussed how much ABA in the xylem stream is actually synthesised in the roots (e.g. McAdam et al., 2016). In some of the first 'root signalling' work, Loveys (1984) quantified the relative proportions of root- and shoot-sourced ABA in the xylem of drought-stressed grapevines, demonstrating that ABA moving from roots to shoots largely originated in the leaves. Some of this ABA, moving to the roots in the phloem, is retained in the roots, while the rest is released to the xylem and transported to the shoots, constituting a 'root signal' (Slovik et al, 1995). Root/soil conditions determine how much ABA recirculates to the shoot and how much is lost to the soil. Once in the xylem, root-sourced (but not necessarily root-synthesised) ABA can be depleted/augmented as a function of pH gradients between the xylem and the xylem parenchyma and by ABA synthesis in the surrounding cells in the stele (Hartung et al., 2002; Okamoto et al., 2009). Four different anatomical levels regulate the intensity of the root to shoot ABA signal: (1) the rhizosphere, (2) the root cortex, (3) the stem and (4) the leaves (Jiang and Hartung, 2008). Relevant processes include (1) root–soil interactions including soil pH and growth-promoting rhizobacteria (Belimov et al., 2014); (2) anatomical features of the root cortex (e.g. Casparian bands) that can constitute a barrier to ABA transport (3) apoplastic pH regulation in the stem and the leaf which will significantly impact fluxes of ABA into and out of the transport pathway (Hartung et al., 2002) and (4) the action of apoplastic and cytosolic β-glucosidases in the leaves. While the conjugate ABA-GE could be a perfect long-distance signal *in planta* because it is translocated without loss to the surrounding tissues (Jiang and Hartung, 2008), ABA catabolism in both vascular tissues and guard cells contributes to systemic ABA action that controls stomatal movement (Okamoto et al., 2009).

A number of hormones can operate as an integrative 'hormonal signal' (see e.g. Belimov et al., 2009) that regulates vegetative plant growth and development. The model proposed by Pierik et al. (2006) focuses on ethylene, where many hormones affect photosynthesis and cell expansion to influence plant growth by interacting with various other signal transduction pathways. Models proposed by Tardieu and Davies (1992), Tardieu et al. (2015) and Tardieu (2016) assess the control of plant functioning as a function of chemical, hydraulic and combined signals.

The significance of chemical signalling should be assessed in relation to the reproductive development of droughted plants. The hormonal balance of developing reproductive structures can affect yield of both wheat and rice (Yang and Zhang, 2006). In the following section, we test the hypothesis that quantitatively assessing plant hormone accumulation in an appropriate tissue location and at an appropriate stage of development can help explain stress resilience and plant productivity. This may provide breeders with a basis to develop improved crop plants for hot and dry regions.

3 Hormonal regulation of crop productivity and resilience

3.1 Yield components and score index

As discussed previously, yield components are determined during different phenological stages. Defining a narrow window of time during plant development can help us understand plant sensitivity to drought stress in terms of grain number. For example, identifying new traits to aid breeding for SF should focus on the rapid spike-growth phase until shortly after anthesis when grain number per spike, a highly complex and polygenic trait, is determined (Slafer, 2012). Score indices could be applied to this specific yield component and be divided into a resilient and productive sub-component (Thiry, 2017).

3.2 Hormones vs stress resilience. Where and when hormones should be measured?

As previously mentioned, the stem elongation phase/late reproductive stage (from the terminal spikelet stage to anthesis) is a critical period to determine floret potential number, viability and SF (González et al., 2011; Guo et al., 2018; Miralles et al., 2000; Slafer, 2012). Floret development determines wheat potential grain number from terminal spikelet to booting stage, and up to 50% of the floret primordia can be lost from booting to heading/anthesis, thereby reducing yield potential (Guo et al., 2018; Slafer, 2012).

The post-anthesis phase (from anthesis to physiological maturity) determines individual grain weight (Guo et al., 2018). Grain weight potential (grain size) is determined from the start of anthesis (when ovary development starts) until the start of the grain filling (Brinton and Uauy, 2019; Calderini et al., 1999, 2001; Ugarte et al., 2007). Floret and plant growth are affected by both genetic and environmental factors (Guo et al., 2018). A better understanding of pollen viability (physiological and molecular processes) may increase the grain number under stress conditions (Dolferus et al., 2011; Gietler et al., 2020; Li et al., 2019; Zheng et al., 2019). Studies should focus on a narrow time window during the stem elongation phase (from the green anthers phase to anthesis) when both anther and ovary size and survival influence final grain number (Dwivedi et al., 2017; Guo et al., 2018).

From studies of this kind, we have a clear idea of when hormone accumulation and effects should be assessed as reproductive structures develop. We have discussed earlier how a root signal might integrate drought stress effects and such a signal is not hard to measure, assuming xylem sap can be readily collected from the crop of interest (Jackson, 1993) by using/adapting new improved and faster method for xylem sap collection (Kan et al., 2019). Hormone assessments in developing reproductive structures are complicated by their size and location (Westgate et al., 1996), and some dissection of these structures from surrounding tissues may be necessary. Since these measurements are not straightforward in field crops, under certain circumstances phenotyping in controlled environments may be considered (Cobb et al., 2013) to minimise the effects of pests and diseases and pesticide use on hormone accumulation (Abeles et al., 1992). Since hormone concentrations will also be under circadian regulation, sampling times must be consistent (Kobayashi and Saka, 2000; Levinsh and Kreicbergs, 1992; McClung, 2000). For example, ethylene emission often peaks mid-photoperiod (Thain et al., 2004) and ABA synthesis is also under circadian control (Khan et al., 2010).

ABA and ethylene are two major plant hormones linked to many plant drought stress responses. Pollen sterility is associated with higher spike ABA concentrations in sensitive wheat genotypes (Gietler et al., 2020; Ji et al., 2010, 2011), and during early development, external application of high ethylene doses induces anther sterility (Campbell et al., 2001). While ethylene does not induce ovary sterility (Rowell and Miller, 1971), stress ethylene can negatively affect grain-filling rate and grain abortion, while promoting leaf senescence (Wilkinson et al., 2012). Several plant drought stress responses such as reduced gs, leaf growth and grain-filling rate are best explained in terms of the interactions between ABA and ethylene (Davies et al., 2013; Müller, 2021; Peleg and Blumwald, 2011; Tanaka et al., 2005; Valluru et al., 2014, 2017; Wilkinson and Davies, 2009; Yang and Zhang, 2006). ABA–ethylene concentration balance has been suggested as a novel target for plant breeders to identify adapted and/or highly productive genotypes under drought stress (Müller, 2021; Wilkinson et al., 2012; Yang et al., 2007).

3.3 Measuring hormones under field conditions

High-throughput methods of sampling and hormone analysis, focusing on specific phenological stages, are needed. Most plant hormones (except ethylene) can be analysed through a multi-analyte physicochemical technique using HPLC/MS analysis (Albacete et al., 2008). When costs of such analyses are prohibitive, radioimmunoassay for ABA (RIA-ABA) and assessing ethylene with a laser-based photoacoustic ethylene detector (ETD-300, Sensor Sense B.V., Nijmegen, The Netherlands) (Thiry, 2017; Thiry et al., 2017) are both relatively rapid and cost-effective measurement methods. Developing a rapid, simple sampling method to quantify ethylene emission rate under field conditions, or identifying a surrogate for this variable, is a priority as all samples should be collected during a 90-min window to maintain similar incubation conditions between the first and the last plot (such as temperature, light, etc.) (Thiry et al., 2017). Although ABA sampling under field conditions is a faster process (excising plant tissues and freezing them in liquid nitrogen), ABA determination through RIA-ABA requires more laboratory time to freeze-dry and grind tissues before sample extraction and RIA analysis *per se*. After preliminary sample preparation, a single operator can analyse *circa* 160–240 samples for RIA-ABA and 100–150 for ethylene in a single day (Thiry, 2017; Thiry et al., 2017).

3.4 ABA/ethylene ratio as a new trait – a case study

Better understanding the effects of abiotic stress on reproductive plant development will drive the development of more reliable screening methods (Ahmadizadeh et al., 2016; Dolferus et al., 2011). Focusing on variation in grain number per m^2 (G#/m^2; the best component to explain wheat yield variations under stress – Fischer, 1985) and SF (a major component of G#/m^2) (Abbate et al., 1998; Fischer, 1984; Martino et al., 2015) is important, but both these traits are negatively related to grain weight (Martino et al., 2015).

Our field experiment quantified leaf ABA concentration and ethylene emission rate at four different phenological stages of wheat during the pre-anthesis period from booting to anthesis. Data were collected on eight genotypes from CIMCOG-ROOT (a panel of genotypes selected from the CIMMYT core germplasm collection) subjected to drought in Mexico during the 2013–2014 cropping season. Studying different yield components using the score index (Thiry et al., 2016) showed that the resilience index of grain number per m^2 (Grain#/spike-RCI) was highly positively related to grain number per m^2 (Table 3). Surprisingly, grain number per m^2 was not correlated to yield under stress (Ys), likely due to the sowing method (Table 4). Plots were sown with a fixed seed weight (5 g/m^2), not a fixed number of seeds per plot. Thus, genotypes with lower grain weight will have more seeds per plot affecting tiller development (Valério et al., 2013) and inevitably the yield component of spike number/m^2. Removing the two genotypes with lower grain weight re-establishes the expected relationship between grain#/m^2 and yield (Fischer, 1984) (Table 4). Table 5 shows the score values of the Resilience Capacity Index of grain number per spike (Grain#/spike-RCI), the productivity Capacity Index of grain number per spike (Grain#/spike-PCI) and the combination of both indices, the grain number per spike stress score index (Grain#/spike-SSI). The original value of grain number per spike is highly positively correlated with Grain#/spike-SSI ($r = 0.97$, $P < 0.01$) (Table 5).

Contrasting values of Grain#/spike-RCI within the 8 genotypes (Table 5) indicate genotypes 1, 2, 4, 8 and 10 are more resilient and genotypes 5, 6 and 9 are more sensitive to drought. ABA and ETH were quantified in two single

Table 3 Pearson correlation coefficient between the yield components and yield components score indices directly related to the grain setting per area or plants (Grain#/m^2, Grain#/spike, respectively) and the original value of yield under drought stress and grain number per metre square obtained in the field, data from the CIMCOG-ROOT trial under drought stress (eight genotypes) during the crop season 2013–2014. Drought stress means that the last water supply was applied when 50% of the plot has reached emergence. (Grain#/m^2: number of grains per square metre, Grain#/m^2-SSI: grains number per square metre stress score index, Grain#/m^2-PCI: grains number per square metre productivity capacity index, Grain#/m^2-RCI: grains number per square metre resilient capacity index, Grain#/spk: number of grains per spike, Grain#/spike-SSI: Grain number per spike stress score index, Grain#/spike-PCI: grains number per square metre productivity capacity index, Grain#/spike-RCI: grains number per square metre resilient capacity index)

	Ys	Grain#/m^2
Ys	1.00[b]	0.10
Grain#/m^2	0.10	1.00[b]
Grain#/m^2-SSI	0.24	0.97[b]
Grain#/m^2-PCI	−0.56	0.70
Grain#/m^2-RCI	0.83[a]	0.38
Grain#/spike	−0.18	0.69
Grain#/spike-SSI	−0.09	0.81[a]
Grain#/spike-PCI	−0.30	0.39
Grain#/spike-RCI	0.12	0.94[b]

[a] $P < 0.05$.
[b] $P < 0.01$.

plants per plot for eight genotypes with three field replicates (plots) during critical phenological stages (from booting to heading). Foliar ethylene emission rate and whole leaf [ABA] generally do not discriminate between traits, with only some phenological stages showing significant genetic variability under yield potential (Yp) conditions and/or under drought (D) (Fig. 2).

Drought-induced increases in ABA generally result in the hormone ratio ABA/ETH under drought stress exceeding that observed under irrigated conditions (Yp). As we've discussed, ethylene concentrations in plant parts may or may not increase as tissue water potentials decline. Importantly, the hormone ratio ABA/ETH shows significant genotypic variation at all developmental stages under both environments (Fig. 3), with discrimination highest during booting, late booting and heading stages.

These results show the relevance of considering the phenological stage of the plant when considering plant hormones as a potential trait. It is also important to consider the plant tissue that may be crucial in the development of yield during development, as well as the time of day when measurements might be made. This is because plant hormone accumulation will vary with time of day (e.g. Thiry et al., 2017).

Table 4 Pearson correlation coefficient between the yield components and yield components score indices directly related to grain number production per area or plants (Grain#/m^2, Grain#/spike, respectively) and the original value of yield under drought stress and grain number per metre square obtained in the field, data from the CIMCOG-ROOT trial under drought stress (six genotypes with similar grain weight) during the crop season 2013–2014. Drought stress means that the last water supply was applied when 50% of the plot has reached emergence. (Grain#/m^2: number of grains per square metre, Grain#/m^2-SSI: grains number per square metre stress score index, Grain#/m^2-PCI: grains number per square metre productivity capacity index, Grain#/m^2-RCI: grains number per square metre resilient capacity index, Grain#/spk: number of grains per spike, Grain#/spike-SSI: Grain number per spike stress score index, Grain#/spike-PCI: grains number per square metre productivity capacity index, Grain#/spike-RCI: grains number per square metre resilient capacity index)

	Ys	Grain#/m^2-YS
Ys	1.00[b]	0.81[a]
Grain#/m^2-YS	0.81[a]	1.00[b]
Grain#/m^2-SSI	0.91[a]	0.98[b]
Grain#/m^2-PCI	0.13	0.66
Grain#/m^2-RCI	0.97[b]	0.77[a]
Grain#/spike-YS	0.10	0.59
Grain#/spike-SSI	0.30	0.75[a]
Grain#/spike-PCI	−0.20	0.30
Grain#/spike-RCI	0.73[a]	0.96[b]

[a] $P < 0.05$.
[b] $P < 0.01$.

3.5 Hormones and/or their balance: a trait for drought stress resilience

At the late booting stage, the resilience component of thousand grain weight (TGW-RCI) was significantly positively correlated with leaf ABA concentration (r = 0.85, $P < 0.05$) (Table 6), while the resilience component of grain number per spike (Grain#/spk-RCI) was significantly positively correlated with leaf ethylene emission (r = 0.83, $P < 0.05$) (Table 6). At this time, leaf hormone ratio ABA/ETH was significantly negatively related to the resilience component of grain number per spike (Grain#/spk-RCI) (r = −0.95, $P < 0.01$), the best correlation observed (Table 6). Figure 4 illustrates the linear regression at this stage (late-booting) of G#/spk-RCI plotted vs (1) leaf ABA accumulation, (2) leaf ethylene emission and 3) leaf hormone ratio (ABA/ETH). Although this CIMCOG-ROOT trial was not specifically selected for contrasting genotypic responses to drought stress, using the index score method with yield and yield components data identified 1 sensitive and 2 resilient genotypes, based on contrasting resilience score index of grain per m^2 and grain number per spike which is

Table 5 Values of the Resilience Capacity Index of grain number per spike (Grain#/spike-RCI), the Productivity Capacity Index of grain number per spike (Grain#/spike-PCI), and the result of their combination as the grain number per spike stress score index (Grain#/spike-SSI) which is highly correlated with the real value of grain number per spike under drought stress (Grain#/spike-D). Values are calculated using the grain number per spike data from the CIMCOG-ROOT trial (eight genotypes) under drought stress, for the 2013–14 cropping season (Y13–14)

Genotypes	Grain#/spike-RCI	Grain#/spike-PCI	Grain#/spike-SSI	Grain#/spike-D	Pearson correlation (Grain#/spike-SSI vs Grain#/spike-YS)
1	9	8	8.5	50.9	0.97[a]
2	7	8	7.5	49.13	
4	6	8	7	49.42	
5	1	4	2.5	41.28	
6	4	1	2.5	39.3	
8	10	6	8	49.31	
9	4	8	6	48.21	
10	6	10	8	51.85	

[a] $P < 0.01$.

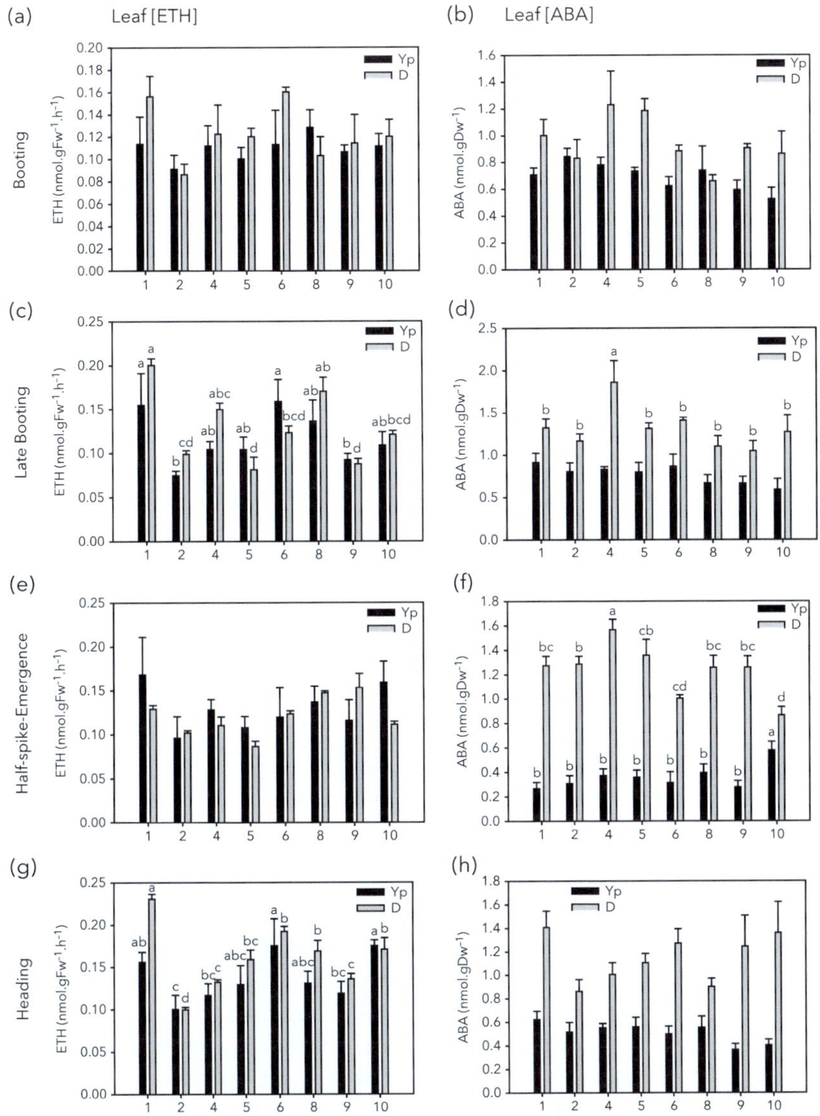

Figure 2 Genotypic variation on flag leaf ethylene emission rate (nmol.gFw^{-1}.h^{-1}) (a, c, e, g) and leaf ABA concentrations (nmol.g Dw^{-1}) (b, d, f, h) under different environments (irrigated (Yp, dark bar) and drought stress (d, grey bar)) at different phenological stages: Booting (a and b), Late Booting (c and d), Half-spike-Emergence (e and f) and Heading (g and h). Data were collected on eight genotypes from CIMCOG-ROOT during the crop season 2013–2014. Columns and bars are means ± standard error of six replicates with different letters indicating significant ($P < 0.05$) difference among genotypes within an environment and only when significant differences exist between the genotypes.

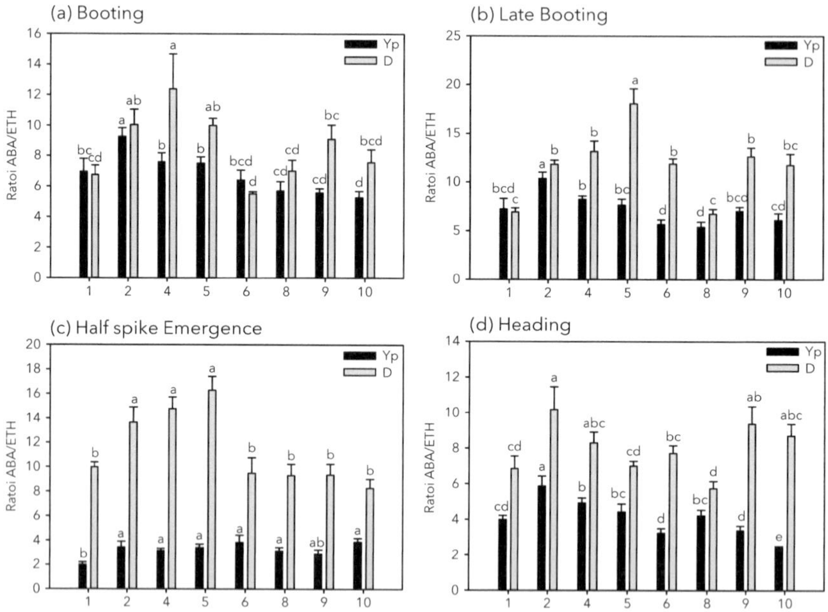

Figure 3 Genotypic variation of the ratio ABA/ETH in leaf tissues and their changes under different treatments yield potential (Yp, dark bar) and drought (d, grey bar) along different phenological stages, Booting (a), Late Booting (b), Half spike Emergence (c) and Heading (d). Data were collected on eight genotypes from CIMCOG-ROOT during the crop season 2013–2014. Columns and bars are means ± standard error of six replicates, with different letters indicating significant ($P < 0.05$) difference among the genotypes within an environment and only when significant differences exist between the genotypes.

a much more specific contrast than the conventional tolerant vs susceptible genotype description base usually on yield response.

Compared to quantification of production of individual hormones, hormone ratios (ABA/ETH) better differentiated genotypic variation in stress resilience response. Late booting stage could be a sensitive stage for the development of basal and terminal spikelets. At meiosis, stress-induced reduction in the availability of carbon and nitrogen can be critical for yield development (Acevedo et al., 2002) and can result in the death of florets from the terminal and basal spikelets (Oosterhuis and Cartwright, 1983). Kobayashi and Saka (2000) concluded that sucrose content, which is critical for grain development, positively regulates the rate of ethylene evolution in the rice seedling leaf, and therefore, it could be hypothesized that through this mechanism higher leaf ethylene production may be then associated with more grains per spike.

Additionally, it has been observed that stomatal closure induced by ABA can be restricted in the presence of ethylene, thereby positively impacting the photosynthetic rate (Wilkinson and Davies, 2009, 2010). All genotypes show

Table 6 Table of Pearson coefficient correlation between the resilient capacity index (RCI) calculated on yield and yield components vs mean flag leaf ABA concentration, mean ethylene emission rate and the mean ratio ABA/ETH under drought stress. RCI: resilient capacity index, Y: yield, TGW: thousand grain weight, Grain#/m^2: grain number per m^2, Spk#/m^2 spike number per square metre, Grain#/spk: grain number per spike, at the different stages (booting, late-booting, half-emergence and heading) where genotypes have shown significant differences with the leaf hormone accumulation (ABA, ethylene) and hormone ratio ABA/ETH

Hormone	Stages	Y-RCI	TGW-RCI	Grain#/m^2-RCI	Spk#/m^2-RCI	Grain#/spk-RCI
ABA Leaf	L. Booting	0.32	0.85[a]	−0.16	−0.22	−0.22
	Half-Ear-Emergence	−0.09	0.57	−0.37	−0.28	−0.11
ETH Leaf	L. Booting	0.14	−0.03	0.10	−0.74[a]	0.83[a]
	Heading	−0.36	−0.38	−0.24	−0.57	0.50
ABA/ETH Leaf	Booting	−0.10	0.68[a]	−0.40	0.08	−0.56
	L. Booting	−0.21	0.44	−0.41	0.48	−0.95[b]
	Half-Ear-Emergence	−0.28	0.64[a]	−0.57	−0.11	−0.51
	Heading	0.09	0.13	0.06	0.44	−0.26

[a] $P < 0.05$.
[b] $P < 0.01$.

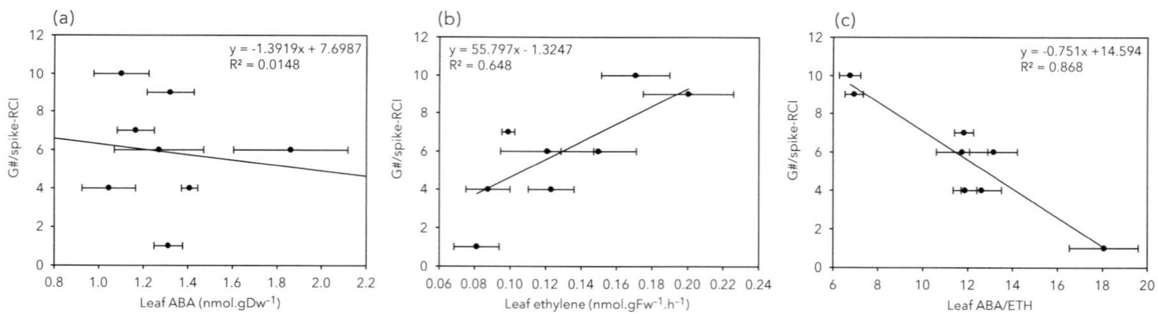

Figure 4 Linear regression between the resilient capacity index of grain number per spike (G#/spk-RCI) (y-axis) vs (a) mean leaf ABA accumulation (nmol.g Dw-1), (b) mean leaf ethylene emission (nmol.gFw^{-1}.h^{-1}) and (c) mean leaf hormone ratio ABA/ETH at late booting stage. Data were collected on CIMCOG-ROOT trial (eight genotypes) during the cropping season 2013-2014.

an increased concentration of ABA in the leaf with no significant differences between genotypes. We suggest that at this stage when the spike reaches the maximum growth rate and competes with vegetative growth for carbohydrates, the higher ethylene emission, observed with genotypes with higher G#/spike-RCI, could decelerate shoot growth increasing resource availability for reproductive development.

Where a high ethylene emission rate at the late booting stage from leaf tissue under field conditions is apparently a determining factor for SF, this might be useful as a new trait for selection for resilience of SF to severe drought stress. However, we have to consider that these genotypes did not show a significant variation in ABA accumulation. Hormone ratio showed better discrimination along the growth stages suggesting that it may be a better trait for selection than single hormone accumulation. Additionally, the interactive effect of ABA and ethylene on carbon availability (Wilkinson and Davies, 2009, 2010) might explain the difficulty in clearly identifying an individual PGR ideotype for ABA (e.g. Blum, 2015).

There is no doubt that an increase in [ABA] is associated with an increased closure of stomata in the leaf and such a response can often reduce growth. However, the role of ethylene in the regulation of plant development is not so clear. An increased emission rate of this hormone has been reported to both promote (this study) and restrict growth, development and yield. For example, Hays et al. (2007) have concluded that increased ethylene accumulation can reduce yield by inducing kernel abortion under heat stress. Ethylene can also reduce grain-filling rate under drought stress (Yang et al., 2006a). From this work, we highlight the relevance of phenological stages, as a key factor in determining the effects of stress on yield. The work of the Hays and Yang groups focused on the post-anthesis period of development while the work here focused on pre-anthesis and specifically on the spike growth stage and this likely explains some differences in reported ethylene effects. It is important also to consider factors such as the kind of stress applied (short time vs long stress, field vs controlled environment), the type of tissues studied (spike, kernel or leaf) and daytime (morning, noon, evening) (Thiry, 2017). Focusing on responses to the application of hormones along the stages of development would likely be an effective way to understand the role of particular hormones, but the assessment of endogenous concentrations in different plant compartments would be important. Application of inhibitors or promoters of a specific hormone will often yield interesting data highlighting the relevance of hormone concentration as demonstrated by Pierik et al. (2006) in the case of ethylene.

3.6 The score index: a new screening method

Score indices (RCI and PCI) have been shown to identify those rare genotypes that do not show negative relationships between yield stability and high productivity (Thiry et al., 2016 and Fig. 1b) and also to identity contrasting genotypes or parents for further crosses (i.e. high resilient × high productive genotypes). By focusing on resilience and productivity, this method can considerably decrease the number of selected genotypes, saving breeders time and money. This novel classification using the score index provides much more information about genotype responses to the stress environment than the usual stress-related terminology (such as tolerant/resilient and sensitive/susceptible genotypes) which does not always describe the basis of these classifications.

If breeders are looking for a highly productive and resilient genotype under stress, we propose an initial focus on productivity performance (high PCI), then reduce the selection by selecting more resilient genotypes (high RCI). However, using the index method requires caution. Score index interpretations should be derived from a group of scores (e.g. grouping the three highest and/or the three lowest scores or selecting genotypes with a score superior or equal to the control cultivar) and the scores should ideally be estimated for a large population as they are based on comparing the mean responses of the full population trial. Evaluating a substantial population increases the genotypic diversity and allows the researcher or breeder to avoid complications of working with genotypes presenting an intermediate stress response. This quantitative method identifies contrasted genotypes in terms of yield components RCI and PCI and allows us to highlight hormonal variation that might be used as a physiological trait. This approach can be beneficial in discovering the identification of novel physiological traits specifically related to resilience or production capacity in terms of yield or yield components. This work could be extended with QTL analysis to identify the genetic component associated with the single trait under consideration. Moreover, using a uniform criterion (PCI and RCI) in fundamental research ensures more meaningful comparisons between research results obtained across a selection panel where there may only be a hazy understanding of potential selection criteria.

The benefits of our proposed approach to plant improvement become more accessible as we develop a new methodology for the identification and quantification of potential regulators of growth and development. Decades of research on hormones have produced convincing data on the effects of a range of regulators on plant growth, development and functioning, but thus far we have largely failed to fully exploit, for the benefit of society, our understanding in this active research area which can be essential as global challenges to our way of life become more and more pressing. Modern genetics opens new opportunities for exploitation through plant breeding and the evolution of new cropping systems. The studies reported upon here show that as we narrow down our targets for manipulation the more likely it is that genetic modifications in the biology of hormones will reap benefits for society.

4 Future trends in research

- Changing agronomic practices and field management such as applying agro-chemical products (fungicides and/or pesticides) or irrigation might disturb the hormone balance and modify or confound the interpretation of resilience response of the genotype to drought stress during selection trials. For example, applying triazole fungicide can alter the hormone balance by inhibiting the gibberellin synthesis, reducing ethylene evolution and increasing cytokinin levels (Rajalekshmi et al., 2009). These variations of ethylene and cytokinin production could have a beneficial effect on the grain-filling rate (based on Christopher et al., 2016; Deveshwar et al., 2020; Reynolds et al., 2021; Thomas and Ougham, 2014, 2015; Yang et al., 2006a,b) but could negatively affect spikelet formation (Wang et al., 2018) and grain number if applied from early to late booting under drought stress, based on this work.
- Development- and tissue-specific promoters provide further opportunities to generate variation in plant hormone status in a controlled manner, while selecting based on the score index may identify natural allelic variation in hormone-related genes. In addition, progress in genomics and phenomics allows for accurate and comprehensive characterization of the QTLs that govern the sort of variation proposed here for targeted breeding programmes (Salvi and Tuberosa, 2015). This is important in the context of the programme proposed here. Alleles at major QTLs have already been used in marker-assisted breeding.
- Further elucidation (within the framework provided by the score index) of the effects of different hormone interactions on plant growth development and functioning. These studies are supported by advances in the techniques available for quantifying plant hormones.
- Further elucidation of xylem-mobile hormone signals as a potential integrator of the effects on growth, development and yield of the plant's above-ground and below-ground environments. Studies of this kind can help optimise interactions between genotypic improvements and agronomic manipulations aimed at optimising water productivity of crops in increasingly dry environments.

5 Where to look for further information

The review shows the progress of the understanding of different PGRs on plant tolerance to abiotic stress such as heat, drought, salinity, and flood.

- Sabagh, A. E. L., Mbarki, S., Hossain, A., Iqbal, M. A., Islam, M. S., Raza, A., Llanes, A., Reginato, M., Rahman, M. A., Mahboob, W., Singhal, R. K., Kumari, A., Rajendran, K., Wasaya, A., Javed, T., Shabbir, R., Rahim, J., Barutçular, C., Habib Ur Rahman, M., Raza, M. A., Ratnasekera, D., Konuskan I, Ö., Hossain, M. A., Meena, V. S., Ahmed, S., Ahmad, Z., Mubeen, M., Singh, K., Skalicky, M., Brestic, M., Sytar, O., Karademir, E., Karademir, C., Erman, M. and Farooq, M. 2021. Potential Role of Plant Growth Regulators in Administering Crucial Processes Against Abiotic Stresses. Front. Agron. 3, 1-28. https://doi.org/10.3389/fagro.2021.648694.

The next review discusses how basic and new knowledge can be applied for breeding and agronomy practices to improve productivity. It is exposed that filling the gap of knowledge on hormone crosstalk in response to stress environment can have favorable impacts on improving crop yield.

- Reynolds, M., Atkin, O. K., Bennett, M., Cooper, M., Dodd, I. C., Foulkes, M. J., Frohberg, C., Hammer, G., Henderson, I. R., Huang, B., Korzun, V., McCouch, S. R., Messina, C. D., Pogson, B. J., Slafer, G. A., Taylor, N. L. and Wittich, P. E. 2021. Addressing Research Bottlenecks to Crop Productivity. Trends Plant Sci. https://doi.org/10.1016/j.tplants.2021.03.011.

More information about wheat and maize breeding:

- Reynolds, M. P. and Braun, H.-J. (eds) 2022. Wheat Improvement. Springer, Cham. https://doi.org/10.1007/978-3-030-90673-3.
- www.cimmyt.org - International Maize and Wheat Improvement Center.

Information about rice breeding:

- www.irri.org - International Rice Research Institute.

6 Acknowledgements

Initial development of the score index and the first relationships between hormones and spike fertility were supported by the project Modernizacion de la Agricultura Tradicional of the Mexican Government (MasAgro), the International Wheat Yield Partnership (IWYP) and by the CGIAR research project in wheat (CRP Wheat). Further development of the score index method and hormone relationships was supported by an Institutional Links grant ID 414238654 and a Newton Fund Impact Scheme ID 630222342 under the Newton-Paulet Fund partnership. The latter awards are funded by the UK Department for Business, Energy and Industrial Strategy and CONCYTEC and delivered by the British Council.

7 References

Abbate, P. E., Andrade, F. H., Lázaro, L., Bariffi, J. H., Berardocco, H. G., Inza, V. H. and Marturano, F. 1998. Grain yield increase in recent argentine wheat cultivars. *Crop Science* 38(5), 1203–1209. DOI: 10.2135/cropsci1998.0011183X003800050015x.

Abbate, P. E., Pontaroli, A. C., Lázaro, L. and Gutheim, F. 2013. A method of screening for spike fertility in wheat. *The Journal of Agricultural Science* 151(3), 322–330. DOI: 10.1017/S0021859612000068.

Abeles, F. B., Morgan, P. W. and Saltveit, M. E., Jr. 1992. *Ethylene in Plant Biology* (2nd edn.). New York: Elsevier.

Abley, K., Formosa-Jordan, P., Tavares, H., Chan, E. Y. T., Afsharinafar, M., Leyser, O. and Locke, J. C. W. 2021. An Aba-Ga bistable switch can account for natural variation in the variability of Arabidopsis seed germination time. *eLife* 10. DOI: 10.7554/eLife.59485.

Acevedo, E., Silva, P. and Silva, H. 2002. Wheat growth and physiology, development. In: Curtis, B. C., Rajaram, S. S. and Macpherson, H. G. (Eds.), *Bread Wheat Improvement and Production, Plant Production and Protection Series* (Vol. 30). Rome, Italy: Food and Agriculture Organization. Available at: http://www.fao.org/3/y4011e/y4011e00.htm.

Ahmadizadeh, M., Vispo, N. A., Calapit-Palao, C. D. O., Pangaan, I. D., Viña, C. D. and Singh, R. K. 2016. Reproductive stage salinity tolerance in rice: A complex trait to phenotype. *Indian Journal of Plant Physiology* 21(4), 528–536. DOI: 10.1007/s40502-016-0268-6.

Albacete, A., Ghanem, M. E., Martínez-Andújar, C., Acosta, M., Sánchez-Bravo, J., Martínez, V., Lutts, S., Dodd, I. C. and Pérez-Alfocea, F. 2008. Hormonal changes in relation to biomass partitioning and shoot growth impairment in salinized tomato (*Solanum lycopersicum* L.) plants. *Journal of Experimental Botany* 59(15), 4119–4131. DOI: 10.1093/jxb/ern251.

Araus, J. L. and Cairns, J. E. 2014. Field high-throughput phenotyping: The new crop breeding frontier. *Trends in Plant Science* 19(1), 52–61. DOI: 10.1016/j.tplants.2013.09.008.

Araus, J. L., Slafer, G. A., Reynolds, M. P. and Royo, C. 2004. Physiology of yield and adaptation in wheat and barley breeding. In: Nguyen, H. T. and Blum, A. (Eds.), *Physiology and Biotechnology Integration for Plant Breeding*. New York: Taylor and Francis: 1–49.

Ashikari, M., Sakakibara, H., Lin., S., Yamamoto, T., Takashi, T., Nishimura, A., Angeles, E. R., Qian, Q., Kitano, H. and Matsuoka, M. 2005. Cytokinin oxidase regulates rice grain production. *Science* 309(5735), 741–745. DOI: 10.1126/science.1113373.

Bao, Y., Aggarwal, P., Robbins, N. E., Sturrock, C. J., Thompson, M. C., Tan, H. Q., Tham, C., Duan, L., Rodriguez, P. L., Vernoux, T., Mooney, S. J., Bennett, M. J. and Dinneny, J. R. 2014. Plant roots use a patterning mechanism to position lateral root branches toward available water. *Proceedings of the National Academy of Sciences of the United States of America* 111(25), 9319–9324. DOI: 10.1073/pnas.1400966111.

Belimov, A. A., Dodd, I. C., Hontzeas, N., Theobald, J. C., Safronova, V. I. and Davies, W. J. 2009. Rhizosphere bacteria containing ACC deaminase increase yield of plants grown in drying soil via both local and systemic hormone signalling. *New Phytologist* 181(2), 413–423. DOI: 10.1111/j.1469-8137.2008.02657.x.

Belimov, A. A., Dodd, I. C., Safronova, V. I., Dumova, V. A., Shaposhnikov, A. I., Ladatko, A. G. and Davies, W. J. 2014. Plant Physiology and Biochemistry Abscisic acid metabolizing rhizobacteria decrease ABA concentrations in planta and alter plant growth. *Plant Physiology and Biochemistry* 74, 84–91. DOI: 10.1016/j.plaphy.2013.10.032.

Bennani, S., Nsarellah, N., Jlibene, M. and Tadesse, W. 2017. Efficiency of drought tolerance indices under different stress severities for bread wheat selection. *Australian Journal of Crop Science* 11(4), 395–405. DOI: 10.21475/ajcs.17.11.04.pne272.

Blackman, P. G. and Davies, W. J. 1983. The effects of cytokinins and ABA on stomatal behaviour of maize and Commelina. *Journal of Experimental Botany* 34(12), 1619–1626. DOI: 10.1093/jxb/34.12.1619.

Blum, A. 1996. Yield potential and drought resistance: Are they mutually exclusive? In: Reynolds, M. P., Rajaram, S. and McNab, A. (Eds.), *Increasing Yield Potential in Wheat: Breaking the Barriers*. Mexico: CIMMYT: 90–100.

Blum, A. 2005. Drought resistance, water-use efficiency, and yield potential - Are they compatible, dissonant, or mutually exclusive? *Australian Journal of Agricultural Research* 56(11), 1159–1168. DOI: 10.1071/AR05069.

Blum, A. 2011. Drought resistance - Is it really a complex trait? *Functional Plant Biology* 38(10), 753–757. DOI: 10.1071/FP11101.

Blum, A. 2015. Towards a conceptual ABA ideotype in plant breeding for water limited environments. *Functional Plant Biology* 42(6), 502–513. DOI: 10.1071/FP14334.

Blum, A., Shpiler, L., Golan, G. and Mayer, J. 1989. Yield stability and canopy temperature of wheat genotypes under drought-stress. *Field Crops Research* 22(4), 289–296. DOI: 10.1016/0378-4290(89)90028-2.

Blum, A. and Tuberosa, R. 2018. Dehydration survival of crop plants and its measurement. *Journal of Experimental Botany* 69(5), 975–981. DOI: 10.1093/jxb/erx445.

Borrás, L., Slafer, G. A. and Otegui, M. E. 2004. Seed dry weight response to source–sink manipulations in wheat, maize and soybean: A quantitative reappraisal. *Field Crops Research* 86(2–3), 131–146. DOI: 10.1016/j.fcr.2003.08.002.

Bourgault, M., Webber, H. A., Chenu, K., O'Leary, G. J., Gaiser, T., Siebert, S., Dreccer, F., Huth, N., Fitzgerald, G. J., Tausz, M. and Ewert, F. 2020. Early vigour in wheat: Could it lead to more severe terminal drought stress under elevated atmospheric [CO2] and semi-arid conditions? *Global Change Biology* 26(7), 4079–4093. DOI: 10.1111/gcb.15128.

Braun, H. J. and Payne, T. 2012. Mega-environment breeding. In: Reynolds, M. R., Pask, A. J. D. and Mullan, D. M. (Eds.), *Physiological Breeding I: Interdisciplinary Approaches to Improve Crop Adaptation. CIMMYT*, pp. 6–17.

Brenner, M. L. and Cheikh, N. 1995. The role of hormones in photosynthate partitioning and seed filling. In: Davies, P. J. (Ed.), *Plant Hormones, Physiology, Biochemistry and Molecular Biology*. Dordrecht, The Netherlands: Kluwer Academic Publishers: 649–670.

Brinton, J. and Uauy, C. 2019. A reductionist approach to dissecting grain weight and yield in wheat. *Journal of Integrative Plant Biology* 61(3), 337–358. DOI: 10.1111/jipb.12741.

Calderini, D. F., Abeledo, L. G., Savin, R. and Slafer, G. A. 1999. Effect of temperature and carpel size during pre-anthesis on potential grain weight in wheat. *The Journal of Agricultural Science* 132(4), 453–459. DOI: 10.1017/S0021859699006504.

Calderini, D. F., Savin, R., Abeledo, L. G., Reynolds, M. P. and Slafer, G. A. 2001. The importance of the period immediately preceding anthesis for grain weight determination in wheat. *Euphytica* 119(1/2), 199–204. DOI: 10.1023/A:1017597923568.

Campbell, W. F., Salisbury, F. B., Bugbee, B., Klassen, S., Naegle, E., Strickland, D. T., Bingham, G. E., Levinskikh, M., Iljina, G. M., Veselova, T. D., Sytchev, V. N., Podolsky, I., McManus, W. R., Bubenheim, D. L., Stieber, J. and Jahns, G. 2001. Comparative floral development of Mir-grown and ethylene-treated, earth-grown Super Dwarf wheat. *Journal of Plant Physiology* 158(8), 1051–1060. DOI: 10.1078/S0176-1617(04)70129-7.

Cattivelli, L., Rizza, F., Badeck, F.-W., Mazzucotelli, E., Mastrangelo, A. M., Francia, E., Marè, C., Tondelli, A. and Stanca, A. M. 2008. Drought tolerance improvement in crop plants: An integrated view from breeding to genomics. *Field Crops Research* 105(1–2), 1–14. DOI: 10.1016/j.fcr.2007.07.004.

Ceccarelli, S. and Grando, S. 1991. Selection environment and environmental sensitivity in barley. *Euphytica* 57(2), 157–167. DOI: 10.1007/BF00023074.

Chávez-Dulanto, P. N., Thiry, A. A. A., Glorio-Paulet, P., Vögler, O. and Carvalho, F. P. 2021. Increasing the impact of science and technology to provide more people with healthier and safer food. *Food and Energy Security* 10(1), e259. DOI: 10.1002/FES3.259.

Chenu, K., Deihimfard, R. and Chapman, S. C. 2013. Large-scale characterization of drought pattern: A continent-wide modelling approach applied to the Australian wheat belt-spatial and temporal trends. *New Phytologist* 198(3), 801–820. DOI: 10.1111/nph.12192.

Christopher, J. T., Christopher, M. J., Borrell, A. K., Fletcher, S. and Chenu, K. 2016. Stay-green traits to improve wheat adaptation in well-watered and water-limited environments. *Journal of Experimental Botany* 67(17), 5159-5172. DOI: 10.1093/jxb/erw276.

Cobb, J. N., DeClerck, G., Greenberg, A., Clark, R. and McCouch, S. 2013. Next-generation phenotyping: Requirements and strategies for enhancing our understanding of genotype-phenotype relationships and its relevance to crop improvement. *TAG: Theoretical and Applied Genetics. Theoretische und Angewandte Genetik* 126(4), 867–887. DOI: 10.1007/s00122-013-2066-0.

Collins, N. C., Tardieu, F. and Tuberosa, R. 2008. Quantitative trait loci and crop performance under abiotic stress: where do we stand? *Plant Physiology* 147, 469–486. DOI: 10.1104/pp.108.118117.

Condon, A. G., Richards, R. A., Rebetzke, G. J. and Farquhar, G. D. 2004. Breeding for high water-use efficiency. *Journal of Experimental Botany* 55(407), 2447–2460. DOI: 10.1093/jxb/erh277.

Cox, R. C., Snaith, P. J. and Mansfield, T. A. 1985. The significance of natural and synthetic auxins in the control of stomatal movements. *Acta Horticulturae* 171(171), 247–254. DOI: 10.17660/ActaHortic.1985.171.22.

Davies, W. J., Dodd, I. C. and Wilkinson, S. 2013. Improving spike fertility by understanding and modifying its sensitivity to environmental cues. In: *Proceedings of the 3rd International Workshop of the Wheat Yield Consortium*, CIM, *MYT*, pp. 36–38.

Desikan, R., Last, K., Harrett-Williams, R., Tagliavia, C., Harter, K., Hooley, R., Hancock, J. T. and Neill, S. J. 2006. Ethylene-induced stomatal closure in Arabidopsis occurs via AtrbohF-mediated hydrogen peroxide synthesis. *Plant Journal: For Cell and Molecular Biology* 47(6), 907–916.

Deveshwar, P., Prusty, A., Sharma, S. and Tyagi, A. K. 2020. Phytohormone-mediated molecular mechanisms involving multiple genes and QTL govern grain number in rice. *Frontiers in Genetics* 11, 586462. DOI: 10.3389/FGENE.2020.586462.

Dewdney, S. J. and McWha, J. A. 1979. Abscisic acid and the movement of photosynthetic assimilates towards developing wheat (*Triticum aestivum* L.) grains. *Z. Pflanzenphysiologie* 92, 186–193. DOI: 10.1111/j.1469-8137.2006.01753.x.

Dodd, I. C. and Kudoyarova, G. 2021. Water uptake in drying soil: Balancing supply and demand. In: Rengel, Z. (Ed.), *The Root System in Agricultural Intensification*. New York: John Wiley & Sons.

Dohlman, F. G. and Long, S. P. 2009. More productive than maize in the midwest: How does Miscanthus do it? *Plant Physiology* 150(4), 2104–2115. DOI: 10.1104/pp.109.139162.

Dolferus, R., Ji, X. and Richards, R. A. 2011. Abiotic stress and control of grain number in cereals. *Plant Science* 181(4), 331–341. DOI: 10.1016/j.plantsci.2011.05.015.

Dong, B., Zheng, X., Liu, H., Able, J. A., Yang, H., Zhao, H., Zhang, M., Qiao, Y., Wang, Y. and Liu, M. 2017. Effects of drought stress on pollen sterility, grain yield, abscisic acid and protective enzymes in two winter wheat cultivars. *Frontiers in Plant Science* 8, 1008. DOI: 10.3389/fpls.2017.01008.

Dorion, S., Lalonde, S. and Saini, H. S. 1996. Induction of male sterility in wheat by meiotic-stage water deficit is preceded by a decline in invertase activity and changes in carbohydrate metabolism in anthers. *Plant Physiology* 111(1), 137–145. DOI: 10.1104/pp.111.1.137.

Draye, X., Kim, Y., Lobet, G. and Javaux, M. 2010. Model-assisted integration of physiological and environmental constraints affecting the dynamic and spatial patterns of root water uptake from soils. *Journal of Experimental Botany* 61(8), 2145–2155. DOI: 10.1093/jxb/erq077.

Dwivedi, S. K., Basu, S., Kumar, S., Kumar, G., Prakash, V., Kumar, S., Mishra, J. S., Bhatt, B. P., Malviya, N., Singh, G. P. and Arora, A. 2017. Heat stress induced impairment of starch mobilisation regulates pollen viability and grain yield in wheat: Study in eastern Indo-Gangetic Plains. *Field Crops Research* 206, 106–114. DOI: 10.1016/j.fcr.2017.03.006.

Farshadfar, E. and Elyasi, P. 2012. Screening quantitative indicators of drought tolerance in bread wheat (*Triticum aestivum* L.) landraces. *European Journal of Experimental Biology* 2, 577–584.

Feng, B., Zhang, C., Chen, T., Zhang, X., Tao, L. and Fu, G. 2018. Salicylic acid reverses pollen abortion of rice caused by heat stress. *BMC Plant Biology* 18(1), 245. DOI: 10.1186/s12870-018-1472-5.

Fernandez, G. C. 1992. Effective selection criteria for assessing plant stress tolerance. In: *Proceedings of the International Symposium on 'Adaptation of Vegetables and Other Food Crops in Temperature and Water Stress'*, Tainan. Taiwan: AVRDC Publication: 257–270.

Fernández, M. A., Gaggioli, C. and Repollo, R. 2019. The effect of the sowing date, genotype and the fertility on grain yield of durum wheat in the dry-subhumid pampas region. *Semiárida: Revista de la Facultad de Agronomía UNLPam* 29(2), 51–60. DOI: 10.19137/semiarida.2019(02).51-60.

Fischer, R. A. 1984. Wheat. In: Smith, W. H. and Banta, J. (Eds.), *Potential Productivity of Field Crops Under Different Environments*. Los Baños: IRRI: 129–153.

Fischer, R. A. 1985. Number of kernels in wheat crops and the influence of solar radiation and temperature. *The Journal of Agricultural Science* 105(2), 447–461. DOI: 10.1017/S0021859600056495.

Fischer, R. A. and Maurer, R. 1978. Drought resistance in spring wheat cultivars. I. Grain yield responses. *Australian Journal of Agricultural Research* 29(5), 897–912. DOI: 10.1071/AR9780897.

Fischer, R. A. and Edmeades, G. O. 2010. Breeding and cereal yield progress. *Crop Science* 50, S85. DOI: 10.2135/cropsci2009.10.0564.

Flohr, B. M., Hunt, J. R., Kirkegaard, J. A., Rheinheimer, B., Swan, T., Goward, L., Evans, J. R. and Bullock, M. 2020. Deep soil water-use determines the yield benefit of long-cycle wheat. *Frontiers in Plant Science* 11, 548. DOI: 10.3389/fpls.2020.00548.

Fleury, D., Jefferies, S., Kuchel, H. and Langridge, P. 2010. Genetic and genomic tools to improve drought tolerance in wheat. *Journal of Experimental Botany* 61(12), 3211-3222. DOI: 10.1093/jxb/erq152.

Foulkes, M. J., Slafer, G. A., Davies, W. J., Berry, P. M., Sylvester-Bradley, R., Martre, P., Calderini, D. F., Griffiths, S. and Reynolds, M. P. 2011. Raising yield potential of wheat. III. Optimizing partitioning to grain while maintaining lodging resistance. *Journal of Experimental Botany* 62(2), 469–486. DOI: 10.1093/jxb/erq300.

Gasparis, S., Kała, M., Przyborowski, M., Łyznik, L. A., Orczyk, W. and Nadolska-Orczyk, A. 2018. A simple and efficient CRISPR/Cas9 platform for induction of single and multiple, heritable mutations in barley (*Hordeum vulgare* L.). *Plant Methods* 14, 111. DOI: 10.1186/S13007-018-0382-8.

Gietler, M., Fidler, J., Labudda, M. and Nykiel, M. 2020. Abscisic acid–Enemy or savior in the response of cereals to abiotic and biotic stresses? *International Journal of Molecular Sciences* 21(13), 1–29. DOI: 10.3390/IJMS21134607.

Gleason, S. M., Barnard, D. M., Green, T. R., Scott Mackay, D., Wang, D. R., Ainsworth, E. A., Altenhofen, J., Brodribb, T. J., Cochard, H., Comas, L. H. and Cooper, M. 2022. Physiological trait networks enhance understanding of crop growth and water use in contrasting environments. *Plant, Cell and Environment*. DOI: 10.1101/2022.03.11.482897.

Gomez-Cadenas, A., Tadeo, F. R., Talon, M. and Primo-Millo, E. 1996. Leaf abscission induced by ethylene in water-stressed intact seedlings of Cleopatra mandarin requires previous abscisic acid accumulation in roots. *Plant Physiology* 112(1), 401-408. DOI: 10.1104/pp.112.1.401.

González, F. G., Miralles, D. J. and Slafer, G. A. 2011. Wheat floret survival as related to pre-anthesis spike growth. *Journal of Experimental Botany* 62(14), 4889–4901. DOI: 10.1093/jxb/err182.

Griffiths, C. A., Reynolds, M. P. and Paul, M. J. 2020. Combining yield potential and drought resilience in a spring wheat diversity panel. *Food and Energy Security* 9(4), e241. DOI: 10.1002/fes3.241.

Guo, Z., Chen, D. and Schnurbusch, T. 2018. Plant and floret growth at distinct developmental stages during the stem elongation phase in wheat. *Frontiers in Plant Science* 9, 330. DOI: 10.3389/fpls.2018.00330.

Habben, J. E., Bao, X., Bate, N. J., DeBruin, J. L., Dolan, D., Hasegawa, D., Helentjaris, T. G., Lafitte, R. H., Lovan, N., Mo, H., Reimann, K. and Schussler, J. R. 2014. Transgenic alteration of ethylene biosynthesis increases grain yield in maize under field drought-stress conditions *plant. Plant Biotechnology Journal* 12(6), 685-693. DOI: 10.1111/pbi.12172.

Haberer, G. and Kieber, J. J. 2002. Cytokinins: New insights into a classic phytohormone. *Plant Physiology* 128(2), 354–362. DOI: 10.1104/pp.010773.

Halperin, O., Gebremedhin, A., Wallach, R. and Moshelion, M. 2017. High-throughput physiological phenotyping and screening system for the characterization of plant-environment interactions. *Plant Journal: For Cell and Molecular Biology* 89(4), 839-850. DOI: 10.1111/tpj.13425.

Hartung, W., Radin, J. W. and Hendrix, D. L. 1988. Abscisic acid movement into the apoplastic solution of water-stressed cotton leaves, role apoplastic pH. *Plant Physiology* 86(3), 908–913. DOI: 10.1104/pp.86.3.908.

Hartung, W., Sauter, A. and Hose, E. 2002. Abscisic acid in the xylem: Where does it come from, where does it go to? *Journal of Experimental Botany* 53(366), 27-32. DOI: 10.1093/jexbot/53.366.27.

Hays, D. B., Do, J. H., Mason, R. E., Morgan, G. and Finlayson, S. A. 2007. Heat stress induced ethylene production in developing wheat grains induces kernel abortion and increased maturation in a susceptible cultivar. *Plant Science* 172(6), 1113–1123. DOI: 10.1016/j.plantsci.2007.03.004.

Holubova, K., Hensel, G., Vojta, P., Tarkowski, P., Bergougnoux, V. and Galuszka, P. 2018. Modification of barley plant productivity through regulation of cytokinin content by reverse-genetics approaches. *Frontiers in Plant Science* 9, 1676. DOI: 10.3389/fpls.2018.01676.

Huang, Y., Guo, Y., Liu, Y., Zhang, F., Wang, Z., Wang, H., Wang, F., Li, D., Mao, D., Luan, S., Liang, M. and Chen, L. 2018. 9-cis-epoxycarotenoid dioxygenase 3 regulates plant growth and enhances multi-abiotic stress tolerance in rice. *Frontiers in Plant Science* 9, 162. DOI: 10.3389/fpls.2018.00162.

Huang, H., Zhang, J., Xu, D., Zhou, Y., Luo, J., Li, M., Chen, S. and Wang, L. 2014. Determination of 21 plant growth regulator residues in fruits by QuEChERS-high performance liquid chromatography-tandem mass spectrometry. *Chinese Journal of Chromatography* 32(7), 707–716. DOI: 10.3724/SP.J.1123.2014.03003.

Hughes, J., Hepworth, C., Dutton, C., Dunn, J. A., Hunt, L., Stephens, J., Waugh, R., Cameron, D. D. and Gray, J. E. 2017. Reducing stomatal density in barley improves drought tolerance without impacting on yield. *Plant Physiology* 174(2), 776–787. DOI: 10.1104/pp.16.01844.

Hunt, J. R., Lilley, J. M., Trevaskis, B., Flohr, B. M., Peake, A., Fletcher, A., Zwart, A. B., Gobbett, D. and Kirkegaard, J. A. 2019. Early sowing systems can boost Australian wheat yields despite recent climate change. *Nature Climate Change* 9(3), 244–247. DOI: 10.1038/s41558-019-0417-9.

Ihsan, M. Z., El-Nakhlawy, F. S., Ismail, S. M., Fahad, S. and Daur, I. 2016. Wheat phenological development and growth studies as affected by drought and late season high temperature stress under arid environment. *Frontiers in Plant Science* 7, 795. doi: 10.3389/FPLS.2016.00795.

IPCC. 2021. The physical science basis. Contribution of working group I to the sixth assessment report of the intergovernmental panel on climate change [Masson-Delmotte, V., Zhai, P., Pirani, A., Connors, S. L., Péan, C., Berger, S., Caud, N., Chen, Y., Goldfarb, L., Gomis, M. I., Huang, M., Leitzell, K., Lonnoy, E., Matthews, J. B. R., Maycock, T. K., Waterfield, T., Yelekçi, O., Yu, R. and Zhou, B. (eds.)]. Summary for policymakers. In: *Climate Change*. Cambridge: Cambridge University Press.

Israeli, Y. and Lahav, E. 2017. Banana. In: Murphy, D. and Murray, B. (Eds.), *Encyclopedia of Applied Plant Sciences* (2nd edn.). London: Elsevier: 363–381.

Jackson, M. B. 1993. Are plant hormones involved in root to shoot communication? *Advances in Botanical Research* 19, 103–187. DOI: 10.1016/S0065-2296(08)60204-9.

Jagadish, S. V. K. 2020. Heat stress during flowering in cereals – Effects and adaptation strategies. *New Phytologist* 226(6), 1567–1572. DOI: 10.1111/NPH.16429.

Jagadish, S. V. K., Craufurd, P. Q. and Wheeler, T. R. 2007. High temperature stress and spikelet fertility in rice (*Oryza sativa* L.). *Journal of Experimental Botany* 58(7), 1627–1635. DOI: 10.1093/jxb/erm003.

Ji, X., Dong, B., Shiran, B., Talbot, M. J., Edlington, J. E., Hughes, T., White, R. G., Gubler, F. and Dolferus, R. 2011. Control of abscisic acid catabolism and abscisic acid homeostasis is important for reproductive stage stress tolerance in cereals. *Plant Physiology* 156(2), 647–662. DOI: 10.1104/pp.111.176164.

Ji, X., Shiran, B., Wan, J., Lewis, D. C., Jenkins, C. L., Condon, A. G., Richards, R. A. and Dolferus, R. 2010. Importance of pre-anthesis anther sink strength for maintenance of grain number during reproductive stage water stress in wheat. *Plant, Cell and Environment* 33(6), 926–942. DOI: 10.1111/j.1365-3040.2010.02130.x.

Jiang, F. and Hartung, W. 2008. Long-distance signalling of abscisic acid (ABA): The factors regulating the intensity of the ABA signal. *Journal of Experimental Botany* 59(1), 37–43. DOI: 10.1093/jxb/erm127.

Kan, M., Yamazaki, K., Fujiwara, T. and Kamiya, T. 2019. A simple and high-throughput method for xylem sap collection. *BioTechniques* 67(5), 242–245. DOI: 10.2144/btn-2019-0046.

Kant, S., Burch, D., Badenhorst, P., Palanisamy, R., Mason, J. and Spangenberg, G. 2015. Regulated expression of a cytokinin biosynthesis gene IPT delays leaf senescence and improves yield under rainfed and irrigated conditions in canola (*Brassica napus* L.). *PLoS ONE* 10(1), e0116349. DOI: 10.1371/journal.pone.0116349.

Khan, S., Rowe, S. C. and Harmon, F. G. 2010. Coordination of the maize transcriptome by a conserved circadian clock. *BMC Plant Biology* 10, 126. DOI: 10.1186/1471-2229-10-126.

Khayatnezhad, M., Zaeifizadeh, M. and Gholamin, R. (Eds.). 2010. Investigation and selection index for drought stress. *Australian Journal of Basic and Applied Sciences* 4, 4815–4822.

Khodarahmpour, Z., Choukan, R., Bihamta, M. R. and Majidi, H. E. 2011. Determination of the best heat stress tolerance indices in maize (*Zea mays* L.) inbred lines and hybrids under Khuzestan Province conditions. *Journal of Agricultural Science and Technology* 13(1), 111–121.

Kholova, J., Murugesan, T.,Kaliamoorthy, S., Malayee, S., Baddam, R., Hammer, G. L., McLean, G., Deshpande, S., Hash, C. T., Craufurd, P. Q. and Vadez, V. 2014. Modelling the effect of plant water use traits on yield and stay-green expression in sorghum. *Functional Plant Biology* 41(11), 1019–1034. DOI: 10.1071/FP13355.

Kirkegaard, J. A. and Lilley, J. M. 2021. Using systems ecology to exploit deeper roots in crops. In: Burridge, J. G. (Ed.), *Understanding and Improving Crop Root Function*. Burleigh Dodds.

Kirkegaard, J. A., Lilley, J. M., Howe, G. N. and Graham, J. M. 2007. Impact of subsoil water use on wheat yield. *Australian Journal of Agricultural Research* 58(4), 303–315. DOI: 10.1071/AR06285.

Kobayashi, H. and Saka, H. 2000. Relationship between ethylene evolution and sucrose content in excised leaf blades of rice. *Plant Production Science* 3(4), 398–403. DOI: 10.1626/pps.3.398/pdf/10.1626/pps.3.398.

Kuanar, S. R., Molla, K. A., Chattopadhyay, K., Sarkar, R. K. and Mohapatra, P. K. 2019. Introgression of Sub1 (SUB1) QTL in mega rice cultivars increases ethylene production to the detriment of grain-filling under stagnant flooding. *Scientific Reports* 9(1), 18567. DOI: 10.1038/s41598-019-54908-2.

Langridge, P. and Reynolds, M. 2021. Breeding for drought and heat tolerance in wheat. *Theoretical and Applied Genetics* 134(6), 1753–1769. DOI: 10.1007/s00122-021-03795-1.

Langridge, P. and Reynolds, M. P. 2015. Genomic tools to assist breeding for drought tolerance. *Current Opinion in Biotechnology* 32, 130–135. DOI: 10.1016/j.copbio.2014.11.027.

Leach, K. A., Hejlek, L. G., Hearne, L. B., Nguyen, H. T., Sharp, R. E. and Davis, G. L. 2011. Primary root elongation rate and abscisic acid levels of maize in response to water stress. *Crop Science* 51(1), 157–172. DOI: 10.2135/CROPSCI2009.12.0708.

Levinsh, G. and Kreicbergs, O. 1992. Endogenous rhythmicity of ethylene production in growing intact cereal seedlings. *Plant Physiology* 100(3), 1389–1391. DOI: 10.1104/pp.100.3.1389.

Levitt, J. 1980. Seccion I: Stress concepts. In: Levitt, J. (Ed.), *Responses of Plants to Environmental Stresses, Water, Radiation, Salt, and Other Stresses* (2nd edn., Vol. II). New York: Academic Press: 1–21.

Li, J., Zhang, J., Li, H., Niu, H., Xu, Q., Jiao, Z., An, J., Jiang, Y., Li, Q. and Niu, J. 2019. The major factors causing the microspore abortion of genic male sterile mutant NWMS1 in wheat (*Triticum aestivum* L.). *International Journal of Molecular Sciences* 20(24), NWMS1. DOI: 10.3390/IJMS20246252.

Li, X., Chen, L., Forde, B. G. and Davies, W. J. 2017. The biphasic root growth response to abscisic acid in Arabidopsis involves interaction with ethylene and auxin signalling pathways. *Frontiers in Plant Science* 8, 1493. DOI: 10.3389/fpls.2017.01493.

Li, Y., Yuan, W., Li, L., Dai, H., Dang, X., Miao, R., Baluška, F., Kronzucker, H. J., Lu, C., Zhang, J. and Xu, W. 2020. Comparative analysis reveals gravity is involved in the MIZ1-regulated root hydrotropism. *Journal of Experimental Botany* 71(22), 7316–7330. DOI: 10.1093/jxb/eraa409.

Lilley, J. M. and Kirkegaard, J. A. 2016. Farming system context drives the value of deep wheat roots in semi-arid environments. *Journal of Experimental Botany* 67(12), 3665–3681. DOI: 10.1093/jxb/erw093.

Lopes, M. S. and Reynolds, M. P. 2010. Partitioning of assimilates to deeper roots is associated with cooler canopies and increased yield under drought in wheat. *Functional Plant Biology* 37(2), 147–156. DOI: 10.1071/FP09121.

López, J., Way, D. A. and Sadok, W. 2021. Systemic effects of rising atmospheric vapor pressure deficit on plant physiology and productivity. *Global Change Biology* 27(9), 1704–1720. DOI: 10.1111/gcb.15548.

López-Calcagno, P. E., Brown, K. L., Simkin, A. J., Fisk, S. J., Vialet-Chabrand, S., Lawson, T. and Raines, C. A. 2020. Stimulating photosynthetic processes increases productivity and water-use efficiency in the field. *Nature Plants* 6(8), 1054–1063. DOI: 10.1038/s41477-020-0740-1.

Loveys, B. R. 1984. Diurnal changes in water relations and abscisic acid in field-grown *Vitis vinifera* cultivars. III. The influence of xylem-derived abscisic acid on leaf gas exchange. *New Phytologist* 98(4), 563–573. DOI: 10.1111/j.1469-8137.1984.tb04149.x.

Lu, G., Coneva, V., Casaretto, J. A., Ying, S., Mahmood, K., Liu, F., Nambara, E., Bi, Y. M. and Rothstein, S. J. 2015. OsPIN5b modulates rice (*Oryza sativa*) plant architecture and yield by changing auxin homeostasis, transport and distribution. *Plant Journal: For Cell and Molecular Biology* 83(5), 913–925. DOI: 10.1111/tpj.12939.

Ludlow, M. M. and Muchow, R. C. 1990. A critical evaluation of traits for improving crop yields in water-limited environments. *Advances in Agronomy* 43, 107–153. DOI: 10.1016/S0065-2113(08)60477-0.

Luo, X., Chen, Z., Gao, J. and Gong, Z. 2014. Abscisic acid inhibits root growth in Arabidopsis through ethylene biosynthesis. *Plant Journal: For Cell and Molecular Biology* 79(1), 44–55. DOI: 10.1111/TPJ.12534.

Lynch, J. P. 2019. Root phenotypes for improved nutrient capture: An underexploited opportunity for global agriculture. *New Phytologist* 223(2), 548–564. DOI: 10.1111/nph.15738.

Martino, D. L., Abbate, P. E., Cendoya, M. G., Gutheim, F., Mirabella, N. E. and Pontaroli, A. C. 2015. Wheat spike fertility: Inheritance and relationship with spike yield components in early generations. *Plant Breeding* 134(3), 264–270. DOI: 10.1111/pbr.12262.

Masia, A., Pitacco, A., Braggio, L. and Giulivo, C. 1994. Hormonal responses to partial drying of the root system of *Helianthus annuus*. *Journal of Experimental Botany* 45(1), 69–76. DOI: 10.1093/jxb/45.1.69.

Maurel, C., Boursiac, Y., Luu, D. T., Santoni, V., Shahzad, Z. and Verdoucq, L. 2015. Aquaporins in plants. *Physiological Reviews* 95(4), 1321–1358. DOI: 10.1152/physrev.00008.2015.

Maurel, C. and Nacry, P. 2020. Root architecture and hydraulics converge for acclimation to changing water availability. *Nature Plants* 6(7), 744–749. DOI: 10.1038/s41477-020-0684-5.

McAdam, S. A., Manzi, M., Ross, J. J., Brodribb, T. J. and Gómez-Cadenas, A. 2016. Uprooting an abscisic acid paradigm: Shoots are the primary source. *Plant Signaling and Behavior* 11(6), e1169359. DOI: 10.1080/15592324.2016.1169359.

McClung, C. R. 2000. Circadian rhythms in plants: A millennial view. *Physiologia Plantarum* 109(4), 359–371. DOI: 10.1034/j.1399-3054.2000.100401.x.

Messina, C. D., Sinclair, T. R., Hammer, G. L., Curan, D., Thompson, J., Oler, Z., Gho, C. and Cooper, M. 2015. Limited-transpiration trait may increase maize drought tolerance in the US Corn Belt. *Agronomy Journal* 107(6), 1978–1986. DOI: 10.2134/agronj15.0016.

Miralles, D. J., Richards, R. A. and Slafer, G. A. 2000. Duration of the stem elongation period influences the number of fertile florets in wheat and barley. *Functional Plant Biology* 27(10), 931–940. DOI: 10.1071/PP00021.

Mohammadi, M., Karimizadeh, R. and Abdipour, M. 2011. Evaluation of drought tolerance in bread wheat genotypes under dryland and supplemental irrigation conditions. *Australian Journal of Crop Science* 5(4), 487–493.

Monneveux, P., Jing, R. and Misra, S. C. 2012. Phenotyping for drought adaptation in wheat using physiological traits. *Frontiers in Physiology* 3, 429. DOI: 10.3389/fphys.2012.00429.

Morgan, P. W. and Drew, M. C. 1997. Ethylene and plant responses to stress. *Physiologia Plantarum* 100(3), 620–630. DOI: 10.1111/j.1399-3054.1997.tb03068.x.

Moriwaki, T., Miyazawa, Y., Kobayashi, A. and Takahashi, H. 2013. Molecular mechanisms of hydrotropism in seedling roots of *Arabidopsis thaliana* (Brassicaceae). *American Journal of Botany* 100(1), 25–34. DOI: 10.3732/ajb.1200419.

Müller, M. 2021. Foes or friends: ABA and ethylene interaction under abiotic stress. *Plants* 10(3), 1–7. DOI: 10.3390/plants10030448.

Nouri, A., Etminan, A., Teixeira da Silva, J. A. and Mohammadi, R. 2011. Assessment of yield, yield-related traits and drought tolerance of durum wheat genotypes (*Triticum turjidum* var. durum Desf.). *Australian Journal of Crop Science* 5(1), 8–6.

Ober, E. S. and Sharp, R. E. 1994. Proline accumulation in maize (*Zea mays* L.) primary roots at low water potentials. I. Requirement for increased levels of ABA. *Plant Physiology* 105(3), 981–987. DOI: 10.1104/pp.105.3.981.

Ogunkanmi, A. B., Tucker, D. J. and Mansfield, T. A. 1973. An improved bioassay for abscisic acid and other antitranspirants. *New Phytologist* 72(2), 277-282. DOI: 10.1111/j.1469-8137.1973.tb02033.x.

Ogura, T., Goeschl, C., Filiault, D., Mirea, M., Slovak, R., Wolhrab, B., Satbhai, S. B. and Busch, W. 2019. Root system depth in Arabidopsis is shaped by EXOCYST70A3 via the dynamic modulation of auxin transport. *Cell* 178(2), 400–412.e16. DOI: 10.1016/j.cell.2019.06.021.

Okamoto, M., Tanaka, Y., Abrams, S. R., Kamiya, Y., Seki, M. and Nambara, E. 2009. High humidity induces abscisic acid 8'-hydroxylase in stomata and vasculature to regulate local and systemic abscisic acid responses in Arabidopsis. *Plant Physiology* 149(2), 825–834. DOI: 10.1104/pp.108.130823.

Oosterhuis, D. M. and Cartwright, P. M. 1983. Spike differentiation and floret survival in semidwarf spring wheat as affected by water stress and photoperiod 1. *Crop Science* 23(4), 711–717. DOI: 10.2135/cropsci1983.0011183X002300040026x.

Ort, D. R. and Long, S. P. 2014. Limits on yields in the Corn Belt. *Science* 344(6183), 484-485. DOI: 10.1126/science.1253884.

Pandey, B. K., Huang, G., Bhosale, R., Hartman, S., Sturrock, C. J., Jose, L., Martin, O. C., Karady, M., Voesenek, L. A. C. J., Ljung, K., Lynch, J. P., Brown, K. M., Whalley, W. R., Mooney, S. J., Zhang, D. and Bennett, M. J. 2021. Plant roots sense soil compaction through restricted ethylene diffusion. *Science* 371(6526), 276–280. DOI: 10.1126/science.abf3013.

Pantin, F., Monnet, D., Janneau, J. M., Costa, J. M., Renaud, J., Muller, B., Simonneau, T. And Genty, B. 2012. The dual effect of abscisic acid on stomata. *New Phytologist* 197, 65–72. DOI: 10.1111/nph.12013.

Parry, M. A. J., Reynolds, M., Salvucci, M. E., Raines, C., Andralojc, P. J., Zhu, X. G., Price, G. D., Condon, A. G. and Furbank, R. T. 2011. Raising yield potential of wheat. II. Increasing photosynthetic capacity and efficiency. *Journal of Experimental Botany* 62(2), 453–467. DOI: 10.1093/jxb/erq304.

Passioura, J. B. 1977. Grain yield, harvest index, and water use of wheat. *Journal of the Australian Institute of Agricultural Science* 43, 117–120. Available at: http://hdl.handle.net/102.100.100/302777?index=1.

Passioura, J. B. and Angus, J. F. 2010. Improving productivity of crops in water-limited environments. *Advances in Agronomy* 106, 37–75. DOI: 10.1016/S0065-2113(10)06002-5.

Peleg, Z. and Blumwald, E. 2011. Hormone balance and abiotic stress tolerance in crop plants. *Current Opinion in Plant Biology* 14(3), 290–295. DOI: 10.1016/j.pbi.2011.02.001.

Peltonen-Sainio, P., Kangas, A., Salo, Y. and Jauhiainen, L. 2007. Grain number dominates grain weight in temperate cereal yield determination: Evidence based on 30 years of multi-location trials. *Field Crops Research* 100(2-3), 179–188. DOI: 10.1016/j.fcr.2006.07.002.

Pérez-Pérez, J. G., Puertolas, J., Albacete, A. and Dodd, I. C. 2020. Alternation of wet and dry sides during partial rootzone drying irrigation enhances leaf ethylene evolution. *Environmental and Experimental Botany* 176, 104095. DOI: 10.1016/j.envexpbot.2020.104095.

Pierik, R., Tholen, D., Poorter, H., Visser, E. J. and Voesenek, L. A. 2006. The Janus face of ethylene: Growth inhibition and stimulation. *Trends in Plant Science* 11(4), 176-183. DOI: 10.1016/j.tplants.2006.02.006.

Pinto, R. S., Reynolds, M. P., Mathews, K. L., McIntyre, C. L., Olivares-Villegas, J. J. and Chapman, S. C. 2010. Heat and drought adaptive QTL in a wheat population designed to minimize confounding agronomic effects. *TAG: Theoretical and Applied Genetics. Theoretische und Angewandte Genetik* 121(6), 1001–1021. DOI: 10.1007/s00122-010-1351-4.

Pu, C. H., Lin, S. K., Chuang, W. C. and Shyu, T. H. 2018. Modified QuEChERS method for 24 plant growth regulators in grapes using LC-MS/MS. *Journal of Food and Drug Analysis* 26(2), 637–648. DOI: 10.1016/j.jfda.2017.08.001.

Rajalekshmi, K. M., Jaleel, C. A., Azooz, M. M. and Panneerselvam, R. 2009. Effect of triazole growth regulators on growth and pigment contents in *Plectranthus aromaticus* and *Plectranthus vettiveroids*. *Advances in Biological Research* 3(3-4), 117–122.

Ramireddy, E., Hosseini, S. A., Eggert, K., Gillandt, S., Gnad, H., von Wirén, N. and Schmülling, T. 2018. Root engineering in barley: Increasing cytokinin degradation produces a larger root system, mineral enrichment in the shoot and improved drought tolerance. *Plant Physiology* 177(3), 1078–1095. DOI: 10.1104/pp.18.00199.

Rebetzke, G. J., Chenu, K., Biddulph, B., Moeller, C., Deery, D. M., Rattey, A. R., Bennett, D., Barrett-Lennard, E. G. and Mayer, J. E. 2013. A multisite managed environment facility for targeted trait and germplasm phenotyping. *Functional Plant Biology* 40(1), 1-13. DOI: 10.1071/FP12180.

Reynolds, M., Chapman, S., Crespo-Herrera, L., Molero, G., Mondal, S., Pequeno, D. N. L., Pinto, F., Pinera-Chavez, F. J., Poland, J., Rivera-Amado, C., Saint Pierre, C. and Sukumaran, S. 2020. Breeder friendly phenotyping. *Plant Science* 295, 110396. DOI: 10.1016/j.plantsci.2019.110396.

Reynolds, M. P., Aitkin, O. K., Bennett, M., Cooper, M., Dodd, I. C., Foulkes, M. J., Frohberg, C., Hammer, G., Henderson, I. R., Huang, B., Korzun, V., McCouch, S. R., Messina, C. D., Pogson, B. J., Slafer, G. A., Taylor, N. L. and Wittich, P. E. 2021. Addressing research bottlenecks to crop productivity. *Trends in Plant Science* 26(6), 607–630. DOI: 10.1016/j.tplants.2021.03.011.

Reynolds, M. P., Foulkes, M. J., Slafer, G. A., Berry, P., Parry, M. A., Snape, J. W. and Angus, W. J. 2009. Raising yield potential in wheat. *Journal of Experimental Botany* 60(7), 1899–1918. DOI: 10.1093/jxb/erp016.

Reynolds, M. P. and Tuberosa, R. 2008. Translational research impacting on crop productivity in drought-prone environments. *Current Opinion in Plant Biology* 11(2), 171–179. DOI: 10.1016/j.pbi.2008.02.005.

Rezaul, I. M., Baohua, F., Tingting, C., Weimeng, F., Caixia, Z., Longxing, T. and Guanfu, F. 2019. Abscisic acid prevents pollen abortion under high-temperature stress by mediating sugar metabolism in rice spikelets. *Physiologia Plantarum* 165(3), 644-663. DOI: 10.1111/ppl.12759.

Richards, R. A. 1996. Defining selection criteria to improve yield under drought. *Plant Growth Regulation* 20(2), 157-166. DOI: 10.1007/BF00024012.

Richards, R. A. 2006. Physiological traits used in the breeding of new cultivars for water-scarce environments. *Agricultural Water Management* 80(1-3), 197-211. DOI: 10.1016/j.agwat.2005.07.013.

Richards, R. A., Rebetzke, G. J., Watt, M., Spielmeyer, W. and Dolferus, R. 2010. Breeding for improved water productivity in temperate cereals: Phenotyping, quantitative trait loci, markers and the selection environment. *Functional Plant Biology* 37(2), 85-97. DOI: 10.1071/FP09219.

Robinson, G. M. 2018. Globalization of agriculture. *Annual Review of Resource Economics* 10(1), 133-160. DOI: 10.1146/annurev-resource-100517-023303.

Rosales, M. A., Maurel, C. and Nacry, P. 2019. Abscisic acid coordinates dose-dependent developmental and hydraulic responses of roots to water deficit. *Plant Physiology* 180(4), 2198-2211. DOI: 10.1104/pp.18.01546.

Rosielle, A. A. and Hamblin, J. 1981. Theoretical aspects of selection for yield in stress and non-stress environments. *Crop Science* 21(6), 943-946. DOI: 10.2135/cropsci1981.0011183X002100060033x.

Rowe, J. H., Topping, J. F., Liu, J. and Lindsey, K. 2016. Abscisic acid regulates root growth under osmotic stress conditions via an interacting hormonal network with cytokinin, ethylene and auxin. *New Phytologist* 211(1), 225-239. DOI: 10.1111/nph.13882.

Rowell, P. L. and Miller, D. G. 1971. Induction of male sterility in wheat with 2-Chloroethylphosphonic acid (ethrel)1. *Crop Science* 11(5), 629-631. DOI: 10.2135/cropsci1971.0011183X001100050007x.

Ryan, A. C., Dodd, I. C., Rothwell, S. A., Jones, R., Tardieu, F., Draye, X. and Davies, W. J. 2016. Gravimetric phenotyping of whole plant transpiration responses to atmospheric vapour pressure deficit identifies genotypic variation in water use efficiency. *Plant Science* 251, 101-109. DOI: 10.1016/j.plantsci.2016.05.018.

Saab, I. N., Sharp, R. E., Pritchard, J. and Voetberg, G. S. 1990. Increased endogenous abscisic acid maintains primary root growth and inhibits shoot growth of maize seedlings at low water potentials. *Plant Physiology* 93(4), 1329-1336. DOI: 10.1104/pp.93.4.1329.

Sakata, T., Oshino, T., Miura, S., Tomabechi, M., Tsunaga, Y., Higashitani, N., Miyazawa, Y., Takahashi, H., Watanabe, M. and Higashitani, A. 2010. Auxins reverse plant male sterility caused by high temperatures. *Proceedings of the National Academy of Sciences of the United States of America* 107(19), 8569-8574. DOI: 10.1073/pnas.1000869107.

Salvi, S. and Tuberosa, R. 2015. The crop QTLome comes of age. *Current Opinion in Biotechnology* 32, 179-185. DOI: 10.1016/j.copbio.2015.01.001.

Sareen, S., Tyagi, B. S. and Sharma, I. 2012. Response estimation of wheat synthetic lines to terminal heat stress using stress indices. *Journal of Agricultural Science* 4(10), 97-104. DOI: 10.5539/jas.v4n10p97.

Sayre, K. D., Rajaram, S. and Fischer, R. A. 1997. Yield potential progress in short bread wheats in northwest Mexico. *Crop Science* 37(1), 36-42. DOI: 10.2135/cropsci1997.0011183X003700010006x.

Sharp, R. E., Wu, Y., Voetberg, G. S., Saab, I. N. and LeNoble, M. E. 1994. Confirmation that abscisic acid accumulation is required for maize primary root elongation at low water potentials. *Journal of Experimental Botany* 45(Special_Issue), 1743–1751. DOI: 10.1093/jxb/45.Special_Issue.1743.

Sheoran, I. S. and Saini, H. S. 1996. Drought-induced male sterility in rice: Changes in carbohydrate levels and enzyme activities associated with the inhibition of starch accumulation in pollen. *Sexual Plant Reproduction* 9(3), 161–169. DOI: 10.1007/BF02221396.

Shi, C., Qi, C., Ren, H., Huang, A., Hei, S. and She, X. 2015. Ethylene mediates brassinosteroid-induced stomatal closure via Gα protein-activated hydrogen peroxide and nitric oxide production in Arabidopsis. *Plant Journal: For Cell and Molecular Biology* 82(2), 280–301. DOI: 10.1111/tpj.12815.

Slafer, G. A. 2003. Genetic basis of yield as viewed from a crop physiologist's perspective. *Annals of Applied Biology* 142(2), 117–128. DOI: 10.1111/j.1744-7348.2003.tb00237.x.

Slafer, G. A. 2005. Physiology of determination of major wheat yield components. In: Buck, H. T., Nisi, J. E. and Salomón, N. (Eds.), *Proceedings of the 7th International Wheat Conference*, 27 November, Mar del Plata, Argentina (Developments in Plant Breeding). Springer, pp. 557–565.

Slafer, G. A. 2012. Wheat development: Its role adaptation, phenotyping and improving crop. In: Reynolds, M. P., Pask, A. and Mullan, D. (Eds.), *Physiological Breeding I: Interdisciplinary Approaches to Improve Crop. CIMMYT*, pp. 109–123.

Slafer, G. A. and Araus, J. L. 2007. Physiological traits for improving wheat yield under a wide range of conditions. In: Spiertz, J. H. J., Struik, P. C. and van Laar, H. H. (Eds.), *Scale and Complexity in Plant Systems Research: Gene-Plant-Crop Relations*. Dordrecht: Springer Netherlands: 147–156.

Slovik, S., Daeter, W. and Hartung, W. 1995. Compartmental distribution and redistribution of abscisic acid (ABA) in roots as influenced by environmental changes: A biomathematical model. *Journal of Experimental Botany* 46(8), 881–894. DOI: 10.1093/jxb/46.8.881.

South, P. F., Cavanagh, A. P., Liu, H. W. and Ort, D. R. 2019. Synthetic glycolate metabolism pathways stimulate crop growth and productivity in the field. *Science* 4, 363. DOI: 10.1126/science.aat9077.

Spollen, W. G., LeNoble, M. E., Samuels, T. D., Bernstein, N. and Sharp, R. E. 2000. Abscisic acid accumulation maintains maize primary root elongation at low water potentials by restricting ethylene production. *Plant Physiology* 122(3), 967–976. DOI: 10.1104/pp.122.3.967.

Spollen, W. G., Sharp, R. E. and Saab, I. N. 1993. Regulation of cell expansion in roots and shoots at low water potentials. In: Smith, J. A. C. and Griffiths, H. (Eds.), *Water Deficits. Plant Responses From Cell to Community*. Oxford: Bios Scientific Publishers: 37–52.

Steffens, B., Wang, J. and Sauter, M. 2006. Interactions between ethylene, gibberellin and abscisic acid regulate emergence and growth rate of adventitious roots in deepwater rice. *Planta* 223(3), 604–612. DOI: 10.1007/S00425-005-0111-1.

Talebi, R., Fayaz, F. and Naji, A. M. 2009. Effective selection criteria for assessing drought stress tolerance in durum wheat (*Triticum durum* Desf.). *General and Applied Plant Physiology* 35(1–2), 64–74.

Tanaka, Y., Sano, T., Tamaoki, M., Nakajima, N., Kondo, N. and Hasezawa, S. 2005. Ethylene inhibits abscisic acid-induced stomatal closure in Arabidopsis. *Plant Physiology* 138(4), 2337–2343. DOI: 10.1104/pp.105.063503.

Tang, N., Shahzad, Z., Lonjon, F., Loudet, O., Vailleau, F. and Maurel, C. 2018. Natural variation at XND1 impacts root hydraulics and trade-off for stress responses in Arabidopsis. *Nature Communications* 9(1), 3884. DOI: 10.1038/s41467-018-06430-8.

Tardieu, F. 2012. Any trait or trait-related allele can confer drought tolerance: Just design the right drought scenario. *Journal of Experimental Botany* 63(1), 25–31. DOI: 10.1093/jxb/err269.

Tardieu, F. 2016. Too many partners in root-shoot signals. Does hydraulics qualify as the only signal that feeds back over time for reliable stomatal control? *New Phytologist* 212(4), 802–804. DOI: 10.1111/nph.14292.

Tardieu, F. and Davies, W. J. 1992. Stomatal response to abscisic-acid is a function of current plant water status. *Plant Physiology* 98(2), 540–545. DOI: 10.1104/pp.98.2.540.

Tardieu, F., Draye, X. and Javau, M. 2017. Root water uptake and ideotypes of the root system: Whole-plant controls matter. *Vadose Zone Journal* 16(9), 1–10. DOI: 10.2136/vzj2017.05.0107.

Tardieu, F., Parent, B. and Simmoneau, T. 2010a. Control of leaf growth by abscisic acid: Hydraulic or non-hydraulic processes? *Plant, Cell and Environment* 33(4), 636–647. DOI: 10.1111/j.1365-3040.2009.02091.x.

Tardieu, F., Tuberosa, R., Varshney, R. K. and Cook, D. R. 2010b. Dissection and modelling of abiotic stress tolerance in plants. *Current Opinion in Plant Biology* 13(2), 206–212. DOI: 10.1016/J.PBI.2009.12.012.

Tardieu, F., Simmoneau, T. and Parent, B. 2015. Modelling the coordination of the controls of stomatal aperture, transpiration, leaf growth, and abscisic acid: Update and extension of the Tardieu–Davies model. *Journal of Experimental Botany* 66(8), 2227–2237. DOI: 10.1093/jxb/erv039.

Tardieu, F., Zhang, J., Katerji, N., Bethenod, O., Palmer, S. and Davies, W. J. 1992. Xylem ABA controls the stomatal conductance of field-grown maize subjected to soil compaction or soil drying. *Plant, Cell and Environment* 15(2), 193–197. DOI: 10.1111/j.1365-3040.1992.tb01473.x.

Tattaris, M., Reynolds, M. P. and Chapman, S. C. 2016. A direct comparison of remote sensing approaches for high-throughput phenotyping in plant breeding. *Frontiers in Plant Science* 7, 1131. DOI: 10.3389/fpls.2016.01131.

Thain, S. C., Vandenbussche, F., Laarhoven, L. J. J., Dowson-Day, M. J., Wang, Z. Y., Tobin, E. M., Harren, F. J., Millar, A. J. and Van Der Straeten, D. 2004. Circadian rhythms of ethylene emission in Arabidopsis. *Plant Physiology* 136(3), 3751–3761. DOI: 10.1104/pp.104.042523.

Thiry, A. A. 2017. *Assessment of Plant Hormone Quantification as a New Physiological Trait for Crop Breeding Using a New Classification Method for Crop Performance under Drought*. Lancaster: Lancaster University. DOI: 10.17635/lancaster/thesis/221.

Thiry, A. A., Chavez-Dulanto, P. N., Reynolds, M. P. and Davies, W. J. 2016. How can we improve crop genotypes to increase stress resilience and productivity in a future climate? A new crop screening method based on productivity and resistance to abiotic stress. *Journal of Experimental Botany* 67(19), 5593–5603. DOI: 10.1093/jxb/erw330.

Thiry, A. A., Reynolds, M. P. and Davies, W. J. 2017. ABA and ethylene hormone balance and its diurnal variation as indicators of plant drought stress resilience. In: *Proceedings of the 3rd International TRIGO (Wheat) Yield Potential Workshop*, Cd. Obregon. Sonora, Mexico: CIMMYT, pp. 119–128.

Thomas, H. and Ougham, H. 2014. The stay-green trait. *Journal of Experimental Botany* 65(14), 3889–3900. DOI: 10.1093/jxb/eru037.

Thomas, H. and Ougham, H. 2015. Senescence and crop performance. In: Sadras, V. O. and Calderini, D. F. (Eds.), *Crop Physiology: Applications for Genetic Improvement, Agronomy and Farming Systems* (2nd edn.). New York: Academic Press: 223–249.

Tsenov, N., Gubatov, T., Raykov, G., Ivanova, A., Chamurliiski, P., Holding, I., Preslavsky, K., Dobrich, C. 2017. New approaches for evaluation of the grain yield of winter wheat. *International Journal of Current Research* 9, 44487–44495.

Tuberosa, R. 2012. Phenotyping for drought tolerance of crops in the genomics era. *Frontiers in Physiology* 3, 347. DOI: 10.3389/fphys.2012.00347.

Turner, N. C. 1979. Drought resistance and adaptation to water deficits in crop plants. In: Mussell, H. and Staples, R. C. (Eds.), *Stress Physiology in Crop Plants*. New York: John Wiley and Sons: 343–372.

Turner, N. C. 1986. Adaptation to water deficits: A changing perspective. *Functional Plant Biology* 13(1), 175–190. DOI: 10.1071/PP9860175.

Uga, Y., Sugimoto, K., Ogawa, S., Rane, J., Ishitani, M., Hara, N., Kitomi, Y., Inukai, Y., Ono, K., Kanno, N., Inoue, H., Takehisa, H., Motoyama, R., Nagamura, Y., Wu, J., Matsumoto, T., Takai, T., Okuno, K. and Yano, M. 2013. Control of root system architecture by DEEPER ROOTING 1 increases rice yield under drought conditions. *Nature Genetics* 45(9), 1097–1102. DOI: 10.1038/ng.2725.

Ugarte, C., Calderini, D. F. and Slafer, G. A. 2007. Grain weight and grain number responsiveness to pre-anthesis temperature in wheat, barley and triticale. *Field Crops Research* 100(2–3), 240–248. DOI: 10.1016/j.fcr.2006.07.010.

Valério, I. P., Carvalho, F. I. Fd, Benin, G., Silveira, Gd, Silva, J. A. Gd, Nornberg, R., Hagemann, T., Luche, HdS. and Oliveira, A. Cd 2013. Seeding density in wheat: The more, the merrier? *Scientia Agricola* 70(3), 176–184. DOI: 10.1590/S0103-90162013000300006.

Valluru, R., Reynolds, M. P., Davies, W. J. and Sukumaran, S. 2017. Phenotypic and genome-wide association analysis of spike ethylene in diverse wheat genotypes under heat stress. *New Phytologist* 214(1), 271–283. DOI: 10.1111/nph.14367.

Valluru, R., Thiry, A., Wilkinson, S., Reynolds, M. P. and Davies, W. J. 2014. Phenotypic selection for spike ethylene. In: *Proceedings of the 4th International Workshop of the Wheat Yield Consortium. Ciudad Obregon, Cd. Obregon*. Sonora, Mexico: CIMMYT, pp. 109–113.

Wang, C., Feng, L., Wu, L., Cheng, C., Li, Y., Yan, J., Gao, J. and Chen, F. 2020. Assessment of genotypes and management strategies to improve resilience of winter wheat production. *Sustainability* 12(4), 1–21. DOI: 10.3390/su12041474.

Wang, Z., Zhang, W. and Yang, J. 2018. Physiological mechanism underlying spikelet degeneration in rice. *Journal of Integrative Agriculture* 17(7), 1475–1481. DOI: 10.1016/S2095-3119(18)61981-1.

Wardlaw, I., Dawson, I., Munibi, P. and Fewster, R. 1989. The tolerance of wheat to high temperatures during reproductive growth. I. Survey procedures and general response patterns. *Australian Journal of Agricultural Research* 40(1), 1–13. DOI: 10.1071/AR9890001.

Wardlaw, I. and Wrigley, C. 1994. Heat tolerance in temperate cereals: An overview. *Functional Plant Biology* 21(6), 695–703. DOI: 10.1071/PP9940695.

Westgate, M. E., Passioura, J. B. and Munns, R. 1996. Water status and ABA content of floral organs in drought stressed wheat. *Functional Plant Biology* 23(6), 763–772. DOI: 10.1071/PP9960763.

Willett, W., Rockström, J., Loken, B., Springmann, M., Lang, T., Vermeulen, S., Garnett, T., Tilman, D., DeClerck, F., Wood, A., Jonell, M., Clark, M., Gordon, L. J., Fanzo, J., Hawkes, C., Zurayk, R., Rivera, J. A., De Vries, W., Majele Sibanda, L., Afshin, A., Chaudhary, A., Herrero, M., Agustina, R., Branca, F., Lartey, A., Fan, S., Crona, B., Fox, E., Bignet, V., Troell, M., Lindahl, T., Singh, S., Cornell, S. E., Srinath Reddy, K., Narain, S., Nishtar, S. and Murray, C. J. L. 2019. Food in the Anthropocene: the EAT–Lancet Commission on healthy diets from sustainable food systems. *Lancet*. DOI: 10.1016/S0140-6736(18)31788-4.

Wilkinson, S. and Davies, W. J. 2009. Ozone suppresses soil drying- and abscisic acid (ABA)-induced stomatal closure via an ethylene-dependent mechanism. *Plant, Cell and Environment* 32(8), 949–959. DOI: 10.1111/j.1365-3040.2009.01970.x.

Wilkinson, S. and Davies, W. J. 2010. Drought, ozone, ABA and ethylene: New insights from cell to plant to community. *Plant, Cell and Environment* 33(4), 510–525. DOI: 10.1111/j.1365-3040.2009.02052.x.

Wilkinson, S., Kudoyarova, G. R., Veselov, D. S., Arkhipova, T. N. and Davies, W. J. 2012. Plant hormone interactions: Innovative targets for crop breeding and management. *Journal of Experimental Botany* 63(9), 3499–3509. DOI: 10.1093/jxb/ers148.

Wright, S. T. C. 1977. The relationship between leaf water potential and the levels of abscisic acid and ethylene in excised wheat leaves. *Planta* 134(2), 183–189. DOI: 10.1007/BF00384969.

Xu, W., Jia, L., Shi, W., Liang, J., Zhou, F., Li, Q. and Zhang, J. 2013. Abscisic acid accumulation modulates auxin transport in the root tip to enhance proton secretion for maintaining root growth under moderate water stress. *New Phytologist* 197(1), 139–150. DOI: 10.1111/nph.12004.

Yang, J. and Zhang, J. 2006. Grain filling of cereals under soil drying. *New Phytologist* 169(2), 223–236. DOI: 10.1111/j.1469-8137.2005.01597.x.

Yang, J., Zhang, J., Liu, K., Wang, Z. and Liu, L. 2006a. Abscisic acid and ethylene interact in wheat grains in response to soil drying during grain filling. *New Phytologist* 171(2), 293–303. DOI: 10.1111/j.1469-8137.2006.01753.x.

Yang, J., Zhang, J., Wang, Z., Liu, K. and Wang, P. 2006b. Post-anthesis development of inferior and superior spikelets in rice in relation to abscisic acid and ethylene. *Journal of Experimental Botany* 57(1), 149–160. DOI: 10.1093/jxb/erj018.

Yang, J., Zhang, J., Liu, K., Wang, Z. and Liu, L. 2007. Abscisic acid and ethylene interact in rice spikelets in response to water stress during meiosis. *Journal of Plant Growth Regulation* 26(4), 318–328. DOI: 10.1007/s00344-007-9013-8.

Yang, J., Zhang, J., Wang, Z., Zhu, Q. and Wang, W. 2001. Remobilization of carbon reserves in response to water deficit during grain filling of rice. *Field Crops Research* 71(1), 47–55. DOI: 10.1016/S0378-4290(01)00147-2.

Yang, J. C., Zhang, J. H., Ye, Y. X., Wang, Z. Q., Zhu, Q. S. and Liu, L. J. 2004. Involvement of abscisic acid and ethylene in the responses of rice grains to water stress during filling. *Plant, Cell and Environment* 27(8), 1055–1064. DOI: 10.1111/j.1365-3040.2004.01210.x.

Young, T. E., Meeley, R. B. and Gallie, D. R. 2004. ACC synthase expression regulates leaf performance and drought tolerance in maize. *Plant Journal: For Cell and Molecular Biology* 40(5), 813–825. DOI: 10.1111/j.1365-313X.2004.02255.x.

Zeevaart, J. A. D. and Creelman, R. A. 1988. Metabolism and physiology of abscisic acid. *Annual Review of Plant Physiology and Plant Molecular Biology* 39(1), 439–473.

Zenda, T., Liu, S. and Duan, H. 2021. Adapting cereal grain crops to drought stress: 2020 and Beyond. In: Fahad, S., Saud, S., Chen, Y., Wu, C. and Wang, D. (Eds.), *Abiotic Stress in Plants*. IntechOpen. DOI: 10.5772/intechopen.93845.

Zeng, Z. and King, R. 1986. Regulation of grain number in wheat: Changes in endogenous levels of abscisic acid. *Functional Plant Biology* 13(3), 347–352. DOI: 10.1071/PP9860347.

Zhang, C., Li, G., Chen, T., Feng, B., Fu, W., Yan, J., Islam, M. R., Jin, Q., Tao, L. and Fu, G. 2018. Heat stress induces spikelet sterility in rice at anthesis through inhibition of pollen tube elongation interfering with auxin homeostasis in pollinated pistils. *Rice* 11, 1–12. DOI: 10.1186/S12284-018-0206-5/FIGURES/6.

Zhang, L., Richards, R. A., Condon, A. G., Liu, D. C. and Rebetzke, G. J. 2015. Recurrent selection for wider seedling leaves increases early biomass and leaf area in wheat (*Triticum aestivum* L.). *Journal of Experimental Botany* 66(5), 1215–1226. DOI: 10.1093/jxb/eru468.

Zhang, Q., Yuan, W., Wang, Q., Cao, Y., Xu, F., Dodd, I. C. and Xu, W. 2022. ABA regulation of root growth during soil drying and recovery can involve auxin response. *Plant, Cell and Environment* 45(3), 871–883. DOI: 10.1111/pce.14137.

Zhao, Z., Wang, E., Kirkegaard, J. A. and Rebetzke, G. J. 2022. Novel wheat varieties facilitate deep sowing to beat the heat of changing climates. *Nature Climate Change* 12(3), 291–296. DOI: 10.1038/s41558-022-01305-9.

Zhao, Z., Rebetzke, G. J., Zheng, B., Chapman, S. C. and Wang, E. 2019. Modelling impact of early vigour on wheat yield in dryland regions. *Journal of Experimental Botany* 70(9), 2535–2548. DOI: 10.1093/jxb/erz069.

Zheng, S., Li, J., Ma, L., Wang, H., Zhou, H., Ni, E., Jiang, D., Liu, Z. and Zhuang, C. 2019. OsAGO2 controls ROS production and the initiation of tapetal PCD by epigenetically regulating OsHXK1 expression in rice anthers. *Proceedings of the National Academy of Sciences of the United States of America* 116(15), 7549–7558. DOI: 10.1073/PNAS.1817675116.

Zhou, M., Wani, S. H., Shavrukov, Y., Shavrukov, Y., Kurishbayev, A., Jatayev, S., Shvidchenko, V., Zotova, L., Koekemoer, F., de Groot, S., Soole, K. and Langridge, P. 2017. Early flowering as a drought escape mechanism in plants: How can aid wheat production? *Frontiers in Plant Science* 8, 1950. doi: 10.3389/fpls.2017.01950.

Chapter 3

Genetics of drought tolerance in cereals

Ambra Viviani, Roberto Tuberosa and Marco Maccaferri, Department of Agricultural and Food Sciences (DISTAL) – University of Bologna, Italy

1 Introduction: drought stress in cereal crops

Cereal crops are a major source of food for humans and livestock around the world (FAO statistics, 2021). However, grain production is at risk from global warming (Bailey-Serres et al., 2019; Cohen et al., 2021; Kumar et al., 2022; Ostmeyer et al., 2020). Climate change has contributed to extreme drought, high temperatures, and flooding events affecting many regions of the world (Lesk et al., 2016). Drought has been seen as the most critical of these abiotic stresses affecting crop growth and leading to severe yield losses (Thanmalagan et al., 2022). Over the past decade, global economic losses in agriculture due to drought have been estimated at about US$37 billion (FAO report, 2021). Water demand for agriculture is predicted to increase twofold by 2050, while freshwater availability will decrease by up to 50% due to the effects of global warming. There is an urgent need to develop high-yielding crops that are climate resilient and more efficient in water use (Baldoni et al., 2021; Langridge et al., 2022; Lesk et al., 2016).

Plants adopt various mechanisms to cope with drought (Fig. 1):

http://dx.doi.org/10.19103/AS.2022.0109.15

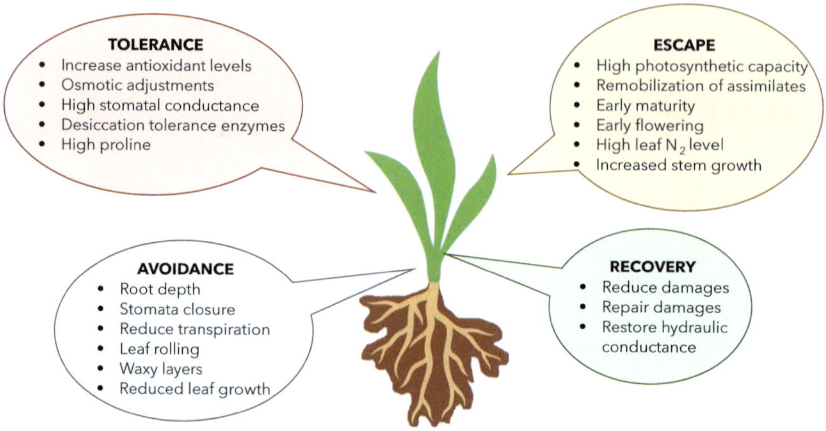

Figure 1 Mechanisms for drought resistance in plants: escape, tolerance, avoidance, and recovery. A particularly important role is played by the rhizosphere and in particular by the interaction between plants and rhizospheric microbiomes, highlighted in the figure with the color brown.

- Drought escape – the ability of plants to complete their life cycle before the onset of drought stress so that they do not experience drought stress.
- Drought avoidance – the ability of plants, under mild or moderate drought stress conditions, to access and/or store as much water as possible to maintain a water status sufficient to support the basic metabolic processes that enable them to survive unfavorable conditions. To this end, plants adopt several strategies: active osmotic adjustment (Blum, 2017) rapid closure of stomata, rolling up of leaves (to reduce the leaf area exposed to incident radiation), accumulation of wax on the cell surface (which reflects sunlight and prevents excessive transpiration), increased rooting at the expense of shoots, increased root depth, and increased water storage capacity (Ludlow and Muchow, 1990; Tuberosa, 2012; Varshney et al., 2018).
- Drought tolerance – the ability of plants to withstand severe drought by maintaining a low tissue water content and sustaining a minimum level of physiological activity through the regulation of thousands of genes and a range of metabolic pathways to reduce or repair resulting stress damage (Blum, 2009; Blum and Tuberosa, 2018).
- Drought recovery – the ability of plants to recover from a state of dehydration after being exposed to prolonged drought stress and low water status (Carraro and Di Iorio, 2022; Fang and Xiong, 2015).

A key role in plant resistance to drought stress is played by roots. Roots are critical for the uptake of water and nutrients from the soil and can undergo changes to adapt to different abiotic stresses. A particularly important role is played by beneficial rhizosphere microbes associated with roots which protect plants from pathogens and promote nutrient uptake efficiency. Beneficial microbiomes can also improve plant performance and phytohormone-mediated resistance to abiotic stresses. Recent studies suggest that microbiomes can influence plant morphological traits and phenology, suggesting a clear role of this beneficial interaction on plant resistance to stresses (Aslam et al., 2022).

The identification of genomic regions for traits associated with drought tolerance and heat stress is a key step in maker-assisted selection (MAS) (Fig. 2). The use of MAS includes (Hasan et al., 2021; Maccaferri et al., 2022):

- Marker-assisted backcrossing
- Marker-assisted recurrent selection
- Marker-assisted gene pyramiding.

Marker-assisted backcrossing is used for introgression of important loci (preferably with each one accounting for more than 10% of phenotypic variation)

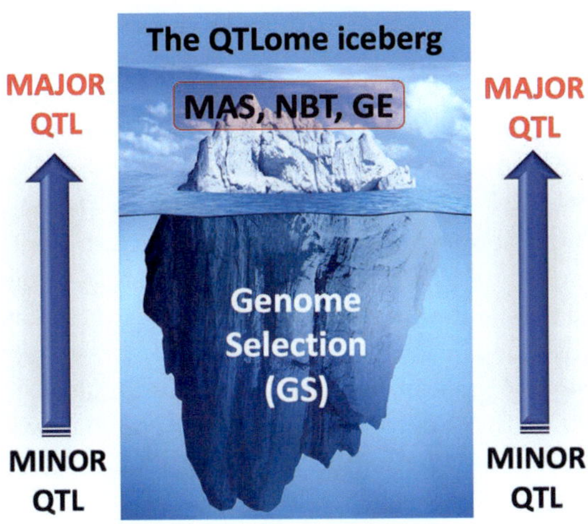

Figure 2 The QTLome iceberg represents the portion of QTLs with a strong enough additive effect to allow for their mapping and selection of the beneficial alleles by MAS. Only a small fraction of these major QTLs has been cloned, thus enabling the application of new breeding technologies (NBT; e.g. gene editing) and/or genetic engineering (GE). Most QTLs have additive effects too small to allow for their mapping, so their selection is only possible indirectly through genome selection (GS) (Maccaferri et al., 2022).

into elite but otherwise susceptible cultivars (or deficient in a particular trait). Marker-assisted gene pyramiding is used to introgress one or more genes, mainly major genes or quantitative trait loci (QTLs), into a common background to enhance resistance to drought and/or other abiotic/biotic stresses.

Complex traits such as drought and heat stress tolerance, which are regulated by countless minor QTLs, can be improved effectively through marker-assisted recurrent selection aimed at the 'tip of the QTLome iceberg' and also by genomic selection (GS) approaches targeting the prevailing underwater and unresolved submerged portion of the iceberg depicted in Fig. 2 (Choudhary et al., 2023; Maccaferri et al., 2022).

A genetic locus or more precisely a gene and/or noncoding sequence (haplotype) affecting natural variation in a quantitative trait is known as a QTL. An important approach to increasing drought tolerance is based on genetic mapping of major QTLs with significant and predictable effects in a given environment. Determining how loci (genes and/or QTLs) contribute to the genotype x environment x management interaction (GxExM) is the 'holy grail' for plant breeders in identifying the most informative phenotypic traits and context-depending alleles for a specific environment and agronomic practices (Christopher et al., 2013; Collins et al., 2008; Cooper and Messina, 2023; Lasky et al., 2023; Maccaferri et al., 2016). Genetic mapping provides a basis for identifying genomic regions involved in tolerance to drought and other abiotic stresses, disease resistance, yield, and quality traits. These complex traits are controlled by polygenes and minor genes (Tuberosa et al., 2002; Varshney et al., 2021). The hypothesis that phenotypic variation in complex traits may be linked to several structural variants has characterized genetic mapping studies in plants (Carlborg and Haley, 2004). The identification of QTLs linked to natural phenotypic variation is the main goal of genetic mapping by assigning loci on a chromosome and estimate their distance from each other. Linkage mapping routinely locates loci in biparental populations.

Genes/QTLs can also be identified by extensive genotyping and phenotyping of a sufficiently large (at least 150) panel of unrelated germplasm accessions representing phenotypic variability in the environments under investigation, with the goal of revealing/confirming marker-trait association (Salvi et al., 2007). This approach is known as genome-wide association mapping (GWAS) (Fig. 3) (Xu et al., 2017; Zia et al., 2022).

Great progress has been made in the past decade in understanding drought resistance, thanks to new discoveries and the development of new approaches and methodologies (Fig. 4) including meta-QTL analysis (Arriagada et al., 2022; Martinez et al., 2016) as well as understanding the potential role of microRNA and methylation on gene expression (Singroha et al., 2021; Sun

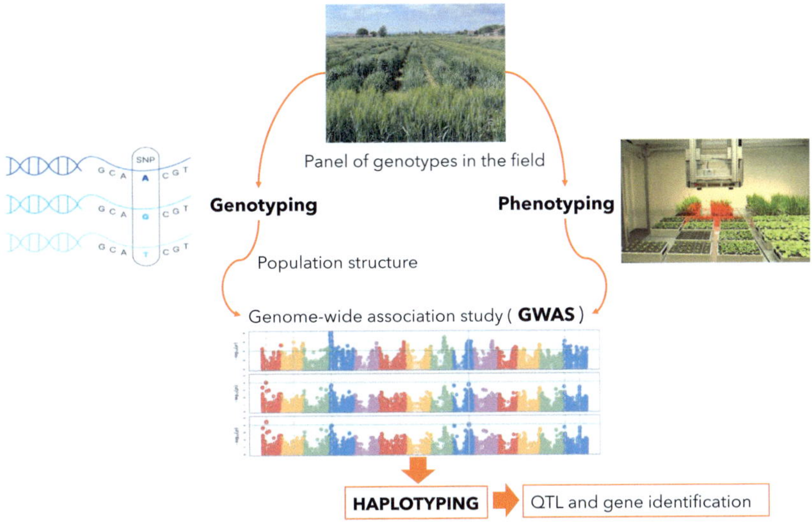

Figure 3 The main steps of a QTL analysis experiment. It starts with a panel of genotypes (field or controlled conditions), for which good phenotyping and genotyping is required. GWAS analysis is performed through different statistical models leading to possible QTL loci to be analyzed to find genes associated with the trait of interest. To increase the power of GWAS, haplotypes can be constructed between neighboring SNPs on a chromosome. Haplotypes are specific sets of alleles observed on a single chromosome and thus are inherited together with little chance of simultaneous recombination. GWAS analysis with haplotypes may be superior to single marker-based analysis in terms of statistical significance (better *P* values) and estimation of allelic effects (Sehgal et al., 2020).

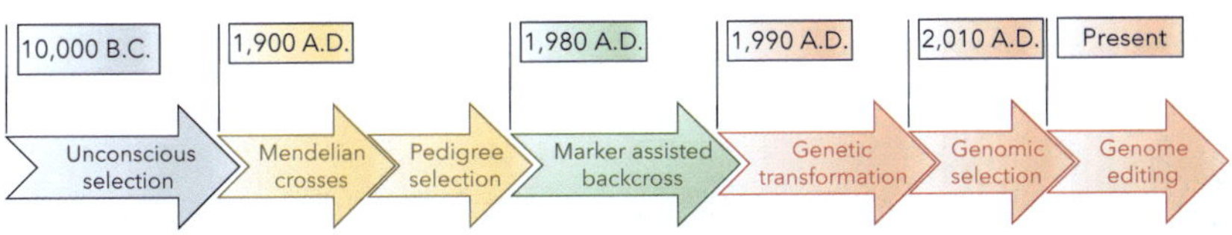

Figure 4 Timeline of plant breeding phases. Source: Modified from Ramstein et al. (2019).

et al., 2022). Together with selection for mono- and oligogenic traits (e.g. disease resistance), selection for quantitative traits has been and will be the driver behind the significant gains necessary in crop yields achieved by modern breeding and is likely to continue to be so.

With the rapid development of modern biological theories and technologies (such as genomics, proteomics, metabolomics, and other 'omics' technologies), researchers have also made great strides in elucidating the functional and signaling networks involved in drought responses. However, the sophisticated underlying mechanisms that differentiate resistance from susceptibility within a species, especially for crops, still remain unclear, particularly for the following topics:

- How drought resistance in crops evolved during domestication
- How to identify and validate drought resistance genes and evaluate their potential in selection
- How to bridge the gap between basic science and field breeding.

To answer these questions, we must better dissect the genetic, transcriptomic, metabolomic, and epigenetic basis of drought resistance traits in crops so we can best apply this knowledge in crop breeding (Fang and Xiong, 2015; Luo et al., 2019).

The following sections discuss some of the latest findings related to drought resistance of the world's most widely grown cereals. We start with rice because of its key role as an ancestral founder species and phylogenetically pivotal role for cereal evolution, and because of its quite small genome comprised of only 430 million base pairs (Mb) (Fig. 5) *Oryza sativa* ssp. *japonica* cv. Nipponbare, which was the first to be sequenced among cereals (Goff et al., 2002). We discuss barley, second in phylogeny after rice, with a genome of about 5.3 billion base pairs (Gb), durum wheat with a genome of about 10 Gb, bread wheat with a genome of about 17 Gb, sorghum with a genome of about 750 Mb, and maize with a genome of about 2.4 Gb (Fig. 5).

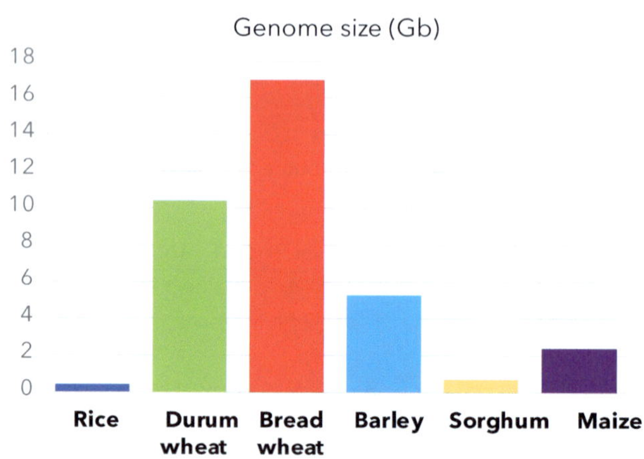

Figure 5 Comparison of genome sizes of the Poaceae considered in this chapter. As the figure shows, the bread wheat genome is the largest, followed by that of durum wheat, barley, maize, sorghum, and rice.

2 Rice: understanding and optimizing drought tolerance

Rice (*Oryza. sativa* L.) is the most widely consumed cereal by humans as a staple food worldwide, especially in Asian countries. Half of humankind relies on rice as a staple food; globally a total of 160 million hectares (Mha) are sown with this crop, with around 40% cultivated in Asian countries (Prasad et al., 2017). Rice belongs to the Poaceae family and the Oryzeae tribe and currently includes 22 wild species but only two principal cultivated species. With a genome of 430 Mb (Fig. 5) (Goff et al., 2002), and phylogenetically with a key and unique role, rice is the ideal cereal model crop. Drought affects rice production in various ways. In rain-fed areas, drought causes paddy fields to dry up, while in irrigated areas it causes groundwater depletion. Rice cultivation usually requires seasonal flooding, but this is no longer sustainable given increasing competition for surface water use. This was illustrated in the loss of rice production in South and Southeast Asia due to severe drought stress in the 2000s (Kumar, 2018). One way to increase productivity is to develop drought-tolerant and high-productivity rice varieties (Sahebi et al., 2018; Satrio et al., 2023).

Of particular importance in developing drought-tolerant rice cultivars is the ability to understand and leverage the molecular mechanisms underlying drought resistance (Mehana et al., 2021). The underlying molecular mechanisms which reflect the haplotype structure need to be better understood at a holistic level using omics techniques (Zargar et al., 2021; Satrio et al., 2023).

Recent advancements in molecular genetics and genotyping have led to the identification of major effect QTLs related to key abiotic stresses. Muthu et al. (2020) e.g. used MAS to introgress major QTLs for tolerance to drought ($qDTY_{1.1}$, $qDTY_{2.1}$), salinity (*Saltol*), and submergence (*Sub1*) into a popular rice variety (IWP). Backcrossed inbred lines exhibited enhanced performance against dehydration, salinity, and submergence stress compared with recurrent parent IWP (Muthu et al., 2020).

Among different DNA-based molecular markers, both first-generation markers (RAPD, RFLP, and AFLP) and second-generation markers (SSR and SNP) are closely related to the structural identity of their host genes but are not affected by environmental factors and plant developmental stages. With recently available microarray and RNA-Seq data, several functional markers have been identified. These markers derive from functionally characterized sequence motifs linked to the phenotypic trait of interest (Salgotra and Stewart, 2020; Samanta and Dey, 2023). Samanta and Dey (2023) showed that miRNA-based markers can be used for screening a rice population under drought stress. Genetic mapping has been combined with image acquisition to try to increase phenotyping precision. Image-based traits can be used in GWAS studies to find genes related to drought resistance (Guo et al., 2018b; Jiang et al., 2021).

QTL analysis e.g. revealed that *OsRINGzf1* (a novel E3 ligase that mediates aquaporin degradation in rice) positively regulates drought resistance by improving water retention capacity (Chen et al., 2022). Satrio et al. (2023) used a recombinant inbred line population to detect the QTL associated with agro-physiological traits under drought stress and found 154 QTLs distributed into 66 regions with LOD ranging from 2.0 to 38.7, including a large QTL cluster on rice chromosome 8 called 'hotspot QTL' that is strongly associated with drought tolerance.

In a study of 114 rice genotypes evaluating various root and shoot traits under water stress conditions, a total of 11 significant marker-trait associations were detected for various root, shoot, and drought tolerance traits with the coefficient of determination (R^2) ranging from 18.9% to 53.4%. A novel major QTL was detected for root length associated with marker *RM127*, explaining 19.3% of variation. Marker alleles linked with loci with notable phenotypic effects for root and drought tolerance traits can be exploited for marker-assisted selection (Verma and Sarma, 2021).

Another approach to obtain a drought-resistant plant is to modify genes of interest. A pivotal role is played by phytohormones which regulate plant response to stresses including drought. They regulate root growth by mediating specific growth processes. Abscisic acid (ABA) in particular has biphasic effects on primary root growth in rice through an auxin biosynthesis-mediated process, causing suppression of root elongation and promotion of root swelling in response to soil compaction. Enhancing ABA induced expression of auxin biosynthesis genes and auxin accumulation in roots. Conversely, blocking auxin biosynthesis reduced ABA sensitivity in roots, showing longer and thinner primary roots with larger root meristem size and smaller root diameter (Qin et al., 2023). ABA in rice is activated by osmotic stress/ABA-activated protein kinases (SAPKs). *OsSAPK3* plays a particularly important role in both ABA-dependent and ABA-independent drought stress responses. *OsSAPK3* could improve rice yield by indirectly regulating tiller number and grain size. These findings provide new insights for the development of drought-resistant rice (Lou et al., 2023).

Another interesting kinase in rice is *OsESG1* which can be activated by abiotic stress. Analysis of transgenic plants showed that *OsESG1* affects crown root development by controlling auxin distribution and transport and modulates drought stress response by regulating the antioxidative system and expression of stress-related genes (Pan et al., 2020). Another study involving a gene (*OsFBX257*) encoding for an F-box protein showed that this gene is co-expressed in a network involving protein kinases and phosphatases. OsFBX257 modulates root architecture and drought-stress tolerance in rice.

OsFBX257 knockdown lines show reduced total root length and depth, crown root number, panicle size, and survival under stress. In contrast, overexpression increases root depth, leaf and grain length, number of panicles, and grain yield (Sharma et al., 2022a).

Lateral root (LR) plasticity is a key trait for plant adaptation to fluctuating soil moisture stressed environments. It describes the plant's ability to alter its root phenotype in response to changing environmental conditions (O'Toole and Bland, 1987). Root plasticity in rice promotes LR development in response to drought stress (Kano-Nakata et al., 2011). A number of backcrossed F_2 lines derived from Nipponbare x chromosome segment substituted at line number 47 (Nipponbare/Kasalath) were used to analyze ten substituted chromosome regions with the 'Kasalath' allele associated with root plasticity. Two QTLs associated with LR plasticity on chromosome 12 significantly contributed to greater root system development and maintenance of total dry matter production under soil moisture fluctuation (SMF) stress (Niones et al., 2015). In addition to kinases, transcription factors also help activate the ABA pathway in rice seedlings, increasing drought tolerance. A particular transcription factor (encoded by the *IPA1* gene) has been shown to improve root system development, reduce stomatal opening, and increase ABA content in seedlings (Zhu et al., 2022). Improved stress tolerance resulting from *OsDSR2* gene deletion could be related to ABA and IAA signaling pathways, emphasizing the important role of phytohormones (ABA in particular) in plant drought resistance (Luo et al., 2023). Xu et al. (2020) found that a transcription factor *SiMYB56* may enhance the drought resistance of transgenic rice plants by regulating lignin biosynthesis and ABA signaling pathway, providing another candidate gene for drought resistance improvement (Xu et al., 2020).

Zinc finger proteins also influence gene expression. Tandem CCCH zinc finger (TZF) in particular plays critical roles in growth development and stress response through posttranscriptional regulation of target mRNA. Guo et al. (2022) showed that overexpression of *OsTZF7* (a gene encoding for a novel tandem zinc finger protein) increased drought tolerance and enhanced sensitivity to exogenous ABA in rice. RNA-seq analysis revealed that genes related to response to stress, ABA signaling, and methylated histone binding are regulated by *OsTZF7* (Guo et al., 2022).

Histone methyltransferases and demethylases also regulate gene expression relating to drought tolerance. A histone demethylase from rice is the Jumonji C domain-containing protein (JMJ710). This is a negative regulator of drought stress-responsive genes which blocks activation of drought response under normal conditions but allows for rapid activation during drought stress (Zhao et al., 2022b).

Table 1 reports other interesting work on how to increase drought resistance of rice.

Table 1 References to studies on drought resistance of rice

Locating genes associated with root morphology and drought avoidance in rice via linkage to molecular markers	Champoux et al. (1995)
Mapping genes controlling root morphology and root distribution in a doubled-haploid population of rice	Yadav et al. (1997)
Mapping QTLs associated with drought avoidance in upland rice	Courtois et al. (2000)
Deep rooting conferred by *DEEPER ROOTING 1* enhances rice yield in paddy fields	Arai-Sanoh et al. (2014)
Physiological mechanisms contributing to the QTL-combination effects on improved performance of IR64 rice NILs under drought	Henry et al. (2015)
QTLs underlying natural variation of root growth angle among rice cultivars with the same functional allele of *DEEPER ROOTING 1*	Kitomi et al. (2015)
qRT9, a quantitative trait locus controlling root thickness and root length in upland rice	Li et al. (2015)
Transcriptome profiling of drought-responsive noncoding RNAs and their target genes in rice	Chung et al. (2016)
Identification of candidate network hubs involved in metabolic adjustments of rice under drought stress by integrating transcriptome data and genome-scale metabolic network	Mohanty et al. (2016)
Discovery of QTLs for water mining and water-use efficiency traits in rice under water-limited condition through association mapping	Raju et al. (2016)
Detection of QTL for exudation rate at ripening stage in rice and its contribution to hydraulic conductance	Yamamoto et al. (2016)
Genetic control of the root system in rice under normal and drought stress conditions by genome-wide association study (GWAS)	Li et al. (2017)
Association between root growth angle and root length density of a near-isogenic line of IR64 rice with *DEEPER ROOTING 1* under different levels of soil compaction	Ramalingam et al. (2017)
Genetic analysis of yield and agronomic traits under reproductive-stage drought stress in rice using a high-resolution linkage map	Bhattarai and Subudhi (2018)
RSL Class II transcription factors guide the nuclear localization of RHL1 to regulate root hair development	Moon et al. (2019)
Tuning a timing device that regulates lateral root development in rice	Acevedo et al. (2019)
Introgression of root and water-use efficiency traits enhances water productivity: an evidence for physiological breeding in rice (*O. sativa* L.)	Dharmappa et al. (2019)
Genome-wide identification of SNAC1-targeted genes involved in drought response in rice	Li et al. (2019b)
Involvement of *OsGF14b* adaptation in the drought resistance of rice plants	Liu et al. (2019)
The interaction of genes controlling root traits is required for the developmental acquisition of deep and thick root traits and improving root architecture in response to low water or nitrogen content in rice (*O. sativa* L.)	Nada et al. (2019)

(*Continued*)

Table 1 (*Continued*)

A coumarin analog NFA from endophytic *Aspergillus fumigatus* improves drought resistance in rice as an antioxidant	Qin et al. (2019)
Genetic mapping identifies consistent quantitative trait loci for yield traits of rice under greenhouse drought conditions	Baisakh et al. (2020)
Expression of the CCCH-tandem zinc finger protein gene *OsTZF5* under a stress-inducible promoter mitigates the effect of drought stress on rice grain yield under field conditions	Selvaraj et al. (2020)
Natural variation of *DROT1* confers drought adaptation in upland rice	Sun et al. (2022)
OsGRP3 enhances drought resistance by altering phenylpropanoid biosynthesis pathway in rice (*Oryza. sativa* L.)	Xu et al. (2022)
IPA1 improves drought tolerance by activating *SNAC1* in rice	Chen et al. (2023)

3 Barley: understanding and optimizing drought tolerance

Barley (*Hordeum vulgare* L.) is one of the oldest domesticated crops with a genome of about 5.3 Gb (Fig. 5) (Mascher et al., 2017). It is widely used in human and animal nutrition as well as malt production. As a rich source of soluble dietary fiber, it reduces the risk of illnesses such as type II diabetes, colorectal cancer, and cardiovascular disease (Prasadi and Joye, 2020). Genetic and archaeological evidence suggest that barley is a mosaic crop (Poets et al., 2015; Pankin et al., 2018) developed from different populations and possibly without a single origin (Elakhdar et al., 2022; Haas et al., 2019). Barley is important in the study of drought resistance because of its ability to tolerate abiotic stresses such as drought (Elakhdar et al., 2023; Forster et al., 2000; Sallam et al., 2019; Talamè et al., 2007). Its close relationship with the Triticeae family also makes it possible to transfer developments in barley to wheat (Fusi et al., 2022; Kirschner et al., 2021).

Secondary messengers such as calcium play a role in drought resistance in barley. Calcium ions (Ca^{2+}) play a central role in the perception and transduction of signals in response to biotic and abiotic stress, particularly in response to osmotic changes. Calcium channels are also involved in growth regulation and biosynthesis (She et al., 2022). Osmosensors related to the hyperosmolality-gated calcium-permeable channel gene family (*OSCA*) may be involved in abiotic stress regulation (She et al., 2022). These *HvOSCAs* include cis-regulatory elements such as drought-responsive elements (MBS) and anaerobic induction elements (ARE). All 14 *HvOSCAs* are induced by drought and other abiotic stresses, given their role in osmotic regulation (Cai et al., 2022). Another transporter is the K^+ channel. Feng et al. (2020) have studied two important K^+ transporters, HvAKT2 and HvHAK1, involved in drought tolerance of barley. Overexpression of *HvAKT2* and *HvHAK1* enhanced K^+ uptake and H^+ homoeostasis leading to drought tolerance in barley transgenic lines. MicroRNAs (miRNAs) are also involved in gene expression. Yuan et al. (2019) showed that over-expression of miR393 enhanced drought tolerance via ABA biosynthesis.

Numerous drought stress-related genes have been identified and characterized by cloning techniques. Transcriptional regulation of gene expression has been used to characterize stress-related genes and their pathways e.g. using transcriptomic, metabolic, and proteomic approaches (Harb et al., 2020; Javadi et al., 2021; Ozturk et al., 2002). Wang et al. (2023) used

Table 2 References to studies on drought resistance of barley

Dehydration-induced transcriptomic responses in two Tibetan hulless barley (*Hordeum vulgare* var. *nudum*) accessions distinguished by drought tolerance	Liang et al. (2017)
Root architectural traits and yield: exploring the relationship in barley breeding trials	Robinson et al. (2018)
Foliar application of betaine improves water-deficit stress tolerance in barley (*Hordeum vulgare* L.)	Wang et al. (2019)
Drought-induced anatomical modifications of barley (*Hordeum vulgare* L.) leaves: An allometric perspective	Wyka et al. (2019)
Genetic components of root architecture and anatomy adjustments to water-deficit stress in spring barley	Oyiga et al. (2020)
Proline-mediated drought tolerance in the barley (*Hordeum vulgare* L.) isogenic line is associated with lateral root growth at the early seedling stage	Frimpong et al. (2021)
Association analysis of physiological traits in spring barley (*H. vulgare* L.) under water-deficit conditions	Jabbari et al. (2021)
Identification of *HvLRX*, a new dehydration and light responsive gene in Tibetan hulless barley (*H. vulgare* var. *nudum*)	Liang et al. (2021)
Interactive effects of hydrogen sulfide and silicon enhance drought and heat tolerance by modulating hormones, antioxidant defense enzymes, and redox status in barley (*H. vulgare* L.)	Naz et al. (2022)
Seed endophyte bacteria enhance drought stress tolerance in *H. vulgare* by regulating, physiological characteristics, antioxidants, and mineral uptake	Abideen et al. (2022)
Most drought-induced DNA methylation changes switched to pre-stress state after re-irrigation in barley (*H. vulgare* L.) cultivars	Falahi et al. (2022)
Identification of DNA markers associated with phenological traits in spring barley (*H. vulgare* L.) under drought stress conditions	Jabbari et al. (2022)
Mycorrhizal biofertilization improves grain yield and quality of hulless barley (*H. vulgare* ssp. *nudum* L.) under water stress conditions	Jerbi et al. (2022)
Identification of *PP2C* genes in Tibetan hulless barley (*H. vulgare* var. *nudum*) under dehydration stress and initiatory expression and functional analysis of HvPP2C59	Liang et al. (2022)
QTLs controlling physiological and morphological traits of barley (*H. vulgare* L.) seedlings under salinity, drought, and normal conditions	Makhtoum et al. (2022)

DAP-seq-based transcriptomic analysis to identify novel transcription factors (TFs) in barley affecting drought resistance. Transcriptome RT-qPCR analysis showed that TF HOVUSG2784400 regulated gene expression under drought stress (Wang et al., 2023). Another TF belonging to MYB family (*HvMYB1*) is reported to be involved in protecting plants against drought by acting as a mediator of ABA (Alexander et al., 2019).

Four major QTLs related to drought tolerance have been detected for spike diameter, awn weight, and flag leaf weight (*qPPT*-7, *qAWW*-7, *qFLW*-4a, and *qFLW*-5) (Makhtoum et al., 2022). Kandemir and Saygili (2023) transferred drought tolerance QTL on chromosome 2 from the drought-tolerant cultivar 'Tadmor' to drought-susceptible 'Baronesse' and 'Aydanhanım'. This QTL resulted in 5.0% and 9.1% improvement in leaf relative water content of 'Baronesse' and 'Aydanhanım' cultivars, respectively. This QTL region increased grain yields by 1.1 t/ha and 0.6 t/ha in 'Baronesse' and 'Aydanhanım', respectively, mainly by increasing the number of tillers (Kandemir and Saygili, 2023).

Table 2 reports other interesting work on how to increase drought resistance in barley.

4 Durum (tetraploid) wheat: understanding and optimizing drought tolerance

Durum (tetraploid) wheat (*Triticum turgidum* L.) has a key role within the genus *Triticum*. Its many subspecies, including wild forms, exhibit many desirable agronomic traits for improvement of durum and bread wheat. The tetraploid durum wheat genome is about 10.45 Gb (Fig. 5) (Maccaferri et al., 2019) and includes two subgenomes: A and B hexaploid bread wheat evolved through two distinct past hybridization events that included three diploid progenitor species. The wheat genome thus includes three subgenomes: A, B, and D (2n = 6x = 42). The A wheat genome was donated by *Triticum urartu* (AA), while *Aegilops speltoides* (BB) is believed to be the donor of the B genome. The D genome was donated by *Aegilops tauschii* (DD) (Yadav et al., 2023). As shown in Fig. 6, wheat underwent a first domestication, hence a strong biodiversity (i.e. haplotype diversity) reduction bottleneck when wild emmer was domesticated (Maccaferri et al., 2019) and, subsequently, when bread (hexaploid) wheat went through the second domestication bottleneck. Haplotype analysis can identify native haplotypes mostly absent in modern wheat.

Durum wheat is primarily cultivated in the semiarid Mediterranean Basin where moisture is mostly provided through rain, increasing vulnerability to drought stress. Some of the physiological mechanisms of drought adaptation of durum wheat have been established (Royo et al., 2010; Graziani et al., 2014). This helps in selecting varieties with morphological and physiological adaptation mechanisms such as increased water use efficiency and yield stability

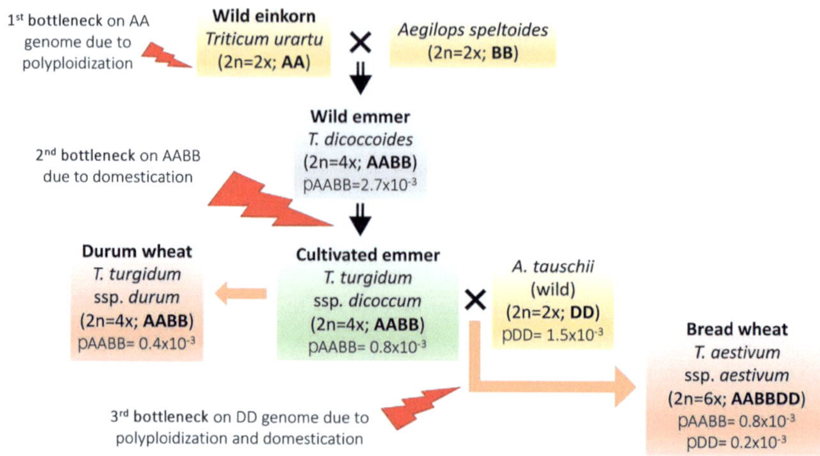

Figure 6 Origin and evolution of wheat. In the figure, the three bottlenecks that occurred during the domestication of wheat are highlighted with a red flash. The three bottlenecks vary in magnitude, but it is likely that the most important, at the level of domestication, was the second, in which wild emmer became cultivated. π = nucleotide diversity (Haudry et al., 2007; Akhunov et al., 2010).

in water-limited environments. Among novel traits providing an integrated view of the plant response to drought, chlorophyll fluorescence (ChlF) is a powerful noninvasive technique for probing photosynthesis in physiological breeding under drought. Zendonadi dos Santos et al. (2020) showed that the light-induced fluorescence transient (LIFT) sensor detected significant differences in ChlF and photosynthetic performance in a large panel of durum wheat genotypes of the Global Durum Panel (Mazzucotelli et al., 2021) under progressive drought. The same panel revealed significant differences among genotypes for carbon isotope composition ($\delta^{13}C$) at the leaf level, a highly heritable trait that captures well the plastic response of the plant to drought (Reynolds and Tuberosa, 2008).

Numerous wheat wild relatives or landraces provide a genetic resource for the improvement of both durum and bread wheat (Yadav et al., 2023). A study of the Italian durum wheat (*Triticum durum* Desf.) landrace Saragolle Lucana e.g. shows increased rhizosheath formation during drought stress despite a smaller root system, potentially optimizing water take-up and retention (Bochicchio et al., 2022). Wild emmer wheat [*Triticum turgidum* ssp. *dicoccoides* (Körn.) Thell], the direct allotetraploid (2n = 4x = 28; genome BBAA) progenitor of domesticated wheat, exhibits robust vegetative growth under arid conditions and can serve as a potential source for enhancing drought adaptation in durum wheat (Bacher et al., 2021). Licaj et al. (2023) showed that the ancient cultivar Saragolla is more resistant to drought stress than the modern Svevo cultivar widely used for drought-tolerant studies (Maccaferri et al., 2008, 2011).

'Omics' approaches can help identify stress-related genes as potential markers of drought tolerance in *Triticum* species (Condorelli et al., 2022). A consensus map with different markers has been developed to explore genetic diversity in *Triticum durum*, providing nearly complete genome coverage (Maccaferri et al., 2014). The consensus map provides the basis for identifying high-density single-nucleotide polymorphic (SNP) markers in durum wheat with the potential to integrate tetraploid and hexaploid wheat genomics (Maccaferri et al., 2015).

Crop modeling also has a role to play (Hammer et al., 2008; Tardieu and Tuberosa, 2010). Dettori et al. (2022) e.g. used the CERES-Wheat crop model to assess how dehydration responsive element-binding *(DREB)* genes in three durum wheat cultivars (Creso, Duilio, and Simeto) grown in Southern Sardinia (Italy) might improve drought stress response.

The greatest yield losses in wheat usually occur when drought stress takes place during flowering and seed filling. Plants must have a good root system to capture water both early in the season and moisture in deeper soil layers toward the end of the season (Hassouni et al., 2018; Pour-Aboughadareh et al., 2020). Root system architecture (RSA) is thus critical to drought tolerance. Plants with a narrow root angle can grow roots deeper and can thus access water and nutrients in deeper soil layers.

Maccaferri et al. (2016) used linkage and association mapping for RSA in two recombinant inbred line populations and using an association mapping panel of elite durum wheat (*Triticum turgidum* L. var. *durum* Desf.) accessions to reveal 20 QTL clusters for root length and number, as well as 30 QTLs for root growth angle (RGA). A GWAS for seminal root angle (SRA) identified seven marker-trait associations clustered on chromosome 6A, representing a major QTL (qSRA-6A) with high levels of pairwise LD (r^2 = 0.67). Subsequent haplotype analysis revealed loci related to gravitropism, polar growth, and hormone signaling. This study revealed that there is a genetic predisposition for deep root growth in conditions of water limitation, though root systems show plasticity in root growth in response to water availability in upper soil layers (Alahmad et al., 2019).

Another approach to increase tolerance to drought in wheat through the root system is to avoid soil acidity. Subsoil acidity with high Al^{3+} concentration inhibits growth, development, and grain yield of durum wheat lines lacking Al^{3+} tolerance genes. Introgression of the Al^{3+}-tolerant *TaMATE1B* gene into durum wheat improved resistance to terminal drought by allowing root growth and proliferation in deep layers of acidic soil with Al^{3+} (Liu et al., 2022a).

Meta-QTL analysis is a useful approach for analyzing the genetic basis of complex traits to identify putative molecular markers for marker-assisted selection. A meta-QTL analysis by Soriano et al. (2021) assessed 45 durum wheat traits, including abiotic stress-related traits. They found a total of 368 QTLs distributed on all 14 chromosomes of genomes A and B: 171 corresponded to quality-related traits, 127 to abiotic stress, and 71 to biotic stress. This is particularly useful in identifying major loci for quantitative traits and, by increasing map resolution, in identifying candidate genes that control polygenic traits (Soriano et al., 2021).

An important component of drought resistance in crops is osmotic adjustment (OA). The genetic basis of OA in wheat and other crops remains largely unknown. A study of 248 field-grown durum wheat elite accessions found high variability but also high trait repeatability in OA (h^2 = 72.3%) (Condorelli et al., 2022). A GWAS revealed 15 significant QTLs for OA (global R^2 = 63.6%), as well as eight major QTL hotspots/clusters on chromosome arms 1BL, 2BL, 4AL, 5AL, 6AL, 6BL, and 7BS. Higher OA capacity was positively associated with leaf-relative water content and/or chlorophyll content associated with water retention. Gene content analysis of cluster regions revealed the presence of several candidate genes for OA, including the bidirectional sugar transporter SWEET, rhomboid-like protein, the *S*-adenosyl-L-methionine-dependent methyltransferases superfamily protein, and DREB1.

Genes activated by drought include those coding for proteins acting as regulators of signal transduction. Gullì et al. (2022) have characterized a *Triticum durum* gene, *TdDRG1*, in two commercial cultivars, Primadur and Svevo. *TdDRG1* codes for a putative transcription factor belonging to the VPS72/YL-1 family. This gene is mostly expressed in roots of the resistant cultivar Svevo compared to Primadur.

Table 3 References to studies on drought resistance of durum wheat

Ancestral QTL alleles from wild emmer wheat improve drought resistance and productivity in modern wheat cultivars	Merchuk-Ovnat et al. (2016)
Genetic variation for aerenchyma and other root anatomical traits in durum wheat (*T. durum* Desf.)	Nazemi et al. (2016)
Identification and characterization of durum wheat microRNAs in leaf and root tissues	Fileccia et al. (2017)
Ancestral QTL alleles from wild emmer wheat enhance root development under drought in modern wheat	Merchuk-Ovnat et al. (2017)
Yield of chromosomally engineered durum wheat-*Thinopyrum ponticum* recombinant lines in a range of contrasting rain-fed environments.	Kuzmanović et al. (2018)
Overexpression of a wheat aquaporin gene, *TdPIP2;1*, enhances salt and drought tolerance in transgenic durum wheat cv. Maali	Ayadi et al. (2019)
A novel aspect of essential oils: coating seeds with thyme essential oil induces drought resistance in wheat	Ben-Jabeur et al. (2019)
Genetic dissection of the seminal root system architecture in Mediterranean durum wheat landraces by GWAS	Roselló et al. (2019)
Screening for durum wheat (*T. durum* Desf.) cultivar resistance to drought stress using an integrated physiological approach	Guellim et al. (2020)
Gene co-expression network analysis to identify critical modules and candidate genes of drought-resistance in wheat	Lv et al. (2020)

Table 3 reports other interesting work on drought resistance in durum wheat.

5 Bread (hexaploid) wheat: understanding and optimizing drought tolerance

Hexaploid (bread) wheat is a staple cereal grown on ca. 220 million hectares to meet the food needs of 2.5 billion people globally (FAO, 2022). It has been estimated that current global production of 763 Mt needs to increase to at least ~860 Mt by 2050 (United Nations, World population prospect, 2022). Drought stress has been recognized as a major cause of wheat yield loss. If drought occurs early, poor rooting of the crop can result in severe and even complete yield loss, while later drought can lead to drastic reductions in harvest index and lower yields (Langridge and Reynolds, 2021; Sehgal et al., 2023).

Bread wheat is one of the most widely studied hexaploid crops, particularly at the level of cytogenetics and genetics, with three closely related sub-genomes ($2n = 6x = 42$, AABBDD) and a genome size of about 17 Gb (Fig. 5) (International Wheat Genome Sequencing Consortium (IWGSC), 2018). The hexaploid nature of bread wheat allows it to tolerate major structural and numerical changes in the composition of its chromosome (Gupta et al., 2020; Zhao et al., 2022a).

Khan et al. (2023) has evaluated 25 wheat genotypes to identify genetic variability in relation to drought stress. Tolerance traits reflect the ability of wheat plants to withstand prolonged drought stress, while recovery trait measure the ability of plants to recover after prolonged drought stress (Sallam et al., 2022). A meta-QTL (MQTL) has identified a total of 854 individual QTLs related to drought and heat stress, including 502 associated with drought stress (Acuña-Galindo et al., 2015). These include stay-green QTLs co-located with QTLs for seminal root angle, a highly heritable trait first investigated and described in durum wheat (Sanguineti et al., 2007) and later shown to be related to grain yield as affected by available soil moisture (Maccaferri et al., 2016; Alahmad et al., 2019; Condorelli et al., 2022).

Relatively few studies have considered the role of root system architecture (RSA) traits in stress response. Each stage of root initiation, growth, and development is genetically controlled and responsive to the environment, resulting in a complex QTLome (Salvi and Tuberosa, 2015) with underutilized opportunities for genetic improvement (Ober et al., 2021). Heat-stressed plants increase transpiration rate to allow canopy cooling. This requires increased stomatal conductance and vascular capacity, including roots. Low canopy temperatures in heat stress conditions have been associated with increased plant access to water through deeper roots (Reynolds and Tuberosa, 2008).

In water stress conditions, COOL genotypes show a deeper root system than HOT genotypes, allowing extraction of 35% more water from the 30–90 cm soil profile (Pinto and Reynolds, 2015). One study, which mapped QTL in the *D_GA8KES401CIKOJ-160-BS00067285_51* marker interval for root length (co-localized with a QTL for grain weight per spike), identified three QTLs related to improved grain yield under drought conditions, including the *Q.Gy-2D* haplotype, proving a potential target for breeding drought-tolerant varieties (Salarpour et al., 2020).

The symbiotic relationship between wheat and arbuscular mycorrhizae can also enhance drought stress tolerance in wheat. Rhizobacteria have a direct role in producing phytohormones and making nutrients more available and an indirect role is suppressing pathogens and reducing disease pressure (Lucas et al., 2014). Two genome regions on chromosomes 3D and 7D have been found to be significantly associated with mycorrhizal response under drought stress conditions (Lehnert et al., 2018). Drought-stressed wheat seedlings inoculated with the arbuscular fungus *Funneliformis mosseae* were found to have higher root and stem biomass with overexpression of three aquaporin-related genes *TaNIP1-10*, *TaNIP3-3*, and *TaNIP3-4* (Asadollahi et al., 2023).

Like mycorrhizae, other beneficial microbes can improve drought resistance. Wheat seedling exposure to the bioprotectant bacterium *Streptomyces pactum* Act12 enhanced plant growth under drought stress, with increases in shoot fresh weight and length as well as root length. Further analysis revealed that Act12 treatment increased foliar abscisic acid content and upregulated expression of drought resistance-related genes such as *EXPA2*, *EXPA6*, *P5CS,* and *SnRK2* (Li et al., 2020).

Transcriptome profiling of the genotype of drought-resistant bread wheat landraces has shown upregulation of genes involved in oxidoreduction processes, secondary metabolite biosynthesis, abiotic stress response, transferase activity, and heat shock proteins, indicating that landrace drought tolerance mainly involves avoidance strategies (Chaichi et al., 2019). The WRKY family of transcription factors is associated with responses to biotic and abiotic stresses. *TaWRKY1-2D* may play important roles in wheat drought tolerance through posttranslational regulation of *TaDHN3* (a gene encoding for a dehydrin protein), making it a promising candidate gene for selecting drought-tolerant wheat varieties (Yu et al., 2023).

Rye (*Secale cereale* L.), a close relative of wheat, is more tolerant than wheat to water shortage and has been reported to have a more robust root system (Howell et al., 2019; Sharma et al., 2022b). Introgression of the 1RS arm of rye into wheat has been associated with increased drought resistance since root biomass is increased (Howell et al., 2019). D-genome introgressions

Table 4 References to studies on drought resistance of bread wheat

Detection of two major grain yield QTL in bread wheat (*T. aestivum* L.) under heat, drought, and high-yield potential environments	Bennett et al. (2012)
Tolerance of wheat to vegetative stage soil waterlogging is conditioned by both constitutive and adaptive QTL	Ballesteros et al. (2015)
High-resolution mapping of traits related to whole-plant transpiration under increasing evaporative demand in wheat	Schoppach et al. (2016)
Identification and validation of root length QTLs for water stress resistance in hexaploid wheat (*T. aestivum* L.)	Ayalew et al. (2017)
Drought response in wheat: key genes and regulatory mechanisms controlling root system architecture and transpiration efficiency	Kulkarni et al. (2017)
QTL mapping for seedling morphology under drought stress in wheat cross synthetic (W7984)/Opata	Khalid et al. (2018)
Comparative analysis of root transcriptome profiles between drought-tolerant and susceptible wheat genotypes in response to water stress	Hu et al. (2018)
Association study reveals novel genomic regions for grain yield and yield-related traits in drought-stressed synthetic hexaploid wheat	Bhatta et al. (2018)
QTL for stay-green traits in wheat in well-watered and water-limited environments	Christopher et al. (2018)
Loci and candidate genes controlling root traits in wheat seedlings - a wheat root GWAS	Beyer et al. (2019)
Comparative metabolomic profiling in the roots and leaves in contrasting genotypes reveals complex mechanisms involved in post-anthesis drought tolerance in wheat	Kang et al. (2019)
Overexpression of *AtWRKY30* transcription factor enhances heat and drought stress tolerance in wheat (*T. aestivum* L.)	El-Esawi et al. (2019)
Transcriptome analysis of osmotic-responsive genes in ABA-dependent and ABA-independent pathways in wheat (*T. aestivum* L.) roots	Li et al. (2019a)
The *LATERAL ROOT DENSITY* gene regulates root growth during water stress in wheat	Placido et al. (2020)
Change in the activity of genes of transcription factors *TaNAC69*, *TaDREB1*, and *TabZIP60* in bread wheat plants with water deficiency and hypothermia	Zaikina et al. (2022)
Impact of ethylene and abscisic acid on programmed cell death in endosperm and grain filling of wheat (*T. aestivum* L.)	Li et al. (2023)
Could buckwheat and spelt be alternatives to wheat under future environmental conditions? Study of their physiological response to drought	Martínez-Goñi et al. (2023)
Effects of elevated CO_2 on the photosynthesis, chlorophyll fluorescence, and yield of two wheat cultivars (*T. aestivum* L.) under persistent drought stress	Yang et al. (2023)

from *Aegilops tauschii* have been shown to improve drought-adaptive traits, particularly MNH5 and MSD345, suggesting these multiple synthetic derivative lines (MSDLs) could be explored for breeding drought-resistant wheat lines (Itam et al., 2020).

Table 4 reports other interesting work on how to increase drought resistance in bread wheat.

6 Sorghum: understanding and optimizing drought tolerance

Sorghum evolved in Africa after a split with rice ca. 50–70 million years ago (Wolfe et al., 1989). It has a genome size of about 730 Mb (Fig. 5) (Paterson et al., 2009). Sorghum is a relatively drought-tolerant C4 species that can cultivated in marginal environments. It is a staple crop in semiarid environments of sub-Saharan Africa and west-central India (Ari Akin et al., 2022; I. Mohamed et al., 2022). However, average sorghum yields in semiarid regions has plateaued at around 1.0 ton/ha compared to the global average of 2.5 ton/ha, mainly due to recurrent droughts and heat-stress episodes (Yahaya and Shimelis, 2022). There are two types of drought stress in sorghum: pre-flowering and post-flowering stress. In the case of pre-flowering stress, sorghum plants show leaf curling, discoloration, and reduction in grain yield of potentially over 40% (Souza et al., 2020). Post-flowering drought stress, which occurs during anthesis and seed set, can lead to complete crop failure (Yahaya and Shimelis, 2022).

GWAS has identified 33 candidate genes with potential functions related to stress signaling, oxidative stress protection, hormonal response to stress, and dehydration protection, helping to understand genetic control of sorghum photosynthetic response to drought (Ortiz and Salas-Fernandez, 2022). A total of 154 QTLs were identified for eight drought-related traits, 64.9% of which overlapped with previously identified QTLs for traits related to stay-green, the flowering regulator *Ma6* and the drought-resistant gene *P5CS2* (Dong et al., 2022). Abdel-Ghany et al. (2020) has located around 180 mainly upregulated genes linked to drought stress response. Among these, about 70 genes are novel with no known function, while the remaining are transcription factors, signaling proteins and stress-related proteins implicated in drought tolerance in other crops.

Another trait that confers adaptation to drought in sorghum is the stay-green trait. Borrell et al. (2014) showed that stay-green sorghum plants exhibit greener leaves and stems during the grain-filling period under water-limited conditions, resulting in increased grain yield, grain mass, and lodging resistance. Four QTLs (*Stg1*, *Stg2*, *Stg3,* and *Stg4*) in particular were involved in regulation of canopy size, leaf anatomy, and root growth. The generic physiological mechanisms underlying the stay-green trait suggest that similar *Stg* ortholog

QTLs could also improve post-anthesis drought adaptation in maize, wheat, and rice (Borrell et al., 2014; Faye et al., 2022).

Research has shown that rhizobacterial isolates can protect sorghum from drought stress (Carlson et al., 2020). Root-tissue colonizing bacteria have been found to upregulate drought-responsive genes such as *sbP5CS2* and *sbP5CS1* in drought-stressed plants (Govindasamy et al., 2020). Transcription factors also play an important role in sorghum drought tolerance. Yang et al. (2020)

Table 5 References to studies on drought resistance of maize and sorghum

Quantitative trait locus mapping reveals regions of the maize genome controlling root system architecture	Zurek et al. (2015)
Genetic dissection of maize seedling root system architecture traits using an ultra-high density bin-map and a recombinant inbred line population	Song et al. (2016)
Synergy between root hydrotropic response and root biomass in maize (*Zea mays* L.) enhances drought avoidance	Eapen et al. (2017)
The maize ABA receptors *ZmPYL8*, *9*, and *12* facilitate plant drought resistance	He et al. (2018)
QTL-by-environment interaction in the response of maize root and shoot traits to different water regimes	Li et al. (2018)
Maize lateral root developmental plasticity induced by mild water stress. I: Genotypic variation across a high-resolution series of water potentials	Dowd et al. (2019)
Carbon isotope composition, water-use efficiency, and drought sensitivity are controlled by a common genomic segment in maize	Avramova et al. (2019)
Genome-scale mining of root-preferential genes from maize and characterization of their promoter activity	Li et al. (2019c)
Transcription factors as molecular switches to regulate drought adaptation in maize	Leng and Zhao (2020)
Nucleotide diversity and association analysis of *ZmMADS60* with root length in the maize seedling stage	Li et al. (2020)
A comprehensive transcriptomics analysis reveals long noncoding RNA to be involved in the key metabolic pathway in response to waterlogging stress in maize	Yu et al. (2020)
Genome-wide selection and genetic improvement during modern maize breeding	Wang et al. (2020)
QTL for nodal root angle in sorghum (*Sorghum bicolor* L. Moench) colocate with QTL for traits associated with drought adaptation	Mace et al. (2012)
Studying the genetic basis of drought tolerance in sorghum by managed stress trials and adjustments for phenological and plant height differences	Sabadin et al. (2012)
QTLs associated with crown root angle, stomatal conductance, and maturity in sorghum	Lopez et al. (2017)
Influence of abiotic stresses on morphophysiological characteristics and biological value of grain *Sorghum bicolor* (L.) Moench	Kibalnik et al. (2023)

identified a WRKY transcription factor, *SbWRKY30,* induced by drought stress, suggesting it might be a candidate gene for breeding drought-tolerant varieties (Yang et al., 2020). Another transcription factor belonging to NAC family, the *SbNAC9* gene, is also involved in drought tolerance in sorghum (Jin et al., 2023).

Table 5 reports other interesting work on how to increase drought resistance in sorghum.

7 Maize: understanding and optimizing drought tolerance

Maize (*Zea mays* L.) or corn has an annual output of 1.15 billion tons from an area of about 197 million hectares (Choudhary et al., 2023). Maize and sorghum diverged from a common ancestor around 12 million years ago. Approximately 60% of annotated genes belong to conserved syntenic regions from both species (Zheng et al., 2020). Maize has a large genome of about 2.4 Gb (Fig. 5) (Schnable et al., 2009).

Drought tolerance in maize is a complex trait that can be improved by multitrait targeted selection (Beissinger et al., 2018). About 52 QTLs have been identified affecting plant height, ear height, anthesis-silking interval, ear weight, panicle weight, 100-grain weight, and ear length in two $F_{2:3}$ populations under drought conditions (Zhao et al., 2018). RNA-seq analyses of 224 maize accessions identified 97 genes associated with drought tolerance (Liu et al., 2020). However, GWAS analysis found only a few overlapping candidate genes associated with the same drought-related traits under different conditions (Khan et al., 2022).

Meta-QTL analysis has identified five maize genes affecting root development and root response to abiotic stress in maize (Guo et al., 2018a). Three QTLs have been found controlling seminal root number (SRN) (Salvi et al., 2016). PTF1, a basic helix-loop-helix (bHLH) transcription factor has also been found to contribute to root development (Li et al., 2019). Liu et al. (2022b) found coexpression of two genes (*ZmVPP1* and *ZmNAC111*) conferred drought resistance in maize seedlings with higher grain yields under water-deficit conditions in field trials.

Research has identified a network of small RNA transcriptomes (sRNAomes) mediated by transposable elements and inverted repeats (TE-IR) that balance plant drought tolerance with yield-related traits, providing potential genomic targets for selection of crops with enhanced stress tolerance without compromising yield (Sun et al., 2023).

Poly-γ-glutamic acid (γ-PGA) is a nontoxic, water-soluble, biodegradable biopolymer composed of D/L-glutamic acid monomers and fermented by *Bacillus subtilis*. γ-PGA has received increasing attention in enhancing plant take-up of fertilizers. Ma et al. (2022) have shown that application of γ-PGA significantly increased dry weight of maize, ABA, soluble sugar, proline, chlorophyll, and photosynthetic rate under severe drought stress. RNA-seq data suggests that γ-PGA may enhance drought resistance by affecting expression of ABA biosynthesis, signal transduction, photosynthesis-related genes, and other stress-responsive genes. In another study, melatonin treatment of maize seedlings significantly improved photosynthetic gas exchange and stomatal behavior, maintaining chlorophyll content and photosynthesis in drought stress conditions (Li et al., 2021b). Melatonin also stimulated the antioxidant system, improving elimination of reactive oxygen species and preventing severe damage from drought.

Inoculation of maize with arbuscular mycorrhizal fungi (AMF) has been found to reduce the effects of drought-induced oxidative damage through elimination of reactive oxygen species (ROS) (Begum et al., 2019). Flavonoids also contribute in ROS removal and alleviate stress-induced oxidative damage (Li et al., 2021a). However, it should be noted that the capacity to accumulate ABA in maize has decreased (Sanguineti et al., 2006) with the development of modern maize varieties (Duvick, 2005).

Table 5 reports other interesting work on how to increase drought resistance in maize.

8 Conclusion and future trends

This chapter provides a glimpse of how circular genomics is increasingly coming of age in more effectively dissecting the genetic basis of drought tolerance in cereals (Langridge et al., 2022; Maccaferri et al., 2022; Adel and Carels, 2023). Among the many drought-adaptive traits considered in this chapter, RSA-related traits are receiving increasing attention (Alemu et al., 2021; Kirschner et al., 2021; Ober et al., 2021; Tuberosa et al., 2021; Fusi et al., 2022; Gregory, 2022; Langridge et al., 2022). Field phenotyping, heritability, and accurate envirotyping play a key role to accurately identify the most reliable and important candidates at the QTLome level (Salvi and Tuberosa, 2015).

Since a single reference genome does not allow for accurate exploration of haplotype diversity within a species, the concept of pangenome has been developed. A pangenome is a collection of all sequences available for a species, including consensus sequences, large structural variations as well as small variations such as SNPs and insertions/deletions (Wang et al., 2023). Graph-based pangenomes have been proposed to expand linear reference genomes to include additional genetic diversity regions to enhance breeding opportunities and effectiveness. Increasingly affordable sequencing costs allow for pangenome assembly in cereals even with the large Triticeae genomes (Chen et al., 2023).

Ultimately, closer integration between advanced genomics and bioinformatics with sound phenotyping (Tuberosa, 2012) and envirotyping, including the water status of both the plant and soil, still poorly understood, coupled with modelling (Cooper and Messina, 2023) will allow for more effective identification and selection of beneficial haplotypes able to mitigate drought effects while contributing to a lower environmental footprint of cereal production to combat global warming.

9 Where to look for further information

The following link provides useful information:

- PlantStress: https://plantstress.com.

10 References

Abdel-Ghany, S. E., Ullah, F., Ben-Hur, A. and Reddy, A. S. N. (2020). Transcriptome analysis of drought-resistant and drought-sensitive sorghum (*Sorghum bicolor*) genotypes in response to PEG-induced drought stress, *International Journal of Molecular Sciences* 21(3), 772.

Abideen, Z., Cardinale, M., Zulfiqar, F., Koyro, H. W., Rasool, S. G., Hessini, K., Darbali, W., Zhao, F. and Siddique, K. H. M. (2022). Seed endophyte bacteria enhance drought stress tolerance in Hordeum vulgare by regulating, physiological characteristics, antioxidants and minerals uptake, *Frontiers in Plant Science* 13, 980046.

Acevedo, L. A., Korson, N. E., Williams, J. M. and Nicholson, L. K. (2019). Tuning a timing device that regulates lateral root development in rice, *Journal of Biomolecular NMR* 73(8-9), 493–507.

Acuña-Galindo, M. A., Mason, R. E., Subramanian, N. K. and Hays, D. B. (2015). Meta-analysis of wheat QTL regions associated with adaptation to drought and heat stress, *Crop Science* 55(2), 477–492.

Adel, S. and Carels, N. (2023). Plant tolerance to drought stress with emphasis on wheat, *Plants* 12(11), 2170.

Akhunov, E. D., Akhunova, A. R., Anderson, O. D., Anderson, J. A., Blake, N., Clegg, M. T., Coleman-Derr, D., Conley, E. J., Crossman, C. C., Deal, K. R., Dubcovsky, J., Gill, B. S., Gu, Y. Q., Hadam, J., Heo, H., Huo, N., Lazo, G. R., Luo, M. C., Ma, Y. Q., Matthews, D. E., McGuire, P. E., Morrell, P. L., Qualset, C. O., Renfro, J., Tabanao, D., Talbert, L. E., Tian, C., Toleno, D. M., Warburton, M. L., You, F. M., Zhang, W. and Dvorak, J. (2010). Nucleotide diversity maps reveal variation in diversity among wheat genomes and chromosomes, *BMC Genomics* 11(702), 702.

Alahmad, S., El Hassouni, K., Bassi, F. M., Dinglasan, E., Youssef, C., Quarry, G., Aksoy, A., Mazzucotelli, E., Juhász, A., Able, J. A., Christopher, J., Voss-Fels, K. P. and Hickey, L. T. (2019). A major root architecture QTL responding to water limitation in durum wheat, *Frontiers in Plant Science* 10, 436.

Alemu, A., Feyissa, T., Maccaferri, M., Sciara, G., Tuberosa, R., Ammar, K., Badebo, A., Acevedo, M., Letta, T. and Abeyo, B. (2021). Genome-wide association analysis unveils novel QTLs for seminal root system architecture traits in Ethiopian durum wheat, *BMC Genomics* 22(1), 1-16.

Alexander, R. D., Wendelboe-Nelson, C. and Morris, P. C. (2019). The barley transcription factor *HvMYB1* is a positive regulator of drought tolerance, *Plant Physiology and Biochemistry* 142, 246–253.

Ari Akin, P., Demirkesen, I., Bean, S. R., Aramouni, F. and Boyaci, I. H. (2022). Sorghum flour application in bread: Technological challenges and opportunities, *Foods* 11(16), 2466.

Arriagada, O., Gadaleta, A., Marcotuli, I., Maccaferri, M., Campana, M., Reveco, S., Alfaro, C., Matus, I. and Schwember, A. R. (2022). A comprehensive meta-QTL analysis for yield-related traits of durum wheat (*Triticum turgidum* L. var. *durum*) grown under different water regimes, *Frontiers in Plant Science* 13, 984269.

Arai-Sanoh, Y., Takai, T., Yoshinaga, S., Nakano, H., Kojima, M., Sakakibara, H., Kondo, M. and Uga, Y. (2014). Deep rooting conferred by *DEEPER ROOTING 1* enhances rice yield in paddy fields, *Scientific Reports* 4(1), 5563.

Asadollahi, M., Iranbakhsh, A., Ahmadvand, R., Ebadi, M. and Mehregan, I. (2023). *Synergetic Effect of Water Deficit and Arbuscular Mycorrhizal Symbiosis on the Expression of Aquaporins in Wheat (Triticum aestivum L.) Roots: Insights from NGS RNA-Sequencing*. Physiology and Molecular Biology of Plants.

Aslam, M. M., Okal, E. J., Idris, A. L., Qian, Z., Xu, W., Karanja, J. K., Wani, S. H. and Yuan, W. (2022). Rhizosphere microbiomes can regulate plant drought tolerance, *Pedosphere* 32(1), 61–74.

Avramova, V., Meziane, A., Bauer, E., Blankenagel, S., Eggels, S., Gresset, S., Grill, E., Niculaes, C., Ouzunova, M., Poppenberger, B., Presterl, T., Rozhon, W., Welcker, C., Yang, Z., Tardieu, F. and Schön, C. C. (2019). Carbon isotope composition, water use efficiency, and drought sensitivity are controlled by a common genomic segment in maize, *Theoretical and Applied Genetics* 132(1), 53–63.

Ayadi, M., Brini, F. and Masmoudi, K. (2019). Overexpression of a wheat aquaporin gene, *TdPIP2;1*, enhances salt and drought tolerance in transgenic durum wheat cv. Maali, *International Journal of Molecular Sciences* 20(10), 2389.

Ayalew, H., Liu, H. and Yan, G. (2017). Identification and validation of root length QTLs for water stress resistance in hexaploid wheat (*Titicum aestivum* L.), *Euphytica* 213(6).

Bacher, H., Zhu, F., Gao, T., Liu, K., Dhatt, B. K., Awada, T., Zhang, C., Distelfeld, A., Yu, H., Peleg, Z. and Walia, H. (2021). Wild emmer introgression alters root-to-shoot growth dynamics in durum wheat in response to water stress, *Plant Physiology* 187(3), 1149–1162.

Bailey-Serres, J., Parker, J. E., Ainsworth, E. A., Oldroyd, G. E. D. and Schroeder, J. I. (2019). Genetic strategies for improving crop yields, *Nature* 575(7781), 109–118.

Baisakh, N., Yabes, J., Gutierrez, A., Mangu, V., Ma, P., Famoso, A. and Pereira, A. (2020). Genetic mapping identifies consistent quantitative trait loci for yield traits of rice under greenhouse drought conditions, *Genes* 11(1), 62.

Baldoni, E., Frugis, G., Martinelli, F., Benny, J., Paffetti, D. and Buti, M. (2021). A comparative transcriptomic meta-analysis revealed conserved key genes and regulatory networks involved in drought tolerance in cereal crops, *International Journal of Molecular Sciences* 22(23), 13062.

Ballesteros, D. C., Mason, R. E., Addison, C. K., Andrea Acuña, M., Nelly Arguello, M., Subramanian, N., Miller, R. G., Sater, H., Gbur, E. E., Miller, D., Griffey, C. A., Barnett, M. D. and Tucker, D. (2015). Tolerance of wheat to vegetative stage soil waterlogging is conditioned by both constitutive and adaptive QTL, *Euphytica* 201(3), 329–343.

Begum, N., Ahanger, M. A., Su, Y., Lei, Y., Mustafa, N. S. A., Ahmad, P. and Zhang, L. (2019). Improved drought tolerance by AMF inoculation in maize (*Zea mays*) involves physiological and biochemical implications, *Plants* 8(12), 579.

Beissinger, T., Kruppa, J., Cavero, D., Ha, N. T., Erbe, M. and Simianer, H. (2018). A simple test identifies selection on complex traits, *Genetics* 209(1), 321–333.

Ben-Jabeur, M., Vicente, R., López-Cristoffanini, C., Alesami, N., Djébali, N., Gracia-Romero, A., Serret, M. D., López-Carbonell, M., Araus, J. L. and Hamada, W. (2019). A novel aspect of essential oils: Coating seeds with thyme essential oil induces drought resistance in wheat, *Plants* 8(10), 371.

Bennett, D., Reynolds, M., Mullan, D., Izanloo, A., Kuchel, H., Langridge, P. and Schnurbusch, T. (2012). Detection of two major grain yield QTL in bread wheat (*Triticum aestivum* L.) under heat, drought and high yield potential environments, *Theoretical and Applied Genetics* 125(7), 1473–1485.

Beyer, S., Daba, S., Tyagi, P., Bockelman, H., Brown-Guedira, G. and Mohammadi, M. (2019). Loci and candidate genes controlling root traits in wheat seedlings—a wheat root GWAS, *Functional and Integrative Genomics* 19(1), 91–107.

Bhatta, M., Morgounov, A., Belamkar, V. and Baenziger, P. S. (2018). Genome-wide association study reveals novel genomic regions for grain yield and yield-related traits in drought-stressed synthetic hexaploid wheat, *International Journal of Molecular Sciences* 19(10), 3011.

Bhattarai, U. and Subudhi, P. K. (2018). Genetic analysis of yield and agronomic traits under reproductive-stage drought stress in rice using a high-resolution linkage map, *Gene* 669, 69–76.

Blum, A. (2009). Effective use of water (EUW) and not water-use efficiency (WUE) is the target of crop yield improvement under drought stress, *Field Crops Research* 112(2–3), 119–123.

Blum, A. (2017). Osmotic adjustment is a prime drought stress adaptive engine in support of plant production, *Plant, Cell and Environment* 40(1), 4–10.

Blum, A. and Tuberosa, R. (2018). Dehydration survival of crop plants and its measurement, *Journal of Experimental Botany* 69(5), 975–981.

Bochicchio, R., Labella, R., Rossi, R., Perniola, M. and Amato, M. (2022). Effects of soil water shortage on seedling shoot and root growth of Saragolle lucana tetraploid wheat (*Triticum durum* Desf.) landrace, *Plants* 11(24), 3492.

Borrell, A. K., Mullet, J. E., George-Jaeggli, B., van Oosterom, E. J., Hammer, G. L., Klein, P. E. and Jordan, D. R. (2014). Drought adaptation of stay-green sorghum is associated with canopy development, leaf anatomy, root growth, and water uptake, *Journal of Experimental Botany* 65(21), 6251–6263.

Cai, Q., Wang, Y., Ni, S., Mu, J., Liu, M., Wang, Y. and Zhao, Y. (2022). Genomewide identification and analysis of the OSCA gene family in barley (*Hordeum vulgare* L.), *Journal of Genetics* 101, 34.

Carlborg, O. and Haley, C. S. (2004). Epistasis: Too often neglected in complex trait studies? *Nature Reviews: Genetics* 5(8), 618-625.

Carlson, R., Tugizimana, F., Steenkamp, P. A., Dubery, I. A., Hassen, A. I. and Labuschagne, N. (2020). Rhizobacteria-induced systemic tolerance against drought stress in *Sorghum bicolor* (L.) Moench, *Microbiological Research* 232, 126388.

Carraro, E. and Di Iorio, A. (2022). Eligible strategies of drought response to improve drought resistance in woody crops: A mini-review, *Plant Biotechnology Reports* 16(3), 265–282.

Chaichi, M., Sanjarian, F., Razavi, K. and Gonzalez-Hernandez, J. L. (2019). Analysis of transcriptional responses in root tissue of bread wheat landrace (*Triticum aestivum* L.) reveals drought avoidance mechanisms under water scarcity, *PLoS ONE* 14(3), e0212671.

Champoux, M. C., Wang, G., Sarkarung, S., Mackill, D. J., O'Toole, J. C., Huang, N. and Mccouch, S. R. (1995). Locating genes associated with root morphology and drought avoidance in rice via linkage to molecular markers, *Theoretical and Applied Genetics*, 90(7–8), 969–981.

Chen, F., Zhang, H., Li, H., Lian, L., Wei, Y., Lin, Y., Wang, L., He, W., Cai, Q., Xie, H., Zhang, H. and Zhang, J. (2023). *IPA1* improves drought tolerance by activating *SNAC1* in rice, *BMC Plant Biology* 23(1), 55.

Chen, S., Xu, K., Kong, D., Wu, L., Chen, Q., Ma, X., Ma, S., Li, T., Xie, Q., Liu, H. and Luo, L. (2022). Ubiquitin ligase *OsRINGzf1* regulates drought resistance by controlling the turnover of *OsPIP2;1*, *Plant Biotechnology Journal* 20(9), 1743–1755.

Chen, Y., Guo, Y., Xie, X., Wang, Z., Miao, L., Yang, Z., Jiao, Y., Xie, C., Liu, J., Hu, Z., Xin, M., Yao, Y., Ni, Z., Sun, Q., Peng, H. and Guo, W. (2023). Pangenome-based trajectories of intracellular gene transfers in Poaceae unveil high cumulation in Triticeae, *Plant Physiology* 193(1), kiad319.

Choudhary, M., Kumar, P., Kumar, P., Sheoran, S., Zunjare, R. U. and Jat, B. S. (2023). Chapter 4 - Molecular breeding for drought and heat stress in maize: Revisiting the progress and achievements. In: Wani, S. H., Wang, D. and Pratap Singh, G. (Eds). *QTL Mapping in Crop Improvement*. Boston: Academic Press, pp. 57–74.

Christopher, J., Christopher, M., Jennings, R., Jones, S., Fletcher, S., Borrell, A., Manschadi, A. M., Jordan, D., Mace, E. and Hammer, G. (2013). QTL for root angle and number in a population developed from bread wheats (*Triticum aestivum*) with contrasting adaptation to water-limited environments, *Theoretical and Applied Genetics* 126(6), 1563–1574.

Christopher, M., Chenu, K., Jennings, R., Fletcher, S., Butler, D., Borrell, A. and Christopher, J. (2018). QTL for stay-green traits in wheat in well-watered and water-limited environments, *Field Crops Research* 217, 32–44.

Chung, P. J., Jung, H., Jeong, D. H., Ha, S. H., Choi, Y. D. and Kim, J. K. (2016). Transcriptome profiling of drought responsive noncoding RNAs and their target genes in rice, *BMC Genomics* 17(1), 563.

Cohen, I., Zandalinas, S. I., Huck, C., Fritschi, F. B. and Mittler, R. (2021). Meta-analysis of drought and heat stress combination impact on crop yield and yield components, *Physiologia Plantarum* 171(1), 66–76.

Condorelli, G. E., Newcomb, M., Groli, E. L., Maccaferri, M., Forestan, C., Babaeian, E., Tuller, M., White, J. W., Ward, R., Mockler, T., Shakoor, N. and Tuberosa, R. (2022). Genome wide association study uncovers the QTLome for osmotic adjustment and related drought adaptive traits in durum wheat, *Genes* 13(2), 293.

Collins, N. C., Tardieu, F. and Tuberosa, R. (2008). Quantitative trait loci and crop performance under abiotic stress: Where do we stand? *Plant Physiology* 147(2), 469-486.

Cooper, M. and Messina, C. D. (2023). Breeding crops for drought-affected environments and improved climate resilience, *The Plant Cell* 35(1), 162-186.

Courtois, B., Mclaren, G., Sinha, P. K., Prasad, K., Yadav, R. and Shen, L. (2000). Mapping QTLs associated with drought avoidance in upland rice, *Molecular Breeding* 6(1), 55-66.

Dettori, M., Cesaraccio, C., Duce, P. and Mereu, V. (2022). Performance prediction of durum wheat genotypes in response to drought and heat in climate change conditions, *Genes* 13(3), 488.

Dharmappa, P. M., Doddaraju, P., Malagondanahalli, M. V., Rangappa, R. B., Mallikarjuna, N. M., Rajendrareddy, S. H., Ramanjinappa, R., Mavinahalli, R. P., Prasad, T. G., Udayakumar, M. and Sheshshayee, S. M. (2019). Introgression of root and water use efficiency traits enhances water productivity: An evidence for physiological breeding in rice (*Oryza sativa* L.), *Rice* 12(1), 14.

Dong, H., Birhan, T., Abajebel, N., Wakjira, M., Mitiku, T., Lemke, C., Vadez, V., Paterson, A. H. and Bantte, K. (2022). Natural variation further increases resilience of sorghum bred for chronically drought-prone environments, *Journal of Experimental Botany* 73(16), 5730-5744.

Dowd, T. G., Braun, D. M. and Sharp, R. E. (2019). Maize lateral root developmental plasticity induced by mild water stress. I: Genotypic variation across a high-resolution series of water potentials, *Plant, Cell and Environment* 42(7), 2259-2273.

Duvick, D. N. (2005). The contribution of breeding to yield advances in maize (*Zea mays* L.). In: *Advances in Agronomy*. Academic Press, pp. 83-145.

Eapen, D., Martínez-Guadarrama, J., Hernández-Bruno, O., Flores, L., Nieto-Sotelo, J. and Cassab, G. I. (2017). Synergy between root hydrotropic response and root biomass in maize (*Zea mays* L.) enhances drought avoidance, *Plant Science* 265, 87-99.

El-Esawi, M. A., Al-Ghamdi, A. A., Ali, H. M. and Ahmad, M. (2019). Overexpression of AtWRKY30 transcription factor enhances heat and drought stress tolerance in wheat (*Triticum aestivum* L.), *Genes* 10(2), 163.

Elakhdar, A., Elakhdar, I., Desouky, S. E. and Khaidr, R. (2023). Chapter 14 - Recent progress in molecular breeding approaches to improve drought tolerance in barley. In: Wani, S. H., Wang, D. and Pratap Singh, G. (Eds). *QTL Mapping in Crop Improvement*. Boston: Academic Press, pp. 275-309.

Elakhdar, A., Solanki, S., Kubo, T., Abed, A., Elakhdar, I., Khedr, R., Hamwieh, A., Capo-chichi, L. J. A., Abdelsattar, M., Franckowiak, J. D. and Qualset, C. O. (2022). Barley with improved drought tolerance: Challenges and perspectives, *Environmental and Experimental Botany* 201, 104965.

Falahi, A., Zarei, L. and Cheghamirza, K. (2022). Most drought-induced DNA methylation changes switched to pre-stress state after re-irrigation in barley (*Hordeum vulgare* L.) cultivars, *Cereal Research Communications* 50(3), 429-438.

Fang, Y. and Xiong, L. (2015). General mechanisms of drought response and their application in drought resistance improvement in plants, *Cellular and Molecular Life Sciences* 72(4), 673-689.

FAO (2021). *The Impact of Disasters and Crises on Agriculture and Food Security* (Vol. 2021). Rome. https://doi.org/10.4060/cb3673en.

FAO (2022). Available at: https://www.fao.org/faostat/en/#data.

FAO statistics, 2021. https://doi.org/10.4060/cb3673en.

Faye, J. M., Akata, E. A., Sine, B., Diatta, C., Cisse, N., Fonceka, D. and Morris, G. P. (2022). Quantitative and population genomics suggest a broad role of stay-green loci in the drought adaptation of sorghum, *The Plant Genome* 15(1), e20176.

Feng, X., Liu, W., Qiu, C. W., Zeng, F., Wang, Y., Zhang, G., Chen, Z. H. and Wu, F. (2020). HvAKT2 and HvHAK1 confer drought tolerance in barley through enhanced leaf mesophyll H^+ homoeostasis, *Plant Biotechnology Journal* 18(8), 1683–1696.

Fileccia, V., Bertolini, E., Ruisi, P., Giambalvo, D., Frenda, A. S., Cannarozzi, G., Tadele, Z., Crosatti, C. and Martinelli, F. (2017). Identification and characterization of durum wheat microRNAs in leaf and root tissues, *Functional and Integrative Genomics* 17(5), 583–598.

Forster, B. P., Ellis, R. P., Thomas, W. T. B., Newton, A. C., Tuberosa, R., This, D., El-Enein, R. A., Bahri, M. H. and Ben Salem, M. (2000). The development and application of molecular markers for abiotic stress tolerance in barley, *Journal of Experimental Botany* 51(342), 19–27.

Frimpong, F., Anokye, M., Windt, C. W., Naz, A. A., Frei, M., Dusschoten, D. V. and Fiorani, F. (2021). Proline-mediated drought tolerance in the barley (*Hordeum vulgare* L.) isogenic line is associated with lateral root growth at the early seedling stage, *Plants* 10(10), 2177.

Fusi, R., Rosignoli, S., Lou, H., Sangiorgi, G., Bovina, R., Pattem, J. K., Borkar, A. N., Lombardi, M., Forestan, C., Milner, S. G., Davis, J. L., Lale, A., Kirschner, G. K., Swarup, R., Tassinari, A., Pandey, B. K., York, L. M., Atkinson, B. S., Sturrock, C. J., Mooney, S. J., Hochholdinger, F., Tucker, M. R., Himmelbach, A., Stein, N., Mascher, M., Nagel, K. A., De Gara, L., Simmonds, J., Uauy, C., Tuberosa, R., Lynch, J. P., Yakubov, G. E., Bennett, M. J., Bhosale, R. and Salvi, S. (2022). Root angle is controlled by *EGT1* in cereal crops employing an antigravitropic mechanism, *Proceedings of the National Academy of Sciences of the United States of America* 119(31), e2201350119.

Goff, S. A., Ricke, D., Lan, T. H., Presting, G., Wang, R., Dunn, M., Glazebrook, J., Sessions, A., Oeller, P., Varma, H., Hadley, D., Hutchison, D., Martin, C., Katagiri, F., Lange, B. M., Moughamer, T., Xia, Y., Budworth, P., Zhong, J., Miguel, T., Paszkowski, U., Zhang, S., Colbert, M., Sun, W. L., Chen, L., Coope, B., Park, S., Wood, T. C., Mao, L., Quail, P., Wing, R., Dean, R., Yu, Y., Zharkikh, A., Shen, R., Sahasrabudhe, S., Thomas, A., Cannings, R., Gutin, A., Pruss, D., Reid, J., Tavtigian, S., Mitchell, J., Eldredge, G., Scholl, T., Miller, R. M., Bhatnagar, S., Adey, N., Rubano, T., Tusneem, N., Robinson, R., Feldhaus, J., Macalma, T., Oliphant, A. and Briggs, S. (2002). A draft sequence of the rice genome (*Oryza sativa* L. ssp. *japonica*), *Science* 296(5565), 92–100.

Govindasamy, V., George, P., Kumar, M., Aher, L., Raina, S. K., Rane, J., Annapurna, K. and Minhas, P. S. (2020). Multi-trait PGP rhizobacterial endophytes alleviate drought stress in a senescent genotype of sorghum [*Sorghum bicolor* (L.) Moench], *3 Biotech* 10(1), 13.

Graziani, M., Maccaferri, M., Royo, C., Salvatorelli, F. and Tuberosa, R. (2014). QTL dissection of yield components and morpho-physiological traits in a durum wheat elite population tested in contrasting thermo-pluviometric conditions, *Crop and Pasture Science* 65(1), 80–95.

Gregory, P. J. (2022). RUSSELL REVIEW Are plant roots only "in" soil or are they "of" it? Roots, soil formation and function, *European Journal of Soil Science* 73(1), e13219.

Guellim, A., Hirel, B., Chabrerie, O., Catterou, M., Tetu, T., Dubois, F., Ahmed, H. B. and Kichey, T. (2020). Screening for durum wheat (*Triticum durum* Desf.) cultivar resistance to drought stress using an integrated physiological approach, *Journal of Crop Science and Biotechnology* 23(4), 355–365.

Gullì, M., De Pascali, M., Perrotta, C. and Rampino, P. (2022). A stress-related transcription factor belonging to the YL-1 family is differently regulated in durum wheat cultivars differing in drought sensitivity, *Plant Physiology and Biochemistry* 170, 307–315.

Guo, C., Chen, L., Cui, Y., Tang, M., Guo, Y., Yi, Y., Li, Y., Liu, L. and Chen, L. (2022). RNA binding protein OsTZF7 traffics between the nucleus and processing bodies/stress granules and positively regulates drought stress in rice, *Frontiers in Plant Science* 13, 802337.

Guo, J., Chen, L., Li, Y., Shi, Y., Song, Y., Zhang, D., Li, Y., Wang, T., Yang, D. and Li, C. (2018a). Meta-QTL analysis and identification of candidate genes related to root traits in maize, *Euphytica* 214(12), 223.

Guo, Z., Yang, W., Chang, Y., Ma, X., Tu, H., Xiong, F., Jiang, N., Feng, H., Huang, C., Yang, P., Zhao, H., Chen, G., Liu, H., Luo, L., Hu, H., Liu, Q. and Xiong, L. (2018b). Genome-wide association studies of image traits reveal genetic architecture of drought resistance in rice, *Molecular Plant* 11(6), 789–805.

Gupta, P. K., Balyan, H. S., Sharma, S. and Kumar, R. (2020). Genetics of yield, abiotic stress tolerance and biofortification in wheat (*Triticum aestivum* L.), *Theoretical and Applied Genetics* 133(5), 1569–1602.

Haas, M., Schreiber, M. and Mascher, M. (2019). Domestication and crop evolution of wheat and barley: Genes, genomics, and future directions, *Journal of Integrative Plant Biology* 61(3), 204–225.

Hammer, G., Cooper, M., Tardieu, F., Welch, S., Walsh, B., van Eeuwijk, F., Chapman, S. and Podlich, D. (2006). Models for navigating biological complexity in breeding improved crop plants, *Trends in Plant Science* 11(12).

Harb, A., Simpson, C., Guo, W., Govindan, G., Kakani, V. G. and Sunkar, R. (2020). The effect of drought on transcriptome and hormonal profiles in barley genotypes with contrasting drought tolerance, *Frontiers in Plant Science* 11, 618491.

Hasan, N., Choudhary, S., Naaz, N., Sharma, N. and Laskar, R. A. (2021). Recent advancements in molecular marker-assisted selection and applications in plant breeding programmes, *Journal, Genetic Engineering and Biotechnology* 19(1), 128.

Hassouni, K., Alahmad, S., Belkadi, B., Filali-Maltouf, A., Hickey, L. T. and Bassi, F. M. (2018). Root system architecture and its association with yield under different water regimes in durum wheat, *Crop Science* 58(6), 2331–2346.

Haudry, A., Cenci, A., Ravel, C., Bataillon, T., Brunel, D., Poncet, C., Hochu, I., Poirier, S., Santoni, S., Glémin, S. and David, J. (2007). Grinding up wheat: A massive loss of nucleotide diversity since domestication, *Molecular Biology and Evolution* 24(7), 1506–1517.

He, Z., Zhong, J., Sun, X., Wang, B., Terzaghi, W. and Dai, M. (2018). The maize ABA receptors *ZmPYL8*, 9, and 12 facilitate plant drought resistance, *Frontiers in Plant Science* 9, 422.

Henry, A., Swamy, B. P., Dixit, S., Torres, R. D., Batoto, T. C., Manalili, M., Anantha, M. S., Mandal, N. P. and Kumar, A. (2015). Physiological mechanisms contributing to the QTL-combination effects on improved performance of IR64 rice NILs under drought, *Journal of Experimental Botany* 66(7), 1787–1799.

Howell, T., Moriconi, J. I., Zhao, X., Hegarty, J., Fahima, T., Santa-Maria, G. E. and Dubcovsky, J. (2019). A wheat/rye polymorphism affects seminal root length and yield across different irrigation regimes, *Journal of Experimental Botany* 70(15), 4027–4037.

Hu, L., Xie, Y., Fan, S., Wang, Z., Wang, F., Zhang, B., Li, H., Song, J. and Kong, L. (2018). Comparative analysis of root transcriptome profiles between drought-tolerant and susceptible wheat genotypes in response to water stress, *Plant Science* 272, 276–293.

International Wheat Genome Sequencing Consortium (IWGSC) (2018). Shifting the limits in wheat research and breeding using a fully annotated reference genome, *Science* 361(6403), 1–13.

Itam, M., Abdelrahman, M., Yamasaki, Y., Mega, R., Gorafi, Y., Akashi, K. and Tsujimoto, H. (2020). *Aegilops tauschii* introgressions improve physio-biochemical traits and metabolite plasticity in bread wheat under drought stress, *Agronomy* 10(10), 1588.

Jabbari, M., Fakheri, B. A., Aghnoum, R., Darvishzadeh, R., Mahdi Nezhad, N., Ataei, R., Koochakpour, Z. and Razi, M. (2021). Association analysis of physiological traits in spring barley (*Hordeum vulgare* L.) under water-deficit conditions, *Food Science and Nutrition* 9(3), 1761–1779.

Jabbari, M., Fakheri, B. A., Aghnoum, R., Darvishzadeh, R., Mahdi Nezhad, N., Ataei, R., Koochakpour, Z. and Razi, M. (2022). Identification of DNA markers associated with phenological traits in spring barley (*Hordeum vulgare* L.) under drought stress conditions, *Cereal Research Communications* 50(2), 171–178.

Javadi, S. M., Shobbar, Z. S., Ebrahimi, A. and Shahbazi, M. (2021). New insights on key genes involved in drought stress response of barley: Gene networks reconstruction, hub, and promoter analysis, *Journal, Genetic Engineering and Biotechnology* 19(1), 2.

Jerbi, M., Labidi, S., Laruelle, F., Tisserant, B., Ben Jeddi, F. and Lounès-Hadj Sahraoui, A. (2022). Mycorrhizal biofertilization improves grain yield and quality of hulless barley (*Hordeum vulgare* ssp. *nudum* L.) under water stress conditions, *Journal of Cereal Science* 104, 103436.

Jiang, Z., Tu, H., Bai, B., Yang, C., Zhao, B., Guo, Z., Liu, Q., Zhao, H., Yang, W., Xiong, L. and Zhang, J. (2021). Combining UAV-RGB high-throughput field phenotyping and genome-wide association study to reveal genetic variation of rice germplasms in dynamic response to drought stress, *New Phytologist* 232(1), 440–455.

Jin, X., Zheng, Y., Wang, J., Chen, W., Yang, Z., Chen, Y., Yang, Y., Lu, G. and Sun, B. (2023). SbNAC9 improves drought tolerance by enhancing scavenging ability of reactive oxygen species and activating stress-responsive genes of sorghum, *International Journal of Molecular Sciences* 24(3), 2401.

Kandemir, N. and Saygili, I. (2023). Evaluation of a barley (*Hordeum vulgare*) chromosome 2 drought tolerance quantitative trait locus in genomic backgrounds of two cultivars, *Plant Breeding* 142(2), 211–222.

Kang, Z., Babar, M. A., Khan, N., Guo, J., Khan, J., Islam, S., Shrestha, S. and Shahi, D. (2019). Comparative metabolomic profiling in the roots and leaves in contrasting genotypes reveals complex mechanisms involved in post-anthesis drought tolerance in wheat, *PLoS ONE* 14(3), e0213502.

Kano-Nakata, M., Inukai, Y., Wade, L. J., Siopongco, J. D. L. C. and Yamauchi, A. (2011). Root development and water uptake, and shoot dry matter production under water deficit conditions in twoCSSLs of rice: Functional roles of root plasticity, *Plant Production Science* 14, 329–339.

Khalid, M., Gul, A., Amir, R., Ali, M., Afzal, F., Quraishi, U. M., Ahmed, Z. and Rasheed, A. (2018). QTL mapping for seedling morphology under drought stress in wheat cross synthetic (W7984)/Opata, *Plant Genetic Resources: Characterization and Utilization* 16(4), 359–366.

Khan, M. A., Waseem Akram, M., Iqbal, M., Ghulam Muhu-Din Ahmed, H., Rehman, A., Arslan Iqbal, H. S. M. and Alam, B. (2023). Multivariate and association analyses of quantitative attributes reveal drought tolerance potential of wheat (*Triticum aestivum* L.) genotypes, *Communications in Soil Science and Plant Analysis* 54(2), 178–195.

Khan, S. U., Zheng, Y., Chachar, Z., Zhang, X., Zhou, G., Zong, N., Leng, P. and Zhao, J. (2022). Dissection of maize drought tolerance at the flowering stage using genome-wide association studies, *Genes* 13(4), 564.

Kibalnik, O. P., Sazonova, I. A., Bochkareva, Y. V., Bychkova, V. V. and Semin, D. S. (2023). Influence of abiotic stresses on morphophysiological characteristics and biological value of grain Sorghum *bicolor* (L.) Moench, *International Journal of Plant Biology* 14(1), 150–161.

Kirschner, G. K., Rosignoli, S., Guo, L., Vardanega, I., Imani, J., Altmüller, J., Milner, S. G., Balzano, R., Nagel, K. A., Pflugfelder, D., Forestan, C., Bovina, R., Koller, R., Stöcker, T. G., Mascher, M., Simmonds, J., Uauy, C., Schoof, H., Tuberosa, R., Salvi, S. and Hochholdinger, F. (2021). ENHANCED GRAVITROPISM 2 encodes a STERILE ALPHA MOTIF-containing protein that controls root growth angle in barley and wheat, *Proceedings of the National Academy of Sciences* 118(35), e2101526118.

Kitomi, Y., Kanno, N., Kawai, S., Mizubayashi, T., Fukuoka, S. and Uga, Y. (2015). QTLs underlying natural variation of root growth angle among rice cultivars with the same functional allele of *DEEPER ROOTING 1*, *Rice* 8(1), 16.

Kulkarni, M., Soolanayakanahally, R., Ogawa, S., Uga, Y., Selvaraj, M. G. and Kagale, S. (2017). Drought response in wheat: Key genes and regulatory mechanisms controlling root system architecture and transpiration efficiency, *Frontiers in Chemistry* 5, 106.

Kumar, P., Choudhary, M., Halder, T., Prakash, N. R., Singh, V., Vineeth, T. V., Sheoran, S., Ravikiran, K. T., Longmei, N., Rakshit, S. and Siddique, K. H. M. (2022). Salinity stress tolerance and omics approaches: Revisiting the progress and achievements in major cereal crops, *Heredity* 128(6), 497–518.

Kumar, R. (2018). Development of drought resistance in rice, *International Journal of Current Microbiology and Applied Sciences* 7(5), 1439–1456.

Kuzmanović, L., Ruggeri, R., Able, J. A., Bassi, F. M., Maccaferri, M., Tuberosa, R., De Vita, P., Rossini, F. and Ceoloni, C. (2018). Yield of chromosomally engineered durum wheat-*Thinopyrum ponticum* recombinant lines in a range of contrasting rain-fed environments, *Field Crops Research* 228, 147–157.

Langridge, P. and Reynolds, M. (2021). Breeding for drought and heat tolerance in wheat, *Theoretical and Applied Genetics* 134(6), 1753–1769.

Langridge, P., Alaux, M., Almeida, N. F., Ammar, K., Baum, M., Bekkaoui, F., Bentley, A. R., Beres, B. L., Berger, B., Braun, H., Brown-Guedira, G., Burt, C. J., Caccamo, M. J., Cattivelli, L., Charmet, G., Civáň, P., Cloutier, S., Cohan, J., Devaux, P. J., Doohan, F. M., Dreccer, M. F., Ferrahi, M., Germán, S. E., Goodwin, S. B., Griffiths, S., Guzmán, C., Handa, H., Hawkesford, M. J., He, Z., Huttner, E., Ikeda, T. M., Kilian, B., King, I. P., King, J., Kirkegaard, J. A., Lage, J., Le Gouis, J., Mondal, S., Mullins, E., Ordon, F., Ortiz-Monasterio, J. I., Özkan, H., Öztürk, İ., Pereyra, S. A., Pozniak, C. J., Quesneville, H., Quincke, M. C., Rebetzke, G. J., Christoph Reif, J., Saavedra-Bravo, T., Schurr,

U., Sharma, S., Singh, S. K., Singh, R. P., Snape, J. W., Tadesse, W., Tsujimoto, H., Tuberosa, R., Willis, T. G. and Zhang, X. (2022). Meeting the challenges facing wheat production: The strategic research agenda of the global wheat initiative, *Agronomy* 12(11), 2767.

Lasky, J. R., Josephs, E. B. and Morris, G. P. (2023). Genotype-environment associations to reveal the molecular basis of environmental adaptation, *The Plant Cell* 35(1), 125–138.

Lehnert, H., Serfling, A., Friedt, W. and Ordon, F. (2018). Genome-wide association studies reveal genomic regions associated with the response of wheat (*Triticum aestivum* L.) to mycorrhizae under drought stress conditions, *Frontiers in Plant Science* 9, 1728.

Leng, P. and Zhao, J. (2020). Transcription factors as molecular switches to regulate drought adaptation in maize, *Theoretical and Applied Genetics* 133(5), 1455–1465.

Lesk, C., Rowhani, P. and Ramankutty, N. (2016). Influence of extreme weather disasters on global crop production, *Nature* 529(7584), 84–87.

Li, B., Fan, R., Sun, G., Sun, T., Fan, Y., Bai, S., Guo, S., Huang, S., Liu, J., Zhang, H., Wang, P., Zhu, X. and Song, C.-P. (2021a). Flavonoids improve drought tolerance of maize seedlings by regulating the homeostasis of reactive oxygen species, *Plant and Soil* 461(1–2), 389–405.

Li, C., Zhang, W., Yuan, M., Jiang, L., Sun, B., Zhang, D., Shao, Y., Liu, A., Liu, X. and Ma, J. (2019a). Transcriptome analysis of osmotic-responsive genes in ABA-dependent and -independent pathways in wheat (*Triticum aestivum* L.) roots, *PeerJ* 7, e6519.

Li, C., Zhang, X., Guo, W., Fu, K., Li, C. and Li, C. (2023). Impact of ethylene and abscisic acid on programmed cell death in endosperm and grain filling of wheat (*Triticum aestivum* L.), *Journal of Plant Growth Regulation* 42(7), 4275–4289.

Li, J., Han, Y., Liu, L., Chen, Y., Du, Y., Zhang, J., Sun, H. and Zhao, Q. (2015). qRT9, a quantitative trait locus controlling root thickness and root length in upland rice, *Journal of Experimental Botany* 66(9), 2723–2732.

Li, P., Zhang, Y., Yin, S., Zhu, P., Pan, T., Xu, Y., Wang, J., Hao, D., Fang, H., Xu, C. and Yang, Z. (2018). QTL-by-environment interaction in the response of maize root and shoot traits to different water regimes, *Frontiers in Plant Science* 9, 229.

Li, X., Chang, Y., Ma, S., Shen, J., Hu, H. and Xiong, L. (2019b). Genome-wide identification of SNAC1-targeted genes involved in drought response in rice, *Frontiers in Plant Science* 10, 982.

Li, X., Guo, Z., Lv, Y., Cen, X., Ding, X., Wu, H., Li, X., Huang, J. and Xiong, L. (2017). Genetic control of the root system in rice under normal and drought stress conditions by genome-wide association study, *PLoS Genetics* 13(7), e1006889.

Li, Y., Liu, X., Chen, R., Tian, J., Fan, Y. and Zhou, X. (2019c). Genome-scale mining of root-preferential genes from maize and characterization of their promoter activity, *BMC Plant Biology* 19(1), 584.

Li, Z., Liu, C., Zhang, Y., Wang, B., Ran, Q. and Zhang, J. (2019). The bHLH family member *ZmPTF1* regulates drought tolerance in maize by promoting root development and abscisic acid synthesis, *Journal of Experimental Botany* 70(19), 5471–5486.

Li, Z., Su, X., Chen, Y., Fan, X., He, L., Guo, J., Wang, Y. and Yang, Q. (2021b). Melatonin improves drought resistance in maize seedlings by enhancing the antioxidant system and regulating abscisic acid metabolism to maintain stomatal opening under PEG-induced drought, *Journal of Plant Biology* 64(4), 299–312.

Liang, J., Chen, X., Deng, G., Pan, Z., Zhang, H., Li, Q., Yang, K., Long, H. and Yu, M. (2017). Dehydration induced transcriptomic responses in two Tibetan hulless barley

(*Hordeum vulgare* var. *nudum*) accessions distinguished by drought tolerance, *BMC Genomics* 18(1), 775.

Liang, J., Yi, L., Li, L., Zhang, H., Zhang, Y., Deng, G., Long, H. and Yu, M. (2022). Identification of *PP2C* genes in Tibetan hulless barley (*Hordeum vulgare* var. *nudum*) under dehydration stress and initiatory expression and functional analysis of HvPP2C59, *Plant Molecular Biology Reporter* 40(4), 611–627.

Liang, J., Zhang, H., Yi, L., Tang, Y., Long, H., Yu, M. and Deng, G. (2021). Identification of *HvLRX*, a new dehydration and light responsive gene in Tibetan hulless barley (*Hordeum vulgare* var. *nudum*), *Genes and Genomics* 43(12), 1445–1461.

Licaj, I., Di Meo, M. C., Fiorillo, A., Samperna, S., Marra, M. and Rocco, M. (2023). Comparative analysis of the response to polyethylene glycol-simulated drought stress in roots from seedlings of "modern" and "ancient" wheat varieties, *Plants* 12(3), 428.

Liu, J., Sun, X., Liao, W., Zhang, J., Liang, J. and Xu, W. (2019). Involvement of *OsGF14b* adaptation in the drought resistance of rice plants, *Rice* 12(1), 82.

Liu, L., Bai, C., Chen, Y., Palta, J. A., Delhaize, E. and Siddique, K. H. M. (2022a). Durum wheat with the introgressed *TaMATE1B* gene shows resistance to terminal drought by ensuring deep root growth in acidic and Al^{3+}-toxic subsoils, *Plant and Soil* 478(1–2), 311–324.

Liu, S., Li, C., Wang, H., Wang, S., Yang, S., Liu, X., Yan, J., Li, B., Beatty, M., Zastrow-Hayes, G., Song, S. and Qin, F. (2020). Mapping regulatory variants controlling gene expression in drought response and tolerance in maize, *Genome Biology* 21(1), 163.

Liu, S., Liu, X., Zhang, X., Chang, S., Ma, C. and Qin, F. (2022b). Co-expression of *ZmVPP1* with *ZmNAC111* confers robust drought resistance in maize, *Genes* 14(1), 8.

Lopez, J. R., Erickson, J. E., Munoz, P., Saballos, A., Felderhoff, T. J. and Vermerris, W. (2017). QTLs associated with crown root angle, stomatal conductance, and maturity in sorghum, *The Plant Genome* 10(2), 1–12.

Lou, D., Lu, S., Chen, Z., Lin, Y., Yu, D. and Yang, X. (2023). Molecular characterization reveals that *OsSAPK3* improves drought tolerance and grain yield in rice, *BMC Plant Biology* 23(1), 53.

Luo, C., Akhtar, M., Min, W., Alam, Y., Ma, T., Shi, Y., She, Y. and Lu, X. (2023). The suppressed expression of a stress responsive gene "*OsDSR2*" enhances rice tolerance in drought and salt stress, *Journal of Plant Physiology* 282, 153927.

Lv, L., Zhang, W., Sun, L., Zhao, A., Zhang, Y., Wang, L., Liu, Y., Li, Z., Li, H. and Chen, X. (2020). Gene co-expression network analysis to identify critical modules and candidate genes of drought-resistance in wheat, *PLoS ONE* 15(8), e0236186.

Ma, H., Li, P., Liu, X., Li, C., Zhang, S., Wang, X. and Tao, X. (2022). Poly-γ-glutamic acid enhanced the drought resistance of maize by improving photosynthesis and affecting the rhizosphere microbial community, *BMC Plant Biology* 22(1), 11.

Maccaferri, M., Bruschi, M. and Tuberosa, R. (2022). Sequence-based marker assisted selection in wheat. In: *Wheat Improvement: Food Security in a Changing Climate'*. Cham: Springer International Publishing, pp. 513–538.

Maccaferri, M., Cane, M. A., Sanguineti, M. C., Salvi, S., Colalongo, M. C., Massi, A., Clarke, F., Knox, R., Pozniak, C. J., Clarke, J. M., Fahima, T., Dubcovsky, J., Xu, S., Ammar, K., Karsai, I., Vida, G. and Tuberosa, R. (2014). A consensus framework map of durum wheat (*Triticum durum* Desf.) suitable for linkage disequilibrium analysis and genome-wide association mapping, *BMC Genomics*, 15(1), 873.

Maccaferri, M., El-Feki, W., Nazemi, G., Salvi, S., Canè, M. A., Colalongo, M. C., Stefanelli, S. and Tuberosa, R. (2016). Prioritizing quantitative trait loci for root system architecture in tetraploid wheat, *Journal of Experimental Botany* 67(4), 1161–1178.

Maccaferri, M., Harris, N. S., Twardziok, S. O., Pasam, R. K., Gundlach, H., Spannagl, M., Ormanbekova, D., Lux, T., Prade, V. M., Milner, S. G., Himmelbach, A., Mascher, M., Bagnaresi, P., Faccioli, P., Cozzi, P., Lauria, M., Lazzari, B., Stella, A., Manconi, A., Gnocchi, M., Moscatelli, M., Avni, R., Deek, J., Biyiklioglu, S., Frascaroli, E., Corneti, S., Salvi, S., Sonnante, G., Desiderio, F., Marè, C., Crosatti, C., Mica, E., Özkan, H., Kilian, B., De Vita, P., Marone, D., Joukhadar, R., Mazzucotelli, E., Nigro, D., Gadaleta, A., Chao, S., Faris, J. D., Melo, A. T. O., Pumphrey, M., Pecchioni, N., Milanesi, L., Wiebe, K., Ens, J., MacLachlan, R. P., Clarke, J. M., Sharpe, A. G., Koh, C. S., Liang, K. Y. H., Taylor, G. J., Knox, R., Budak, H., Mastrangelo, A. M., Xu, S. S., Stein, N., Hale, I., Distelfeld, A., Hayden, M. J., Tuberosa, R., Walkowiak, S., Mayer, K. F. X., Ceriotti, A., Pozniak, C. J. and Cattivelli, L. (2019). Durum wheat genome highlights past domestication signatures and future improvement targets, *Nature Genetics* 51(5), 885–895.

Maccaferri, M., Ricci, A., Salvi, S., Milner, S. G., Noli, E., Martelli, P. L., Casadio, R., Akhunov, E., Scalabrin, S., Vendramin, V., Ammar, K., Blanco, A., Desiderio, F., Distelfeld, A., Dubcovsky, J., Fahima, T., Faris, J., Korol, A., Massi, A., Mastrangelo, A. M., Morgante, M., Pozniak, C., N'Diaye, A., Xu, S. and Tuberosa, R. (2015). A high-density, SNP-based consensus map of tetraploid wheat as a bridge to integrate durum and bread wheat genomics and breeding, *Plant Biotechnology Journal* 13(5), 648–663.

Maccaferri, M., Sanguineti, M. C., Corneti, S., Ortega, J. L., Salem, M. B., Bort, J., DeAmbrogio, E., del Moral, L. F., Demontis, A., El-Ahmed, A., Maalouf, F., Machlab, H., Martos, V., Moragues, M., Motawaj, J., Nachit, M., Nserallah, N., Ouabbou, H., Royo, C., Slama, A. and Tuberosa, R. (2008). Quantitative trait loci for grain yield and adaptation of durum wheat (*Triticum durum* Desf.) across a wide range of water availability, *Genetics* 178(1), 489–511.

Maccaferri, M., Sanguineti, M. C., Demontis, A., El-Ahmed, A., Garcia del Moral, L., Maalouf, F., Nachit, M., Nserallah, N., Ouabbou, H., Rhouma, S., Royo, C., Villegas, D. and Tuberosa, R. (2011). Association mapping in durum wheat grown across a broad range of water regimes, *Journal of Experimental Botany* 62(2), 409–438.

Mace, E. S., Singh, V., Van Oosterom, E. J., Hammer, G. L., Hunt, C. H. and Jordan, D. R. (2012). QTL for nodal root angle in sorghum (*Sorghum bicolor* L. Moench) co-locate with QTL for traits associated with drought adaptation, *Theoretical and Applied Genetics* 124(1), 97–109.

Makhtoum, S., Sabouri, H., Gholizadeh, A., Ahangar, L. and Katouzi, M. (2022). QTLs controlling physiological and morphological traits of barley (*Hordeum vulgare* L.) seedlings under salinity, drought, and normal conditions, *Biotech* 11(3), 26.

Martinez, A. K., Soriano, J. M., Tuberosa, R., Koumproglou, R., Jahrmann, T. and Salvi, S. (2016). Yield QTLome distribution correlates with gene density in maize, *Plant Science* 242, 300–309.

Martínez-Goñi, X. S., Miranda-Apodaca, J. and Pérez-López, U. (2023). Could buckwheat and spelt be alternatives to wheat under future environmental conditions? Study of their physiological response to drought, *Agricultural Water Management* 278, 108176.

Mascher, M., Gundlach, H., Himmelbach, A., Beier, S., Twardziok, S. O., Wicker, T., Radchuk, V., Dockter, C., Hedley, P. E., Russell, J., Bayer, M., Ramsay, L., Liu, H., Haberer, G., Zhang, X. Q., Zhang, Q., Barrero, R. A., Li, L., Taudien, S., Groth, M., Felder, M., Hastie,

A., Šimková, H., Staňková, H., Vrána, J., Chan, S., Muñoz-Amatriaín, M., Ounit, R., Wanamaker, S., Bolser, D., Colmsee, C., Schmutzer, T., Aliyeva-Schnorr, L., Grasso, S., Tanskanen, J., Chailyan, A., Sampath, D., Heavens, D., Clissold, L., Cao, S., Chapman, B., Dai, F., Han, Y., Li, H., Li, X., Lin, C., McCooke, J. K., Tan, C., Wang, P., Wang, S., Yin, S., Zhou, G., Poland, J. A., Bellgard, M. I., Borisjuk, L., Houben, A., Doležel, J., Ayling, S., Lonardi, S., Kersey, P., Langridge, P., Muehlbauer, G. J., Clark, M. D., Caccamo, M., Schulman, A. H., Mayer, K. F. X., Platzer, M., Close, T. J., Scholz, U., Hansson, M., Zhang, G., Braumann, I., Spannagl, M., Li, C., Waugh, R. and Stein, N. (2017). A chromosome conformation capture ordered sequence of the barley genome, *Nature* 544(7651), 427–433.

Mehana, M., Abdelrahman, M., Emadeldin, Y., Rohila, J. S. and Karthikeyan, R. (2021). Impact of genetic improvements of rice on its water use and effects of climate variability in Egypt, *Agriculture* 11(9), 865.

Merchuk-Ovnat, L., Fahima, T., Ephrath, J. E., Krugman, T. and Saranga, Y. (2017). Ancestral QTL alleles from wild emmer wheat enhance root development under drought in modern wheat, *Frontiers in Plant Science* 8, 703.

Merchuk-Ovnat, L., Barak, V., Fahima, T., Ordon, F., Lidzbarsky, G. A., Krugman, T. and Saranga, Y. (2016). Ancestral QTL alleles from wild emmer wheat improve drought resistance and productivity in modern wheat cultivars, *Frontiers in Plant Science* 7, 452.

Mohamed, H., Fawzi, M., Basit, A., Kaleemullah, Lone, R. and Sofy, M. R. (2022). Sorghum: Nutritional factors, bioactive compounds, pharmaceutical and application in food systems: A review, *Phyton* 91(7), 1303–1325.

Mohanty, B., Kitazumi, A., Cheung, C. Y. M., Lakshmanan, M., de Los Reyes, B. G., Jang, I. C. and Lee, D. Y. (2016). Identification of candidate network hubs involved in metabolic adjustments of rice under drought stress by integrating transcriptome data and genome-scale metabolic network, *Plant Science* 242, 224–239.

Moon, S., Cho, L. H., Kim, Y. J., Gho, Y. S., Jeong, H. Y., Hong, W. J., Lee, C., Park, H., Jwa, N. S., Dangol, S., Chen, Y., Park, H., Cho, H. S., An, G. and Jung, K. H. (2019). RSL Class II transcription factors guide the nuclear localization of *RHL1* to regulate root hair development, *Plant Physiology* 179(2), 558–568.

Muthu, V., Abbai, R., Nallathambi, J., Rahman, H., Ramasamy, S., Kambale, R., Thulasinathan, T., Ayyenar, B. and Muthurajan, R. (2020). Pyramiding QTLs controlling tolerance against drought, salinity, and submergence in rice through marker assisted breeding, *PLoS ONE* 15(1), e0227421.

Nada, R. M., Abo-Hegazy, S. E., Budran, E. G. and Abogadallah, G. M. (2019). The interaction of genes controlling root traits is required for the developmental acquisition of deep and thick root traits and improving root architecture in response to low water or nitrogen content in rice (*Oryza sativa* L.) cultivars, *Plant Physiology and Biochemistry* 141, 122–132.

Naz, R., Gul, F., Zahoor, S., Nosheen, A., Yasmin, H., Keyani, R., Shahid, M., Hassan, M. N., Siddiqui, M. H., Batool, S., Anwar, Z., Ali, N. and Roberts, T. H. (2022). Interactive effects of hydrogen sulphide and silicon enhance drought and heat tolerance by modulating hormones, antioxidant defence enzymes and redox status in barley (*Hordeum vulgare* L.), *Plant Biology* 24(4), 684–696.

Nazemi, G., Valli, F., Ferroni, L., Speranza, M., Maccaferri, M., Tuberosa, R. and Salvi, S. (2016). Genetic variation for aerenchyma and other root anatomical traits in durum wheat (*Triticum durum* Desf.), *Genetic Resources and Crop Evolution* 63(5), 771–779.

Niones, J. M., Inukai, Y., Suralta, R. R. and Yamauchi, A. (2015). QTL associated with lateral root plasticity in response to soil moisture fluctuation stress in rice, *Plant and Soil* 391(1-2), 63-75.

Ober, E. S., Alahmad, S., Cockram, J., Forestan, C., Hickey, L. T., Kant, J., Maccaferri, M., Marr, E., Milner, M., Pinto, F., Rambla, C., Reynolds, M., Salvi, S., Sciara, G., Snowdon, R. J., Thomelin, P., Tuberosa, R., Uauy, C., Voss-Fels, K. P., Wallington, E. and Watt, M. (2021). Wheat root systems as a breeding target for climate resilience, *Theoretical and Applied Genetics* 134(6), 1645-1662.

Ortiz, D. and Salas-Fernandez, M. G. (2022). Dissecting the genetic control of natural variation in sorghum photosynthetic response to drought stress, *Journal of Experimental Botany* 73(10), 3251-3267.

Ostmeyer, T., Parker, N., Jaenisch, B., Alkotami, L., Bustamante, C. and Jagadish, S. V. K. (2020). Impacts of heat, drought, and their interaction with nutrients on physiology, grain yield, and quality in field crops, *Plant Physiology Reports* 25(4), 549-568.

Oyiga, B. C., Palczak, J., Wojciechowski, T., Lynch, J. P., Naz, A. A., Léon, J. and Ballvora, A. (2020). Genetic components of root architecture and anatomy adjustments to water-deficit stress in spring barley, *Plant, Cell and Environment* 43(3), 692-711.

Ozturk, Z. N., Talamé, V., Deyholos, M., Michalowski, C. B., Galbraith, D. W., Gozukirmizi, N., Tuberosa, R. and Bohnert, H. J. (2002). Monitoring large-scale changes in transcript abundance in drought- and salt-stressed barley, *Plant Molecular Biology* 48(5), 551-573.

O'Toole, J. C. and Bland, W. L. (1987). Genotypic variation in crop plant root systems, *Advances in Agronomy* 41, 91-145.

Pan, J., Li, Z., Wang, Q., Yang, L., Yao, F. and Liu, W. (2020). An S-domain receptor-like kinase, *OsESG1*, regulates early crown root development and drought resistance in rice, *Plant Science* 290, 110318.

Pankin, A., Altmüller, J., Becker, C. and Von Korff, M. (2018). Targeted resequencing reveals genomic signatures of barley domestication, *New Phytologist* 218(3), 1247-1259.

Paterson, A. H., Bowers, J. E., Bruggmann, R., Dubchak, I., Grimwood, J., Gundlach, H., Haberer, G., Hellsten, U., Mitros, T., Poliakov, A., Schmutz, J., Spannagl, M., Tang, H., Wang, X., Wicker, T., Bharti, A. K., Chapman, J., Feltus, F. A., Gowik, U., Grigoriev, I. V., Lyons, E., Maher, C. A., Martis, M., Narechania, A., Otillar, R. P., Penning, B. W., Salamov, A. A., Wang, Y., Zhang, L., Carpita, N. C., Freeling, M., Gingle, A. R., Hash, C. T., Keller, B., Klein, P., Kresovich, S., McCann, M. C., Ming, R., Peterson, D. G., Mehboob-ur-Rahman, Ware, D., Westhoff, P., Mayer, K. F., Messing, J. and Rokhsar, D. S. (2009). The Sorghum bicolor genome and the diversification of grasses, *Nature* 457(7229), 551-556.

Pinto, R. S. and Reynolds, M. P. (2015). Common genetic basis for canopy temperature depression under heat and drought stress associated with optimized root distribution in bread wheat, *Theoretical and Applied Genetics* 128(4), 575-585.

Placido, D. F., Sandhu, J., Sato, S. J., Nersesian, N., Quach, T., Clemente, T. E., Staswick, P. E. and Walia, H. (2020). The *LATERAL ROOT DENSITY* gene regulates root growth during water stress in wheat, *Plant Biotechnology Journal* 18(9), 1955-1968.

PlantStress Website. Available at: https://plantstress.com/.

Poets, A. M., Fang, Z., Clegg, M. T. and Morrell, P. L. (2015). Barley landraces are characterized by geographically heterogeneous genomic origins, *Genome Biology* 16(1), 173.

Pour-Aboughadareh, A., Mohammadi, R., Etminan, A., Shooshtari, L., Maleki-Tabrizi, N. and Poczai, P. (2020). Effects of drought stress on some agronomic and morpho-physiological traits in durum wheat genotypes, *Sustainability* 12(14), 5610.

Prasad, R., Shivay, Y. S. and Kumar, D. (2017). Current status, challenges, and opportunities in rice production. In: Chauhan, B. S., Jabran, K. and Mahajan, G. (Eds). *Rice Production Worldwide*. Cham: Springer International Publishing, pp. 1-32.

Prasadi, N. P. V. and Joye, I. J. (2020). Dietary fibre from whole grains and their benefits on metabolic health, *Nutrients* 12(10), 3045.

Qin, H., Wang, J., Zhou, J., Qiao, J., Li, Y., Quan, R. and Huang, R. (2023). Abscisic acid promotes auxin biosynthesis to inhibit primary root elongation in rice, *Plant Physiology* 191(3), 1953–1967.

Qin, W., Liu, C., Jiang, W., Xue, Y., Wang, G. and Liu, S. (2019). A coumarin analogue NFA from endophytic *Aspergillus fumigatus* improves drought resistance in rice as an antioxidant, *BMC Microbiology* 19(1), 1–11.

Raju, B. R., Mohankumar, M. V., Sumanth, K. K., Rajanna, M. P., Udayakumar, M., Prasad, T. G. and Sheshshayee, M. S. (2016). Discovery of QTLs for water mining and water use efficiency traits in rice under water-limited condition through association mapping, *Molecular Breeding* 36(3), 35.

Ramalingam, P., Kamoshita, A., Deshmukh, V., Yaginuma, S. and Uga, Y. (2017). Association between root growth angle and root length density of a near-isogenic line of IR64 rice with *DEEPER ROOTING 1* under different levels of soil compaction, *Plant Production Science* 20(2), 162–175.

Ramstein, G. P., Jensen, S. E. and Buckler, E. S. (2019). Breaking the curse of dimensionality to identify causal variants in breeding 4, *Theoretical and Applied Genetics* 132(3), 559–567.

Reynolds, M. and Tuberosa, R. (2008). Translational research impacting on crop productivity in drought-prone environments, *Current Opinion in Plant Biology* 11(2), 171–179.

Robinson, H., Kelly, A., Fox, G., Franckowiak, J., Borrell, A. and Hickey, L. (2018). Root architectural traits and yield: Exploring the relationship in barley breeding trials, *Euphytica* 214(9), 151.

Roselló, M., Royo, C., Sanchez-Garcia, M. and Soriano, J. M. (2019). Genetic dissection of the seminal root system architecture in Mediterranean durum wheat landraces by genome-wide association study, *Agronomy* 9(7), 364.

Royo, C., Maccaferri, M., Álvaro, F., Moragues, M., Sanguineti, M. C., Tuberosa, R., Maalouf, F., del Moral, L. F. G. Demontis, A., Rhouma, S., Nachit, M., Nserallah, N. and Villegas, D. (2010). Understanding the relationships between genetic and phenotypic structures of a collection of elite durum wheat accessions, *Field Crops Research* 119(1), 91–105.

Sabadin, P. K., Malosetti, M., Boer, M. P., Tardin, F. D., Santos, F. G., Guimarães, C. T., Gomide, R. L., Andrade, C. L. T., Albuquerque, P. E. P., Caniato, F. F., Mollinari, M., Margarido, G. R. A., Oliveira, B. F., Schaffert, R. E., Garcia, A. A. F., Van Eeuwijk, F. A. and Magalhaes, J. V. (2012). Studying the genetic basis of drought tolerance in sorghum by managed stress trials and adjustments for phenological and plant height differences, *Theoretical and Applied Genetics* 124(8), 1389–1402.

Sahebi, M., Hanafi, M. M., Rafii, M. Y., Mahmud, T. M. M., Azizi, P., Osman, M., Abiri, R., Taheri, S., Kalhori, N., Shabanimofrad, M., Miah, G. and Atabaki, N. (2018). Improvement of drought tolerance in rice (*Oryza sativa* L.): Genetics, genomic tools, and the WRKY gene family. *BioMed Research International* 2018, 1-20.

Salarpour, M., Pakniyat, H., Abdolshahi, R., Heidari, B., Razi, H. and Afzali, R. (2020). Mapping QTL for agronomic and root traits in the Kukri/RAC875 wheat (*Triticum aestivum* L.) population under drought stress conditions, *Euphytica* 216(7), 105.

Salgotra, R. K. and Stewart, C. N. (2020). Functional markers for precision plant breeding, *International Journal of Molecular Sciences* 21(13), 4792.

Sallam, A., Alqudah, A. M., Dawood, M. F. A., Baenziger, P. S. and Börner, A. (2019). Drought stress tolerance in wheat and barley: Advances in physiology, breeding and genetics research, *International Journal of Molecular Sciences* 20(13), 3137.

Sallam, A., Eltaher, S., Alqudah, A. M., Belamkar, V. and Baenziger, P. S. (2022). Combined GWAS and QTL mapping revealed candidate genes and SNP network controlling recovery and tolerance traits associated with drought tolerance in seedling winter wheat, *Genomics* 114(3), 110358.

Salvi, S., Giuliani, S., Ricciolini, C., Carraro, N., Maccaferri, M., Presterl, T., Ouzunova, M. and Tuberosa, R. (2016). Two major quantitative trait loci controlling the number of seminal roots in maize co-map with the root developmental genes *rtcs* and *rum1*, *Journal of Experimental Botany* 67(4), 1149–1159.

Salvi, S., Sponza, G., Morgante, M., Tomes, D., Niu, X., Fengler, K. A., Meeley, R., Ananiev, E. V., Svitashev, S., Bruggemann, E., Li, B., Hainey, C. F., Radovic, S., Zaina, G., Rafalski, J. A., Tingey, S. V., Miao, G. H., Phillips, R. L. and Tuberosa, R. (2007). Conserved noncoding genomic sequences associated with a flowering-time quantitative trait locus in maize, *Proceedings of the National Academy of Sciences of the United States of America* 104(27), 11376–11381.

Salvi, S. and Tuberosa, R. (2015). The crop QTLome comes of age, *Current Opinion in Biotechnology* 32, 179–185.

Samanta, P. and Dey, N. (2023). microRNA-marker based genetic diversity analysis for drought tolerance in rice (*Oryza sativa* L.), *Plant Physiology Reports* 28(1), 43-52.

Sanguineti, M. C., Duvick, D. N., Smith, J., Landi, P. and Tuberosa, R. (2006). Effects of long-term selection on seedling traits and ABA accumulation in commercial maize hybrids, *Maydica* 51, 329–338.

Sanguineti, M. C., Li, S., Maccaferri, M., Corneti, S., Rotondo, F., Chiari, T. and Tuberosa, R. (2007). Genetic dissection of seminal root architecture in elite durum wheat germplasm, *Annals of Applied Biology* 151(3), 291–305.

Satrio, R. D., Fendiyanto, M. H., Supena, E. D. J., Suharsono, S. and Miftahudin, M. (2023). Mapping and identification of QTL for agro-physiological traits in rice (*Oryza sativa* L.) under drought stress, *Plant Gene* 33, 100397.

Schoppach, R., Taylor, J. D., Majerus, E., Claverie, E., Baumann, U., Suchecki, R., Fleury, D. and Sadok, W. (2016). High resolution mapping of traits related to whole-plant transpiration under increasing evaporative demand in wheat, *Journal of Experimental Botany* 67(9), 2847–2860.

Schnable, P. S., Ware, D., Fulton, R. S., Stein, J. C., Wei, F., Pasternak, S., Liang, C., Zhang, J., Fulton, L., Graves, T. A., Minx, P., Reily, A. D., Courtney, L., Kruchowski, S. S., Tomlinson, C., Strong, C., Delehaunty, K., Fronick, C., Courtney, B., Rock, S. M., Belter, E., Du, F., Kim, K., Abbott, R. M., Cotton, M., Levy, A., Marchetto, P., Ochoa, K., Jackson, S. M., Gillam, B., Chen, W., Yan, L., Higginbotham, J., Cardenas, M., Waligorski, J., Applebaum, E., Phelps, L., Falcone, J., Kanchi, K., Thane, T., Scimone, A., Thane, N., Henke, J., Wang, T., Ruppert, J., Shah, N., Rotter, K., Hodges, J., Ingenthron, E.,

Cordes, M., Kohlberg, S., Sgro, J., Delgado, B., Mead, K., Chinwalla, A., Leonard, S., Crouse, K., Collura, K., Kudrna, D., Currie, J., He, R., Angelova, A., Rajasekar, S., Mueller, T., Lomeli, R., Scara, G., Ko, A., Delaney, K., Wissotski, M., Lopez, G., Campos, D., Braidotti, M., Ashley, E., Golser, W., Kim, H., Lee, S., Lin, J., Dujmic, Z., Kim, W., Talag, J., Zuccolo, A., Fan, C., Sebastian, A., Kramer, M., Spiegel, L., Nascimento, L., Zutavern, T., Miller, B., Ambroise, C., Muller, S., Spooner, W., Narechania, A., Ren, L., Wei, S., Kumari, S., Faga, B., Levy, M. J., McMahan, L., Van Buren, P., Vaughn, M. W., Ying, K., Yeh, C. T., Emrich, S. J., Jia, Y., Kalyanaraman, A., Hsia, A. P., Barbazuk, W. B., Baucom, R. S., Brutnell, T. P., Carpita, N. C., Chaparro, C., Chia, J. M., Deragon, J. M., Estill, J. C., Fu, Y., Jeddeloh, J. A., Han, Y., Lee, H., Li, P., Lisch, D. R., Liu, S., Liu, Z., Nagel, D. H., McCann, M. C., SanMiguel, P., Myers, A. M., Nettleton, D., Nguyen, J., Penning, B. W., Ponnala, L., Schneider, K. L., Schwartz, D. C., Sharma, A., Soderlund, C., Springer, N. M., Sun, Q., Wang, H., Waterman, M., Westerman, R., Wolfgruber, T. K., Yang, L., Yu, Y., Zhang, L., Zhou, S., Zhu, Q., Bennetzen, J. L., Dawe, R. K., Jiang, J., Jiang, N., Presting, G. G., Wessler, S. R., Aluru, S., Martienssen, R. A., Clifton, S. W., McCombie, W. R., Wing, R. A. and Wilson, R. K. (2009). The B73 maize genome: Complexity, diversity, and dynamics, *Science* 326(5956), 1112–1115.

Sehgal, D., Dhakate, P., Ambreen, H., Shaik, K. H. B., Rathan, N. D., Anusha, N. M., Deshmukh, R. and Vikram, P. (2023). Wheat omics: Advancements and opportunities, *Plants* 12(3), 426.

Sehgal, D., Mondal, S., Crespo-Herrera, L., Velu, G., Juliana, P., Huerta-Espino, J., Shrestha, S., Poland, J., Singh, R. and Dreisigacker, S. (2020). Haplotype-based, genome-wide association study reveals stable genomic regions for grain yield in CIMMYT spring bread wheat, *Frontiers in Genetics* 11, 589490.

Selvaraj, M. G., Jan, A., Ishizaki, T., Valencia, M., Dedicova, B., Maruyama, K., Ogata, T., Todaka, D., Yamaguchi-Shinozaki, K., Nakashima, K. and Ishitani, M. (2020). Expression of the CCCH-tandem zinc finger protein gene *OsTZF5* under a stress-inducible promoter mitigates the effect of drought stress on rice grain yield under field conditions, *Plant Biotechnology Journal* 18(8), 1711–1721.

Sharma, E., Bhatnagar, A., Bhaskar, A., Majee, S. M., Kieffer, M., Kepinski, S., Khurana, P. and Khurana, J. P. (2022a). Stress-induced F-Box protein-coding gene *OsFBX257* modulates drought stress adaptations and ABA responses in rice, *Plant, Cell and Environment* 46(4), 1207–1231.

Sharma, P., Chaudhary, H. K., Kapoor, C., Manoj, N. V., Singh, K. and Sood, V. K. (2022b). Molecular cytogenetic analysis of novel wheat-rye translocation lines and their characterization for drought tolerance and yellow rust resistance, *Cereal Research Communications* 50(4), 655–665.

She, K., Pan, W., Yan, Y., Shi, T., Chu, Y., Cheng, Y., Ma, B.and Song, W. (2022). Genome-wide identification, evolution and expressional analysis of OSCA gene family in barley (Hordeum vulgare L.), *International Journal of Molecular Sciences* 23(21), 13027.

Singroha, G., Sharma, P. and Sunkur, R. (2021). Current status of microRNA-mediated regulation of drought stress responses in cereals, *Physiologia Plantarum* 172(3), 1808–1821.

Song, W., Wang, B., Hauck, A. L., Dong, X., Li, J. and Lai, J. (2016). Genetic dissection of maize seedling root system architecture traits using an ultra-high density bin-map and a recombinant inbred line population, *Journal of Integrative Plant Biology* 58(3), 266–279.

Soriano, J. M., Colasuonno, P., Marcotuli, I. and Gadaleta, A. (2021). Meta-QTL analysis and identification of candidate genes for quality, abiotic and biotic stress in durum wheat, *Scientific Reports* 11(1), 11877.

Souza, A. A. D., Carvalho, A. J. D., Bastos, E. A., Portugal, A. F., Torres, L. G., Batista, P. S. C., Julio, M. P. M., Julio, B. H. M. and Menezes, C. B. D. (2020). Grain sorghum grown under drought stress at pre- and post-flowering in semiarid environment, *Journal of Agricultural Science* 12(4), 97.

Sun, X., Xiang, Y., Dou, N., Zhang, H., Pei, S., Franco, A. V., Menon, M., Monier, B., Ferebee, T., Liu, T., Liu, S., Gao, Y., Wang, J., Terzaghi, W., Yan, J., Hearne, S., Li, L., Li, F. and Dai, M. (2023). The role of transposon inverted repeats in balancing drought tolerance and yield-related traits in maize, *Nature Biotechnology* 41(1), 120–127.

Sun, X., Xiong, H., Jiang, C., Zhang, D., Yang, Z., Huang, Y., Zhu, W., Ma, S., Duan, J., Wang, X., Liu, W., Guo, H., Li, G., Qi, J., Liang, C., Zhang, Z., Li, J., Zhang, H., Han, L., Zhou, Y., Peng, Y. and Li, Z. (2022). Natural variation of *DROT1* confers drought adaptation in upland rice, *Nature Communications* 13(1), 4265.

Talamè, V., Ozturk, N. Z., Bohnert, H. J. and Tuberosa, R. (2007). Barley transcript profiles under dehydration shock and drought stress treatments: A comparative analysis, *Journal of Experimental Botany* 58(2), 229–240.

Tardieu, F. and Tuberosa, R. (2010). Dissection and modelling of abiotic stress tolerance in plants, *Current Opinion in Plant Biology* 13(2), 206–212.

Thanmalagan, R. R., Jayaprakash, A., Roy, A., Arunachalam, A. and Lakshmi, P. T. V. (2022). A review on applications of plant network biology to understand the drought stress response in economically important cereal crops, *Plant Gene* 29, 100345.

Tuberosa, R. (2012). Phenotyping for drought tolerance of crops in the genomics era, *Frontiers in Physiology* 3, 347.

Tuberosa, R., Frascaroli, E., Maccaferri, M. and Salvi, S. (2021). Understanding and exploiting the genetics of plant root traits. In: Gregory, P. J. (Ed). *Understanding and improving crop root function*. Cambridge: Burleigh Science Publishing, pp. 395–466.

Tuberosa, R., Gill, B. S. and Quarrie, S. A. (2002). Cereal genomics: Ushering in a brave new world, *Plant Molecular Biology* 48(5–6), 445–449.

United Nations. World population prospect 2022. Available at: https://population.un.org/wpp/.

Varshney, R. K., Barmukh, R., Roorkiwal, M., Qi, Y., Kholova, J., Tuberosa, R., Reynolds, M. P., Tardieu, F. and Siddique, K. H. M. (2021). Breeding custom-designed crops for improved drought adaptation, *Advanced Genetics* 2(3), e202100017.

Varshney, R. K., Tuberosa, R. and Tardieu, F. (2018). Progress in understanding drought tolerance: From alleles to cropping systems, *Journal of Experimental Botany* 69(13), 3175–3179.

Verma, H. and Sarma, R. N. (2021). Identification of markers for root traits related to drought tolerance using traditional rice germplasm, *Molecular Biotechnology* 63(12), 1280–1292.

Wang, B., Lin, Z., Li, X., Zhao, Y., Zhao, B., Wu, G., Ma, X., Wang, H., Xie, Y., Li, Q., Song, G., Kong, D., Zheng, Z., Wei, H., Shen, R., Wu, H., Chen, C., Meng, Z., Wang, T., Li, Y., Li, X., Chen, Y., Lai, J., Hufford, M. B., Ross-Ibarra, J., He, H. and Wang, H. (2020). Genome-wide selection and genetic improvement during modern maize breeding, *Nature Genetics* 52(6), 565–571.

Wang, N., Cao, F., Richmond, M. E. A., Qiu, C. and Wu, F. (2019). Foliar application of betaine improves water-deficit stress tolerance in barley (*Hordeum vulgare* L.), *Plant Growth Regulation* 89(1), 109–118.

Wang, S., Qian, Y. Q., Zhao, R. P., Chen, L. L. and Song, J. M. (2023). Graph-based pan-genomes: Increased opportunities in plant genomics, *Journal of Experimental Botany* 74(1), 24–39.

Wang, Y., Li, H., Zhao, C., Yang, C., Xu, Q., Yuan, H., Yang, H. and Zeng, X. (2023). Identification of a novel transcription factor under long-term drought resistance in highland barley: A DNA affinity purification sequencing-based transcriptomic analysis, *Chemical and Biological Technologies in Agriculture* 10(1), 1.

Wolfe, K. H., Sharp, P. M. and Li, W.-H. (1989). Rates of synonymous substitution in plant nuclear genes, *Journal of Molecular Evolution* 29(3), 208–211.

Wyka, T. P., Bagniewska-Zadworna, A., Kuczyńska, A., Mikołajczak, K., Ogrodowicz, P., Żytkowiak, M., Surma, M. and Adamski, T. (2019). Drought-induced anatomical modifications of barley (*Hordeum vulgare* L.) leaves: An allometric perspective, *Environmental and Experimental Botany* 166, 103798.

Xu, Y. (2016). Envirotyping for deciphering environmental impacts on crop plants, *Theoretical and Applied Genetics*, 129, 653–673.

Xu, W., Dou, Y., Geng, H., Fu, J., Dan, Z., Liang, T., Cheng, M., Zhao, W., Zeng, Y., Hu, Z. and Huang, W. (2022). *OsGRP3* enhances drought resistance by altering phenylpropanoid biosynthesis pathway in rice (*Oryza sativa* L.), *International Journal of Molecular Sciences* 23(13), 7045.

Xu, W., Tang, W., Wang, C., Ge, L., Sun, J., Qi, X., He, Z., Zhou, Y., Chen, J., Xu, Z., Ma, Y. Z. and Chen, M. (2020). *SiMYB56* confers drought stress tolerance in transgenic rice by regulating lignin biosynthesis and ABA signaling pathway, *Frontiers in Plant Science* 11, 785.

Xu, Y., Li, P., Yang, Z. and Xu, C. (2017). Genetic mapping of quantitative trait loci in crops, *The Crop Journal* 5(2), 175–184.

Yadav, R., Courtois, B., Huang, N. and Mclaren, G. (1997). Mapping genes controlling root morphology and root distribution in a doubled-haploid population of rice, *Theoretical and Applied Genetics* 94(5), 619–632.

Yadav, I. S., Singh, N., Wu, S., Raupp, J., Wilson, D. L., Rawat, N., Gill, B. S., Poland, J. and Tiwari, V. K. (2023). Exploring genetic diversity of wild and related tetraploid wheat species *Triticum turgidum* and *Triticum timopheevii*, *Journal of Advanced Research* 48, 47–60.

Yahaya, M. A. and Shimelis, H. (2022). Drought stress in sorghum: Mitigation strategies, breeding methods and technologies–a review, *Journal of Agronomy and Crop Science* 208(2), 127–142.

Yamamoto, T., Suzuki, T., Suzuki, K., Adachi, S., Sun, J., Yano, M., Ookawa, T. and Hirasawa, T. (2016). Detection of QTL for exudation rate at ripening stage in rice and its contribution to hydraulic conductance, *Plant Science* 242, 270–277.

Yang, Q., Li, P., Zhang, D., Lin, W., Hao, X. and Zong, Y. (2023). Effects of elevated CO_2 on the photosynthesis, chlorophyll fluorescence and yield of two wheat cultivars (*Triticum aestivum* L.) under persistent drought stress, *Sustainability* 15(2), 1593.

Yang, Z., Chi, X., Guo, F., Jin, X., Luo, H., Hawar, A., Chen, Y., Feng, K., Wang, B., Qi, J., Yang, Y. and Sun, B. (2020). SbWRKY30 enhances the drought tolerance of plants and regulates a drought stress-responsive gene, *SbRD19*, in sorghum, *Journal of Plant Physiology* 246–247, 153142.

Yu, F., Tan, Z., Fang, T., Tang, K., Liang, K. and Qiu, F. (2020). A comprehensive transcriptomics analysis reveals long non-coding RNA to be involved in the key metabolic pathway in response to waterlogging stress in maize, *Genes* 11(3), 267.

Yu, Y., Song, T., Wang, Y., Zhang, M., Li, N., Yu, M., Zhang, S., Zhou, H., Guo, S., Bu, Y., Wang, T., Xiang, J. and Zhang, X. (2023). The wheat WRKY transcription factor TaWRKY1-2D confers drought resistance in transgenic Arabidopsis and wheat (*Triticum aestivum* L.), *International Journal of Biological Macromolecules* 226, 1203–1217.

Yuan, W., Suo, J., Shi, B., Zhou, C., Bai, B., Bian, H., Zhu, M. and Han, N. (2019). The barley miR393 has multiple roles in regulation of seedling growth, stomatal density, and drought stress tolerance, *Plant Physiology and Biochemistry* 142, 303–311.

Zaikina, E. A., Musin, K. G., Kuluev, A. R., Nikonov, V. I., Dmitriev, A. M. and Kuluev, B. R. (2022). Change in the activity of genes of transcription factors *TaNAC69*, *TaDREB1*, and *TabZIP60* in bread wheat plants with water deficiency and hypothermia, *Russian Journal of Plant Physiology* 69(3), 1–9.

Zargar, S. M., Mir, R. A., Ebinezer, L. B., Masi, A., Hami, A., Manzoor, M., Salgotra, R. K., Sofi, N. R., Mushtaq, R., Rohila, J. S. and Rakwal, R. (2021). Physiological and multi-omics approaches for explaining drought stress tolerance and supporting sustainable production of rice, *Frontiers in Plant Science* 12, 803603.

Zhao, J., Zheng, X., Qiao, L., Yang, C., Wu, B., He, Z., Tang, Y., Li, G., Yang, Z., Zheng, J. and Qi, Z. (2022a). Genome-wide association study reveals structural chromosome variations with phenotypic effects in wheat (*Triticum aestivum* L.), *The Plant Journal* 112(6), 1447–1461.

Zhao, W., Wang, X., Zhang, Q., Zheng, Q., Yao, H., Gu, X., Liu, D., Tian, X., Wang, X., Li, Y. and Zhu, Z. (2022b). H3K36 demethylase JMJ710 negatively regulates drought tolerance by suppressing *MYB48-1* expression in rice, *Plant Physiology* 189(2), 1050–1064.

Zhao, X., Peng, Y., Zhang, J., Fang, P. and Wu, B. (2018). Identification of QTLs and meta-QTLs for seven agronomic traits in multiple maize populations under well-watered and water-stressed conditions, *Crop Science* 58(2), 507–520.

Zheng, Z., Hey, S., Jubery, T., Liu, H., Yang, Y., Coffey, L., Miao, C., Sigmon, B., Schnable, J. C., Hochholdinger, F., Ganapathysubramanian, B. and Schnable, P. S. (2020). Shared genetic control of root system architecture between *Zea mays* and *Sorghum bicolor*, *Plant Physiology* 182(2), 977–991.

Zhu, M., He, Y., Zhu, M., Ahmad, A., Xu, S., He, Z., Jiang, S., Huang, J., Li, Z., Liu, S., Hou, X. and Zhang, Z. (2022). *ipa1* improves rice drought tolerance at seedling stage mainly through activating abscisic acid pathway, *Plant Cell Reports* 41(1), 221–232.

Zia, M. A. B., Yousaf, M. F., Asim, A. and Naeem, M. (2022). An overview of genome-wide association mapping studies in Poaceae species (model crops: Wheat and rice), *Molecular Biology Reports* 49(12), 12077–12090.

Zurek, P. R., Topp, C. N. and Benfey, P. N. (2015). Quantitative trait locus mapping reveals regions of the maize genome controlling root system architecture, *Plant Physiology* 167(4), 1487–1496.

Chapter 4

The role of drought-induced proteins in regulating drought tolerance in cereals

Mitiku A. Mengistu and John C. Cushman, University of Nevada – Reno, USA

1 Introduction

A growing human population estimated to reach more than 9 billion by 2050 (United Nations, 2019) will increase global food demand above current production levels by 33–62% between 2010 and 2050, depending upon dietary requirements, actual population growth, and the impacts of global climate change (van Dijk et al., 2021). Adopting plant-based diets for humans is seen as beneficial in terms of caloric conversion efficiency and food security (Shepon et al., 2016), environmental sustainability and land-carrying capacity (Peters et al., 2016), and the health and longevity of the human population (Hever, 2016; Kahleova et al., 2017). In addition to food for human consumption, if global meat and egg production increases 1.6-fold by 2050, as projected by the UN Food and Agriculture Organization (FAO) (Alexandratos and Bruinsma, 2012), then substantially more cereal production will be needed to feed livestock as well.

Cereals are named from the Latin word *cerealis*, which means 'grain', and are derived primarily from monocot grasses within the Poaceae. True cereal grains come from members of the Gramineae and include maize (*Zea mays*), wheat (*Triticum aestivum*), spelt (*Triticum spelta*), rice (*Oryza sativa*), African rice (*Oryza glaberrima*), wild rice (*Zizania palustris*), barley (*Hordeum vulgare*), sorghum (*Sorghum bicolor*), oat (*Avena sativa*), rye (*Secale cereale*), pearl

http://dx.doi.org/10.19103/AS.2022.0109.04

millet (*Cenchrus americanus*), finger millet (*Eleusine coracana*), foxtail millet (*Sertaria italica*), teff (*Eragrostis tef*), Japanese millet (*Echinochloa esculenta*), proso millet (*Panicum millaceum*), and fonio (*Digitaria exilis*), among others (Sarwar et al., 2013; Pardo and VanBuren, 2021). Related species that produce edible grains within the eudicots, such as buckwheat (*Fagopyrum esculentum*), quinoa (*Chenopodium quinoa*), and chia (*Salvia hispanica*), are referred to as pseudocereals. Current annual cereal production is estimated at 2.7 M Mg worldwide (FAO, 2021). More than 50% of the global daily caloric intake is derived from direct consumption of cereal grains (Awika, 2011). The percentage of caloric intake varies from about 30% in developed countries to 60% in developing countries to more than 80% in the poorest countries of the world (Awika, 2011).

However, global warming and associated climate changes represent major threats to global agricultural production systems that will challenge current efforts to meet future crop demands. The major impact of rising global temperatures is greater soil drying from longer and more severe drought events that limit water uptake and root health and increase vapor-pressure deficits experienced by plant leaves that ultimately constrains photosynthetic processes and net primary productivity (Zhao and Running, 2009; Naumann et al., 2018). Global climate change will likely result in a global expansion of drylands (Huang et al., 2015), adding to the greater than 50% global land area currently classified as dry subhumid, semiarid, and arid (Zika and Erb, 2009). Higher temperatures are responsible for an estimated 57% increase in atmospheric thirst or reference evapotranspiration (ETo) (loss of water from the Earth's surface to the atmosphere) in the western continental United States over the last 40 years and will likely continue current trends of water supply shortages, increased droughts, and wildfires in the future (Albano et al., 2022). Multiple large ensemble simulations showed increased risk of spatially compounded droughts of ~40–60%, due to a combination of hot and dry conditions, by the middle and late twenty-first century, respectively, driven by the El Niño–Southern Oscillation (ENSO) with a disproportionate increase in risk across the Amazon Basin and North America (Singh et al., 2022a).

The negative impacts of climate change on global crop production are becoming increasingly evident. Future crop production gains will likely fail to meet projected demands by 2050 (Ray et al., 2013), thus creating major negative impacts on food security, especially in low- and middle-income countries (Praveen and Sharma, 2019; Kumar et al., 2021). Cereal grain production has already declined 9–10% between 1964 and 2007 due to droughts and extreme heat events (Lesk et al., 2016). Crop productivity is expected to decrease in many regions of the globe in response to warming and drying trends (Wheeler and Von Braun, 2013). Crop yield simulations in both temperate and tropical regions for three major cereal crops (maize, rice, and wheat) predict that declines in aggregate crop production will accelerate during the latter half of the twenty-first century with increasing temperatures (Challinor et al., 2014).

Using four different assessment methods and models, each degree-Celsius rise in temperature is predicted to reduce global yields by 6.0%, 3.2%, 7.4%, and 3.1%, for wheat, rice, maize, and soybean, respectively (Zhao et al., 2017). Ensemble modeling has also projected that yield loss risks will increase 9–19% for maize, rice, wheat, and soybeans by the end of the twenty-first century due to global warming trends (Leng and Hall, 2019). An ensemble of 18 different global circulation models predicted that decadal rainfed maize yields will decline 2.2% to up to 21.5% depending upon the Representative Concentration Pathway (RCP) used (Irmak et al., 2022). Although projected irrigated yields declined less, in the range of 3.7–15.6% depending upon the RCP, and were less variable than rainfed yield, projected losses in productivity accelerated from 2070 to 2090 despite supplemental irrigation (Irmak et al., 2022). Moreover, some cereals such as wheat and barley are expected to be highly impacted by climate change, whereas others, such as sorghum and teff, are likely to be less affected (Gebresamuel et al., 2022). To meet these future needs for food and fiber, improvements in drought tolerance through genetic engineering or genome-editing approaches and agronomic management practices, such as minimizing input factors, including fertilizer and irrigation to maximize crop yields, are needed to increase production trends into the future, (Gupta et al., 2020).

2 Drought tolerance mechanisms

The general term of drought resistance refers to the ability of plants to grow productively when exposed to water deficits (Clarke et al., 1992; Mitra, 2001). Drought resistance adaptations are complex and based upon various biochemical, physiological, morphological, and anatomical mechanisms (Fang and Xiong, 2015). Examples include, but are not limited to root traits, manipulating carbon allocation, cuticular wax production, stomatal responses, leaf area, and delayed senescence (Stay-green) (Martignago et al., 2020; Yahaya and Shimelis, 2022).

In many cases, larger, longer, or deeper roots, steeper or wider root angles, and increased root number and root volume are traits associated with greater drought resistance, which allows plants to maintain higher water potential and continue transpiration under water-deficit conditions (Girma et al., 2020; Siddiqui et al., 2021; Kiran et al., 2022; Yahaya and Shimelis, 2022). For example, enhanced drought resistance was engineered in rice by the overexpression of a NAC family transcription factor (*Os*NAC9), which resulted in increased root diameter and enlarged cells within the root xylem and cortex (Redillas et al., 2012).

The climate resilience of many cereals arises, in part, from the evolution of C_4 photosynthesis, a drought-avoidance adaptation that reduces photorespiration and improves water-use efficiency (WUE) through reductions in stomatal conductance (Christin et al., 2013). A majority of cereals belong to two evolutionary and phenotypically distinct evolutionary clades designated as (Pardo and VanBuren, 2021):

- the Bambusoideae, Oryzoideae, and Pooideae (BOP) subfamilies; and
- the Panicoideae, Arundinoideae, Chloridoideae, Micrairoideae, Aristidoideae, and Danthonioideae (PACMAD) subfamilies.

Members of the BOP clade, including rice, wheat, barley, and rye, generally perform C_3 photosynthesis and display lower WUE than C_4 photosynthesis species. Members of the PACMAD clade within the Panicoideae are typically the NADP-ME type of C_4 photosynthesis, whereas members of the Chloridoideae are typically the NAD-ME type of C_4 photosynthesis, reflective of their respective evolutionary histories (Pardo and VanBuren, 2021).

In general, the panicoid grasses, which include the major crops of maize, sorghum, and sugarcane, are taller and larger but less drought-stress-tolerant than the shorter and smaller, yet more drought-stress-resilient chloridoid grasses, which include teff, finger millet, and the desiccation tolerant grasses *Sporobolus stapfianus* and *Oropetium capense* (Gaff and McGregor, 1979; Gaff et al., 2009). Chloroidoid grass lineages, in which C_4 photosynthesis evolved earlier, tend to exhibit lower leaf hydraulic conductance and higher WUE than panicoid grasses (Liu and Osborne, 2015; Zhou et al., 2021). The chloridoid grasses also typically exhibit lower productivity, making them less attractive as crop species, which is linked to a trade-off between productivity and stress tolerance especially evident among desiccation tolerant species (Alpert, 2006; da Silva et al., 2019).

Plants have evolved a number of strategies to overcome drought stress (Fang and Xiong, 2015; Rane et al., 2021):

- Avoidance;
- Escape;
- Tolerance; and
- Recovery.

Drought avoidance refers to the ability of a plant to conduct normal physiological processes under mild or moderate water-deficit stress conditions mainly by maintaining high water potentials within its tissues (Mitra, 2001; Basu et al., 2016). Water loss prevention can be accomplished through rapid stomatal closure of dumbbell-shaped stomatal guard cells accompanied by two subsidiary cells (McAusland et al., 2016; Lawson and Vialet-Chabrand,

2019), stomatal size, distribution, and density and vertical leaf angles (Pardo and VanBuren, 2021), leaf rolling (Liao et al., 2019; Zhang et al., 2021), and the accumulation of epicuticular waxes to limit non-stomatal transpiration (Carvalho et al., 2020; Wang et al., 2021). As with the drought resistance mechanisms mentioned earlier, water loss can be offset by enhanced water uptake from well-developed fibrous root systems or water storage within root structures, thereby avoiding the negative consequences of water loss (Fry et al., 2018).

Drought escape is related to drought avoidance, but refers more specifically to the plant's ability to complete its ontology rapidly through early seedling vigor and often involves accelerating flowering or early maturity to avoid terminal drought typically associated with the end of the growing season particularly in Mediterranean climates (Shavrukov et al., 2017; Dorrani-Nejad et al., 2022; Yahaya and Shimelis, 2022). Drought escape is also associated with avoiding drought stress during critical reproductive stages such as flowering by maintaining pollen viability and increasing the rate of grain filling (Mitra, 2001; Ndlovu et al., 2021). Drought escape can also be accomplished by shifting the planting season earlier to take advantage of winter moisture and to avoid terminal drought or by adjusting the crop cycle to seasonal rainfall patterns to increase water availability during grain filling (Araus et al., 2002). However, early planting requires that germplasm possess a greater degree of chilling tolerance (Raymundo et al., 2021).

Tolerating drought through maintaining physiological and biochemical activities within the plant involves various mechanisms associated with protecting the intracellular environment from various perturbations arising from the detrimental effects of drought or combinations of heat and drought (Rane et al., 2021). Drought tolerance involves a wide range of adaptations, including osmotic adaptation, protection against oxidative stress, phytohormone signaling responses, accumulation of photoprotective pigments, transpirational cooling, and circadian clock control of early morning pollination and fertilization (Fang and Xiong, 2015; Rane et al., 2021). Osmotic adjustment involves maintaining turgor pressure within the cell through the synthesis, transport, and accumulation of cellular compatible solutes, osmolytes, or osmoprotective metabolites such as amino acids, betaines, polyols, non-reducing sugars, and polyamines among others, and is generally associated with sustaining crop yield under drought stress conditions (Blum, 2017; Turner, 2018). The ability to maintain leaf turgor is governed by leaf osmotic adjustment and tissue flexibility with greater cell elasticity associated with a greater capacity to maintain leaf integrity or a more negative turgor loss point under water-deficit conditions (Bartlett et al., 2012; Liu and Osborne, 2015). Drought stress also leads to an over-accumulation of reactive oxygen species (ROS), which can damage membrane lipids, proteins and enzymes, and nucleic acids leading to cellular deterioration and death (Cruz de Carvalho, 2008; Verma et al., 2019). To overcome this damage, plants respond through the production of ROS scavenging enzymes and compounds to detoxify ROS and ameliorate such oxidative damage (Miller et al., 2010; Choudhury et al., 2017).

Drought recovery refers to the ability of a crop to resume growth and continue to accrue biomass or grain production following an episode of severe drought (Fang and Xiong, 2015). The ability to maintain high photosynthetic activity as well as other cellular functions during drought stress and to recover rapidly following rewatering leads to less severe yield losses (Abid et al., 2018). The functional basis of drought avoidance, escape, tolerance, and recovery lies with the protein expression patterns and adaptive proteins and enzymes synthesized by the plant in response to drought (and heat), adaptation, and recovery. Thus, exploration of the proteomic responses of cereal crops to drought (and heat) can provide crucial insights into key adaptations that can be exploited to improve cereal crop performance following drought and the climate resilience of grain production in the future.

3 Drought-related proteins and proteomic studies

High-throughput plant phenotyping is useful for identifying and selecting specific traits and advancing germplasm for enhanced breeding of climate-resilient cereals (Ul Islam et al., 2021). However, omics-based phenotyping approaches, such as transcriptomics, proteomics, metabolomics, and fluxomics, can complement these approaches. Proteomics or protein profiling and abundance quantitation approaches convey direct information about the protein complement of the plant in response to developmental cues and environmental stimuli (Thelen and Peck, 2007). Proteomic studies can not only survey relative protein abundance patterns, but also investigate posttranslational modifications of selected proteins (Kwon et al., 2006).

Many studies have catalogued proteins expressed in major (cereal) crops subjected to abiotic stresses, including salinity, cold, heat, and drought, imposed under controlled growth chamber or greenhouse conditions or under less controlled field conditions (Kosová et al., 2011; Roy et al., 2011; Hossain et al., 2012; Barkla et al., 2013a,b; Ahmad et al., 2016; Wang et al., 2017; Gautam et al., 2020). The role of proteomic (and multi-omic) studies in cereals (and other species) exposed to drought stress have been reviewed recently (Ngara and Ndimba, 2014; Wang et al., 2016a; Ghatak et al., 2017; Wu et al., 2017; Singh et al., 2022b). Such approaches are particularly relevant to understanding plant responses to environmental stresses and can provide information about associated candidate genes for marker-assisted breeding programs, quantitative trait loci (QTL), and genome-wide association studies (GWAS), as well as candidate genes for gain-of-function and loss-of-function testing using transgenesis and genome-editing approaches.

Researchers now have a near-comprehensive understanding of the range of proteins expressed during drought stress responses. While proteomic studies have revealed a large number of candidate proteins with potential roles in drought adaptation, relatively few proteins have undergone rigorous functional analysis using loss-of-function or gain-of-function transgenic (or genome-editing) approaches.

However, recent efforts have focused on understanding the specific roles of putative drought-adaptive proteins (Priya et al., 2019; Rodziewicz et al., 2019; Singh et al., 2022b). Major targets have included many different families of:

- transcription factors (TFs) controlling the rate of transcription of genetic information from DNA to messenger RNA (examples: basic leucine zippers [bZIP], Petunia NAM and Arabidopsis ATAF1/2 and CUC2 proteins [NAC], conserved amino acid sequence motif WRKYGQK [WRKY], v-myb avian myeloblastosis viral oncogene homolog [MYB], APETALA2/ethylene response factor (AP2/ERF), ABA-responsive element (ABRE), ABRE-binding protein/ABRE-binding factors [AREB/ABFs], and heat shock transcription factors [HSFs]);
- stress sensors and receptors - components of stress-adaptive and phytohormone signal transduction pathways (examples: phytohormone abscisic acid [ABA] receptors, G-protein-coupled receptors [GPCRs], receptor-like kinases [RLK], and calcium sensors);
- transporters (examples: aquaporins, ion, and solute transporters);
- signal transducers (examples: MAP kinases and calcium-dependent protein kinases [CDPKs]);
- regulators and enzymes that modify/catalyze mRNA and protein stability and turnover (examples: multiple micro RNAs [miRNAs], RNA-binding proteins to the 3′-UTR of *ZmRtn16*);
- anatomical/structural traits such as stomatal density, cuticular wax formation, and leaf curling (examples: epidermal patterning factor [EPF] signaling peptides, and SHINE/WAX INDUCER1 [SHN1/WIN1]); and
- proteins with protective functions for protein and membrane stability (examples: heat shock proteins [HSPs], late embryogenesis abundant [LEA] proteins, saposins/sphingolipid activator proteins [SAPs], dehydrins, osmotins, and antifreeze proteins).

This provides a context for a number of drought-adaptive proteins discussed in the following section. The use of protein functionality to improve plant performance is exemplified by Moroney and colleagues who have used transporter-based approaches (envelope-based, thylakoid-based, or carboxysome-based approaches) and structural approaches (protopyrenoid or pyrenoid-based approaches to optimize pyrenoid structure) to improve photosynthetic function (Moroney et al., 2022).

4 Analysis of drought-adaptive proteins using transgenesis

In comparison to transgenesis and genome-editing approaches, relatively few examples to improve drought tolerance of cereals have been documented using induced mutagenesis approaches (Zenda et al., 2021). In this section we summarize recent advances in major cereal crops, including maize, rice, and wheat.

4.1 Maize

Maize is among the most intensively studied cereal grains in terms of drought stress or drought combined with other stress treatments such as heat stress. Proteomic studies, in addition to other omics technologies, can help to improve marker-assisted breeding strategies and multi-omics studies aimed at developing drought-tolerant maize hybrids and cultivars (Maazou et al., 2016; Nepolean et al., 2018; Blein-Nicolas et al., 2020; Liu and Qin, 2021). One interesting example of a GWAS revealed a natural variant with improved drought tolerance due to a 366-bp insertion containing three MYB cis-elements, which conferred drought-inducible expression of a gene *ZmVPP1* encoding a vacuolar H^+ pyrophosphatase, which resulted in enhanced photosynthetic efficiency and root development (Wang et al., 2016b). This phenotype was validated by transgenic overexpression of *ZmVPP1*, which also resulted in enhanced tolerance to water-deficit stress.

In another example of a natural variant identified by GWAS, the promoter of a *ZmNAC111* TF was found to contain an insertion of a miniature inverted-repeat transposable element, which is associated with drought sensitivity in maize (Mao et al., 2015). As expected, ectopic overexpression of *ZmNAC111* in maize (and *Arabidopsis*) resulted in increased survival rates under water-deficit stress treatments from decreased transpiration rates, reduced stomatal conductance, and improved WUE. Although the mode of action of *ZmNAC111* was unclear, transcriptome profiling revealed the increased expression of several drought-inducible genes and ABA, ethylene, and abiotic stress pathways (Mao et al., 2015). Ectopic overexpression of the rice MYB gene (*OsMYB55*) resulted in improved drought and heat tolerance compared with wild-type control maize lines (Casaretto et al., 2016). Transcriptome analysis revealed the elevated expression of multiple abiotic stress-responsive TFs, including genes involved in hormone signaling, oxidative stress, and defense responses. Ectopic overexpression of a member of the maize bHLH TF family, *ZmPTF1*, which promotes root development and confers low-phosphate tolerance in maize, improved root system size, increased ABA content, and activated the expression of ABA synthesis and signaling components and the expression of other TFs with roles in drought stress responses (Li et al., 2019b). These events resulted in increased water-deficit stress tolerance at both the seedling and flower stages of development.

In addition to TFs, other signaling proteins can affect drought tolerance in maize. For example, GWAS revealed a negative correlation between the expression of *ZmPP2C-A10*, a member of the protein phosphatase 2C family involved in mediating ABA signaling and drought tolerance (Xiang et al., 2017). *ZmPP2C-A10* was shown to be a negative regulator of drought tolerance and its expression was linked with endoplasmic reticulum stress-induced signaling and expression, and a repression of ABA signaling and responses following water-deficit stress treatment. Map-based cloning was used to identify a QTL for lateral root number variation, which led to the identification of the *ZmLRT* gene, which when knocked out led to greater lateral root formation and greater drought tolerance compared to wild-type plants, whereas *ZmLRT*-overexpressing plants had fewer lateral roots (Zhang et al., 2022).

Transgenic strategies that reduce ethylene biosynthesis by silencing ACC synthases, which catalyze the rate-limiting step in ethylene biosynthesis, have been demonstrated to improve grain yield in maize under abiotic stress conditions (Habben et al., 2014). The ARGOS gene family are negative regulators of ethylene signal transduction. Thus, overexpression of *ZmARGO1* and *ZmARGO8* reduced plant sensitivity to ethylene and reduced expression of ethylene-responsive genes, leading to enhanced drought tolerance (Shi et al., 2015). Transgenic maize lines ectopically expressing *ZmARGO1* showed reduced sensitivity to ethylene and greater grain yield than non-transgenic controls under well-watered and drought stress conditions (Shi et al., 2015).

Given its central role in drought stress signaling and response, the ABA pathway has been a key target for improving crop performance under drought stress conditions (Du and Xiong, 2014). The ABA signaling pathway is controlled at several points by ubiquitination. Two members of the RING-H2 domain E3 ligase gene family, *ZmXerico1* and *ZmXerico2*, which are induced by drought stress, confer ABA hypersensitivity and improved WUE when ectopically overexpressed in maize (and *Arabidopsis*) (Brugière et al., 2017). Overexpression of *ZmXerico1* and *ZmXerico2* resulted in increased ABA content and decreased levels of ABA breakdown products, delayed senescence, reduced leaf rolling, and enhanced maize grain yield performance under water-deficit stress conditions.

Reducing the abundance of gibberellic acid (GA) through the ectopic overexpression of *AtG2ox1*, an *Arabidopsis* gene encoding a member of the GA 2-oxidase family, reduced endogenous GA content in maize leaves resulting in improved drought tolerance in addition to the expected reduction in plant stature (Chen et al., 2019). Although the mechanism of action for reducing GA content was unclear, transcriptomic analysis revealed that expression of TFs involved in drought and abiotic stress responses was enhanced leading to increased proline and soluble sugar content and ROS enzyme activities.

Ectopic overexpression of a gene encoding *ZmSDD1*, which encodes a serine proteinase, resulted in a 30% reduction in stomatal density in maize leaves (Liu et al., 2015). As a result, transpiration was dramatically reduced in these transgenic lines, which displayed higher WUE and photosynthetic rates, due in part to increased phosphoenolpyruvate carboxylase (PEPC) activity, CO_2 utilization rates, and lower CO_2 compensation points. Limiting transpiration rates have been suggested to be a useful breeding target to improve maize performance in drought-prone regions (Messina et al., 2015). In addition to reducing stomatal density, changes in the abundance of the cuticular wax composition of the epidermal layer of the plant leaf can result in improved drought tolerance (Xue et al., 2017). In maize, a mutation in the glossy 6 (*Zmgl6*) gene resulted in decreased epicuticular wax deposition, increased cuticular permeability, and reduced seedling water-deficit stress tolerance (Li et al., 2019a). These results showed the importance of epicuticular wax accumulation in drought tolerance and suggested that *ZmGL6* is likely involved in the intracellular trafficking of cuticular waxes.

Transgenic maize expressing a bacterial cold shock protein B (*CspB*) showed increased grain yield under drought conditions on an average of 6% compared with controls (Nemali et al., 2015). Although the exact mechanism of action of the *CspB* gene remains unclear, yield performance increases were associated with higher soil water content during the treatment phase, increased ear growth, decreased leaf area, leaf dry weight, and sap flow during silking, and increased kernel number and harvest index compared with controls. These parameters might be the result of a transient reduction in leaf growth resulting in a lower water use, higher WUE, and lower stress exposure during silking, which in turn would lead to an increase in ear partitioning and ear growth during silking. The larger ears would then be able to attract more assimilations during the kernel filling stage resulting in increased rates of kernel set and grain yield in addition to a greater harvest index.

Multiple miRNAs have been implicated in participating in the regulation of transcript abundance patterns of genes associated with drought tolerance in maize via multiple signaling pathways (Das and Mondal, 2021). Alterations in the rate of transcript turnover can affect drought tolerance in maize. For example, the recent genome sequencing of a prominent drought-resistant genotype, CIMBL55, revealed a 28-bp insertion in the 3′ UTR of *ZmRtn16* (Zm00001d047517), which encodes a reticulon-like protein. The presence of this insertion reduced the transcript stability for this gene, likely due to enhanced transcript turnover in the presence of RNA-binding motifs within the 28-bp insertion, leading to reduced drought tolerance (Tian et al., 2023).

In contrast, overexpression of *ZmRtn16* using a *ZmUbi1* constitutive promoter resulted in improved drought tolerance relative to wild-type plants.

4.2 Rice

Rice is also one of the most intensively studied cereal grains in terms of drought stress responses due to its global importance and has been the target of intensive efforts to improve its drought tolerance through transgenesis (Todaka et al., 2015; Khan et al., 2021; Kumar et al., 2022). As with maize, TFs have been effective target proteins to improve drought tolerance. For example, overexpression of *OsNAC6* TF enhanced drought resistance in rice by increasing root number and root diameter (Lee et al., 2017b). *OsNAC6* functions through increasing the expression of genes involved in membrane modifications and nicotianamine (NA) biosynthesis and accumulation, a Fe chelator, ROS scavengers, and sulfur metabolites, which in turn improved drought tolerance. The ectopic overexpression of the rice NAC gene (*OsNAC14*) resulted in improved drought tolerance at the vegetative stage compared with non-transgenic control lines (Shim et al., 2018). RNA-seq transcriptome analysis and chromatin immunoprecipitation (ChIP) analysis revealed that *OsNAC14* targeted DNA damage and defense responses resulting in improved drought tolerance. Another abiotic stress-responsive NAC family TF from rice, *SNAC3*, conferred drought (and heat and oxidative) stress tolerance in rice (Fang et al., 2015). *SNAC3* overexpressing plants showed enhanced expression of ROS enzymes, reduced levels of ROS (e.g. H_2O_2 and malondialdehyde [MDA]), and electrolyte leakage.

Overexpression of a dehydration-responsive element binding 2 (DREB2)-like protein called *OsDRAP1* conferred drought tolerance in rice (Huang et al., 2018). *OsDRAP1* expression improved water balance, redox homeostasis, and vascular development in rice plants under water-deficit stress through the activation of many downstream drought-tolerance genes. Overexpression of an *Arabidopsis AtWRKY57* TF conferred improved drought tolerance in rice (Jiang et al., 2016). In addition, transformed plants were also more tolerant to osmotic (25% (w/v) PEG6000) and salinity stress due to increased expression of several abiotic stress-responsive genes, increased proline accumulation, and higher ROS scavenging enzyme activities. In an interesting example, overexpression of *OsWOX13*, a homeobox TF, led to both improved drought tolerance and early flowering phenotypes, and might be involved in the regulation of genes involved in drought escape (Minh-Thu et al., 2018). Ectopic overexpression of a MYB family TF, *OsMYB6*, resulted in improved drought and salinity stress tolerance in rice (Tang et al., 2019). *OsMYB6* functions as a stress-responsive TF that increased the expression of abiotic stress-responsive genes, which increased proline accumulation and ROS scavenging enzymes (i.e. CAT, SOD) compared with control lines.

Ectopic overexpression of the novel, drought-inducible gene *OsAHK1*, which encodes an AT-hook motif and PPC domain, improved drought resistance (as imposed artificially by 20% (w/v) PEG) in rice during both seedling and panicle development phases (Zhou et al., 2016). *OsAHK1* directly regulates the control of many genes associated with stress responses, root development during drought stress to enhance drought avoidance, and those that participate in oxidative stress responses and regulate chlorophyll content in leaves. Overexpression of *OsERF71*, a member of the ethylene response factor (ERF) family of TFs, in rice shoots resulted in improved drought stress tolerance by modifying root structures, including a larger aerenchyma and radial root growth (Lee et al., 2017a). *OsERF71*-mediated root modification and drought tolerance were found to arise from a combination of various genes involved in cell wall associated and lignin biosynthesis, which are thought to limit transpiration from plant tissues under drought conditions. Overexpression of the *OsLG3* TF, which also belongs to the ERF family of TFs induced by water-deficit stress, improved the tolerance of rice plants to simulated drought conditions imposed by 20% (w/v) PEG solution or osmotic stress conditions imposed by 200 mM mannitol treatment (Xiong et al., 2018). The overexpressing plants showed less oxidative stress damage, likely the result of increased expression of ROS scavenging enzymes and improved drought stress tolerance. In contrast, some TFs can act as transcriptional repressors. Downregulation of the *OsMADS25* TF via RNAi gene silencing resulted in enhanced tolerance to water-deficit stress in both greenhouse- and field-grown conditions (Khong et al., 2015). *OsMADS25* appears to regulate a wide range of genes involved in biotic and abiotic resistance. Similarly, ectopic over- or underexpression of *OsERF109* significantly reduced and improved drought resistance, respectively, indicating that *OsERF109* functions as a negative regulator in rice by negatively regulating the expression of genes involved in ethylene biosynthesis (Yu et al., 2017b). Overexpression of *OsERF71* also enhanced drought and salinity tolerance in rice by regulating the expression of several ABA-responsive genes and proline biosynthesis genes under drought stress (Li et al., 2018).

Other unknown regulatory proteins in rice have also been discovered that play a role in drought tolerance. For example, *Oryza sativa* Stress tolerance and Grain Length (*OsSGL*), a novel DUF1645 domain-containing protein, conferred improved drought tolerance in rice when overexpressed, presumably through the formation of a more extensive root system, the enhanced accumulation of osmolytes such as proline, mannitol, and soluble sugars, and the elevated expression of other abiotic stress-adaptive genes (Cui et al., 2016).

The ABA signaling pathway also plays an important role in drought stress responses in rice. Overexpression of *OsABF1*, a bZIP TF, conferred drought tolerance in rice (Zhang et al., 2017). *OsABF1* also directly regulates the expression of other PP2C genes and bZIP TFs among many other targets as identified from RNA-seq and ChIP-seq data. Overexpression of one of these direct targets, COR413-TM1, a putative thylakoid membrane protein, also conferred improved drought tolerance when overexpressed (Zhang et al., 2017). Other members of the bZIP transcription family, including *OsbZIP23* and to a lesser extent *OsbZIP45*, have also been implicated in enhancing drought tolerance in rice when overexpressed (Park et al., 2015). *OsbZIP23* is targeted by the rice SUMO protease Overly Tolerant to Salt 1 (*OsOTS1*), for de-SUMOylation, which affects its stability. When *OsOTS1* expression is reduced by RNAi-mediated gene silencing, the abundance of OsbZIP23 is increased leading to increased drought tolerance gene expression and improved drought tolerance (Srivastava et al., 2017). In another bZIP TF example, the overexpression of *OsbZIP42*, which acts as a positive regulator of ABA signaling and drought stress responses, conferred improved drought tolerance in rice (Joo et al., 2019). Ectopic overexpression resulted in ABA hypersensitivity and the elevated expression of ABA-responsive genes such as *LEA3* and *Rab16*. Overexpression of *OsNF-YA7*, a member of the NF-YA TF family, conferred improved drought tolerance in rice (Lee et al., 2015). RNA-seq analysis revealed that *OsNF-YA7* likely mediates drought tolerance in an ABA-independent manner. Ectopic overexpression of *IPA1 (IDEAL PLANT ARCHITECTURE 1)/OsSPL14* TF resulted in improved tolerance in rice seedlings exposed to artificial drought stress treatment (25% (w/v) PEG4000) (Zhu et al., 2022). *IPA1* acts mainly as a positive regulator of the ABA pathway, activating the expression of OsNAC52, a positive regulator of the ABA biosynthetic pathways.

While TFs play a major role in the regulation of transcriptional activation or repression of target genes, few examples of posttranscriptional regulation are apparent in controlling drought tolerance. However, characterization of the rice GLYCINE-RICH PROTEIN 3 (*OsGRP3*) revealed that its overexpression conferred improved drought tolerance by reducing ROS accumulation under conditions of drought stress (Shim et al., 2021). RNA-immunoprecipitation studies suggest that *OsGRP3* associates with transcripts of ROS-related genes and accelerates their decay, which alleviates ROS accumulation and improves drought tolerance.

Auxin signaling also appears to be involved in drought tolerance as the overexpression of the *OsIAA6* gene in rice, a component of the auxin-signaling cascade, resulted in improved drought tolerance apparently through the regulation of auxin biosynthesis genes. Overexpression of *OsSOS2*, a serine-threonine protein kinase involved in the salt overly sensitive pathway, conferred drought (and salinity) tolerance in high-yielding rice at both the seedling and reproductive stages (Kumar et al., 2022).

Other signaling components, such as those involved in phospholipid signaling, have also shown beneficial effects when overexpressed in rice. For example, ectopic overexpression of the *OsPLDα1* gene, which encodes phospholipase Dα1, reduced productivity losses under drought stress conditions compared with non-transgenic control lines (Abreu et al., 2018). *OsPLDα1* plants showed delayed senescence and greater recovery of photosynthesis and lower expression of ROS scavenging enzymes following stress treatment. In addition, the calmodulin-like (CML) gene *OsCML4*, was shown to confer drought tolerance to rice through the expression of ROS scavenging enzymes such as superoxide dismutase (SOD) and catalase (CAT), and also by inducing the expression of abiotic stress-responsive genes in an ABA-independent manner (Yin et al., 2015). Another CML gene, Drought Stress Response-1 from rice (*OsDSR-1*), improved drought tolerance and increased ABA sensitivity (Yin et al., 2017b). *OsDSR-1* causes an increase in ROS scavenging enzymes and increased free proline and soluble sugar contents in the transgenic rice lines compared with untransformed control lines, changes that are expected to decrease oxidative stress damage during drought stress.

As in maize, limiting transpiration has been a useful target for improving drought tolerance in rice. Ectopic overexpression of the rice epidermal patterning factor *OsEPF1*, which encodes a signaling peptide that regulates stomatal density and spacing, resulted in significant reductions in stomatal density and correspondingly low stomatal conductance in rice leaves (Caine et al., 2019) and constitutive formation of root cortical aerenchyma (Mohammed et al., 2019). The low stomatal density lines were able to conserve ~60% more water than control lines, survive acute water-deficit stress better than control lines, and provide equivalent or improved yields, despite reduced rates of photosynthesis. Overexpression of *OsARS5*, which encodes an abscisic acid, stress, and ripening gene, improved the water status of rice under drought stress conditions through stomatal closure (Li et al., 2017). *OsARS5* is thought to play multiple roles leading to improved drought tolerance by regulating ABA biosynthesis, promoting stomatal closure, and acting as a chaperone-like protein to prevent protein inactivation under stress. Expression of a Maize PIF1 in rice has also been shown to improve drought tolerance by inducing stomatal closure (Gao et al., 2018b). *ZmPIF1* overexpression also resulted in enhanced grain yield through increasing tiller and panicle numbers in transgenic rice compared to untransformed controls.

LEA proteins play key roles in desiccation tolerance. Ectopic overexpression of an abiotic stress-inducible group I LEA protein from rice, *OsEm1*, resulted in enhanced ABA sensitivity and osmotic stress tolerance and increased survival at the vegetative stage of development (Yu et al., 2016). *OsEm1* overexpression led to the increased expression of other LEA genes suggesting that *OsEm1* acts as a positive regulator of drought tolerance.

HSPs are induced in response to many abiotic stresses and have been implicated as having protective roles against many such stresses (Timperio et al., 2008). The overexpression of a HSP90 family member, *OsHSP50.2*, provided direct demonstration of the role of HSPs in drought tolerance in rice

(Xiang et al., 2018). Ectopic overexpression of *OsHSP50.2* resulted in reduced water loss and enhanced the ability to tolerate drought and osmotic stress through several likely mechanisms, including increased ROS scavenging and osmotic adjustment from higher proline accumulation and reduced water loss.

Lastly, the overexpression of specific enzymes has been shown to confer improved drought tolerance in rice. For example, transgenic rice expressing maize C_4 photosynthesis PEPC showed improved drought tolerance (Qian et al., 2015). Overexpression of the galactinol synthase gene from *Arabidopsis* improved drought tolerance and increased grain yield in transgenic rice under field conditions (Selvaraj et al., 2017). The improved drought tolerance was associated with a higher relative galactinol content in the leaves along with higher photosynthetic rates, and faster recovery ability from drought stress. In contrast, mutation of the *OsPSL1* gene in rice, which encodes a cell wall-localized polygalacturonase (PG), displayed leaf rolling in response to high light intensity and low humidity, which enhanced the drought tolerance of the mutant relative to wild-type lines presumably through reductions in water loss through modifications in cell wall biosynthesis (Zhang et al., 2021).

4.3 Wheat

Improving the drought tolerance of wheat through transgenic approaches has been reviewed recently (Khan et al., 2019). As with maize and rice, TFs and regulons belonging to five families of TFs (e.g. AP2/ERF, bZIP, MYB/MYC, NAC, and WRKY) have functional roles in conferring improved drought tolerance in wheat (Gahlaut et al., 2016). Constitutive and predominantly root-expression of the NAC TF, *TaRNAC1*, increased drought tolerance under up to 10% (w/v) PEG 8000 stress treatment (Chen et al., 2018). *TaRNAC1* expression resulted in increased root length and significantly increased root biomass, presumably through an increase in active GA production in the roots. A drought-responsive NAC TF from wheat, *TaSNAC8-6A*, was discovered by GWAS of survival rates under severe drought stress at the seedling stage of development (Mao et al., 2020). Ectopic overexpression of *TaSNAC8-6A* in wheat resulted in improved drought tolerance through targeting of auxin- and drought-response signaling pathway components, which stimulated lateral root development and enhanced drought resistance. A different NAC TF, *TaNAC071-A*, was also identified in wheat through GWAS, and when overexpressed also improved drought tolerance (Mao et al., 2022). Drought tolerance was conferred through increased expression of drought-responsive genes and improved WUE. Ectopic overexpression of *TaWRKY1*, a drought- and ABA-induced member of the WRKY TF family, improved drought tolerance compared to control lines (Ding et al., 2016). Transgenic lines exhibited higher biomass accumulation, reduced rates of water loss due to higher stomatal closure rates, more osmolyte accumulation, and higher

ROS scavenging enzyme activities than control lines. Ectopic overexpression of *TaWRKY2* in transgenic wheat also improved drought tolerance with higher survival rates and lower rates of leaf water loss, and greater grain yield and above ground biomass than controls (Gao et al., 2018a). The transgenic lines also displayed higher proline and soluble sugar and chlorophyll content than control plants. Ectopic overexpression of an *Arabidopsis* WRKY30 (*AtWRKY30*) gene in transgenic wheat resulted in improved drought (and heat) tolerance compared to wild-type plants (El-Esawi et al., 2019). Multiple mechanisms contributed to the phenotypes, including expression of stress-responsive genes, higher proline and soluble sugar content, and higher ROS enzyme activities. Lastly, overexpression of an OVATE family protein (OFP) member, *TaOPF29a-A*, which are plant-specific TFs, resulted in improved drought tolerance compared with non-transgenic control lines (Wang et al., 2020).

Like other cereals, the ABA signaling pathway regulates drought stress responses in wheat. Ectopic overexpression of a wheat ABA receptor, *TaPYL4*, increased the plant's sensitivity to ABA and reduced the plant's lifetime water consumption compared to isogenic control lines through reduced stomatal aperture size and transpiration and increased instantaneous WUE (Mega et al., 2019).

As in maize, transgenic wheat expressing a codon-optimized, synthetic (se) version of the bacterial cold shock protein B (*CspB*) showed increased grain yield and 1000-grain weight compared with control genotypes under drought conditions in the field (Yu et al., 2017a). The *SeCspB*-expressing transgenic lines showed stress tolerance features including lower MDA content, lower rates of water loss, and higher chlorophyll and proline content relative to control lines.

Ectopic overexpression of a wheat orthologue of SHN1/WIN1, *TaSHN1*, a member of the AP2/ERF TF family, improved drought tolerance relative to control lines (Bi et al., 2018). *TaSHN1* acted as a transcriptional activator to increase the expression of enzymes involved in epicuticular wax biosynthesis, altered epicuticular wax composition, significantly reduced stomatal density compared with control lines, and increased survival rate following acute and severe water-deficit stress.

Transgenic wheat plants overexpressing a maize C_4 PEPC enzyme improved wheat drought stress tolerance compared with control lines (Qin et al., 2016). The improved drought tolerance was associated with multiple changes, including higher proline, soluble sugar, and soluble protein content in leaves, and higher WUE. The PEPC-expressing lines also showed larger root systems and increased photosynthetic capacity under water-deficit conditions.

As in rice, LEA proteins overexpression can improve drought tolerance in wheat. Ectopic overexpression of a barley *HVA1* gene, a group III LEA protein, improved drought and salinity tolerance in transgenic wheat (Habib et al., 2022). The HVA1 overexpressing wheat lines exhibited improved membrane stability and lower electrolyte leakage after exposure to these stresses.

5 Analysis of drought-adaptive proteins using genome editing

Recent advances in cereal genomics have dramatically improved genomic and transcriptomic sequence information availability for many cereals (Zenda et al., 2021). Such information provides an unprecedented opportunity for crop improvement using genome-editing approaches (Matres et al., 2021; Willmann, 2021). Various examples of genome editing for the major cereal crops, including but not limited to maize, rice, wheat, and barley, have targeted disease resistance, herbicide tolerance, nutritional and grain quality traits, and agronomic traits such as improved yield and drought tolerance (Matres et al., 2021).

Improved drought tolerance in maize was achieved by increasing the expression of the *ARGOS8* gene, which encodes a negative regulator of ethylene responses (Shi et al., 2017). Enhanced expression was achieved by inserting the maize *GOS2* promoter upstream of the *ARGOS8* gene (knock-in) or by replacing the weak endogenous promoter with the stronger *GOS2* promoter (promoter replacement). The genome-edited plants showed higher grain yields of five bushels per acre when drought stress occurred during flowering and no yield reductions under well-watered conditions (Shi et al., 2017). Targeted CRISPR/Cas9 mutagenesis of the rice zinc finger TF *DROUGHT AND SALT TOLERANCE* (*OsDST*) resulted in reduced stomatal densities, which improved WUE (Santosh Kumar et al., 2020). Similarly, targeted mutagenesis of the *SEMI-ROLLED LEAF1,2* (*OsSRL1* and *OsSRL2*) alleles in rice resulted in constitutive leaf rolling until maturity resulting in lower transpiration and stomatal conductance rates, thereby avoiding the effects of water-deficit stress and improving grain filling during severe water-deficit stress conditions (Liao et al., 2019). This mutant phenotype was also associated with the accumulation of a complex set of abiotic stress-responsive proteins.

Lack of overexpression of a rice epidermal patterning factor-like (EPFL) signaling peptide *OsEPFL9* by CRISPR/Cas9 or CRISPR/Cpf1 knock-out resulted in a significant reduction in stomatal density in rice (Yin et al., 2017a). Although the effects of this reduction were not documented, presumably this would have reduced water loss and transpiration resulting in more drought-tolerant rice plants.

CRISPR/Cas9-based mutagenesis of *SEMI-ROLLED LEAF1,2* (*SRL1* and *SRL2*) genes led to loss-of-function mutations resulting in a rolled-leaf phenotype, which showed greater drought tolerance than wild-type control lines in rice (Liao et al., 2019). A proteomic analysis revealed that the expression of many drought-responsive proteins was altered in the rolled-leaf mutant lines.

CRISPR/Cas9-based genome editing has also been applied to wheat in a proof-of-concept study using targeted genome editing of *TaDREB2* and

TaERF3, two genes involved in drought stress tolerance phenotypes (Kim et al., 2018). Transient expression of guide RNAs and Cas9 protein were delivered to protoplasts and editing was confirmed by restriction enzyme digestion, T7 endonuclease assays, and DNA sequencing.

6 Conclusion and future trends

The current global climate crisis is impacting crop productivity currently and will likely continue to negatively impact yields of cereal grains into the future. Among the crops that humankind relies upon for subsistence, the cereal grains provide 30-80% of the human diet. The climate resilience traits needed to sustain productivity in the future are complex and involve avoidance, escape, tolerance, resistance, and recovery strategies. Fortunately, modern omics technologies have provided the ability to characterize in a comprehensive way each of the components of biological systems and to begin to understand the mechanistic details of this complexity through QTL and GWAS and detailed loss-of-function and gain-of-function studies. Transformation technologies for cereal crops have improved significantly in recent decades allowing a much broader range of gene functional testing to be conducted. Major targets have included many different families of TFs, components of stress-adaptive and phytohormone signal transduction pathways, regulators and enzymes that modify mRNA and protein stability and turnover, anatomical traits such as stomatal density, cuticular wax formation, and leaf curling, and proteins with protective functions for protein and membrane stability such as HSPs and LEAs. The functions of many proteins with yet unknown roles in drought tolerance remain to be explored or understood. Such proteins hold the promise of novel strategies to further improve drought tolerance and to create more climate-resilient cereal crops. Furthermore, transformation technologies have also allowed for expanded application of transgenic strategies and CRISPR/Cas genome-editing methods to probe gene function and make changes to the genome without the complication of transgenes, which can present a regulatory burden to the introduction of drought tolerance traits. As DNA-free genome-editing methods become optimized and more widespread, one can expect a rapid expansion of these new tools to a wide range of novel strategies aimed at improving drought avoidance, resistance, and tolerance.

7 Where to look for further information

- For a recent general review about the mechanisms of abiotic stress (drought, salinity, heat, and cold) sensing and signaling, and transcriptional, translational, post-translation, and epigenetic regulation, please see the recent review by Zhang et al., 2022.

- The recent review of drought-stress responses in cereals by Havrlentová et al., 2021 provides an informative overview of how cereal crops cope with drought stress and approaches to improve drought tolerance.
- For a comprehensive review of the diverse set of environmental stresses (drought, salinity, and heat) and their impacts on cereal crop production, transcription factors, stress responses, and transgenic crops developed for these abiotic stresses, please see the recent review by Jeyasri et al., 2021.
- The recent review by Thanmalagan et al., 2022 about the complex regulatory, metabolic, and transcriptional networks involved in perceiving and responding to drought stress in cereals showcases the role of integrative bioinformatics tools and databases to help understand the complexity of multi-omics datasets and leverage them to design well-informed strategies to design more drought tolerant cereal crops.
- The recent monograph entitled 'Omics approaches to manage abiotic stress in cereals' provides an excellent and comprehensive resource for diverse aspects of abiotic stress responses in many different cereal crops (Roychoudhury et al., 2022).
- The Gramene website (https://www.gramene.org/) offers a comprehensive set of data resources for many cereals genomes and pan genomes, metabolic and regulatory pathways (Plant Reactome), as well as data mining tools.
- More information about the negative impacts of drought stress on cereal crop production can be found at the United States Department of Agriculture (USDA) National Drought Mitigation Center website: https://agindrought.unl.edu/RowCrops.aspx.
- Resources for monitoring, managing, and planning for agricultural drought conditions for cereal crops can be found at the USDA National Agricultural Statistics Service website: https://www.nass.usda.gov/Publications/State_Crop_Progress_and_Condition/index.php.

8 Acknowledgements

The authors gratefully acknowledge support from the Nevada Agricultural Experiment Station under projects NEV-00386 and NEV-00391. The authors also thank Lisa M. Petrusa for critical reading of the manuscript.

9 References

Abid, M., Ali, S., Qi, L. K., Zahoor, R., Tian, Z., Jiang, D., Snider, J. L. and Dai, T. (2018). Physiological and biochemical changes during drought and recovery periods at tillering and jointing stages in wheat (*Triticum aestivum* L.). *Scientific Reports* 8(1), 4615.

Abreu, F. R. M., Dedicova, B., Vianello, R. P., Lanna, A. C., de Oliveira, J. A. V., Vieira, A. F., Morais, O. P., Mendonça, J. A. and Brondani, C. (2018). Overexpression of a phospholipase (*OsPLDα1*) for drought tolerance in upland rice (*Oryza sativa* L.). *Protoplasma* 255(6), 1751–1761.

Ahmad, P., Abdel Latef, A. A. H., Rasool, S., Akram, N. A., Ashraf, M. and Gucel, S. (2016). Role of proteomics in crop stress tolerance. *Frontiers in Plant Science* 7, 1336.

Albano, C. M., Abatzoglou, J. T., McEvoy, D. J., Huntington, J. L., Morton, C. G., Dettinger, M. D. and Ott, T. J. (2022). A multidataset assessment of climatic drivers and uncertainties of recent trends in evaporative demand across the continental United States. *Journal of Hydrometeorology* 23(4), 505–519.

Alexandratos, N. and Bruinsma, J. (2012). *World Agriculture Towards* (2012 revision). Rome: Food and Agriculture Organization, p. 2030/2050.

Alpert, P. (2006). Constraints of tolerance: why are desiccation-tolerant organisms so small or rare? *Journal of Experimental Biology* 209(9), 1575–1584.

Araus, J. L., Slafer, G. A., Reynolds, M. P. and Royo, C. (2002). Plant breeding and drought in C_3 cereals: what should we breed for? *Annals of Botany* 89(7) (Spec No), 925–940.

Awika, J. M. (2011). Major cereal grains production and use around the world. In: *Advances in Cereal Science: Implications to Food Processing and Health Promotion*. ACS Publications, pp. 1–13.

Barkla, B. J., Castellanos-Cervantes, T., Diaz de Leon, J. L., Matros, A., Mock, H. P., Perez-Alfocea, F., Salekdeh, G. H., Witzel, K. and Zörb, C. (2013a). Elucidation of salt stress defense and tolerance mechanisms of crop plants using proteomics–current achievements and perspectives. *Proteomics* 13(12–13), 1885–1900.

Barkla, B. J., Vera-Estrella, R. and Pantoja, O. (2013b). Progress and challenges for abiotic stress proteomics of crop plants. *Proteomics* 13(12–13), 1801–1815.

Bartlett, M. K., Scoffoni, C. and Sack, L. (2012). The determinants of leaf turgor loss point and prediction of drought tolerance of species and biomes: A global meta-analysis. *Ecology Letters* 15(5), 393–405.

Basu, S., Ramegowda, V., Kumar, A. and Pereira, A. (2016). Plant adaptation to drought stress. *F1000Research* 5.

Bi, H., Shi, J., Kovalchuk, N., Luang, S., Bazanova, N., Chirkova, L., Zhang, D., Shavrukov, Y., Stepanenko, A., Tricker, P., Langridge, P., Hrmova, M., Lopato, S. and Borisjuk, N. (2018). Overexpression of the *TaSHN1* transcription factor in bread wheat leads to leaf surface modifications, improved drought tolerance, and no yield penalty under controlled growth conditions. *Plant, Cell and Environment* 41(11), 2549–2566.

Blein-Nicolas, M., Negro, S. S., Balliau, T., Welcker, C., Cabrera-Bosquet, L., Nicolas, S. D., Charcosset, A. and Zivy, M. (2020). A systems genetics approach reveals environment-dependent associations between SNPs, protein coexpression, and drought-related traits in maize. *Genome Research* 30(11), 1593–1604.

Blum, A. (2017). Osmotic adjustment is a prime drought stress adaptive engine in support of plant production. *Plant, Cell and Environment* 40(1), 4–10.

Brugière, N., Zhang, W., Xu, Q., Scolaro, E. J., Lu, C., Kahsay, R. Y., Kise, R., Trecker, L., Williams, R. W., Hakimi, S., Niu, X., Lafitte, R. and Habben, J. E. (2017). Overexpression of RING domain E3 ligase ZmXerico1 confers drought tolerance through regulation of ABA homeostasis. *Plant Physiology* 175(3), 1350–1369.

Caine, R. S., Yin, X., Sloan, J., Harrison, E. L., Mohammed, U., Fulton, T., Biswal, A. K., Dionora, J., Chater, C. C., Coe, R. A., Bandyopadhyay, A., Murchie, E. H., Swarup, R., Quick, W. P. and Gray, J. E. (2019). Rice with reduced stomatal density conserves water and has improved drought tolerance under future climate conditions. *New Phytologist* 221(1), 371–384.

Carvalho, H. D. R., Heilman, J. L., McInnes, K. J., Rooney, W. L. and Lewis, K. L. (2020). Epicuticular wax and its effect on canopy temperature and water use of Sorghum. *Agricultural and Forest Meteorology* 284, 107893.

Casaretto, J. A., El-Kereamy, A., Zeng, B., Stiegelmeyer, S. M., Chen, X., Bi, Y. M. and Rothstein, S. J. (2016). Expression of OsMYB55 in maize activates stress-responsive genes and enhances heat and drought tolerance. *BMC Genomics* 17, 312.

Challinor, A. J., Watson, J., Lobell, D. B., Howden, S. M., Smith, D. R. and Chhetri, N. (2014). A meta-analysis of crop yield under climate change and adaptation. *Nature Climate Change* 4(4), 287–291.

Chen, D., Chai, S., McIntyre, C. L. and Xue, G. P. (2018). Overexpression of a predominantly root-expressed NAC transcription factor in wheat roots enhances root length, biomass and drought tolerance. *Plant Cell Reports* 37(2), 225–237.

Chen, Z., Liu, Y., Yin, Y., Liu, Q., Li, N., Li, X., He, W., Hao, D., Liu, X. and Guo, C. (2019). Expression of AtGA2ox1 enhances drought tolerance in maize. *Plant Growth Regulation* 89(2), 203–215.

Choudhury, F. K., Rivero, R. M., Blumwald, E. and Mittler, R. (2017). Reactive oxygen species, abiotic stress and stress combination. *The Plant Journal: for Cell and Molecular Biology* 90(5), 856–867.

Christin, P. A., Osborne, C. P., Chatelet, D. S., Columbus, J. T., Besnard, G., Hodkinson, T. R., Garrison, L. M., Vorontsova, M. S. and Edwards, E. J. (2013). Anatomical enablers and the evolution of C4 photosynthesis in grasses. *Proceedings of the National Academy of Sciences of the United States of America* 110(4), 1381–1386.

Clarke, J. M., DePauw, R. M. and Townley-Smith, T. F. (1992). Evaluation of methods for quantification of drought tolerance in wheat. *Crop Science* 32(3), 723–728.

Cruz de Carvalho, M. H. (2008). Drought stress and reactive oxygen species: production, scavenging and signaling. *Plant Signaling and Behavior* 3(3), 156–165.

Cui, Y., Wang, M., Zhou, H., Li, M., Huang, L., Yin, X., Zhao, G., Lin, F., Xia, X. and Xu, G. (2016). OsSGL, a novel DUF1645 domain-containing protein, confers enhanced drought tolerance in transgenic rice and Arabidopsis. *Frontiers in Plant Science* 7, 2001.

da Silva, A. C., de Freitas Lima, M., Eloy, N. B., Thiebaut, F., Montessoro, P., Hemerly, A. S. and Ferreira, P. C. G. (2019). The Yin and Yang in plant breeding: The trade-off between plant growth yield and tolerance to stresses. *Biotechnology Research and Innovation* 3, 73–79.

Das, R. and Mondal, S. K. (2021). Plant miRNAs: Biogenesis and its functional validation to combat drought stress with special focus on maize. *Plant Gene* 27, 100294.

Ding, W., Fang, W., Shi, S., Zhao, Y., Li, X. and Xiao, K. (2016). Wheat WRKY type transcription factor gene *TaWRKY1* is essential in mediating drought tolerance associated with an ABA-dependent pathway. *Plant Molecular Biology Reporter* 34(6), 1111–1126.

Dorrani-Nejad, M., Kazemipour, A., Maghsoudi-Moud, A. A. and Abdolshahi, R. (2022). Wheat breeding for early heading: Does it improve grain yield under drought stress and well-watered conditions? *Environmental and Experimental Botany* 200, 104902.

Du, H. and Xiong, L. (2014). Improvement of stress tolerance in crops by genetic manipulation of ABA metabolism, signaling, and regulation. In: *Abscisic Acid*. Dordrecht: Springer, pp. 447–465.

El-Esawi, M. A., Al-Ghamdi, A. A., Ali, H. M. and Ahmad, M. (2019). Overexpression of *AtWRKY30* transcription factor enhances heat and drought stress tolerance in wheat (*Triticum aestivum* L.). *Genes* 10(2), 163.

Fang, Y., Liao, K., Du, H., Xu, Y., Song, H., Li, X. and Xiong, L. (2015). A stress-responsive NAC transcription factor SNAC3 confers heat and drought tolerance through modulation of reactive oxygen species in rice. *Journal of Experimental Botany* 66(21), 6803–6817.

Fang, Y. and Xiong, L. (2015). General mechanisms of drought response and their application in drought resistance improvement in plants. *Cellular and Molecular Life Sciences* 72(4), 673–689.

FAO (2021). *Cereal Supply and Demand Brief*. Food and Agriculture Organization of the United Nations.

Firezer, G., Firew, M., Taye, T., Temesgen, M. and Kassahun, B. (2020). Phenotyping sorghum [*Sorghum bicolor* (L.) Moench] for drought tolerance with special emphasis to root angle. *African Journal of Agricultural Research* 16(8), 1213–1222.

Fry, E. L., Evans, A. L., Sturrock, C. J., Bullock, J. M. and Bardgett, R. D. (2018). Root architecture governs plasticity in response to drought. *Plant and Soil* 433(1), 189–200.

Gaff, D. F., Blomstedt, C. K., Neale, A. D., Le, T. N., Hamill, J. D. and Ghasempour, H. R. (2009). *Sporobolus stapfianus*, a model desiccation-tolerant grass. *Functional Plant Biology* 36(7), 589–599.

Gaff, D. F. and McGregor, G. R. (1979). The effect of dehydration and rehydration on the nitrogen content of various fractions from resurrection plants. *Biologia Plantarum* 21(2), 92–99.

Gahlaut, V., Jaiswal, V., Kumar, A. and Gupta, P. K. (2016). Transcription factors involved in drought tolerance and their possible role in developing drought tolerant cultivars with emphasis on wheat (*Triticum aestivum* L.). *TAG. Theoretical and Applied Genetics. Theoretische und Angewandte Genetik* 129(11), 2019–2042.

Gao, H., Wang, Y., Xu, P. and Zhang, Z. (2018a). Overexpression of a WRKY transcription factor *TaWRKY2* enhances drought stress tolerance in transgenic wheat. *Frontiers in Plant Science* 9, 997.

Gao, Y., Wu, M., Zhang, M., Jiang, W., Ren, X., Liang, E., Zhang, D., Zhang, C., Xiao, N., Li, Y., Dai, Y. and Chen, J. (2018b). A maize phytochrome-interacting factors protein ZmPIF1 enhances drought tolerance by inducing stomatal closure and improves grain yield in *Oryza sativa*. *Plant Biotechnology Journal* 16(7), 1375–1387.

Gautam, A., Pandey, P. and Pandey, A. K. (2020). Proteomics in relation to abiotic stress tolerance in plants. In: *Plant Life Under Changing Environment*. Elsevier, pp. 513–541.

Gebresamuel, G., Abrha, H., Hagos, H., Elias, E. and Haile, M. (2022). Empirical modeling of the impact of climate change on altitudinal shift of major cereal crops in South Tigray, Northern Ethiopia. *Journal of Crop Improvement* 36(2), 169–192.

Ghatak, A., Chaturvedi, P. and Weckwerth, W. (2017). Cereal crop proteomics: systemic analysis of crop drought stress responses towards marker-assisted selection breeding. *Frontiers in Plant Science* 8, 757.

Gupta, A., Rico-Medina, A. and Caño-Delgado, A. I. (2020). The physiology of plant responses to drought. *Science* 368(6488), 266–269.

Habben, J. E., Bao, X., Bate, N. J., DeBruin, J. L., Dolan, D., Hasegawa, D., Helentjaris, T. G., Lafitte, R. H., Lovan, N., Mo, H., Reimann, K. and Schussler, J. R. (2014). Transgenic alteration of ethylene biosynthesis increases grain yield in maize under field drought-stress conditions. *Plant Biotechnology Journal* 12(6), 685–693.

Habib, I., Shahzad, K., Rauf, M., Ahmad, M., Alsamadany, H., Fahad, S. and Saeed, N. A. (2022). Dehydrin responsive HVA1 driven inducible gene expression enhanced salt and drought tolerance in wheat. *Plant Physiology and Biochemistry* 180, 124–133.

Havrlentová, M., Kraic, J., Gregusová, V. and Kovácsová, B. (2021). Drought stress in cereals–A review. Agriculture (Pol'nohospodárstvo). *Agriculture* 67(2), 47.

Hever, J. (2016). Plant-based diets: A physician's guide. *The Permanente Journal* 20(3), 15–82.

Hossain, Z., Nouri, M. Z. and Komatsu, S. (2012). Plant cell organelle proteomics in response to abiotic stress. *Journal of Proteome Research* 11(1), 37–48.

Huang, J., Yu, H., G., Wang, G. and Guo, R. (2015). Accelerated dryland expansion under climate change. *Nature Climate Change* 6, 166–171.

Huang, L., Wang, Y., Wang, W., Zhao, X., Qin, Q., Sun, F., Hu, F., Zhao, Y., Li, Z., Fu, B. and Li, Z. (2018). Characterization of transcription factor gene *OsDRAP1* conferring drought tolerance in rice. *Frontiers in Plant Science* 9, 94.

Irmak, S., Sandhu, R. and Kukal, M. S. (2022). Multi-model projections of trade-offs between irrigated and rainfed maize yields under changing climate and future emission scenarios. *Agricultural Water Management* 261, 107344.

Jeyasri, R., Muthuramalingam, P., Satish, L., Pandian, S. K., Chen, J. T., Ahmar, S., Wang, X., Mora-Poblete, F. and Ramesh, M. (2021). An overview of abiotic stress in cereal crops: negative impacts, regulation, biotechnology and integrated omics. *Plants* 10(7), 1472.

Jiang, Y., Qiu, Y., Hu, Y. and Yu, D. (2016). Heterologous expression of *AtWRKY57* confers drought tolerance in Oryza sativa. *Frontiers in Plant Science* 7, 145.

Joo, J., Lee, Y. H. and Song, S. I. (2019). *OsbZIP42* is a positive regulator of ABA signaling and confers drought tolerance to rice. *Planta* 249(5), 1521–1533.

Kahleova, H., Levin, S. and Barnard, N. (2017). Cardio-metabolic benefits of plant-based diets. *Nutrients* 9(8), 848.

Khan, M. I. R., Palakolanu, S. R., Chopra, P., Rajurkar, A. B., Gupta, R., Iqbal, N. and Maheshwari, C. (2021). Improving drought tolerance in rice: ensuring food security through multi-dimensional approaches. *Physiologia Plantarum* 172(2), 645–668.

Khan, S., Anwar, S., Yu, S., Sun, M., Yang, Z. and Gao, Z. Q. (2019). Development of drought-tolerant transgenic wheat: Achievements and limitations. *International Journal of Molecular Sciences* 20(13), 3350.

Khong, G. N., Pati, P. K., Richaud, F., Parizot, B., Bidzinski, P., Mai, C. D., Bès, M., Bourrié, I., Meynard, D., Beeckman, T., Selvaraj, M. G., Manabu, I., Genga, A. M., Brugidou, C., Nang Do, V., Guiderdoni, E., Morel, J. B. and Gantet, P. (2015). OsMADS26 negatively regulates resistance to pathogens and drought tolerance in rice. *Plant Physiology* 169(4), 2935–2949.

Kim, D., Alptekin, B. and Budak, H. (2018). CRISPR/Cas9 genome editing in wheat. *Functional and Integrative Genomics* 18(1), 31–41.

Kiran, B. O., Karabhantanal, S. S., Patil, S. B., Ashwathama, V. H., Jolli, R. B., Tonapi, V. A. (2022). Phenotyping sorghum for drought-adaptive physiological and root architectural traits under water-limited environments. *Cereal Research Communications* 50(4), 885–893.

Kosová, K., Vítámvás, P., Prášil, I. T. and Renaut, J. (2011). Plant proteome changes under abiotic stress–contribution of proteomics studies to understanding plant stress response. *Journal of Proteomics* 74(8), 1301–1322.

Kumar, A., Sengar, R. S., Pathak, R. K. and Singh, A. K. (2023). Integrated approaches to develop drought-tolerant rice: Demand of era for global food security. *Journal of Plant Growth Regulation* 42(1), 96–120.

Kumar, G., Basu, S., Singla-Pareek, S. L. and Pareek, A. (2022). Unraveling the contribution of *OsSOS2* in conferring salinity and drought tolerance in a high-yielding rice. *Physiologia Plantarum* 174(1), e13638.

Kumar, P., Sahu, N. C., Kumar, S. and Ansari, M. A. (2021). Impact of climate change on cereal production: Evidence from lower-middle-income countries. *Environmental Science and Pollution Research International* 28(37), 51597–51611.

Kwon, S. J., Choi, E. Y., Choi, Y. J., Ahn, J. H. and Park, O. K. (2006). Proteomics studies of post-translational modifications in plants. *Journal of Experimental Botany* 57(7), 1547–1551.

Lawson, T. and Vialet-Chabrand, S. (2019). Speedy stomata, photosynthesis and plant water use efficiency. *New Phytologist* 221(1), 93–98.

Lee, D. K., Kim, H. I., Jang, G., Chung, P. J., Jeong, J. S., Kim, Y. S., Bang, S. W., Jung, H., Do Choi, Y. D. and Kim, J. K. (2015). The NF-YA transcription factor *OsNF-YA7* confers drought stress tolerance of rice in an abscisic acid independent manner. *Plant Science* 241, 199–210.

Lee, D. K., Yoon, S., Kim, Y. S. and Kim, J. K. (2017a). Rice *OsERF71*-mediated root modification affects shoot drought tolerance. *Plant Signaling and Behavior* 12(1), e1268311.

Lee, D. K., Chung, P. J., Jeong, J. S., Jang, G., Bang, S. W., Jung, H., Kim, Y. S., Ha, S. H., Choi, Y. D. and Kim, J. K. (2017b). The rice *OsNAC6* transcription factor orchestrates multiple molecular mechanisms involving root structural adaptions and nicotianamine biosynthesis for drought tolerance. *Plant Biotechnology Journal* 15(6), 754–764.

Leng, G. and Hall, J. (2019). Crop yield sensitivity of global major agricultural countries to droughts and the projected changes in the future. *Science of the Total Environment* 654, 811–821.

Lesk, C., Rowhani, P. and Ramankutty, N. (2016). Influence of extreme weather disasters on global crop production. *Nature* 529(7584), 84–87.

Li, J., Guo, X., Zhang, M., Wang, X., Zhao, Y., Yin, Z., Zhang, Z., Wang, Y., Xiong, H., Zhang, H., Todorovska, E. and Li, Z. (2018). OsERF71 confers drought tolerance via modulating ABA signaling and proline biosynthesis. *Plant Science* 270, 131–139.

Li, J., Li, Y., Yin, Z., Jiang, J., Zhang, M., Guo, X., Ye, Z., Zhao, Y., Xiong, H., Zhang, Z., Shao, Y., Jiang, C., Zhang, H., An, G., Paek, N. C., Ali, J. and Li, Z. (2017). *OsASR5* enhances drought tolerance through a stomatal closure pathway associated with ABA and H_2O_2 signalling in rice. *Plant Biotechnology Journal* 15(2), 183–196.

Li, L., Du, Y., He, C., Dietrich, C. R., Li, J., Ma, X., Wang, R., Liu, Q., Liu, S., Wang, G., Schnable, P. S. and Zheng, J. (2019a). Maize glossy6 is involved in cuticular wax deposition and drought tolerance. *Journal of Experimental Botany* 70(12), 3089–3099.

Li, Z., Liu, C., Zhang, Y., Wang, B., Ran, Q. and Zhang, J. (2019b). The bHLH family member ZmPTF1 regulates drought tolerance in maize by promoting root development and abscisic acid synthesis. *Journal of Experimental Botany* 70(5471)_5486, 5471–5486.

Liao, S., Qin, X., Luo, L., Han, Y., Wang, X., Usman, B., Nawaz, G., Zhao, N., Liu, Y. and Li, R. (2019). CRISPR/Cas9-induced mutagenesis of *semi-rolled leaf1, 2* confers curled leaf phenotype and drought tolerance by influencing protein expression patterns and ROS scavenging in rice (*Oryza sativa* L.). *Agronomy* 9(11), 728.

Liu, H. and Osborne, C. P. (2015). Water relations traits of C_4 grasses depend on phylogenetic lineage, photosynthetic pathway, and habitat water availability. *Journal of Experimental Botany* 66(3), 761–773.

Liu, S. and Qin, F. (2021). Genetic dissection of maize drought tolerance for trait improvement. *Molecular Breeding* 41(2), 8.

Liu, Y., Qin, L., Han, L., Xiang, Y. and Zhao, D. (2015). Overexpression of maize SDD1 (ZmSDD1) improves drought resistance in *Zea mays* L. by reducing stomatal density. *Plant Cell, Tissue and Organ Culture* 122(1), 147–159.

Maazou, A.-R. S., Tu, J., Qiu, J. and Liu, Z. (2016). Breeding for drought tolerance in maize (*Zea mays* L.). *American Journal of Plant Sciences* 7, 1858.

Mao, H., Li, S., Chen, B., Jian, C., Mei, F., Zhang, Y., Li, F., Chen, N., Li, T., Du, L., Ding, L., Wang, Z., Cheng, X., Wang, X. and Kang, Z. (2022). Variation in cis-regulation of a NAC transcription factor contributes to drought tolerance in wheat. *Molecular Plant* 15(2), 276–292.

Mao, H., Li, S., Wang, Z., Cheng, X., Li, F., Mei, F., Chen, N. and Kang, Z. (2020). Regulatory changes in *TaSNAC8-6A* are associated with drought tolerance in wheat seedlings. *Plant Biotechnology Journal* 18(4), 1078–1092.

Mao, H., Wang, H., Liu, S., Li, Z., Yang, X., Yan, J., Li, J., Tran, L. S. and Qin, F. (2015). A transposable element in a NAC gene is associated with drought tolerance in maize seedlings. *Nature Communications* 6, 8326.

Martignago, D., Rico-Medina, A., Blasco-Escámez, D., Fontanet-Manzaneque, J. B. and Caño-Delgado, A. I. (2019). Drought resistance by engineering plant tissue-specific responses. *Frontiers in Plant Science* 10, 1676.

Matres, J. M., Hilscher, J., Datta, A., Armario-Nájera, V., Baysal, C., He, W., Huang, X., Zhu, C., Valizadeh-Kamran, R., Trijatmiko, K. R., Capell, T., Christou, P., Stoger, E. and Slamet-Loedin, I. H. (2021). Genome editing in cereal crops: an overview. *Transgenic Research* 30(4), 461–498.

McAusland, L., Vialet-Chabrand, S., Davey, P., Baker, N. R., Brendel, O. and Lawson, T. (2016). Effects of kinetics of light-induced stomatal responses on photosynthesis and water-use efficiency. *New Phytologist* 211(4), 1209–1220.

Mega, R., Abe, F., Kim, J. S., Tsuboi, Y., Tanaka, K., Kobayashi, H., Sakata, Y., Hanada, K., Tsujimoto, H., Kikuchi, J., Cutler, S. R. and Okamoto, M. (2019). Tuning water-use efficiency and drought tolerance in wheat using abscisic acid receptors. *Nature Plants* 5(2), 153–159.

Messina, C. D., Sinclair, T. R., Hammer, G. L., Curan, D., Thompson, J., Oler, Z., Gho, C. and Cooper, M. (2015). Limited-transpiration trait may increase maize drought tolerance in the US Corn Belt. *Agronomy Journal* 107(6), 1978–1986.

Miller, G., Suzuki, N., Ciftci-Yilmaz, S. and Mittler, R. (2010). Reactive oxygen species homeostasis and signalling during drought and salinity stresses. *Plant, Cell and Environment* 33(4), 453–467.

Minh-Thu, P. T., Kim, J. S., Chae, S., Jun, K. M., Lee, G. S., Kim, D. E., Cheong, J. J., Song, S. I., Nahm, B. H. and Kim, Y. K. (2018). A WUSCHEL homeobox transcription factor, OsWOX13, enhances drought tolerance and triggers early flowering in rice. *Molecules and Cells* 41(8), 781–798.

Mitra, J. (2001). Genetics and genetic improvement of drought resistance in crop plants. *Current Science*, 758–763.

Mohammed, U., Caine, R. S., Atkinson, J. A., Harrison, E. L., Wells, D., Chater, C. C., Gray, J. E., Swarup, R. and Murchie, E. H. (2019). Rice plants overexpressing OsEPF1 show reduced stomatal density and increased root cortical aerenchyma formation. *Scientific Reports* 9(1), 5584.

Moroney, J. V., Rai, A. K., Weerasooriya, H., Kasili, R. and Machingura, M. (2022). Improving proteins to optimise photosynthesis. In: *Understanding and improving crop photosynthesis*, Sharwood, R. (ed.). Cambridge, UK: Burleigh Dodds Science Publishing.

Naumann, G., Alfieri, L., Wyser, K., Mentaschi, L., Betts, R. A., Carrao, H., Spinoni, J., Vogt, J. and Feyen, L. (2018). Global changes in drought conditions under different levels of warming. *Geophysical Research Letters* 45(7), 3285–3296.

Ndlovu, E., van Staden, J. and Maphosa, M. (2021). Morpho-physiological effects of moisture, heat and combined stresses on *Sorghum bicolor* [Moench (L.)] and its acclimation mechanisms. *Plant Stress* 2, 100018.

Nemali, K. S., Bonin, C., Dohleman, F. G., Stephens, M., Reeves, W. R., Nelson, D. E., Castiglioni, P., Whitsel, J. E., Sammons, B., Silady, R. A., Anstrom, D., Sharp, R. E., Patharkar, O. R., Clay, D., Coffin, M., Nemeth, M. A., Leibman, M. E., Luethy, M. and Lawson, M. (2015). Physiological responses related to increased grain yield under drought in the first biotechnology-derived drought-tolerant maize. *Plant, Cell and Environment* 38(9), 1866–1880.

Nepolean, T., Kaul, J., Mukri, G. and Mittal, S. (2018). Genomics-enabled next-generation breeding approaches for developing system-specific drought tolerant hybrids in maize. *Frontiers in Plant Science* 9, 361.

Ngara, R. and Ndimba, B. K. (2014). Understanding the complex nature of salinity and drought-stress response in cereals using proteomics technologies. *Proteomics* 14(4–5), 611–621.

Pardo, J. and VanBuren, R. (2021). Evolutionary innovations driving abiotic stress tolerance in C4 grasses and cereals. *The Plant Cell* 33(11), 3391–3401.

Park, S.-H., Jeong, J. S., Lee, K. H., Kim, Y. S., Do Choi, Y. and Kim, J.-K. (2015). *OsbZIP23* and *OsbZIP45*, members of the rice basic leucine zipper transcription factor family, are involved in drought tolerance. *Plant Biotechnology Reports* 9(2), 89–96.

Peters, C. J., Picardy, J., Darrouzet-Nardi, A. F., Wilkins, J. L., Griffin, T. S., Fick, G. W., Kapuscinski, A. R. and Méndez, E. (2016). Carrying capacity of US agricultural land: ten diet scenarios. *Elementa: Science of the Anthropocene* 4, 000116.

Praveen, B. and Sharma, P. (2019). A review of literature on climate change and its impacts on agriculture productivity. *Journal of Public Affairs* 19(4), e1960.

Priya, M., Dhanker, O. P., Siddique, K. H. M., HanumanthaRao, B., Nair, R. M., Pandey, S., Singh, S., Varshney, R. K., Prasad, P. V. V. and Nayyar, H. (2019). Drought and heat stress-related proteins: an update about their functional relevance in imparting stress tolerance in agricultural crops. *TAG. Theoretical and Applied Genetics. Theoretische und Angewandte Genetik* 132(6), 1607–1638.

Qian, B., Li, X., Liu, X., Chen, P., Ren, C. and Dai, C. (2015). Enhanced drought tolerance in transgenic rice over-expressing of maize C_4 phosphoenolpyruvate carboxylase gene via NO and Ca^{2+}. *Journal of Plant Physiology* 175, 9–20.

Qin, N., Xu, W., Hu, L., Li, Y., Wang, H., Qi, X., Fang, Y. and Hua, X. (2016). Drought tolerance and proteomics studies of transgenic wheat containing the maize C4 phosphoenolpyruvate carboxylase (PEPC) gene. *Protoplasma* 253(6), 1503–1512.

Rane, J., Singh, A. K., Kumar, M., Boraiah, K. M., Meena, K. K., Pradhan, A. and Prasad, P. V. V. (2021). The adaptation and tolerance of major cereals and legumes to important abiotic stresses. *International Journal of Molecular Sciences* 22(23), 12970.

Ray, D. K., Mueller, N. D., West, P. C. and Foley, J. A. (2013). Yield trends are insufficient to double global crop production by 2050. *PLoS ONE* 8(6), e66428.

Raymundo, R., Sexton-Bowser, S., Ciampitti, I. A. and Morris, G. P. (2021). Crop modeling defines opportunities and challenges for drought escape, water capture, and yield increase using chilling-tolerant sorghum. *Plant Direct* 5(9), e349.

Redillas, M. C. F. R., Jeong, J. S., Kim, Y. S., Jung, H., Bang, S. W., Choi, Y. D., Ha, S. H., Reuzeau, C. and Kim, J. K. (2012). The overexpression of OsNAC9 alters the root architecture of rice plants enhancing drought resistance and grain yield under field conditions. *Plant Biotechnology Journal* 10(7), 792–805.

Rodziewicz, P., Chmielewska, K., Sawikowska, A., Marczak, Ł., Łuczak, M., Bednarek, P., Mikołajczak, K., Ogrodowicz, P., Kuczyńska, A., Krajewski, P. and Stobiecki, M. (2019). Identification of drought responsive proteins and related proteomic QTLs in barley. *Journal of Experimental Botany* 70(10), 2823–2837.

Roy, A., Rushton, P. J. and Rohila, J. S. (2011). The potential of proteomics technologies for crop improvement under drought conditions. *Critical Reviews in Plant Sciences* 30(5), 471–490.

Roychoudhury, A., Aftab, T. and Acharya, K. (Eds.). (2022). *Omics Approach to Manage Abiotic Stress in Cereals*. Singapore: Springer Nature Publishing, p. 599.

Santosh Kumar, V. V., Verma, R. K., Yadav, S. K., Yadav, P., Watts, A., Rao, M. V. and Chinnusamy, V. (2020). CRISPR-Cas9 mediated genome editing of *drought and salt tolerance* (*OsDST*) gene in indica mega rice cultivar MTU1010. *Physiology and Molecular Biology of Plants* 26(6), 1099–1110.

Sarwar, H., Sarwar, M. F., Sarwar, M., Qadri, N. A. and Moghal, S. (2013). The importance of cereals (Poaceae: Gramineae) nutrition in human health: A review. *Journal of Cereals and Oilseeds* 4(3), 32–35.

Selvaraj, M. G., Ishizaki, T., Valencia, M., Ogawa, S., Dedicova, B., Ogata, T., Yoshiwara, K., Maruyama, K., Kusano, M., Saito, K., Takahashi, F., Shinozaki, K., Nakashima, K. and Ishitani, M. (2017). Overexpression of an *Arabidopsis thaliana* galactinol synthase gene improves drought tolerance in transgenic rice and increased grain yield in the field. *Plant Biotechnology Journal* 15(11), 1465–1477.

Shavrukov, Y., Kurishbayev, A., Jatayev, S., Shvidchenko, V., Zotova, L., Koekemoer, F., De Groot, S., Soole, K. and Langridge, P. (2017). Early flowering as a drought escape mechanism in plants: How can it aid wheat production? *Frontiers in Plant Science* 8, 1950.

Shepon, A., Eshel, G., Noor, E. and Milo, R. (2016). Energy and protein feed-to-food conversion efficiencies in the US and potential food security gains from dietary changes. *Environmental Research Letters* 11(10), 105002.

Shi, J., Gao, H., Wang, H., Lafitte, H. R., Archibald, R. L., Yang, M., Hakimi, S. M., Mo, H. and Habben, J. E. (2017). ARGOS 8 variants generated by CRISPR-Cas9 improve maize grain yield under field drought stress conditions. *Plant Biotechnology Journal* 15(2), 207–216.

Shi, J., Habben, J. E., Archibald, R. L., Drummond, B. J., Chamberlin, M. A., Williams, R. W., Lafitte, H. R. and Weers, B. P. (2015). Overexpression of ARGOS genes modifies plant sensitivity to ethylene, leading to improved drought tolerance in both Arabidopsis and maize. *Plant Physiology* 169(1), 266–282.

Shim, J. S., Oh, N., Chung, P. J., Kim, Y. S., Choi, Y. D. and Kim, J. K. (2018). Overexpression of OsNAC14 improves drought tolerance in rice. *Frontiers in Plant Science* 9, 310.

Shim, J. S., Park, S. H., Lee, D. K., Kim, Y. S., Park, S. C., Redillas, M. C. F. R., Seo, J. S. and Kim, J. K. (2021). The rice glycine-RICH PROTEIN 3 confers drought tolerance by regulating mRNA stability of ROS scavenging-related genes. *Rice* 14(1), 31.

Siddiqui, M. N., Léon, J., Naz, A. A. and Ballvora, A. (2021). Genetics and genomics of root system variation in adaptation to drought stress in cereal crops. *Journal of Experimental Botany* 72(4), 1007–1019.

Singh, J., Ashfaq, M., Skinner, C. B., Anderson, W. B., Mishra, V. and Singh, D. (2022a). Enhanced risk of concurrent regional droughts with increased ENSO variability and warming. *Nature Climate Change* 12(2), 163–170.

Singh, P. K., Indoliya, Y., Agrawal, L., Awasthi, S., Deeba, F., Dwivedi, S., Chakrabarty, D., Shirke, P. A., Pandey, V., Singh, N., Dhankher, O. P., Barik, S. K. and Tripathi, R. D. (2022b). Genomic and proteomic responses to drought stress and biotechnological interventions for enhanced drought tolerance in plants. *Current Plant Biology* 29, 100239.

Srivastava, A. K., Zhang, C., Caine, R. S., Gray, J. and Sadanandom, A. (2017). Rice SUMO protease overly tolerant to salt 1 targets the transcription factor, OsbZIP23 to promote drought tolerance in rice. *The Plant Journal: for Cell and Molecular Biology* 92(6), 1031–1043.

Tang, Y., Bao, X., Zhi, Y., Wu, Q., Guo, Y., Yin, X., Zeng, L., Li, J., Zhang, J., He, W., Liu, W., Wang, Q., Jia, C., Li, Z. and Liu, K. (2019). Overexpression of a MYB family gene, *OsMYB6*, increases drought and salinity stress tolerance in transgenic rice. *Frontiers in Plant Science* 10, 168.

Thanmalagan, R. R., Jayaprakash, A., Roy, A., Arunachalam, A. and Lakshmi, P. T. V. (2022). A review on applications of plant network biology to understand the drought stress response in economically important cereal crops. *Plant Gene* 29, 100345.

Thelen, J. J. and Peck, S. C. (2007). Quantitative proteomics in plants: Choices in abundance. *The Plant Cell* 19(11), 3339–3346.

Tian, T., Wang, S., Yang, S., Yang, Z., Liu, S., Wang, Y., Gao, H., Zhang, S., Yang, X., Jiang, C. and Qin, F. (2023). Genome assembly and genetic dissection of a prominent drought-resistant maize germplasm. *Nature Genetics* 55(3), 496–506.

Timperio, A. M., Egidi, M. G. and Zolla, L. (2008). Proteomics applied on plant abiotic stresses: role of heat shock proteins (HSP). *Journal of Proteomics* 71(4), 391–411.

Todaka, D., Shinozaki, K. and Yamaguchi-Shinozaki, K. (2015). Recent advances in the dissection of drought-stress regulatory networks and strategies for development of drought-tolerant transgenic rice plants. *Frontiers in Plant Science* 6, 84.

Turner, N. C. (2018). Turgor maintenance by osmotic adjustment: 40 years of progress. *Journal of Experimental Botany* 69(13), 3223–3233.

Ul Islam, N., Wani, S. H., Ali, G., Ahmad Dar, Z., Wani, A. and Lone, A. (2021). High-throughput phenotyping for abiotic stress resilience in cereals. *Journal of Cereal Research* 13(1), 1–15.

United Nations (2019). *World Population Prospects 2019: Highlights*. New York: United Nations Department for Economic and Social Affairs.

van Dijk, M., Morley, T., Rau, M. L. and Saghai, Y. (2021). A meta-analysis of projected global food demand and population at risk of hunger for the period 2010–2050. *Nature Food* 2(7), 494–501.

Verma, G., Srivastava, D., Tiwari, P. and Chakrabarty, D. (2019). ROS modulation in crop plants under drought stress. *Reactive Oxygen, Nitrogen and Sulfur Species in Plants: Production, Metabolism, Signaling and Defense Mechanisms*, 311–336.

Wang, D., Cao, Z., Wang, W., Zhu, W., Hao, X., Fang, Z., Liu, S., Wang, X., Zhao, C. and Tang, Y. (2020). Genome-wide characterization of OFP family genes in wheat (*Triticum aestivum* L.) reveals that *TaOPF29a-A* promotes drought tolerance. *BioMed Research International* 2020, 9708324.

Wang, D., Ni, Y., Liao, L., Xiao, Y. and Guo, Y. (2021). *Poa pratensis* ECERIFERUM1 (PpCER1) is involved in wax alkane biosynthesis and plant drought tolerance. *Plant Physiology and Biochemistry* 159, 312–321.

Wang, X., Cai, X., Xu, C., Wang, Q. and Dai, S. (2016a). Drought-responsive mechanisms in plant leaves revealed by proteomics. *International Journal of Molecular Sciences* 17(10), 1706.

Wang, X., Wang, H., Liu, S., Ferjani, A., Li, J., Yan, J., Yang, X. and Qin, F. (2016b). Genetic variation in ZmVPP1 contributes to drought tolerance in maize seedlings. *Nature Genetics* 48(10), 1233–1241.

Wang, X., Xu, C., Cai, X., Wang, Q. and Dai, S. (2017). Heat-responsive photosynthetic and signaling pathways in plants: insight from proteomics. *International Journal of Molecular Sciences* 18(10), 2191.

Wheeler, T. and Von Braun, J. (2013). Climate change impacts on global food security. *Science* 341(6145), 508–513.

Willmann, M. R. (2021). *Genome Editing for Precision Crop Breeding*. Cambridge, UK: Burleigh Dodds Science Publishing.

Wu, S., Ning, F., Zhang, Q., Wu, X. and Wang, W. (2017). Enhancing omics research of crop responses to drought under field conditions. *Frontiers in Plant Science* 8, 174.

Xiang, J., Chen, X., Hu, W., Xiang, Y., Yan, M. and Wang, J. (2018). Overexpressing heat-shock protein OsHSP50. 2 improves drought tolerance in rice. *Plant Cell Reports* 37(11), 1585–1595.

Xiang, Y., Sun, X., Gao, S., Qin, F. and Dai, M. (2017). Deletion of an endoplasmic reticulum stress response element in a ZmPP2C-A gene facilitates drought tolerance of maize seedlings. *Molecular Plant* 10(3), 456–469.

Xiong, H., Yu, J., Miao, J., Li, J., Zhang, H., Wang, X., Liu, P., Zhao, Y., Jiang, C., Yin, Z., Li, Y., Guo, Y., Fu, B., Wang, W., Li, Z., Ali, J. and Li, Z. (2018). Natural variation in OsLG3 increases drought tolerance in rice by inducing ROS scavenging. *Plant Physiology* 178(1), 451–467.

Xue, D., Zhang, X., Lu, X., Chen, G. and Chen, Z. H. (2017). Molecular and evolutionary mechanisms of cuticular wax for plant drought tolerance. *Frontiers in Plant Science* 8, 621.

Yahaya, M. A. and Shimelis, H. (2022). Drought stress in sorghum: Mitigation strategies, breeding methods and technologies–A review. *Journal of Agronomy and Crop Science* 208(2), 127–142.

Yin, X., Biswal, A. K., Dionora, J., Perdigon, K. M., Balahadia, C. P., Mazumdar, S., Chater, C., Lin, H. C., Coe, R. A., Kretzschmar, T., Gray, J. E., Quick, P. W. and Bandyopadhyay, A. (2017a). CRISPR-Cas9 and CRISPR-Cpf1 mediated targeting of a stomatal developmental gene *EPFL9* in rice. *Plant Cell Reports* 36(5), 745–757.

Yin, X., Huang, L., Wang, M., Cui, Y. and Xia, X. (2017b). OsDSR-1, a calmodulin-like gene, improves drought tolerance through scavenging of reactive oxygen species in rice (Oryza sativa L.). *Molecular Breeding* 37(6), 75.

Yin, X. M., Huang, L. F., Zhang, X., Wang, M. L., Xu, G. Y. and Xia, X. J. (2015). OsCML4 improves drought tolerance through scavenging of reactive oxygen species in rice. *Journal of Plant Biology* 58(1), 68–73.

Yu, J., Lai, Y., Wu, X., Wu, G. and Guo, C. (2016). Overexpression of OsEm1 encoding a group I LEA protein confers enhanced drought tolerance in rice. *Biochemical and Biophysical Research Communications* 478(2), 703–709.

Yu, T.-F., Xu, Z.-S., Guo, J.-K., Wang, Y.-X., Abernathy, B., Fu, J.-D., Chen, X., Zhou, Y.-B., Chen, M. and Ye, X.-G. (2017a). Improved drought tolerance in wheat plants overexpressing a synthetic bacterial cold shock protein gene *SeCspA*. *Scientific Reports* 7, 1-1444050.

Yu, Y., Yang, D., Zhou, S., Gu, J., Wang, F., Dong, J. and Huang, R. (2017b). The ethylene response factor *OsERF109* negatively affects ethylene biosynthesis and drought tolerance in rice. *Protoplasma* 254(1), 401–408.

Zenda, T., Liu, S., Dong, A. and Duan, H. (2021). Advances in cereal crop genomics for resilience under climate change. *Life* 11(6), 502.

Zhang, C., Li, C., Liu, J., Lv, Y., Yu, C., Li, H., Zhao, T. and Liu, B. (2017). The OsABF1 transcription factor improves drought tolerance by activating the transcription of COR413-TM1 in rice. *Journal of Experimental Botany* 68(16), 4695–4707.

Zhang, G., Hou, X., Wang, L., Xu, J., Chen, J., Fu, X., Shen, N., Nian, J., Jiang, Z., Hu, J., Zhu, L., Rao, Y., Shi, Y., Ren, D., Dong, G., Gao, Z., Guo, L., Qian, Q. and Luan, S. (2021). PHOTO-SENSITIVE LEAF ROLLING 1 encodes a polygalacturonase that modifies cell wall structure and drought tolerance in rice. *New Phytologist* 229(2), 890–901.

Zhang, H., Zhu, J., Gong, Z. and Zhu, J. K. (2022). Abiotic stress responses in plants. *Nature Reviews. Genetics* 23(2), 104.

Zhang, M., Chen, Y., Xing, H., Ke, W., Shi, Y., Sui, Z., Xu, R., Gao, L., Guo, G. and Li, J. (2022). Positional cloning and characterization reveal the role of a miRNA precursor gene ZmLRT in the regulation of lateral root number and drought tolerance in maize. *Journal of Integrative Plant Biology* 65, 772–790.

Zhao, C., Liu, B., Piao, S., Wang, X., Lobell, D. B., Huang, Y., Huang, M., Yao, Y., Bassu, S., Ciais, P., Durand, J. L., Elliott, J., Ewert, F., Janssens, I. A., Li, T., Lin, E., Liu, Q., Martre, P., Müller, C., Peng, S., Peñuelas, J., Ruane, A. C., Wallach, D., Wang, T., Wu, D., Liu, Z., Zhu, Y., Zhu, Z. and Asseng, S. (2017). Temperature increase reduces global yields of major crops in four independent estimates. *Proceedings of the National Academy of Sciences of the United States of America* 114(35), 9326–9331.

Zhao, M. and Running, S. W. (2010). Drought-induced reduction in global terrestrial net primary production from 2000 through 2009. *Science* 329(5994), 940–943.

Zhou, H., Akçay, E., Edwards, E. and Helliker, B. (2021). The legacy of C_4 evolution in the hydraulics of C_3 and C_4 grasses. *bioRxiv*, 2020.2005.2014.097030.

Zhou, L., Liu, Z., Liu, Y., Kong, D., Li, T., Yu, S., Mei, H., Xu, X., Liu, H., Chen, L. and Luo, L. (2016). A novel gene OsAHL1 improves both drought avoidance and drought tolerance in rice. *Scientific Reports* 6, 30264.

Zhu, M., He, Y., Zhu, M., Ahmad, A., Xu, S., He, Z., Jiang, S., Huang, J., Li, Z., Liu, S., Hou, X. and Zhang, Z. (2022). *ipa1* improves rice drought tolerance at seedling stage mainly through activating abscisic acid pathway. *Plant Cell Reports* 41(1), 221–232.

Zika, M. and Erb, K. (2009). The global loss of net primary production resulting from human-induced soil degradation in drylands. *Ecological Economics* 69(2), 310–318.

Part 2

Techniques for improving resistance

Chapter 5

Advances in phenotyping to identify drought-resistance traits in cereal roots

John (Jack) Christopher, University of Queensland QAAFI, Australia

1 Introduction

There is an urgent need to increase the rate of improvement in crop production in the face of an increasing human population and increasing prosperity, as discussed in earlier chapters (FAO, 2019; Béné et al., 2015; Ray et al., 2013). However, climate change is predicted to worsen the problem of water limitation in many rain-fed cropping areas where it is commonly the major production constraint even in current climates (FAO, 2018; Gupta et al., 2020). In water-limited environments, the crop yield is most commonly limited by the amount of water that the crop can access and the efficiency with which that water can be converted to yield (water use efficiency, WUE; Passioura, 1977). The root system is the most important plant structure for water acquisition. Therefore, improving

http://dx.doi.org/10.19103/AS.2022.0109.06

root adaptation has been proposed as an important target for improving the drought resistance of crops (Lynch et al., 2022; Langridge and Reynolds, 2021; Li et al., 2021; McGrail et al., 2020; Araus et al., 2002, 2008; Lynch, 2007).

Root adaptations could potentially improve drought resistance in several ways, for example, through increased and more efficient soil water acquisition (Langridge and Reynolds, 2021; Li et al., 2021; Watt et al., 2008; Lynch, 2007; Manschadi et al., 2006, 2008). However, selecting root traits that could lead to such improvements requires reliable measurement of the traits which contribute to each adaptation.

2 Root phenotyping strategies and challenges

Choosing the most appropriate root phenotyping strategy will depend on the specific context which determines the importance of the various root types and anatomical structures (Langridge and Reynolds, 2021; Ober et al., 2021). Factors affecting the choice of a strategy include target traits, crop species, crop genotype and the water-stressed environment (Hammer et al., 2020). Soil type is also important as well as the presence or absence of physical constraints (including high bulk density and hard layers) as well as chemical constraints (Dang et al., 2011).

Just as the importance of different components of the root system change depending on the circumstances, so too do the most useful methods of phenotyping. Methods will vary with the growth stage and with the specific components of the root system that need to be measured (Rich et al., 2020; Tracy et al., 2020). It is important to consider which root types are the most relevant to phenotype and at which stage or stages during crop development they should be examined in relation to the target drought environment (Tracy et al., 2020).

Many root traits have been suggested as possible targets for improving drought resistance (Fig. 1). Some are constitutive and expressed throughout development. Others are dynamic, changing during development or in response to environmental conditions such as the timing and distribution of water availability in the soil. Traits may also be expressed only at certain phases of development, for example, in the early vegetative phase or later, between flowering and maturity (Manschadi et al., 2006, 2008).

Unfortunately, measuring root systems presents some serious challenges. Roots have been referred to as the 'hidden half' of the plant as they cannot usually be directly observed below the soil (Eshel and Beeckman, 2013). Phenotyping root traits tends to be labour and resource intensive, with low throughput and high cost, compared to many other more easily observed crop traits. Furthermore, root traits can be highly variable, requiring a high level of replication for reliable measurement. For these reasons, advances in

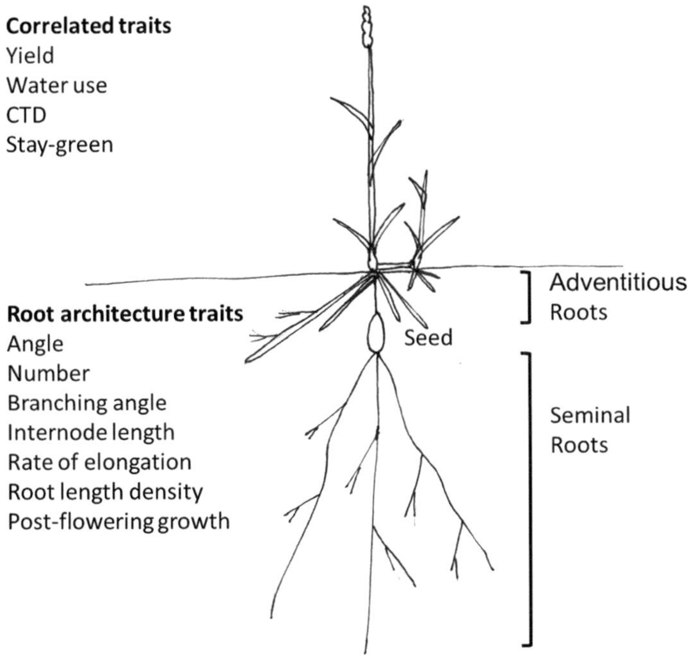

Figure 1 Diagram of a generalized cereal root system showing the seminal root system emanating from the seed and the adventitious root system emanating from the base of the stems and lower nodes. Examples of root architecture traits reported to affect drought resistance are listed to the left. Potentially correlated traits that can be measured above ground are also listed on the left, including canopy temperature depression (CTD).

phenotyping and breeding for root traits have historically been slow compared to aboveground traits. However, more recently, there has been increased interest in this area of research (Lynch et al., 2022; Ober et al., 2021; Tracy et al., 2020).

Methods of phenotyping roots to aid yield improvement vary greatly depending on the trait. Many traits are measured *'ex situ'*, i.e. under controlled conditions when compared to that in the field. Examples include conditions in the laboratory, growth chamber, glasshouse or semi-controlled systems under rainout shelters and gantries in the field (Atkinson et al., 2019). Some of these trait measurement methods are moderately high throughput and some are automated. Most of these methods use homogeneous root growth media or soil that may not necessarily reflect soil conditions in actual cropping fields (Gao et al., 2016; Rich et al., 2020). Methods for detailed measurement of root traits in the field are currently more challenging, requiring high labour input while throughput is generally limited. However, progress is being made with the use of specifically designed equipment and imaging strategies (Wasson et al., 2020, 2016, 2014; Atkinson et al., 2019).

In rain-fed cropping systems, there is usually little scope to increase the water present in the cropping environment. There is some scope for the adoption of management practices such as reduced tillage and stubble retention to improve rainfall infiltration and reduce surface evaporation in order to improve soil water availability (Carter, 2004). To improve the ability of the crop to use the water that is present, it has been proposed that there is potential to increase the ability to extract soil water or to improve the capture of water from small rainfall events. It has also been proposed that optimizing the timing of water extraction during development can increase WUE. For example, in environments with late-season water limitation, reducing water use early in the season can retain soil moisture for use post flowering when the marginal WUE for grain production is high (Kirkegaard et al., 2007; Passioura, 1983).

The effect of most root traits on drought resistance is heavily context dependent, depending particularly on environmental factors and crop management strategies (Hammer et al., 2020). The relationship between root traits measured in controlled conditions compared with the root systems in the field also changes with the level and timing of water stress in the environment as well as with field soil characteristics (Rich et al., 2020). Therefore, it is important to consider how root adaptation might improve water-limited cereal yields in specific target cropping environments (Hammer et al., 2020).

Given that phenotyping root traits tends to be arduous, expensive and variable, particularly in the field, it is important to focus on the traits that are most likely to have the desired impact in a target environment. Computer crop simulation models parameterized to model the effect of root traits can be very useful to examine the value of traits in specific environments (Postma et al., 2017; Chenu et al., 2013; Chapman, 2008; Chapman et al., 2000). Historical climate data from the target cropping region can be used to model the likely effect of changes in root traits in relevant environmental conditions for many previously encountered seasons (Manschadi et al., 2006; Veyradier et al., 2013). They can also model the likely effect of different management strategies for crops with different root traits (Chapman, 2008). Importantly, they can also be used to predict the likely effect of modified root systems in postulated future climates (Collins and Chenu, 2021).

Often, root traits are under complex genetic control, complicating introgression and combining traits in elite genetic backgrounds (Ober et al., 2021). This appears to apply to many of the root traits identified to date. Two strategies that can be used to lower costs and overcome some of the challenges are the use of more readily measured correlated traits and molecular breeding (Ober et al., 2021; Tracy et al., 2020). To identify genetic regions associated with a trait (quantitative trait loci, QTL) and genes for molecular breeding, detailed phenotyping must be conducted for genome-wide association studies (GWASs) and molecular mapping. However, to discover and validate correlated traits and/or QTL, sufficiently large populations of genotypes need to be phenotyped.

Once the earlier factors have been considered, the appropriate root trait or traits must be phenotyped, often *ex situ*. However, the expression of the trait eventually needs to be validated in the field, i.e. in the actual target environment.

3 Root traits to improve drought adaptation in cereals

To determine which root traits to measure, some basic aspects of the cereal root system should be considered. These include the root system architecture (RSA) and the root anatomy. The overall RSA of cereals, which are monocots, differs from that of dicots (Gregory, 2021; Eshel and Beeckman, 2013; Esau, 1977). While dicotyledonous crops tend to develop from an initial central tap root and radiating lateral roots, the cereals tend to have a fibrous and less centralized root system (Eshel and Beeckman, 2013; Watt et al., 2008; Esau, 1977).

The RSA of cereals differs with the developmental stage (Eshel and Beeckman, 2013; Watt et al., 2008; Esau, 1977). There are two major developmental cohorts contributing to the overall RSA (Fig. 1). Firstly, the seminal roots develop, emanating from the embryo. Secondly, the adventitious root system develops later from the base of the stems and from the lower stem nodes. Lateral roots may branch from both the seminal and adventitious roots as the development continues (Eshel and Beeckman, 2013; Watt et al., 2008; Esau, 1977).

The relative importance of water and nutrient acquisition of the seminal and adventitious root systems differs between cereal species. It also differs with environmental factors, including the timing of rainfall and management strategies, as well as with soil properties. In some major cereal crops, including wheat (*Triticum aestivum* l.), rice (*Oryza sativa L.*), barley (*Hordeum vulgare* L.,), oats (*Avena sativa* L.), triticale (×Trichosecale Wittm. ex A. Camus) and rye (*Secale cereale* L.), the seminal roots are important for resource acquisition both early and late in the season. Later in the season, the seminal roots are particularly important for penetrating deep in the soil profile (Watt et al., 2008; Fitter, 2002; Esau, 1977). The later adventitious roots are more important from mid- to late tiller development and for resource acquisition in shallower to moderate soil depths (Watt et al., 2008; Fitter, 2002; Esau, 1977). Although the developmental sequence is similar, in some other major cereals, the seminal root system more often plays a smaller role, while the adventitious root system is more important for resource acquisition through a larger part of the crop cycle. Such examples of cereals include maize (*Zea mays* L.; Mi et al., 2016; Hochholdinger, 2009) and sorghum (*Sorghum bicolor* (L.) Moench.; Singh et al., 2010).

In addition to resource capture, the roots are important for anchoring the crop to the soil. The adventitious root system in many cereals is particularly important in preventing lodging of taller crops, which can lead to lost production. This is true for species such as sorghum and maize (Singh et al., 2010; Hochholdinger, 2009) as well as taller cultivars of wheat and barley (Dreccer et al., 2020; Piñera-Chavez et al., 2016). However, lodging is generally a greater problem in well-watered environments than in drought conditions (Piñera-Chavez et al., 2016). In variable climates, it is important that selection for deep seminal roots to increase deep soil water extraction in dry seasons does not come at the expense of a robust shallow root system in years with higher rainfall.

The internal anatomy of cereal roots also differs widely both within and between species, which may affect resource acquisition (Eshel and Beeckman, 2013). All roots develop from relatively undifferentiated cells in the apical meristem (Esau, 1977). These cells differentiate and develop at various distances behind the advancing meristem to form the main root tissues (Esau, 1977). Tissues include the stele, containing the vascular xylem and phloem tissue, and pericycle (Fig. 2). The stele is surrounded by the endodermis, cortical parenchyma and the outermost epidermis. Single-celled root hairs may extend from the epidermal cells, increasing the potential absorption area (Fig. 2).

Water uptake by roots involves complex interactions between root distribution, root hydraulic properties and the hydraulic properties of soil (reviewed in Ahmed et al., 2021). Water uptake strategies that can improve adaptation to water stress have been shown to include increased ability to extract soil water, conserving water early in development for the critical grain-filling stages and rapid completion of the crop cycle to avoid late drought (Langridge and Reynolds, 2021; Gregory, 2021). The most adaptive strategy depends on the timing and severity of water stress in relation to the crop developmental stage (Chenu et al., 2013).

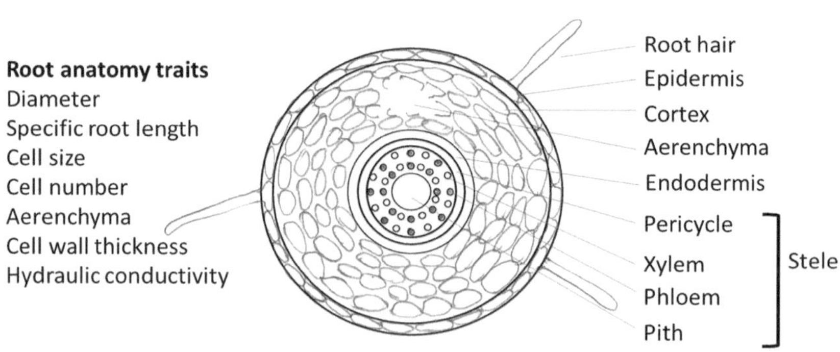

Figure 2 Diagram of a generalized cereal root transverse section showing the main tissues and structures (right). Examples of some root anatomy traits reported to affect drought resistance are listed (left).

Some root system traits that have been proposed to improve crop adaptation to water-limited environments are illustrated in Fig. 1. They range from macroscale changes in the overall RSA to microscopic changes such as changes to the vascular elements. They also include differences in the hormonal control of root growth and the biochemical processes of resource uptake and transfer to the shoots. Changes in the timing of root development may also be important.

3.1 Root system architecture

Traits associated with RSA that have been linked with drought resistance range from those associated with the overall structure and distribution of roots to root branching and fine structures such as root hairs (Li et al., 2021; Tracy et al., 2020). Clearly, the overall final structure can be significantly influenced by the fine structures formed early in development. Many traits have been associated with the development of an overall root structure adapted to water-limited conditions (Fig. 1), including root angle, number, branching angle, internode length, rate of elongation, root length density (RLD) and post-flowering growth (Fig. 1; Manschadi et al., 2010; Lynch, 2018). Architectural traits associated with drought adaptation not shown in Fig. 1 include root length, branching rate, root surface, fine root length (Canales et al., 2019), as well as root system depth, resumption of root growth after rewetting and the number of root tips per root system length (Comas et al., 2013).

For many cropping environments where water occurs at depth, it has been proposed that the overall architecture favouring late, deep root development can improve access to deeper soil water (Li et al., 2021; Ober et al., 2021; Tracy et al., 2020; Manschadi et al., 2006). It has been proposed that genotypes with a longer crop duration or a perennial growth habit have a longer period for root system penetration to greater depth (Clément, 2021; Crews et al., 2018). To achieve deeper roots late in the growing season for annual species with a more conventional crop duration, it has been observed that an overall narrower root system that distributes more resources at depth with less lateral root distribution can be advantageous in wheat (Wasson et al., 2012; Manschadi et al., 2010), sorghum (Singh et al., 2010, 2011), rice (Oyanagi et al., 2001) and maize (Hammer et al., 2009; Nakamoto et al., 1991). Such adaptations may increase water uptake during the crucial grain-filling period when the marginal WUE for converting water to grain is high (Manschadi et al., 2006; Kirkegaard et al., 2007).

In some situations, a narrower overall root system with more roots at depth may not be advantageous, for example when ample water is available in shallow soil layers or where water simply does not accumulate in lower layers. In such cases, a root system extending laterally may be better adapted (Tracy et al., 2020). For taller cereals including sorghum and maize, shallow root development, often of the adventitious roots, is also important to avoid crop losses due to lodging (Piñera-Chavez et al, 2016).

The rate at which water is taken up during crop development is also important. Where water is freely available during the crop cycle, plants that transpire more water can often produce greater biomass that can be potentially converted to yield (Condon et al., 2004; Wright et al., 1994). However, in drought environments, it may be advantageous to limit water use early in the season to retain sufficient soil water for the crucial grain-filling period (Borrell et al., 2014; Wasson et al., 2012; Richards and Passioura, 1981, 1989; Passioura, 1972, 2010). The timing of water uptake may depend largely on shoot demand but can also be modulated by root traits. For example, water uptake can be influenced by the timing of growth and distribution of late, deep roots (Manschadi et al., 2006). Water uptake and transport can also be modulated by aspects of root anatomy affecting uptake and transport to the shoots (Ahmed, et al., 2021).

3.2 Root anatomy

The root anatomy, including the root diameter and cellular structure, affects the efficiency of water uptake and metabolic costs associated with soil resource capture (Fig. 2). Root anatomical traits that have been associated with drought adaptation include root diameter, specific root length (SRL), cell size, cell number, aerenchyma, cell wall thickness and hydraulic conductivity (Fig. 2; Lynch et al., 2014, 2022). Some anatomical traits restrict radial water movement, including root cortical senescence, root cortical aerenchyma and the endodermis (Schneider et al., 2018; Lynch et al., 2014). Other anatomical traits associated with drought tolerance include cortical cell size and cell file number (Schneider et al., 2018).

For example, to explore a larger soil volume using a minimum of energy and resources, narrower roots with greater length per unit of root mass (greater SRL) could theoretically be advantageous (Comas et al., 2013; Klein et al, 2020; Wasaya et al., 2018). Similarly, the production of roots with gas spaces (aerenchyma; Fig. 2), between cells or formed by lysis of cells, can reduce the carbon input required for maintenance and respiration for a given root volume (Armstrong, 1980; Armstrong and Drew, 2002; Mustroph and Albercht, 2003; Postma and Lynch, 2011). However, these traits may not be advantageous in harder-to-penetrate soils with high bulk density, or in soils with hard layers, such as cultivation-induced hard pans. In these conditions, roots of greater diameter, of thicker cell walls and with less flexibility may be superior in accessing water in deeper layers (Klein et al., 2020).

Once the roots have penetrated a given volume of soil, water and nutrients must be taken up and conveyed to the aboveground organs. Water uptake occurs radially from the outside of the root to the stele after which water flows via the axial pathway mainly in the xylem vessels (Eshel and Beeckman, 2013; Ryan et al., 2016). Radial flow is modulated by aquaporin activity, by restrictions imposed by the exodermis (directly below the epidermis) and endodermis as well as by the pathway length (Zarebanadkouki et al., 2013; Steudle and Frensch, 1996). The distance from the outside of the root to the stele is important and is affected by, for example, the cortex tissue volume and the number of cells in the pathway (Bramley et al., 2009; Javot and Maurel, 2002). Aquaporins are membrane channel proteins that are selectively permeable to water and are major contributors to water uptake into the root cells. Aquaporins and radial transport tend to be co-limiting with axial (vascular) limitations (Chaumont and Tyerman, 2017; Frensch and Steudle, 1989). Axial flow is affected by such factors as water demand from the shoots as well as xylem vessel number, diameter, length and hydraulic resistance (reviewed in Bramley et al., 2009). It has been proposed that, in certain environments, restricting axial water transport to the shoots early in the season may allow crops to retain more soil moisture for the high WUE period post flowering and during grain filling (Richards et al., 2010; Richards, 2008; Passioura, 1983; Richards and Passioura, 1989).

3.3 Changes in roots in response to drought

Root systems are dynamic and can respond to many stimuli in the environment, including gravity, water, light and nutrients. Active root responses include gravitropism, hydrotropism and phototropism (Muthert et al., 2020). Gravitropism refers to the degree to which the roots grow downwards in response to gravity. Clearly, the degree to which gravitropism affects the angle of root growth has significant effects on the fine as well as the overall root architecture such as the association between a more vertical root angle and deep root growth (Eshel and Beeckman, 2013; Oyanagi, et al., 2001; Nakamoto et al., 1991). Hydrotropism refers to the degree to which roots respond by growing towards or away from soil volumes with higher water content (Dietrich, 2018). Hydrotropism can affect the ability of roots to seek out soil moisture (Muthert et al., 2020). Phototropism of roots refers to the tendency to grow away from blue or white light and towards red light (Muthert et al., 2020). Aspects of root anatomy may also be modified in response to water limitation, including changes to aerenchyma, cortical cell size, file number as well as xerobranching (Gregory, 2021; Ober at al., 2021; Rich and Watt, 2013; Schneider and Lynch, 2018).

Roots also tend to grow towards favourable concentrations of important soil nutrients such as nitrogen and phosphorus (Muthert et al., 2020). In contrast, root growth can be restricted by toxic concentrations of both nutrients and other chemicals, for example salt (Muthert et al., 2020), which may restrict water uptake (Dang et al., 2016).

Lopes and Reynolds (2010) reported different responses to contrasting patterns of water availability in root architecture and drought resistance. They observed that certain wheat genotypes can produce a greater proportion of late, deep roots where water is limited late in the season but can also produce a greater proportion of shallower roots when water is available in the surface layers. Pinto and Reynolds (2015) reported a common genetic basis for the response in both environments.

Temperature not only affects the rate of root growth and hence, for example, the rate of root system penetration to depth but can also have more subtle effects. Vazquez-Carrasquer et al. (2016) demonstrated that the seminal root angle of wheat tends to increase with an increase in growth temperature from 11°C to 17°C and that this response varied between genotypes.

Responsive traits are often under complex genetic, hormonal and molecular controls, as discussed elsewhere in this volume.

3.4 Combining root traits

The discussions in the previous section suggest that certain root traits can group together to provide adaptation to specific environments. For example, the 'steep, deep and cheap' root system in maize has been proposed to describe a root system with rapid exploitation of deep soil layers providing access to soil moisture by combining a number of different root traits (Lynch, 2013). Recent studies described an advantageous root trait combination for maize, including reduced root maintenance costs, reduced soil moisture uptake early in the season and greater penetration ability of roots in water-stressed environments (Lynch et al., 2022; Klein et al., 2020). Reduced root maintenance costs were related to the production of aerenchyma as well as to the cortical cell size and number. Reduced early water uptake was related to reduced cross-sectional area of the xylem, lower number of xylem vessels and higher hydraulic resistance. Improved soil-penetration ability was associated with smaller outer cortical cells and greater cortex cross-sectional area relative to the stele (Klein et al., 2020).

Any one combination of root traits is likely to be adaptive only for certain environments. However, adaptation to any one environment might also be achieved by more than one combination of traits.

4 Phenotyping in controlled conditions: *ex situ*

The challenges of measuring root traits in the field have led to the development of many methods to phenotype roots in controlled systems. Methods have been developed to measure a range of growth stages from seedlings to mature plants (Li et al, 2021; Ober et al., 2021; Tracy et al., 2020). Measurements can also range from whole root systems and organs to the dissection of the microscopic cellular structures. Specific *ex-situ* phenotyping methods are generally suited to a particular range of developmental stages. Therefore, the methods are discussed in relation to the relevant growth developmental stages.

4.1 Seedlings

Characterizing roots in the early growth stages has the advantage that the roots can be studied after culture for only a few days and large numbers of individual plants can potentially be examined using a relatively small growth space. Some methods of root measurement early in development also have the advantage that they use relatively simple technologies that are more readily applicable in low-income countries.

Some of the simplest and cheapest methods to culture seedling roots use paper to provide support and moisture for growth. It is important to choose a paper suitable for plant growth, as roots are highly sensitive to contaminants that can induce an adverse chemical environment. Papers sold specifically for plant germination or laboratory filter papers are usually free from harmful concentrations of such contamination. More general-purpose papers may be suitable but should be tested before commencing experiments to ensure that they do not inhibit root growth. One of the simplest paper-based methods is the paper roll or 'cigar' method (Merchuk-Ovanat et al., 2017). This allows large numbers of plants to be phenotyped for traits such as comparative initial rate of root growth and the number of seminal roots. There are also commercially available paper systems such as plastic pouches with paper inserts as well commercial germination papers. Examples of crops for which such phenotyping systems have been used include maize (Hund et al., 2009), wheat (Atkinson et al., 2015; Rich et al., 2019) and pearl millet (Passot et al., 2016).

Seedlings have also been cultured in artificial media, including agar gels (Manschadi et al., 2006; Bengough et al., 2004). Other transparent gels include hydrogel systems (Maa et al., 2019; Ahmed, 2015). Transparent gel systems have been used for three-dimensional (3D) imaging of maize (Wang et al., 2015) and rice (Topp et al., 2013; Clark et al., 2011), for example. As with paper products, it is important to choose a highly purified gelling agent, or one specifically recommended for plant growth, to avoid root-inhibiting contaminants. Fungal growth in gel systems can also be a problem. It is usually necessary to sterilize both the medium and the seeds as well as to use sterile technique while mounting seeds. This adds time and expense to the use of gel systems.

Methods to study the angle and number of seedling roots in sand or soil media have proved particularly useful. The growth of rice seedlings in small plastic baskets has been used to identify and characterize rice genotypes with a more vertical root angle, leading to the release of rice varieties with improved drought resistance in water-limited upland systems (Ramalingam et al., 2017; Uga et al., 2013, 2011). A method allowing the study of the seminal root angle and number in soil-filled, transparent pots has also been described (Richard et al., 2015). This allowed up to 600 plants m^{-2} of growth space of wheat or barley to be characterized for root angle after 5 days from sowing and root number at 11 days.

4.2 Early- to mid-growth stages

Studies of cereals during the early-growth stages from early tillering to flowering often use pots or root observation chambers (also called rhizoboxes) of various sizes (Ober et al., 2021; Tracy et al., 2020; Lynch, 2007, 2013). In such systems, it is important to consider the soil volume required to study the root system up to the required growth stage to avoid root structure being restricted by the constraints of the container (Passioura, 2006). Roots are usually grown in sieved and mixed sand or soil, which tends to have lower bulk density and lack the natural structure of field soils. Consequently, roots may grow more rapidly and reach greater length in artificial systems than would be expected in the field. However, the systems can be useful to measure the genetic potential for root growth in the absence of constraints (Singh et al., 2011; Manschadi et al., 2010).

Root observation chambers are generally narrower in one dimension to encourage roots to grow up against a transparent observational surface. This forces a normally 3D root structure to conform to a more 2D growth pattern. These systems can help to detect differences in the initial growth angle and root number during early to mid-development of the root structure but might not faithfully reproduce the structure that would occur in a 3D environment (Ober et al., 2021; Thorup-Kristensen et al., 2020; Tracy et al., 2020; Manschadi et al., 2010; Lynch, 2013, 2007).

4.3 Automated high-throughput systems

Automated and semi-automated systems have been developed to assist with the culture and imaging of roots of seedlings as well as early- to mid-growth stage plants. These systems have the advantage that plant maintenance can be largely automated. Plants can also be mechanically passed through an automatic imaging system at regular intervals, allowing the study of the dynamics of root growth at the observation surface. The Phytomorph system uses agar-filled dishes or trays to measure seedling root traits (Subramanian et al., 2013). GrowScreen-Rhizo has a system of observation chambers for early- to mid-growth stage plants (Nagel et al., 2012). The RADIX rhizoslide system uses plexiglass slides supporting a germination paper growth substrate (Le Marié et al., 2016). In the RhizoTubes system, 18 cm × 50 cm polymethylmethacrylate tubes contain a growing substrate such as soil that is surrounded by a permeable membrane, with plant roots growing between the membrane and the inner surface of the transparent tube (Jeudy et al., 2016). A range of root phenotyping facilities are also incorporated into the Australian Plant Phenomics Facilities in Canberra and Adelaide in Australia. Some automated or semi-automated systems monitor the root development indirectly via infrared imaging, electrical impedance and other measures to estimate changes in the water distribution in the growing medium resulting from root water uptake (Corona-Lopez et al., 2019).

Relatively high-throughput systems of lower cost requiring some manual plant maintenance and/or imaging have also been developed, and the technology has been applied in low-income countries (Singh et al., 2011; Menamo et al., 2017).

4.4 Later growth stages

The major portion of the development of the root system of cereals is thought to be completed by around the time of flowering (Eshel and Beeckman, 2013; Watts et al., 2008). However, root system development can continue post flowering (Manschadi et al., 2006). Continued development and maintenance of roots can have important implications for yield. It has been demonstrated that additional soil water accessed post flowering has a high WUE for conversion to grain yield of wheat of up to 60 kg ha^{-1} mm^{-1} (Manschadi et al., 2006; Kirkegaard et al., 2007). However, phenotyping root systems post flowering requires the extended maintenance and culture of larger plants and larger soil volumes compared to pre-flowering stages. To avoid restricting root growth in the vertical dimension using a manageable container volume, many systems restrict lateral growth. For example, plastic tubes made from various types of plastic pipes, often originally made for use in drainage and construction, have been used to allow root growth to depths of 1.5 m or more. However, most are 0.3 m in diameter or narrower. This restricts the full expression of lateral root growth and can cause the crowding of roots at the sides of the tubes.

This lateral restriction has been partially alleviated by using larger more '2D' chambers. For example, Manschadi et al. (2006) reported the use of chambers that were 2.4 m long, 0.1 m wide and 1.2 m high. This not only reduced the required soil volume but encouraged root growth at the transparent sides to allow imaging. However, this likely tends to induce a greater lateral root spread in the wider dimension. As mentioned earlier, roots can be expected to grow further in sand or soil in containers than in natural field soil conditions. If restricted in one dimension, roots could be anticipated to partially compensate by growing more in another. Observations made in plastic tubes and root observation chambers with restricted size in one or more lateral dimensions can provide useful comparisons between the potential of crop roots to grow in either one or two dimensions, respectively. However, they will likely not faithfully reproduce 3D crop root systems that occur in the field.

Methods to grow roots in more 3D systems have also been devised, although these generally require root washing before imaging and tomography (Dowd et al., 2021).

Tubes have been used in phenotyping systems for late, deep roots with the main aim of measuring root distribution with depth. The materials can be relatively cheap and, depending on the final size and weight, can be readily handled. To make measurements, it is usually necessary to destructively harvest and wash the root system free of soil. Tubes are often cut vertically down two sides to allow extraction of the whole core and root system. Alternatively, they can be cut into sections corresponding to soil depth ranges of interest. For handling and weighing of larger numbers of pipes or for longer and wider tubes, they can be arrayed under a gantry system. Tubes can be arrayed in a pit, which is then filled with sand or soil to maintain a soil temperature closer to the field soil. They may also be arrayed under a rainout shelter to control watering. Such a system has been used to phenotype more than 300 Chinese wheat cultivars and land races, allowing the study of the genetic control of late-season root traits (Li et al., 2019).

Lysimeter systems with automated watering and weighing in large pots allow the measurement of water use and imposition of controlled water stress (Chenu et al., 2018). However, destructive sampling and root washing are usually required to make root measurements.

An interesting semi-hydroponic system using commercially available components has been developed, allowing culture to later growth stages of the broadleaf crop *Lupinus angustifolius* (Chen et al., 2011) and maize (Qiao et al., 2019). The semi-hydroponic system consists of 5-mm thick acrylic panels 260 mm × 480 mm. To support roots, panels are covered in cloth, which is irrigated using a dripper system. The panels are suspended in large, wheeled plastic bins. This technology has potential for use in low-income countries.

4.5 Aeroponic and hydroponic systems

Some culture methods that can be used at a range of growth stages have been developed without direct support from a solid or semi-solid matrix. These include aeroponic and hydroponic systems. In an aeroponic system, the roots are suspended in a chamber below the plant and irrigated regularly by spraying a fine mist of nutrient solution. A high-throughput aeroponic system has been developed, allowing the growth of hundreds of plants to a root depth of 60 cm with automated imaging approximately hourly (https://uclouvain.be/en/research-institutes/eli/elia/rootphair.html). Lower cost and more labour-intensive systems have also been reported (Selvaraj et al., 2019; Tafesse et al., 2021). The largest aeroponic root phenotyping facility known to the author, and possibly the largest root chamber of any kind, is an aeroponics facility located at the Racine Root Laboratory at the Tel Aviv University Botanical Gardens in Israel (https://www.tau.ac.il › lifesci › units › Gardens › roots).

Root structure in aeroponic systems may be at least partially maintained using a 3D scaffold. However, as fine roots grow further from the scaffold, they tend to be weighed down by water droplets, which may cause steeper vertical development than may otherwise occur. The surface tension of water during spray irrigation may also cause roots to clump, making individual roots harder to differentiate in images. Hydroponic systems have also been used. These often also have a 3D scaffold to assist in the partial retention of root architectural structure (Piñeros et al., 2016).

The nutrient solutions in both aeroponic and hydroponic systems can be readily modified to investigate the effects of soil solution composition on RSA and growth. This can also be done in some other *ex-situ* systems irrigated using nutrient solutions, including sand culture and culture using paper supports. Variables of interest may include nutrient deficiencies or toxicities in the solution.

4.6 Phenotyping of root anatomical traits

While phenotyping root architectural traits in large breeding populations can be difficult, phenotyping of root anatomical traits, often at the microscopic level, presents a whole separate range of challenges. However, there has been considerable recent progress using improved measurement techniques combined with automatic trait data extraction software. Some recent reviews provide a useful summary of this work (Ober et al., 2021; Lynch et al., 2022; Lynch et al., 2021).

4.7 Measurements early vs late in development

Measurements at each growth stage are most effectively performed in phenotyping systems adapted for a specific range of growth stages. For example, it may be difficult to observe the initial growth angle and number of seedling seminal roots in a system designed for observation of later root traits such as large pots or large root observation chambers. However, each different phenotyping system presents specific limitations. Additionally, traits measured early in development *ex situ* may not necessarily correlate with later, deeper root development, particularly with development in the field.

Correlation between root traits measured in controlled systems and those measured in the field is often poor (Rich et al., 2020; Watt et al., 2013).

Where significant correlations do occur, they are usually context dependent on environmental variables such as the timing of water stress and soil type. As mentioned earlier, natural soils tend to exhibit variation with depth and over time. For example, bulk density usually increases with depth and hard layers may be present. Oxygen concentration and temperature tend to decrease with depth. It was reported that wheat root biomass decreased more rapidly with depth in field soils than in *ex situ* phenotyping (Gao et al., 2016; Kuijken et al., 2015). Rich et al. (2020) also reported for wheat that observations made in several *ex-situ* phenotyping systems were correlated with root trait measurements only in individual field experiments phenotyped by extensive soil coring (Wasson et al., 2016, 2014). Additionally, *ex-situ* traits correlating with a particular field observed trait in one field experiment were not necessarily correlated in another experiment at a site in the same region or at a similar site in a different season (Rich et al., 2020).

5 Field phenotyping

Plants cultured *ex situ* are usually grown in soil, or other media, that has been processed and mixed to produce a homogeneous growing environment for roots. However, soils in the field are not uniform, either across the field or down the soil profile. They often have distinct horizons. They may have hard layers and other impediments to root growth such as chemical constraints. Therefore, it is important to compare observations made on roots grown *ex situ* with those of roots from actual field soils.

Direct methods suitable for improving cereal drought adaptation include 'shovelomics', minirhizotrons, and extensive soil coring (Trachsel et al., 2011; Rewald and Ephrath, 2013; Wasson et al., 2014S). Root observations made in soil trenches in the field can be informative but cannot be done on a scale useful for cereal improvement programs (Lynch, 2007). Remote methods including ground-penetrating radar have been used to measure the roots of perennial plants and woody species but are not yet at a high enough resolution for measuring cereal roots (Ober et al., 2021; Liu et al., 2016).

The term 'shovelomics' has been used to describe recent improvements to a traditional technique for the study of the crown roots of plants in the field. The crown of the plant is dug from the soil, often to a depth of 20–30 cm and a diameter of about 30 cm depending on the species. The soil is then washed from the crown prior to the measurement of traits such as crown diameter and the outside angle of the roots. The crowns are often composed mainly of adventitious roots. The initial extraction of the crowns was traditionally achieved manually using an implement such as a shovel, leading to the name 'shovelomics'. The crowns can then be imaged, and a number of root traits can be extracted (Colombi et al., 2015). More recent strategies to mechanize and automate the method have been developed to extract, wash and image the crowns (Seethepalli et al., 2020). Machine learning has also been applied to aid the interrogation of the images in conjunction with tomography to extract data on a wider range of root traits, including individual root diameter and angle. However, the data from 'shovelomics' are obviously limited to that section of the root system that can be extracted from near the surface.

Minirhizotrons have been used to study deeper roots in the field (Rewald and Ephrath, 2013; Maeght et al., 2013; Samson and Sinclair, 1994). These use transparent tubes 50 mm in internal diameter that are normally inserted at an angle of 45° and to lengths of up to 2 m or more, thus reaching soil depths of over a metre (Rewald and Ephrath, 2013). A camera is periodically inserted into the tubes where it is rotated as it traverses the length of the tube, imaging the roots exposed at the outside surface of the transparent tubes. This system has the advantage that it allows non-destructive and repeated observation of the root system during crop development to follow root growth. However, the equipment itself and that required to install the tubes can be expensive, especially if large numbers of tubes are required. In some cases, it can be difficult to ensure close contact between the outside of the tube and surrounding soil (Rewald and Ephrath, 2013; Johnson et al., 2001). This can lead to air spaces that may be avoided by roots. Alternatively, smaller gaps which may allow free water movement or accumulation at the outside surface of the tube, and/or a path of reduced soil impedance, may attract a higher density of roots than occur in the main soil volume (Rewald and Ephrath, 2013; Rytter and Rytter, 2012). The insertion of the tubes may also cause smearing of soil onto the outside surface, obscuring roots from the camera. Given the advantages of the method, however, there are many reports of useful characterization of cereal roots using this system both in the field and in large rhizotrons (Merrill et al., 2005).

Soil coring is a useful method to study deeper roots. It is destructive of the roots extracted. However, sampling may be achievable without destroying the crop, for example using machines with high ground clearance. This method can be repeated at different crop developmental stages up to harvest and beyond. Traditionally, roots were extracted from sections of the cores by washing to measure the RLD, root biomass and other parameters. However, more recently, a method to estimate RLD by counting roots at breaks in cores has been devised (Wasson et al., 2014). The method of breaking cores and counting the roots has also been partially automated (Wasson et al., 2016). The method is calibrated by washing a subsample of the core sections, thus greatly reducing labour requirements. The reduced input also potentially allows resources to be diverted to taking a larger number of cores across the field to help account for the high level of spatial variability across the field.

Managed field-based systems have been developed to integrate automation and detailed instrumentation to measure roots as well as water

Figure 3 An example of a field-based system where plants are cultured in tubes in the field and maintained under a rainout shelter showing a view from each end. The PhytoTron is located at the International Centre for Agriculture Research in Dry Areas in Morocco. Source: Photo courtesy: Andrea Visioni, ICARDA.

uptake. These can test plants in the open field environment. However, they generally do not use undisturbed field soils and are restricted to a limited number of locations. Plants are cultured in either large soil-filled tubes or in reconstituted field soil. A novel field-based system called the RadiMax has been developed at the Copenhagen University, allowing the field culture of crops with controlled underground watering at a range of depths and using rainout shelters (Svane et al., 2019). An array of 150 minirhizotron tubes is installed at 40 cm spacings for root observations in the RadiMax. Another field-based array of minirhizotrons and extensive instrumentation has been established in Germany (Cai et al., 2016).

Systems using field gantries with field soil or tubes have also been partially automated. A recently developed example is the PhytoTron facility at the International Centre for Agriculture Research in Dry Areas in Morocco (ICARDA; https://www.icarda.org/media/news/icardas-new-physiotron-facility-ready-unblock-global-phenotyping-efforts; Fig. 3).

6 Three-dimensional imaging and root interactions with the soil

The majority of methods discussed earlier involve removal of roots from the soil by washing or culture in homogenous growth media. They provide little information about either the 3D architecture of the root system in natural field soils or the interactions between roots and the undisturbed soil.

Methods have been developed to characterize the 3D architecture of roots using x-rays and magnetic resonance imaging (MRI), for example (Dowd et al., 2021; Flavel et al., 2017; Piñeros et al., 2016). These have been used to image roots in a variety of growth media. They can also be used to image roots in soil cores extracted from the field (Dowd et al., 2021). This can provide valuable insights that can help to clarify whether architectural traits measured *ex situ* reliably reflect those of the roots in the soil. Images of roots in natural field soil can also provide information on the interactions between the roots and soil (Ober et al., 2021; Lynch et al., 2022; Lynch et al., 2021). However, most of these techniques are restricted to relatively small soil volumes and have low throughput. When soil volumes are increased, imaging times can greatly increase (Metzner et al., 2015). Thus, these techniques are currently more useful for research purposes than for screening large numbers of genotypes in the field.

Root growth in soil cores can be measured at a reasonably small scale using detailed x-ray computer tomography (Flavel et al., 2012, 2017; Mooney et al., 2012; Hainsworth and Aylmore, 1983). MRI using radio waves has also been used (van Dusschoten et al., 2016) and positron emission tomography (PET; Metzner et al., 2015; Atkinson et al., 2019). MRI has also been combined with PET to study the structure and dynamics of carbon transport in maize root systems (Jahnke et al., 2009). Growth media need to be more carefully selected for MRI, which is more sensitive than x-ray computer tomography to ferromagnetic particles in soil (Metzner et al., 2015; Pflugfelder et al., 2017).

X-ray computer tomography can be undertaken for plants cultured in pots or tubes or for soil cores taken in the field. Cores are often a few centimetres in diameter. Larger cores of up to 15–30 cm diameter may also be imaged this way (Atkinson et al., 2019). However, to reach the resolution required to identify cereal roots, scan times must be increased to several hours for larger cores (Metzner et al., 2015). These machines are also large and expensive, requiring highly trained technicians to operate them.

Another system has been developed that uses luminescence reporter genes in *Arabidopsis* to visualize roots in rhizotrons that could be used to measure roots in transgenic crop plants (Rellán-Álvarez et al., 2015).

Laser ablation methods have been developed to study the cellular structure of roots. This method has been reported to also visualize arbuscular *mycorrhiz*al fungi, nematodes and root worm in maize roots (Strock et al., 2019).

Variations in root interactions with the soil may include the tendency of roots to grow within existing voids or on layers of water adhering to soil particles. However, such variations may not be fully expressed in artificial growth media. Roots in soils of high bulk density or with hard pans may be found preferentially in soil channels or pores left by the decay of previous root systems (Watt et al., 2008, 2005; Gao et al., 2016). Interactions with the organisms in the soil near the roots may also be important (microbiome). Imaging methods allowing measurement of roots in small intact soil cores have been recently developed that can elucidate such details (Flaver et al., 2012, 2017).

7 Imaging and trait data extraction

One of the limitations of many of the phenotyping techniques discussed is that they generate a large number of images and large data files. It quickly becomes impractical to analyse images using software requiring user input for individual images or even for small batches of images. Therefore, high-throughput and machine learning techniques have been used and are being developed to interrogate images and extract data on target traits. 'SmartRoot' is a semi-automated system (Lobet et al., 2011). Generator of Root Anatomy in R 'GRANAR' is available for analysis of monocotyledon roots (Heymans et al., 2020). 'RootScan' is a program for semi-automated image analysis of anatomical traits in root cross-sections (Burton et al., 2012). 'Rhizovision' is an integrated system for conducting, imaging and data extraction from the images of root crowns from 'shovelomics' (Seethepalli et al., 2020). Root estimator for shovelomics traits 'REST' is also available for analysing 'shovelomics' images (Colombi et al., 2015). DIRT 'digital imaging of root traits' is also available, with an updated version DIRT3D recently released (Bucksch et al., 2014; Liu et al., 2021). Useful reviews of this area include Dowd et al. (2021), Lynch et al. (2021), Atkinson et al. (2019) and Schulz et al. (2013).

8 Aboveground correlated traits in the field

In the simplest case, an adaptive root system will be correlated with greater yield in the field. However, the many potential confounding factors mean that attribution of a yield advantage to root traits requires validation.

The challenge of directly measuring cereal roots in the field has led to the use of certain traits that may be correlated with desirable root phenotypes in particular field environments (Ober et al., 2021; Tracy et al., 2020). Under these conditions, such traits may also be considered 'proxy traits' for certain root traits (Ober et al., 2021; Tracy et al., 2020). For example, correlated traits can be useful in cropping systems where in-season rainfall is routinely limited, but water remains available late in the season, deep in the soil profile. In such environments, it can be postulated that genotypes with greater ability to access deep soil moisture and maintain transpiration will exhibit certain aboveground traits. For example, they could maintain a lower canopy temperature compared to the ambient air temperature (greater magnitude of canopy temperature depression, CTD). Similarly, they may be able to retain green leaf area for longer than other genotypes after anthesis (stay-green phenotype). Clearly for this to hold, possible confounding factors such as soil diseases, parasitic nematodes and soil constraints must be carefully managed. An increased ability to extract soil moisture is also an indication of an adaptive root system. These correlated traits have the advantage that they can be measured relatively easily, rapidly and non-destructively. This can be done using proximal sensors, unmanned aerial vehicles (UAVs) or satellite imagery.

Genotypes which are transpiring can be differentiated from those which are not transpiring using CTD, under certain field conditions. At close to midday on sunny days when wind speeds remain low, the influence of crop transpiration on the canopy temperature is less variable than at other times and is often close to its greatest magnitude. However, cloud cover as well as higher and more variable wind speeds can cause considerable spatial and temporal variations in the canopy temperature. Traditionally, canopy temperature was measured on the canopy of individual plots using a handheld sensor, which allowed the measurement of one or two plots per minute at the most. Thus, measuring the temperature of a large field trial with hundreds of plots could take periods of hours. The incidental sunlight, cloud cover or wind conditions could change quite significantly during this length of time, introducing high levels of variation in CTD measurements. More recently, UAVs have been used to take infrared images of the canopy. These can capture data from large field trials in periods of minutes rather than hours, reducing the spatial and temporal variations in CTD.

Stay-green phenotype is a term used to describe the ability of a plant to retain green leaf area for longer after anthesis than standard genotypes (Borrell et al., 2014; Thomas and Ougham, 2014; Thomas and Howarth, 2000; Thomas and Smart, 1993). Stay-green expression, measured as leaf or canopy greenness over time after anthesis, is a more stable measure than CTD during each day and can be measured using relatively inexpensive conventional red, green, and blue imaging. It can also be measured using dedicated handheld sensors such as the GreenSeeker® 505 (Trimble Inc., Sunnyvale, CA, USA) (Christopher et al., 2014). The GreenSeeker emits red and infrared light measuring the reflectance back from the canopy. Thus, GreenSeeker measurements are less sensitive to temporal changes in sunlight and atmospheric conditions than sensors measuring the reflectance of ambient light. Multispectral and hyperspectral sensors can also be used to measure stay-green (Gracia-Romero et al., 2019). As with infrared sensors discussed earlier, sensors to measure stay-green may also be mounted on UAVs capable of imaging a trial in a period of minutes. As with CTD, the correlation of stay-green with improved access to soil water depends on the availability of soil water and absence of confounding soil and disease factors. Stay-green can also be associated with other shoot-related factors such as increased transpiration efficiency (TE) and developmental genes directly influencing crop senescence. Thus, correlations between stay-green and root traits usually need to be validated in other root phenotyping systems. An example of this is given in the case study later.

Soil moisture extraction under field plots may also be a useful proxy for the overall depth and soil moisture extraction capability of a crop root system. As with CTD and stay-green, this correlated trait is most useful in environments with limited in-crop rainfall where late-season water limitation regularly occurs. In such situations, soil moisture extraction can be estimated by measuring soil moisture in cores taken near sowing and comparing them to those taken later in the season. As with measurements of RLD and maximum root depth, moisture measurements made using soil cores are subject to a high level of spatial variation.

Methods to study soil moisture without the need for coring or insertion of probes include electrical resistivity (Binley and Kemna, 2005) and electromagnetic induction (EMI; Brevik et al., 2006). Soil water extraction can be measured using EMI over larger soil volumes than those extracted in soil cores (Whalley et al., 2017; Shanahan et al., 2015; Dang et al., 2011). Instruments for EMI measurement in the field include the EM38 (Geonics EM38®, *Mississauga, Ontario*, Canada), the Dualem, which has two receiver coils (Dualem Inc., Milton, Ontario, Canada), and the CMD Explorer and Mini-Explorer, which have three receiver coils (GF Instruments, Brno, Czech Republic). Other systems for proximal soil moisture measurement are also under development (reviewed in Hardie, 2020).

The EM38 in vertical dipole mode, for example, can measure a volume of soil extending roughly 1 m in the lateral dimension and from approximately 0.3 m depth to 1.5 m depth. This soil volume is ideally suited to many small plot crop phenotyping experiments and provides a useful depth to investigate late, deep soil moisture extraction (Dang et al., 2011). For measurements of the top 0.3 m of the profile, the instrument can be used in horizontal dipole mode (Dang et al., 2011). As instruments such as the EM38 are portable and can be used to measure large numbers of plots, it is possible to sample a field site more frequently using this method than repeatedly soil coring. Coring also runs the risk of damaging the crop either directly or by soil compaction even when high ground-clearance machines are used. However, modern global positioning equipment used to accurately position experimental crop rows and to guide soil coring equipment reduces this risk.

Soil moisture can also be measured using probes inserted into access tubes fitted into the soil profile such as neutron soil moisture meters (Li et al., 2003; Silvestri et al., 1991). Access tubes are best fitted either before or shortly after sowing but prior to crop emergence. However, the labour and expense required to fit access tubes, as well as the safety training required for the use of neutron probes, make these methods more expensive than others such as the EMI.

Both coring and non-destructive soil moisture measurements can be confounded by frequent periods of in-crop rainfall between the initial and final measurements. This can reduce the ability to detect the often-small differences between plots of different genotypes compared to the large spatial variation across the field.

9 Trait by environment interactions

The benefit of most of the root traits discussed is dependent on the context of the drought environment. This applies to both constitutive traits and those which can be modulated in response to changes in the environment. A trait that is beneficial in a particular water-stressed environment may have little effect in another environment or may even be deleterious (Hammer et al., 2020). Computer simulation modelling can be an aid to determine the value of individual traits or combinations of traits in the target population of environments (Postma et al., 2017; Casadebaig et al., 2016; Veyradier et al., 2013; Manschadi et al., 2006). To facilitate this, the model needs to be parameterized for the target traits. The model should then be validated and calibrated against field and/or *ex-situ* phenotyping data. An example of this approach is discussed in the case study later.

Computer simulation modelling has the advantage that it can be used to predict crop performance in a wide range of environments prior to the need for extensive field experimentation. Additionally, it can be used to predict performance not only for current environments but also to examine the effect of traits in projected future climates (Collins and Chenu, 2021; Hammer et al., 2020). This is important as these represent the climates likely to affect crop cultivars arising from current research and breeding programs (Collins and Chenu, 2021; Hammer et al., 2020).

10 Phenotyping methods for genomic selection and molecular breeding

For genomic and molecular genetic studies, large numbers of genotypes need to be phenotyped. Many QTL and genes associated with root traits have been reported, as outlined in a separate chapter of this volume. Traits expressed in seedlings and at early growth stages have been characterized using systems described earlier, including rice (Uga et al., 2011), maize (Burton et al., 2014), bread wheat (Christopher et al., 2013), durum (Canè et al., 2014), sorghum (Mace et al., 2012) and barley (Abdel-Ghani et al., 2019; Robinson et al., 2016). However, phenotyping large numbers of lines at later growth stages, particularly in the field, is more difficult and so has been attempted less frequently. In an example of semi-field and field studies, 323 genotypes of wheat, mainly

Chinese, including modern cultivars and landraces, were phenotyped in large tubes (11 cm × 180 cm) arrayed in a pit in the field (Li et al., 2019). Additionally, minirhizotrons were used to measure field-grown plants. This study allowed the identification of QTL for root depth, canopy temperature and yield per plant (Li et al., 2019).

11 Case study: late, deep root development in sub-tropical wheat

Clearly, there are many factors to consider when devising a root phenotyping strategy to assist the development of drought-resistant cereals for a particular environment. Therefore, it can be informative to consider a particular example of where this has been done.

11.1 The target environment

In the north-eastern, sub-tropical cropping region of Australia, spring habit cereals such as wheat and barley are sown as the soil cools in the autumn months, grow during the cooler winter months and are harvested as the season warms again in late spring to early summer. The sub-tropical climate means that these winter-grown crops are heavily reliant on stored soil water from the previous summer rainfall (Chenu et al., 2013). Water is stored in deep, cracking clay soils with high water-holding capacity that predominate in croplands in this region (Isbell, 2021). In-crop rainfall is limited in most years, leading to late-season water stress, adversely affecting yield in 75% of cropping seasons (Chenu et al., 2013). When there is little in-crop rainfall, the soil moisture in the upper layers is depleted as the crop develops, mostly by the seminal root system. This forces the plants to seek moisture deeper in the profile.

11.2 Identifying target traits

Late and deep root development has been shown to be advantageous for water-limited environments in Australia (Hammer et al., 2020). Late, deep roots may arise from the seminal root system as is usually the case for wheat or from the nodal root system as in maize and sorghum, for example. Water accessed between flowering and maturity has been shown to have a high marginal WUE for grain production. Up to 50–60 kg ha^{-1} of extra grain can be produced for each extra millimetre of water extracted from the soil post flowering. This has been demonstrated in field experiments in a temperate Australian cropping region (Kirkegaard et al., 2007). It has also been demonstrated in computer crop simulation modelling studies for the sub-tropical region of Australia (Manschadi et al., 2006).

The computer crop simulation model – 'the agricultural production simulator' (APSIM) – has been used to simulate production in sub-tropical environments using historic climate data from three representative sites and 105 years of daily historical weather data. Crop parameters were modified to simulate the effect of greater RLD at depth. This simulated a genotype with the ability to extract moisture from deeper layers to a lower moisture limit and to do this more rapidly than a standard genotype once roots reached any one soil layer. The differences in these parameters were based on measurements of root distribution and water extraction in large soil-filled root observation chambers, as discussed later (Manschadi et al., 2006).

A similar simulation study was conducted using 123 years of weather data for 61 locations for cropping regions throughout Australia, which included areas of sub-tropical, temperate, and Mediterranean climates (Veyradier et al., 2013). In this study, late, deep soil moisture extraction was predicted to be advantageous in most seasons, to varying degrees, in each of these regions (Veyradier et al., 2013). Importantly, yields for a phenotype with greater late, deep root development were only rarely predicted to be less than those for the standard phenotype in any of the studied regions. This suggests that a phenotype with increased late, deep root development leading to greater deep soil water extraction could be beneficial in many water-limited climates globally.

To study this phenotype in more detail, plants were grown in large root chambers which were 2.4 m long, 0.1 m wide and 1.2 m high with plexiglass sides to allow for observation of the roots during development (Fig. 4). A single plant was grown close to one transparent side of each chamber, and there was a gap of 1.2 m between chambers (Manschadi et al., 2006). The chambers were watered to field capacity at the commencement of the experiment, providing sufficient soil moisture in the large soil volume to allow a single wheat plant to grow to maturity without further watering. This meant that, once moisture had been fully extracted from the soil directly under the plant, the roots had to further explore the soil volume to acquire more water. As the soil was relatively homogeneous throughout the chamber, there was little impedance to root growth, and the genetic potential of each plant to explore the soil, mainly in two dimensions, could be expressed. Two wheat genotypes were compared: a well-characterized and widely grown standard genotype and a genotype which had exhibited greater yield in many rain-fed field experiments in the sub-tropical Australian cropping region (Christopher et al., 2008). Roots at the plexiglass surface were imaged periodically throughout development. Destructive harvests at anthesis and maturity, followed by washing of sections of the soil from the chambers, allowed quantification of root length, RLD and root biomass throughout the chambers (Manschadi et al., 2006).

When harvested at anthesis, the soil volume directly below the plants was nearly fully exploited by both the drought-adapted and the standard

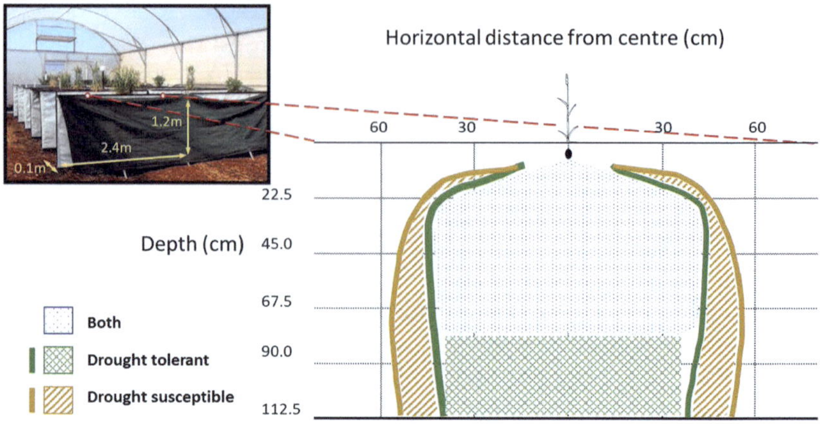

Figure 4 Diagram of part of the transparent side of one of the large root observation chambers shown in the inset photograph. In the photograph, the transparent sides are covered by a white vinyl, light-excluding cover. The maximum extent of the root systems of representative wheat plants is illustrated at maturity showing depth from the soil surface (which was 7.5 cm from the top of the chamber). The drought-tolerant genotype (green lines) had a narrower overall root system than the drought-susceptible genotype (tan lines). Both genotypes had fully exploited a volume of soil directly below the plant by the time of anthesis (blue stippled area). After anthesis, the drought-tolerant genotype developed more roots at depth (green cross-hatched area) than the drought-susceptible genotype, which developed more roots to the sides (tan-hatched area). For further detail, see Manschadi et al. (2006).

genotype. However, there were slightly more roots at depths from 0.7 m to 0.9 m under the drought-adapted genotype, while the standard line had slightly more roots distributed more widely to the side of the plant, at anthesis. These differences were more pronounced at maturity (Fig. 4). Although the root density directly under the plants to a depth of 0.7 m changed little between anthesis and maturity for either genotype, the drought-adapted line produced substantially more roots at depths below 0.8 m than the standard cultivar which produced slightly more at either side of the plant (Fig. 4). Importantly, the two genotypes achieved similar overall biomass in this system for both roots and shoots. The total root length per plant at maturity was also similar. It was the density and the spatial distribution of roots, particularly those developed post anthesis, that differed most between the genotypes.

The increased proportion of late, deep roots produced by the drought-tolerant genotype correlated with significantly more moisture extraction from the chamber post flowering. As water from the top layers was extracted early in crop development by the seminal root system, the development of adventitious roots from the stems and stem nodes was limited. Thus, the difference between genotypes was largely expressed in the seminal root system. These subtle differences in the distribution and timing of root development post flowering would not have been detected by phenotyping during early crop development or by examination of shallow roots in the field.

11.3 Phenotyping mapping and breeding populations

Method development was required to improve throughput to enable phenotyping of mapping and breeding populations. Although they are highly informative, the large chambers are labour intensive with low throughput. A forklift and operator were required to move them once full. A team of four people spent 2 days to assemble and fill 12 chambers. Disassembly also took the team several days at each harvest, and washing the roots from individual sections of the soil took several weeks.

The overall root pattern of the drought-tolerant line was narrower than that of the standard line with more roots at depth. This pattern of root distribution had been previously correlated with a more vertical seminal root angle in seedlings of maize (Nakamoto et al., 1991) and winter habit wheat in Japan (Oyanagi et al., 2001, 1993). Phenotying in small gel-filled chambers confirmed that seedlings of the drought-tolerant wheat line had a narrower seminal root angle as well as a greater number of seminal root (Benough et al., 2004; Maschadi et al., 2006). Phenotyping was then extended to a doubled haploid population made by crossing the drought-tolerant and standard genotypes. This allowed the identification of QTL for seminal root angle and number as well as for stay-green (Christopher et al., 2016, 2013).

Phenotyping of seminal roots in gel-filled chambers was more rapid than growing plants in large root chambers. However, it was still quite labour intensive and slow for characterizing the numbers of lines required for genetic studies. The requirement for sterilization of the materials and for sterile technique while assembling the gel-filled chambers added considerably to the time and labour required.

11.4 Developing tools for selection

To genotype a larger number of breeding lines in a more genetically diverse population, a multireference parent-nested association mapping (MR-NAM) population was developed (Christopher et al., 2021, 2018; Richard, 2017). To develop the MR-NAM, 11 lines donating target root traits and other adaptive traits were crossed to three reference parents producing over 1500 recombinant inbred lines (Christopher et al., 2021; Richard, 2017). To phenotype roots of a large number of lines, a more efficient seedling seminal root trait phenotyping method was developed using soil-filled transparent pots (Richard et al., 2015). This allowed the phenotyping of hundreds of lines in a small area in a growth chamber over only a few days.

The transparent pot system was used to directly select in early, segregating generations for seedling seminal angle (Richard et al., 2018). By selecting single plants for either narrow or wide root angle, both the phenotype and allele frequency were significantly shifted in the desired direction in the derived progeny (Richard et al., 2018).

Narrower overall root architecture was observed in genotypes with narrower seedling seminal root angle. However, phenotyping the top of the root system at the seedling stage might not necessarily translate to a greater RLD at depth, if the overall root mass was less, for example. Therefore, further investigations were conducted using polyvinyl chloride tubes of 1.5 m depth and 0.09 m diameter. These experiments revealed that imposing a moderate level of water stress at or shortly after flowering increased the difference in late, deep root growth (Shazadi et al., 2021).

As a result of these studies, breeding lines with favourable root traits have been selected and provided to wheat breeders in Australia along with information on QTL for the traits. These provide a valuable resource for the development of drought-tolerant wheat cultivars for growers. In addition, the GWAS for the MR-NAM population can potentially be used to devise genomic selection strategies for root traits to improve drought resistance.

12 Conclusion and future trends

12.1 Conclusion and recommendations

The discussion above suggests that researchers should consider some basic steps when devising a phenotyping strategy to identify root traits to select and breed for drought-resistant cereals for target environments. The aim would be to phenotype large numbers of genotypes using high-throughput technologies where possible and conduct detailed field phenotyping only for smaller numbers of selected genotypes in calibration and validation studies. Some basic steps and potential patterns for workflow are indicated below (Fig. 5). These represent a simplified scheme to emphasize the potential role of root phenotyping in breeding. The scheme would not necessarily be followed in the indicated order and is dependent on the target traits, target environments as well as the available genetic sources and prior data. Additionally, the arrows could mostly be followed in either direction.

The first step would be to identify genotypes with potential variations for root traits. For example, these could be lines with contrasting performance in the target environments based on field testing of diverse panels of genotypes, preferably using existing datasets (point 1, Fig. 5). Where root traits are likely contributing to adaptation, further phenotyping would then be conducted to test for variation in the most prospective root traits. Alternatively, genetic variation could be identified while phenotyping diverse genotypes *ex situ*

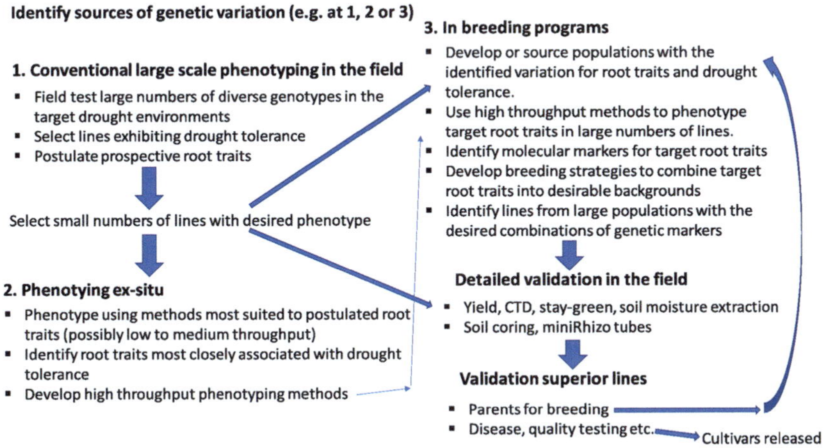

Figure 5 Diagram of some potential workflows for applying root phenotyping in plant breeding programs. It is first necessary to identify genetic variation for root traits, for example at positions 1, 2 or 3.

(point 2) or sourced from populations developed from known sources of genetic diversity for root traits (point 3).

Phenotyping techniques would then be selected based on the postulated target root traits and the developmental stage at which they are most likely to be expressed. Once traits have been identified, computer crop simulation models may potentially be used to identify traits with the most adaptative value in the target environments. However, before this can be done, the model needs to be parameterized for the target traits and carefully calibrated using data from field or *ex-situ* experiments. Methods would then be chosen or developed for high throughput (HTP) testing of the identified root traits. These would likely be *ex-situ* methods and/or field-based correlated traits. The HTP methods would then be used to phenotype large numbers of lines in GWASs and to devise efficient breeding strategies. The lines resulting from the breeding strategy could then be genotypically selected in early generations. Selected lines would then be phenotyped using the HTP strategies to select smaller sets of the most prospective lines to be root phenotyped in detail in the field along with yield testing. Finally, these validated lines would be used as parents in further rounds of breeding. Alternatively, they may be tested for other important traits prior to release as commercial cultivars.

12.2 Future trends

After many decades of slow development compared to shoot phenotyping technologies, root phenotyping has advanced rapidly, particularly in the last decade. Integration of traditional and advanced technologies has been a major contributor. Improvements in technology for automated plant culture and measurement, image capture, data capture and storage as well as analysis of large datasets have each been important.

An extensive array of phenotyping techniques is now available. For the researcher, the most important consideration is to define the research target so as to select the most appropriate and efficient phenotyping techniques. Consideration of the crop species and growth stage is important as are the climatic conditions and management practices encountered by the crop in the target population of environments.

Given the remaining constraints on field phenotyping, it is often still more efficient to conduct studies *ex situ*. This is especially the case for traits expressed early in development. To encompass the complete crop cycle, a range of different *ex-situ* phenotyping methods can be used which are each more suited to a specific range of developmental stages. However, as each method has its own specific limitations, care should be exercised when comparing results between methods and if attempting to integrate the results to explain the development of the root system over the entire crop cycle.

Some of the latest methods use high-cost and specialist instruments also requiring considerable input from highly trained professionals. However, there are methods available that can allow significant research progress even in low-income research settings.

Eventually, adaptations of the root system postulated to improve drought adaptation need to be validated in the field. Non-destructive aboveground phenotyping methods to measure CTD, stay-green and soil moisture extraction can help. These should be combined with focused field root phenotyping.

After a long period of slow progress, it seems that many new technologies are now becoming available. These, combined with innovative new strategies, will greatly increase the rate of advance in root phenotyping techniques in the coming years.

Root phenotyping is now reaching the stage where useful advances in the development of drought-resistant cereals are possible.

13 Where to look for further information

13.1 Further reading

For a good overview:

- Eshel, A. and Beeckman, T. (Eds.). (2013). *Plant Roots*: The Hidden Half. Boca Raton: CRC Press.
- Gregory, P. (Ed.). (2021). *Understanding and Improving Crop Root Function*. Cambridge, UK: Burleigh Dodds Science Publishing.

Recent useful review articles include Lynch (2022); Lynch et al. (2022); Lynch (2021).

13.2 Websites

- Australian Plant Phenomics Facility (APPF); https://www.plantphenomics.org.au › about-u.
- ICARDA, PhysioTron; https://www.icarda.org/media/news/icardas-new-physiotron-facility-ready-unblock-global-phenotyping-efforts.
- Jülich Plant Phenotyping Centre (JPPC); https://www.fz-juelich.de › ibg-2 › backup_research.
- Image analysis software; https://www.quantitative-plant.org.
- Pennsylvania State University: https://plantscience.psu.edu/research/labs/roots/projects/deeper.

13.3 Key journals and conferences

- ISRR; International Society for Root Research conference.
- IPPS; International Plant Phenotyping Symposium.
- Interdrought.
- PAG; Plant and Animal Genome.

13.4 Major international research projects

- IPPN; International Plant Phenotyping Network; http://www.plant-phenotyping.org.
- ARPA; https://arpa-e.energy.gov/technologies/programs/roots.

14 Abbreviations

Australian Plant Phenomics Facilities (APPF); agricultural production simulator, crop simulation model (APSIM); Australian Grains Research and Development Corporation (GRDC); canopy temperature depression (CTD); Centre for International Maize and Wheat Improvement (CIMMYT); electromagnetic induction (EMI); Food and Agriculture Organization of the United Nations (FAO); genome wide association studies (GWAS); ground penetrating radar (GPR); International Centre for Agriculture Research in Dry Areas (ICARDA); quantitative trait loci (QTL); magnetic resonance imaging (MRI); multi reference nested association mapping (MR-NAM); positron emission tomography (PET); Queensland Government, Department of Agriculture and Forestry (QDAF); red, green, blue colour system (RGB); root length density (RLD); root system architecture (RSA); specific root length (SRL); transpiration efficiency (TE);

University of Queensland, Queensland Alliance for Agriculture and Food Innovation (QAAFI); unmanned aerial vehicle (UAV); water use efficiency (WUE).

15 Acknowledgements

The author gratefully acknowledges the resources and funding provided by the University of Queensland, Queensland Alliance for Agriculture and Food Innovation, the Queensland Government Department of Agriculture and Forestry and the Australian Grains Research and Development Corporation (GRDC). Funding and support for the research described in the case study from GRDC and the support of the grain growers of Australia are also gratefully acknowledged. The thoughtful suggestions and editing by the Writers Action Learning group members at the Queensland Government, Leslie Research Facility, are also greatly appreciated. I also wish to thank the anonymous peer reviewers for their generous and thoughtful suggestions. Thanks also to Andrea Visioni for the photographs and information on the PhytoTron facility.

16 References

Abdel-Ghani, A. H., Sharma, R., Wabila, C., Dhanagond, S., Owais, S. J., Duwayri, M. A., Al-Dalain, S. A., Klukas, C., Chen, D., Lübberstedt, T., von Wirén, N., Graner, A., Kilian, B. and Neumann, K. (2019). Genome-wide association mapping in a diverse spring barley collection reveals the presence of QTL hotspots and candidate genes for root and shoot architecture traits at seedling stage, *BMC Plant Biology* 19(1): 216–216. https://doi.org/10.1186/s12870-019-1828-5.

Ahmed, E. M. (2015). Hydrogel: Preparation, characterization, and applications: a review, *Journal of Advanced Research* 6(2): 105–121.

Ahmed, M. A., Vetterlein, D. and Carminati, A. (2021). Advances in understanding plant root water uptake. In: Gregory, P. (Ed.), *Understanding and Improving Crop Root Function*. Cambridge, UK: Burleigh Dodds Science Publishing, pp. 373–392.

APPF. Australian plant phenomics facility. Available at: https://www.plantphenomics.org.au/.

Araus, J. L., Slafer, G. A., Reynolds, M. P. and Royo, C. (2002). Plant breeding and drought in C3 cereals: What should we breed for?, *Annals of Botany* 89(7): 925–940.

Araus, J. L., Slafer, G. A., Royo, C. and Serret, M. D. (2008). Breeding for yield potential and stress adaptation in cereals, *Critical Reviews in Plant Sciences* 27(6): 377–412.

Armstrong, W. (1980). Aeration in higher plants, *Advances in Botanical Research* 7: 225–332.

Armstrong, W. and Drew, M. C. (2002). Root growth and metabolism under oxygen deficiency. In: Eshel, A., A. and Beeckman, T. (Eds.), *Plant Roots: The Hidden Half* (3rd ed.). New York: Marcel Dekker.

Atkinson, J. A., Pound, M. P., Bennett, M. J. and Wells, D. M. (2019). Uncovering the hidden half of plants using new advances in root phenotyping, *Current Opinion in Biotechnology* 55: 1–8. https://doi.org/10.1016/j.copbio.2018.06.002.

Atkinson, J. A., Wingen, L. U., Griffiths, M., Pound, M. P., Gaju, O., Foulkes, M. J., Gouis, J. L., Griffiths, S., Bennett, M. J., King, J. and Wells, D. M. (2015). Phenotyping pipeline reveals major seedling root growth QTL in hexaploid wheat, *Journal of Experimental Botany* 66(8): 2283–2292. https://doi.org/10.1093/jxb/erv006.

Béné, C., Barange, M., Subasinghe, R., Pinstrup-Andersen, P., Merino, G., Hemre, G.-I. and Williams, M. (2015). Feeding 9 billion by 2050 – Putting fish back on the menu, *Food Security* 7(2): 261–274. https://doi.org/10.1007/s12571-015-0427-z.

Bengough, A. G., Gordon, D. C., Al-Menaie, H., Ellis, R. P., Allan, D., Keith, R., Thomas, W. T. B. and Forster, B. P. (2004). Gel observation chamber for rapid screening of root traits in cereal seedlings, *Plant and Soil* 262(1/2): 63–70.

Binley, A. and Kemna, A. (2005). DC resistivity and induced polarization methods. In: Rubin, Y. and Hubbard, S. S. (Eds.), *Hydrogeophysics*. Dordrecht, the Netherlands: Springer, pp. 129–156.

Borrell, A. K., Mullet, J. E., George-Jaeggli, B., van Oosterom, E. J., Hammer, G. L., Klein, P. E. and Jordan, D. R. (2014). Drought adaptation of stay-green cereals associated with canopy development, leaf anatomy, root growth and water uptake, *Journal of Experimental Botany* 65(21): 6251–6263.

Bramley, H., Turner, N. C., Turner, D. W. and Tyerman, S. D. (2009). Roles of morphology, anatomy, and aquaporins in determining contrasting hydraulic behaviour of roots, *Plant Physiology* 150(1): 348–364.

Brevik, E. C., Fenton, T. E. and Lazari, A. (2006). Soil electrical conductivity as a function of soil water content and implications for soil mapping, *Precision Agriculture* 7(6): 393–404. https://doi.org/10.1007/s11119-006-9021-x.

Bucksch, A., Burridge, J., York, L. M., Das, A., Nord, E., Weitz, J. S. and Lynch, J. P. (2014). Image-based high-throughput field phenotyping of crop roots, *Plant Physiology* 166(2): 470–486. https://doi.org/10.1104/pp.114.243519.

Burton, A. L., Johnson, J. M., Foerster, J. M., Hirsch, C. N., Buell, C. R., Hanlon, M. T., Kaeppler, S. M., Brown, K. M. and Lynch, J. P. (2014). QTL mapping and phenotypic variation for root architectural traits in maize (*Zea mays* L.), *TAG: Theoretical and Applied Genetics. Theoretische und Angewandte Genetik* 127(11): 2293–2311. https://doi.org/10.1007/s00122-014-2353-4.

Cai, G., Vanderborght, J., Klotzsche, A., van der Kruk, J., Neumann, J., Hermes, N. and Vereecken, H. (2016). Construction of minirhizotron facilities for investigating root zone processes, *Vadose Zone Journal* 15(9): 1–13. https://doi.org/10.2136/vzj2016.05.0043.

Canales, F. J., Nagel, K. A., Müller, C., Rispail, N. and Prats, E. (2019). Deciphering root architectural traits involved to cope with water deficit in oat, *Frontiers in Plant Science* 10: 1558–1558. https://doi.org/10.3389/fpls.2019.01558.

Canè, M. A., Maccaferri, M., Nazemi, G., Salvi, S., Francia, R., Colalongo, C. and Tuberosa, R. (2014). Association mapping for root architectural traits in durum wheat seedlings as related to agronomic performance, *Molecular Breeding* 34(4): 1629–1645. https://doi-org.ezproxy.library.uq.edu.au/10.1007/s11032-014-0177-1.

Carter, M. R. (2004). "Conservation tillage." Encyclopedia of soils in the environment, *Sciencedirect*. Elsevier B.V.

Casadebaig, P., Zheng, B., Chapman, S., Huth, N., Faivre, R. and Chenu, K. (2016). Assessment of the potential impacts of wheat plant traits across environments by combining crop modeling and global sensitivity analysis, *PLoS ONE* 11(1): e0146385. https://doi.org/10.1371/journal.pone.0146385.

Chapman, S. C. (2008). Use of crop models to understand genotype by environment interactions for drought in real-world and simulated plant breeding trials, *Euphytica* 161(1–2): 195–208. https://doi.org/10.1007/s10681-007-9623-z.

Chapman, S. C., Hammer, G. L., Butler, D. G. and Cooper, M. (2000). Genotype by environment interactions affecting grain sorghum. III. Temporal sequences and spatial patterns in the target population of environments, *Australian Journal of Agricultural Research* 51(2): 223–233. https://doi.org/10.1071/AR99022.

Chaumont, F., Tyerman, S. D. and SpringerLink. (2017). *Plant Aquaporins From Transport to Signaling*. Springer International Publishing.

Chen, Y. L., Dunbabin, V. M., Diggle, A. J., Siddique, K. H. M. and Rengel, Z. (2011). Development of a novel semi-hydroponic phenotyping system for studying root architecture, *Functional Plant Biology* 38(5): 355–363.

Chenu, K., Deihimfard, R. and Chapman, S. C. (2013). Large-scale characterization of drought pattern: A continent-wide modelling approach applied to the Australian wheatbelt – Spatial and temporal trends, *The New Phytologist* 198(3): 801–820. https://doi.org/10.1111/nph.12192.

Chenu, K., Van Oosterom, E. J., McLean, G., Deifel, K. S., Fletcher, A., Geetika, G., Tirfessa, A., Mace, E. S., Jordan, D. R., Sulman, R. and Hammer, G. L. (2018). Integrating modelling and phenotyping approaches to identify and screen complex traits: Transpiration efficiency in cereals, *Journal of Experimental Botany* 69(13): 3181–3194.

Christopher, J. T., Christopher, M. J., Borrell, A. K., Fletcher, S. and Chenu, K. (2016). Stay-green traits to improve wheat adaptation in well-watered and water-limited environments, *Journal of Experimental Botany* 67(17): 5159–5172. https://doi.org/10.1093/jxb/erw276.

Christopher, J. T., Christopher, M. J., Jennings, R., Jones, S., Fletcher, S., Borrell, A. K., Manschadi, A. M., Jordan, D., Mace, E. and Hammer, G. L. (2013). QTL for root angle and number in a population developed from bread wheats (*Triticum aestivum*) with contrasting adaptation to water-limited environments, *TAG: Theoretical and Applied Genetics. Theoretische und Angewandte Genetik* 126(6): 1563–1574.

Christopher, J. T., Manschadi, A. M., Hammer, G. L. and Borrell, A. K. (2008). Developmental and physiological traits associated with high yield and stay-green phenotype in wheat, *Australian Journal of Agricultural Research* 59(4): 354–364.

Christopher, J. T., Veyradier, M., Borrell, A. K., Harvey, G., Fletcher, S. and Chenu, K. (2014). Phenotyping novel stay-green traits to capture genetic variation in senescence dynamics, *Functional Plant Biology* 41(11): 1035–1048.

Christopher, M., Chenu, K., Jennings, R., Fletcher, S., Butler, D., Borrell, A. and Christopher, J. (2018). QTL for stay-green traits in wheat in well-watered and water-limited environments, *Field Crops Research* 217: 32–44. https://doi.org/10.1016/j.fcr.2017.11.003.

Christopher, M., Paccapelo, V., Kelly, A., Macdonald, B., Hickey, L., Richard, C., Verbyla, A., Chenu, K., Borrell, A., Amin, A. and Christopher, J. (2021). QTL identified for stay-green in a multi-reference nested association mapping population of wheat exhibit context dependent expression and parent-specific alleles, *Field Crops Research* 270: 108181.

Clark, R. T., MacCurdy, R. B., Jung, J. K., Shaff, J. E., McCouch, S. R., Aneshansley, D. J. and Kochian, L. V. (2011). Three-dimensional root phenotyping with a novel imaging and software platform, *Plant Physiology* 156(2): 455–465.

Clément, C. (2021). *Deep water uptake of perennial crops. A case study on intermediate wheatgrass and alfalfa*. PhD Thesis, University of Copenhagen, Denmark.
Collins, B. and Chenu, K. (2021). Improving productivity of Australian wheat by adapting sowing date and genotype phenology to future climate, *Climate Risk Management* 32: 100300.
Colombi, T., Kirchgessner, N., Le Marié, C. A., York, L. M., Lynch, J. P. and Hund, A. (2015). Next generation shovelomics: Set up a tent and REST, *Plant and Soil* 388(1-2): 1-20. https://doi.org/10.1007/s11104-015-2379-7.
Comas, L. H., Becker, S. R., Cruz, V. M. V., Byrne, P. F. and Dierig, D. A. (2013). Root traits contributing to plant productivity under drought, *Frontiers in Plant Science* 4: 442-442. https://doi.org/10.3389/fpls.2013.00442.
Condon, A. G., Richards, R. A., Rebetzke, G. J. and Farquhar, G. D. (2004). Breeding for high water-use efficiency, *Journal of Experimental Botany* 55(407): 2447-2460.
Corona-Lopez, D. D. J., Sommer, S., Rolfe, S. A., Podd, F. and Grieve, B. D. (2019). Electrical impedance tomography as a tool for phenotyping plant roots, *Plant Methods* 15: 49-49.
Crews, T. E., Carton, W. and Olsson, L. (2018). Is the future of agriculture perennial? Imperatives and opportunities to reinvent agriculture by shifting from annual monocultures to perennial polycultures, *Global Sustainability* 1: 1-18. https://doi.org/10.1017/sus.2018.11.
Dang, Y. P., Dalal, R. C., Pringle, M. J., Biggs, A. J. W., Darr, S., Sauer, B., Moss, J., Payne, J. and Orange, D. (2011). Electromagnetic induction sensing of soil identifies constraints to the crop yields of north-eastern Australia, *Soil Research* 49(7): 559-571.
Dang, Y. P., Christopher, J. T. and Dalal, R. C. (2016). Genetic diversity in barley and wheat for tolerance to soil constraints, *Agronomy* 6(4): 55. https://doi.org/10.3390/agronomy6040055.
Dietrich, D. (2018). Hydrotropism: How roots search for water, *Journal of Experimental Botany* 69(11): 2759-2771. https://doi.org/10.1093/jxb/ery034.
Dowd, T., McInturf, S., Li, M. and Topp, C. N. (2021). Rated-M for mesocosm: Allowing the multimodal analysis of mature root systems in 3D, *Emerging Topics in Life Sciences* 5(2): 249-260. https://doi.org/10.1042/ETLS20200278.
Dreccer, M. F., Condon, A. G., Macdonald, B., Rebetzke, G. J., Awasi, M.-A., Borgognone, M. G., Peake, A., Piñera-Chavez, F. J., Hundt, A., Jackway, P. and McIntyre, C. L. (2020). Genotypic variation for lodging tolerance in spring wheat: Wider and deeper root plates, a feature of low lodging, high yielding germplasm, *Field Crops Research* 258: 107942. https://doi.org/10.1016/j.fcr.2020.107942.
Esau, K. (1977). *Anatomy of Seed Plants* (2nd ed.). New York: Wiley, pp. 215-239.
Eshel, A. and Beeckman, T. (Eds.). (2013). *Plant Roots: The Hidden Half*. Boca Raton: CRC Press.
FAO (2018). The impact of disasters and crises on agriculture and food security. Available at: http://www.fao.org/documents/card/en/c/cb3673en.
FAO (2019). Proactive approaches to drought preparedness - Where are we now and where do we go from here? Available at: http://www.fao.org/3/ca5794en/ca5794en.pdf.
Fitter, A. (2002). Characterization and function of the root. In: Eshel, A., Beeckman, T. and Dekker, M. (Eds.), *Plant Roots the Hidden Half*. New York, p. 15.
Flavel, R. J., Guppy, C. N., Rabbi, S. M. R. and Young, I. M. (2017). An image processing and analysis tool for identifying and analysing complex plant root systems in 3D soil

using non-destructive analysis: Root, *PLoS ONE* 12(5): e0176433. https://doi.org/10.1371/journal.pone.0176433.

Flavel, R. J., Guppy, C. N., Tighe, M., Watt, M., McNeill, A. and Young, I. M. (2012). Non-destructive quantification of cereal roots in soil using high-resolution X-ray tomography, *Journal of Experimental Botany* 63(7): 2503–2511. https://doi.org/10.1093/jxb/err421.

Frensch, J. and Steudle, E. (1989). Axial and radial hydraulic resistance to roots of maize (*Zea mays* L.), *Plant Physiology* 91(2): 719–726.

Gao, W., Hodgkinson, L., Jin, K., Watts, C. W., Ashton, R. W., Shen, J., Ren, T., Dodd, I. C., Binley, A., Phillips, A. L., Hedden, P., Hawkesford, M. J. and Whalley, W. R. (2016). Deep roots and soil structure, *Plant, Cell and Environment* 39(8): 1662–1668. https://doi.org/10.1111/pce.12684.

Gracia-Romero, A., Kefauver, S. C., Fernandez-Gallego, J. A., Vergara-Díaz, O., Teresa Nieto-Taladriz, M. T. and Araus, J. L. (2019). UAV and ground image-based phenotyping: A proof of concept with durum wheat, *Remote Sensing (Basel, Switzerland)* 11(10): 1244. https://doi.org/10.3390/rs11101244.

Gregory, P. (Ed.). (2021). *Understanding and Improving Crop Root Function*. Cambridge, UK: Burleigh Dodds Science Publishing.

Gupta, A., Rico-Medina, A. and Cano-Delgado, A. I. (2020). The physiology of plant responses to drought, *Science* 368(6488): 266–269. https://doi.org/10.1126/science.aaz7614.

Hainsworth, J. M. and Aylmore, L. A. G. (1983). The use of computer assisted tomography to determine spatial distribution of soil water content, *Soil Research* 21(4): 435–443.

Hammer, G. L., Dong, Z., McLean, G., Doherty, A., Messina, C., Schussler, J., Zinselmeier, C., Paszkiewicz, S. and Cooper, M. (2009). Can changes in canopy and/or root system architecture explain historical maize yield trends in the U. S. corn belt?, *Crop Science* 49(1): 299–312. https://doi.org/10.2135/cropsci2008.03.0152.

Hammer, G. L., McLean, G., von Oosterom, E., Chapman, S., Zheng, B., Wu, A., Doherty, A. and Jordan, D. (2020). Designing crops for adaptation to the drought and high-temperature risks anticipated in future climates, *Crop Science* 60(2): 605–621. https://doi.org/10.1002/csc2.20110.

Hardie, M. (2020). Review of novel and emerging proximal soil moisture sensors for use in agriculture, *Sensors* 20(23): 6934. https://doi.org/10.3390/s20236934.

Heymans, A., Couvreur, V., LaRue, T., Paez-Garcia, A. and Lobet, G. (2020). GRANAR, a computational tool to better understand the functional importance of monocotyledon root anatomy, *Plant Physiology (Bethesda)* 182(2): 707–720. https://doi.org/10.1104/pp.19.00617.

Hochholdinger, F. (2009). The maize root system: Morphology, anatomy, and genetics. In: Bennetzen, J. L. and Hake, S. C. (Eds.), *Handbook of Maize: Its Biology*. New York, NY: Springer. https://doi.org/10.1007/978-0-387-79418-1_8.

Hund, A., Trachsel, S. and Stamp, P. (2009). Growth of axile and lateral roots of maize: I development of a phenotying platform, *Plant and Soil* 325(1–2): 335–349.

ICARDA and PhysioTron. Available at: https://www.icarda.org/media/news/icardas-new-physiotron-facility-ready-unblock-global-phenotyping-efforts.

IPPN, International Plant Phenotyping Network. Available at: http://www.plant-phenotyping.org › Root_phenotyping.

Isbell, R. F. (2021). National Committee on soil terrain, & CSIRO Publishing, issuing body. In: *Australian Soil Classification* (3rd ed.). CSIRO Publishing.

Jahnke, S., Menzel, M. I., van Dusschoten, D., Roeb, G. W., Bühler, J., Minwuyelet, S., Blümler, P., Temperton, V. M., Hombach, T., Streun, M., Beer, S., Khodaverdi, M., Ziemons, K., Coenen, H. H. and Schurr, U. (2009). Combined MRI–PET dissects dynamic changes in plant structures and functions, *The Plant Journal: For Cell and Molecular Biology* 59(4): 634–644. https://doi.org/10.1111/j.1365-313X.2009.03888.x.

Javot, H. and Maurel, C. (2002). The role of aquaporins in root water uptake, *Annals of Botany* 90(3): 301–313.

Jeudy, C., Adrian, M., Baussard, C., Bernard, C., Bernaud, E., Bourion, V., Busset, H., Cabrera-Bosquet, L., Cointault, F., Han, S., Lamboeuf, M., Moreau, D., Pivato, B., Prudent, M., Trouvelot, S., Truong, H. N., Vernoud, V., Voisin, A. S., Wipf, D. and Salon, C. (2016). RhizoTubes as a new tool for high throughput imaging of plant root development and architecture: Test, comparison with pot grown plants and validation, *Plant Methods* 12: 31–31. https://doi.org/10.1186/s13007-016-0131-9.

Johnson, M. G., Tingey, D. T., Phillips, D. L. and Storm, M. J. (2001). Advancing fine root research with minirhizotrons, *Environmental and Experimental Botany* 45(3): 263–289.

Kirkegaard, J. A., Lilley, J. M., Howe, G. N. and Graham, J. M. (2007). Impact of subsoil water use on wheat yield, *Australian Journal of Agricultural Research* 58(4): 303–315.

Klein, S. P., Schneider, H. M., Perkins, A. C., Brown, K. M. and Lynch, J. P. (2020). Multiple integrated root phenotypes are associated with improved drought tolerance, *Plant Physiology* 183(3): 1011–1025. https://doi.org/10.1104/pp.20.00211.

Knodel, P., Silvestri, V., Sarkis, G., Bekkouche, N., Soulie, M. and Tabib, C. (1991). Laboratory and field calibration of a neutron depth moisture gauge for use in high water content soils, *Geotechnical Testing Journal* 14(1): 64–70.

Kuijken, R. C., van Eeuwijk, F. A., Marcelis, L . F. and Bouwmeester, H. J. (2015). Root phenotyping: From component trait in the lab to breeding, *Journal of Experimental Botany* 66(18): 5389–5401.

Langridge, P. and Reynolds, M. (2021). Breeding for drought and heat tolerance in wheat, *TAG. Theoretical and Applied Genetics. Theoretische und Angewandte Genetik* 134(6): 1753–1769. https://doi.org/10.1007/s00122-021-03795-1.

Le Marié, C., Kirchgessner, N., Flütsch, P., Pfeifer, J., Walter, A. and Hund, A. (2016). RADIX: Rhizoslide platform allowing high throughput digital image analysis of root system expansion, *Plant Methods* 12(1): 40. https://doi.org/10.1186/s13007-016-0140-8.

Li, C., Li, L., Reynolds, M. P., Wang, J., Chang, X., Nao, X. and Jing, R. (2021). Recognizing the hidden half: Root system attributes associated with wheat drought tolerance, *Journal of Experimental Botany* 72(14): 5117–5133. https://doi.org/10.1093/jxb/erab124.

Li, J., Smith, D. W., Fityus, S. G. and Sheng, D. (2003). Numerical analysis of neutron moisture probe measurements, *International Journal of Geomechanics* 3(1): 11–20. https://doi.org/10.1061/(ASCE)1532-3641(2003)3:1(11).

Li, L., Peng, Z., Mao, X., Wang, J., Chang, X., Reynolds, M. and Jing, R. (2019). Genome-wide association study reveals genomic regions controlling root and shoot traits at late growth stages in wheat, *Annals of Botany* 124(6): 993–1006. https://doi-org .ezproxy.library.uq.edu.au/10.1093/aob/mcz041Liu.

Liu, S., Barrow, C. S., Hanlon, M., Lynch, J. P. and Bucksch, A. (2021). DIRT/3D: 3D root phenotyping for field-grown maize (*Zea mays*), *Plant Physiology* 187(2): 739–757. https://doi.org/10.1093/plphys/kiab311.

Liu, X., Dong, X. and Leskovar, D. I. (2016). Ground penetrating radar for underground sensing in agriculture: A review, *International Agrophysics* 30(4): 533–543.

Lobet, G., Pagès, L. and Draye, X. (2011). A novel image-analysis toolbox enabling quantitative analysis of root system architecture, *Plant Physiology* 157(1): 29–39. https://doi.org/10.1104/pp.111.179895.

Lopes, M. S. and Reynolds, M. P. (2010). Partitioning of assimilates to deeper roots is associated with cooler canopies and increased yield under drought in wheat, *Functional Plant Biology* 37(2): 147–156.

Lynch, J. P. (2007). Roots of the second green revolution, *Australian Journal of Botany* 55(5): 493–512.

Lynch, J. P. (2013). Steep, cheap and deep: An ideotype to optimize water and N acquisition by maize root systems, *Annals of Botany* 112(2): 347–357. https://doi.org/10.1093/aob/mcs293.

Lynch, J. P. (2018). Rightsizing root phenotypes for drought resistance, *Journal of Experimental Botany* 69(13): 3279–3292. https://doi.org/10.1093/jxb/ery048.

Lynch, J. P. (2021). Root biology in the 21st century: Challenges and opportunities, *Annals of Botany* 128(1): i–ii. https://doi.org/10.1093/aob/mcab062.

Lynch, J. P. (2022). Harnessing root architecture to address global challenges, *The Plant Journal: For Cell and Molecular Biology* 109(2): 415–431. https://doi.org/10.1111/tpj.15560.

Lynch, J. P., Chimungu, J. G. and Brown, K. M. (2014). Root anatomical phenes associated with water acquisition from drying soil: Targets for crop improvement, *Journal of Experimental Botany* 65(21): 6155–6166. https://doi.org/10.1093/jxb/eru162.

Lynch, J. P., Mooney, S. J., Strock, C. F. and Schneider, H. M. (2022). Future roots for future soils, *Plant, Cell and Environment* 45(3): 620–636.

Lynch, J. P., Strock, C. F., Schneider, H. M., Sidhu, J. S., Ajmera, I., Galindo-Castañeda, T., Klein, S. P. and Hanlon, M. T. (2021). Root anatomy and soil resource capture, *Plant and Soil* 466(1–2): 21–63. https://doi.org/10.1007/s11104-021-05010-y.

Maa, L., Shia, Y., Siemianowskia, O., Yuanb, B., Egnerc, T. K., Mirnezamid, S. V., Linda, K. R., Baskar Ganapathysubramani, B., Vendittic, V. and Cademartiria, L. (2019). Hydrogel-based transparent soils for root phenotyping in vivo, *Proceedings of the National Academy of Sciences of the United States of America* 116(22): 11063–11068.

Mace, E. S., Singh, V., Van Oosterom, E. J., Hammer, G. L., Hunt, C. H. and Jordan, D. R. (2012). QTL for nodal root angle in sorghum (*Sorghum bicolor* L. Moench) co-locate with QTL for traits associated with drought adaptation, *TAG. Theoretical and Applied Genetics. Theoretische und Angewandte Genetik* 124(1): 97–109. https://doi.org/10.1007/s00122-011-1690-9.

Maeght, J. L., Rewald, B. and Pierret, A. (2013). How to study deep roots-and why it matters, *Frontiers in Plant Science* 4: 299.

Manschadi, A. M., Christopher, J., deVoil, P. and Hammer, G. L. (2006). The role of root architectural traits in adaptation of wheat to water-limited environments, *Functional Plant Biology: Functional Plant Biology* 33(9): 823–837. https://doi.org/10.1071/FP06055.

Manschadi, A. M., Christopher, J. T., Hammer, G. L. and deVoil, P. (2010). Experimental and modelling studies of drought-adaptive root architectural traits in wheat (*Triticum aestivum* L.), *Plant Biosystems – An International Journal Dealing With All Aspects of Plant Biology* 144(2): 458–462. https://doi.org/10.1080/11263501003731805.

Manschadi, A. M., Hammer, G. L., Christopher, J. T. and deVoil, P. (2008). Genotypic variation in seedling root architectural traits and implications for drought adaptation in wheat (*Triticum aestivum* L.), *Plant and Soil* 303(1–2): 115–129. https://doi.org/10.1007/s11104-007-9492-1.

McGrail, R., Van Sanford, D. and McNear, D. (2020). Trait-based root phenotyping as a necessary tool for crop selection and improvement, *Agronomy* 10(9): 1328.

Menamo, T. M., Bantte, K., Jordan, D., Mace, E., Hunt, C. and Borrell, A. K. (2017). *Characterization of Ethiopian sorghum [*Sorghum bicolor *(L) Moench] germplasm collection for drought adaptation traits associated with roots*. Hyderabad, India: Interdrought-V.

Merrill, S. D., Tanaka, D. L. and Hanson, J. D. (2005). Comparison of fixed-wall and pressurized-wall minirhizotrons for fine root growth measurements in eight crop species, *Agronomy Journal* 97(5): 1367–1373. https://doi.org/10.2134/agronj2004.0209.

Metzner, R., Eggert, A., van Dusschoten, D., Pflugfelder, D., Gerth, S., Schurr, U., Uhlmann, N. and Jahnke, S. (2015). Direct comparison of MRI and X-ray CT technologies for 3D imaging of root systems in soil: Potential and challenges for root trait quantification, *Plant Methods* 11: 17–17. https://doi.org/10.1186/s13007-015-0060-z.

Mi, G., Chen, F., Yuan, L. and Zhang, F. (2016). Chapter 2: Ideotype root system architecture for maize to achieve high yield and resource use efficiency in intensive cropping systems, *Advances in Agronomy* 139: 73–97. https://doi.org/10.1016/bs.agron.2016.05.002.

Mooney, S. J., Pridmore, T. P., Helliwell, J. and Bennett, M. J. (2012). Developing X-ray computed tomography to non-invasively image 3-D root systems architecture in soil, *Plant and Soil* 352(1–2): 1–22.

Mustroph, A. and Albercht, G. (2003). Tolerance of crop plants to oxygen deficiency stress: Fermenative activity and photosynthetic capacity of entire seedlings under hypoxia and anoxia, *Physiologia Plantarum* 117(4): 508–520.

Muthert, L. W. F., Izzo, L. G., van Zanten, M. and Aronne, G. (2020). Root tropisms: Investigations on earth and in space to unravel plant growth direction, *Frontiers in Plant Science* 21. https://doi.org/10.3389/fpls.2019.01807.

Nagel, K. A., Putz, A., Gilmer, F., Heinz, K., Fischbach, A., Pfeifer, J., Faget, M., Blossfeld, S., Ernst, M., Dimaki, C., Kastenholz, B., Kleinert, A. K., Galinski, A., Scharr, H., Fiorani, F. and Schurr, U. (2012). GROWSCREEN-Rhizo is a novel phenotyping robot enabling simultaneous measurements of root and shoot growth for plants grown in soil-filled rhizotrons, *Functional Plant Biology* 39(11): 891–904.

Nakamoto, T., Shimoda, K. and Matsuzaki, A. (1991). Elongation angle of nodal roots and its possible relation to spatial root distribution in maize and foxtail millet, *Japanese Journal of Crop Science* 60(4): 543–549.

Ober, E. S., Alahmad, S., Cockram, J., Forestan, C., Hickey, L. T., Kant, J., Maccaferri, M., Marr, E., Milner, M., Pinto, F., Rambla, C., Reynolds, M., Salvi, S., Sciara, G., Snowdon, R. J., Thomelin, P., Tuberosa, R., Uauy, C., Voss-Fels, K. P., Wallington, E. and Watt, M. (2021). Wheat root systems as a breeding target for climate resilience, *TAG. Theoretical and Applied Genetics. Theoretische und Angewandte Genetik* 134(6): 1645–1662. https://doi.org/10.1007/s00122-021-03819-w.

Oyanagi, A., Kiribuchi-Otobe, C., Yanigisawa, T., Honda, I. and Wada, M. (2001). Selection of wheat experimental lines with deep and shallow root systems based on the growth angle of seminal roots, *Japanese Journal of Crop Science* 70(3): 400–407.

Oyanagi, A., Nakamoto, T. and Wada, M. (1993). Relationship between root growth angle of seedlings and vertical distribution of roots in the field in wheat cultivars, *Japanese Journal of Crop Science* 62(4): 565–570.

Passioura, J. B. (1972). The effect of root geometry on the yield of wheat growing on stored water, *Australian Journal of Agricultural Research* 23(5): 745–752. https://doi.org/10.1071/AR9720745.

Passioura, J. B. (1977). Grain yield, harvest index and water use of wheat, *Journal of the Australian Institute of Agriculture Science* 43: 11–121.

Passioura, J. B. (1983). Roots and drought resistance, *Plant Production and Management Under Drought Conditions* 7: 265–280.

Passioura, J. B. (2006). Viewpoint: The perils of pot experiments, *Functional Plant Biology* 33(12): 1075–1079.

Passioura, J. B. (2010). Scaling up: The essence of effective agricultural research, *Functional Plant Biology* 37(7): 585–591.

Passot, S., Gnacko, F., Moukouanga, D., Lucas, M., Guyomarc'h, S., Ortega, B. M., Atkinson, J. A., Belko, M. N., Bennett, M. J., Gantet, P., Wells, D. M., Guédon, Y., Vigouroux, Y., Verdeil, J. L., Muller, B. and Laplaze, L. (2016). Characterization of pearl millet root architecture and anatomy reveals three types of lateral roots, *Frontiers in Plant Science* 7: 829–829. https://doi.org/10.3389/fpls.2016.00829.

Pflugfelder, D., Metzner, R., van Dusschoten, D., Reichel, R., Siegfried Jahnke, S. and Koller, R. (2017). Non-invasive imaging of plant roots in different soils using magnetic resonance imaging (MRI), *Plant Methods* 13: 102. https://doi.org/10.1186/s13007-017-0252-9.

Piñera-Chavez, F. J., Berry, P. M., Foulkes, M. J., Molero, G. and Reynolds, M. P. (2016). Avoiding lodging in irrigated spring wheat. II. Genetic variation of stem and root structural properties, *Field Crops Research* 196: 64–74.

Piñeros, M. A., Larson, B. G., Shaff, J. E., Schneider, D. J., Falcão, A. X., Yuan, L., Clark, R. T., Craft, E. J., Davis, T. W., Pradier, P. L., Shaw, N. M., Assaranurak, I., McCouch, S. R., Sturrock, C., Bennett, M. and Kochian, L. V. (2016). Evolving technologies for growing, imaging and analyzing 3D root system architecture of crop plants, *Journal of Integrative Plant Biology* 58(3): 230–241. https://doi.org/10.1111/jipb.12456.

Pinto, R. S. and Reynolds, M. P. (2015). Common genetic basis for canopy temperature depression under heat and drought stress associated with optimized root distribution in bread wheat, *TAG. Theoretical and Applied Genetics. Theoretische und Angewandte Genetik* 128(4): 575–585. https://doi.org/10.1007/s00122-015-2453-9.

Postma, J. A., Kuppe, C., Owen, M. R., Mellor, N., Griffiths, M., Bennett, M. J., Lynch, J. P. and Watt, M. (2017). OpenSimRoot: Widening the scope and application of root architectural models, *New Phytologist* 215(3): 1274–1286. Available at: https://rootmodels.gitlab.io/.

Postma, J. A. and Lynch, J. P. (2011). Root cortical aerenchyma enhances the growth of maize on soils with suboptimal availability of nitrogen, phosphorus, and potassium, *Plant Physiology* 156(3): 1190–1201.

Qiao, S., Fang, Y., Wu, A., Xu, B., Zhang, S., Deng, X., Djalovic, I., Siddique, K. H. M. and Chen, Y. (2019). Dissecting root trait variability in maize genotypes using the semi-hydroponic phenotyping platform, *Plant and Soil* 439(1–2): 75–90. https://doi.org/10.1007/s11104-018-3803-6.

Racine Root Laboratory. Tel Aviv University Botanic Gardens. Available at: https://www.tau.ac.il › lifesci › units › Gardens › roots.

Ramalingam, P., Kamoshita, A., Deshmukh, V., Yaginuma, S. and Uga, Y. (2017). Association between root growth angle and root length density of a near-isogenic line of IR64 rice with DEEPER ROOTING 1 under different levels of soil compaction, *Plant Production Science* 20(2): 162–175.

Ray, D. K., Mueller, N. D., West, P. C. and Foley, J. A. (2013). Yield trends are insufficient to double global crop production by 2050, *PLoS ONE* 8(6): e66428. https://doi.org/10.1371/journal.pone.0066428.

Rellán-Álvarez, R., Lobet, G., Lindner, H., Pradier, P. L., Sebastian, J., Yee, M. C., Geng, Y., Trontin, C., LaRue, T., Schrager-Lavelle, A., Haney, C. H., Nieu, R., Maloof, J., Vogel, J. P. and Dinneny, J. R. (2015). GLO-roots: An imaging platform enabling multidimensional characterization of soil-grown root systems, *eLife* 4. https://doi.org/10.7554/eLife.07597.

Rewald, B. and Ephrath, J. (2013). Soil conditions and cereal root architecture: Minirhizotron techniques, *Journal of Experimental Biology* 64: 1193–1208.

Rich, S. M., Christopher, J., Richards, R. and Watt, M. (2020). Root phenotypes of young wheat plants grown in controlled environments show inconsistent correlation with mature root traits in the field, *Journal of Experimental Botany* 71(16): 4751–4762. https://doi.org/10.1093/jxb/eraa201.

Richard, C. (2017). *Breeding wheat for drought adaptation: Development of selection tools for root architectural traits*. Thesis, The University of Queensland, Qld Alliance for Agriculture and Food Innovation. https://doi.org/10.14264/uql.2017.1055.

Richard, C., Christopher, J., Chenu, K., Borrell, A., Christopher, M. and Hickey, L. (2018). Selection in early generations to shift allele frequency for seminal root angle in wheat, *The Plant Genome* 11(2): 170071. https://doi.org/10.3835/plantgenome2017.08.0071.

Richard, C. A., Hickey, L. T., Fletcher, S., Jennings, R., Chenu, K. and Christopher, J. T. (2015). High-throughput phenotyping of seminal root traits in wheat, *Plant Methods* 11: 13.

Richards, R. A. (2008). Genetic opportunities to improve cereal root systems for dryland agriculture, *Plant Production Science* 11(1): 12–16.

Richards, R. A. and Passioura, J. B. (1981). Seminal root morphology and water use of wheat i. environmental effects, *Crop Science* 21(2): 249–252.

Richards, R. A. and Passioura, J. B. (1989). A breeding program to reduce the diameter of the major xylem vessel in the seminal roots of wheat and its effect on grain yield in rain-fed environments, *Australian Journal of Agricultural Research* 40(5): 943–950. https://doi.org/10.1071/AR9890943.

Richards, R. A., Rebetzke, G. J., Watt, M., Condon, A. G., Spielmeyer, W. and Dolferus, R. (2010). Breeding for improved water productivity in temperate cereals: Phenotyping, quantitative trait loci, markers and the selection environment, *Functional Plant Biology* 37(2): 85–97.

Robinson, H., Hickey, L., Richard, C., Mace, E., Kelly, A., Borrell, A., Franckowiak, J. and Fox, G. (2016). Genomic regions influencing seminal root traits in barley, *The Plant Genome* 9(1): 1–13. https://doi.org/10.3835/plantgenome2015.03.0012.

Ryan, P. R., Delhaize, E., Watt, M. and Richardson, A. E. (2016). Plant roots: Understanding structure and function in an ocean of complexity, *Annals of Botany* 118(4): 555–559.

Rytter, R.-M. and Rytter, L. (2012). Quantitative estimates of root densities at minirhizotrons differ from those in the bulk soil, *Plant and Soil* 350(1–2): 205–220. https://doi.org/10.1007/s11104-011-0896-6.

Samson, B. K. and Sinclair, T. R. (1994). Soil core and minirhizotron comparison for the determination of root length density, *Plant and Soil* 161(2): 225–232. https://doi.org/10.1007/BF00046393.

Schneider, M. H. and Lynch, J. P. (2018). Functional implications of root cortical senescence for soil resource capture, *Plant and Soil* 423: 13–26.

Schulz, H., Postma, J. A., van Dusschoten, D., Scharr, H. and Behnke, S. (2013). Plant root system analysis from MRI images. In: Csurka, G., Kraus, M., Laramee, R. S., Richard, P. and Braz, J. (Eds.), *Communications in Computer and Information Science. Computer Vision, Imaging and Computer Graphics. Theory and Application*. Berlin, Heidelberg: Springer, pp. 411–425. https://doi.org/10.1007/978-3-642-38241-3_28.

Seethepalli, A., Guo, H., Liu, X., Griffiths, M., Almtarfi, H., Li, Z., Liu, S., Zare, A., Fritschi, F. B., Blancaflor, E. B., Ma, X. F. and York, L. M. (2020). RhizoVision crown: An integrated hardware and software platform for root crown phenotyping, *Plant Phenomics* 2020: 3074916.

Selvaraj, M. G., Montoya-P, M. E., Atanbori, J., French, A. P. and Pridmore, T. (2019). A low-cost aeroponic phenotyping system for storage root development: Unravelling the below-ground secrets of cassava (*Manihot esculenta*), *Plant Methods* 15: 131. https://doi.org/10.1186/s13007-019-0517-6.

Shanahan, P. W., Binley, A., Whalley, W. R. and Watts, C. W. (2015). The use of electromagnetic induction to monitor changes in soil moisture profiles beneath different wheat genotypes, *Soil Science Society of America Journal* 79(2): 459–466.

Shazadi, K., Chenu, K. and Christopher, J. T. (2021). Late, deep root development varies between wheat cultivars subjected to terminal water stress, Australian Agronomy Conference. Proceedings, Australian Agronomy Society

Singh, V., van Oosterom, E. J., Jordan, D. R., Hunt, C. H. and Hammer, G. L. (2011). Genetic variability and control of nodal root angle in sorghum, *Crop Science* 51(5): 2011–2020.

Singh, V., van Oosterom, E. J., Jordan, D. R., Messina, C. D., Cooper, M. and Hammer, G. L. (2010). Morphological and architectural development of root systems in sorghum and maize, *Plant and Soil* 333(1–2): 287–299. https://doi.org/10.1007/s11104-010-0343-0.

Steudle, E. and Frensch, J. (1996). Water transport in plants: Role of the apoplast, *Plant and Soil* 187(1): 67–79.

Strock, C. F., Schneider, H. M., Galindo-Castañeda, T., Hall, B. T., Van Gansbeke, B., Mather, D. E., Roth, M. G., Chilvers, M. I., Guo, X., Brown, K. and Lynch, J. P. (2019). Laser ablation tomography for visualization of root colonization by edaphic organisms, *Journal of Experimental Botany* 70(19): 5327–5342. https://doi.org/10.1093/jxb/erz271.

Subramanian, R., Spalding, E. P. and Ferrier, N. J. (2013). A high throughput robot system for machine vision based plant phenotype studies, *Machine Vision and Applications* 24(3): 619–636.

Svane, S. F., Jensen, C. S. and Thorup-Kristensen, K. (2019). Construction of a large-scale semi-field facility to study genotypic differences in deep root growth and resources acquisition, *Plant Methods* 15: 26. https://doi.org/10.1186/s13007-019-0409-9.

Tafesse, E. G., Aidoo, M. K., Lazarovitch, N. and Rachmilevitch, S. (2021). Aeroponic systems: A unique tool for estimating plant water relations and NO_3 uptake in response to salinity stress, *Plant Direct* 5(4): e00312. https://doi.org/10.1002/pld3.312.

Thomas, H. and Howarth, C. J. (2000). Five ways to stay green, *Journal of Experimental Botany* 51(Spec No): 329–337. https://doi.org/10.1093/jexbot/51.suppl_1.329.

Thomas, H. and Ougham, H. (2014). The stay-green trait, *Journal of Experimental Botany* 65(14): 3889–3900.

Thomas, H. and Smart, C. M. (1993). Crops that stay green, *Annals of Applied Biology* 123(1): 193–219.

Thorup-Kristensen, K., Halberg, N., Nicolaisen, M. H., Olesen, J. E. and Dresbøll, D. B. (2020). Exposing deep roots: A rhizobox laboratory, *Trends in Plant Science* 25(4): 418–419. https://doi.org/10.1016/j.tplants.2019.12.006.

Topp, C. N., Iyer-Pascuzzi, A. S., Anderson, J. T., Lee, C.-R., Zurek, P. R., Symonova, O., Zheng, Y., Bucksch, A., Mileyko, Y., Galkovskyi, T., Moore, B. T., Harer, J., E., Herbert, M.-O., Thomas, W., Joshua, S. and Benfey, P. N. (2013). 3D phenotyping and quantitative trait locus mapping identify core regions of the rice genome controlling root architecture, *PNAS: Proceedings of the National Academy of Sciences* 110(18): E1695–E1704. https://doi.org/10.1073/pnas.1304354110.

Trachsel, S., Kaeppler, S. M., Brown, K. M. and Lynch, J. P. (2011). Shovelomics: High throughput phenotyping of maize (*Zea mays* L.) root architecture in the field, *Plant and Soil* 341(1-2): 75–87. https://doi.org/10.1007/s11104-010-0623-8.

Tracy, S. R., Nagel, K. A., Postma, J. A., Fassbender, H., Wasson, A. and Watt, M. (2020). Crop improvement from phenotyping roots: Highlights reveal expanding opportunities, *Trends in Plant Science* 25(1): 105–118. https://doi.org/10.1016/j.tplants.2019.10.015.

Uga, Y., Okuno, K. and Yano, M. (2011). Dro1, a major QTL involved in deep rooting of rice under upland field conditions, *Journal of Experimental Botany* 62(8): 2485–2494.

Uga, Y., Sugimoto, K., Ogawa, S., Rane, J., Ishitani, M., Hara, N., Kitomi, Y., Inukai, Y., Ono, K., Kanno, N., Inoue, H., Takehisa, H., Motoyama, R., Nagamura, Y., Wu, J., Matsumoto, T., Takai, T., Okuno, K. and Yano, M. (2013). Control of root system architecture by deeper rooting 1 increases rice yield under drought conditions, *Nature Genetics* 45(9): 1097–1102.

van Dusschoten, D., Metzner, R., Kochs, J., Postma, J. A., Pflugfelder, D., Bühler, J., Schurr, U. and Jahnke, S. (2016). Quantitative 3D analysis of plant roots growing in soil using magnetic resonance imaging, *Plant Physiology* 170(3): 1176–1188. https://doi.org/10.1104/pp.15.01388.

Vazquez-Carrasquer, V., Christopher, J. T. and Chenu, K. (2016). *Impact of Soil Temperature on Seedling Roots in Wheat and Implications for Adaptation to Water-Limited Environments*. Beijing: International Crop Science Congress.

Veyradier, M., Christopher, J. T. and Chenu, K. (2013). Quantifying the potential yield benefit of root traits in a target population of environments. In: Sievänen, R., Nikinmaa, E., Godin, C., Lintunen, A. and Nygren, P. (Eds.), *Proceedings of the 7th International Conference on Functional-Structural Plant Models*, Saariselkä, Finland. Available at: http://www.metla.fi/fspm2013/proceedings.

Wang, Q., Komarov, S., Mathews, A. J., Li, K., Topp, C., O'Sullivan, J. A. and Tai, Y. (2015). *Combined 3D PET and Optical Projection Tomography Techniques for Plant Root Phenotyping, Phys Q-Bio ArXiv150100242*.

Wasaya, A., Zhang, X., Fang, Q. and Yan, Z. (2018). Root phenotyping for drought tolerance: A review, *Agronomy* 8(11): 241. https://doi.org/10.3390/agronomy8110241.

Wasson, A., Bischof, L., Zwart, A. and Watt, M. (2016). A portable fluorescence spectroscopy imaging system for automated root phenotyping in soil cores in the field, *Journal of Experimental Botany* 67(4): 1033–1043. https://doi.org/ 10.1093/jxb/erv570.

Wasson, A. P., Nagel, K. A., Tracy, S. and Watt, M. (2020). Beyond digging: Noninvasive root and rhizosphere phenotyping, *Trends in Plant Science* 25(1): 119–120. https://doi.org/10.1016/j.tplants.2019.10.011.

Wasson, A. P., Rebetzke, G. J., Kirkegaard, J. A., Christopher, J., Richards, R. A. and Watt, M. (2014). Soil coring at multiple field environments can directly quantify variation in deep root traits to select wheat genotypes for breeding, *Journal of Experimental Botany* 65(21): 6231–6249.

Wasson, A. P., Richards, R. A., Chatrath, R., Misra, S. C., Prasad, S. V. S., Rebetzke, G. J., Kirkegaard, J. A., Christopher, J. and Watt, M. (2012). Traits and selection strategies to improve root systems and water uptake in water-limited wheat crops, *Journal of Experimental Botany* 63(9): 3485–3498.

Watt, M., Kirkegaard, J. A. and Rebetzke, G. J. (2005). A wheat genotype developed for rapid leaf growth copes well with the physical and biological constraints of unploughed soil, *Functional Plant Biology* 32(8): 695–706. https://doi.org/10.1071/FP05026.

Watt, M., Magee, L. J. and McCully, M. E. (2008). Types, structure and potential for axial water flow in the deepest roots of field-grown cereals, *New Phytologist* 178(1): 135–146.

Watt, M., Moosavi, S., Cunningham, S. C., Kirkegaard, J. A., Rebetzke, G. J. and Richards, R. A. (2013). A rapid, controlled-environment seedling root screen for wheat correlates well with rooting depths at vegetative, but not reproductive, stages at two field sites, *Annals of Botany* 112(2): 447–455.

Whalley, W. R., Binley, A., Watts, C. W., Shanahan, P., Dodd, I. C., Ober, E. S., Ashton, R. W., Webster, C. P., White, R. P. and Hawkesford, M. J. (2017). Methods to estimate changes in soil water for phenotyping root activity in the field, *Plant and Soil* 415(1): 407–422.

Wright, G. C., Nageswara Rao, R. C. N. and Farquhar, G. D. (1994). Water-use efficiency and carbon isotope discrimination in peanut under water deficit conditions, *Crop Science* 34(1): 92–97. https://doi.org/10.2135/cropsci1994.0011183X003400010016x.

Zarebanadkouki, M., Kim, Y. X. and Carminati, A. (2013). Where do roots take up water? Neutron: Radiography of water flow into the roots of transpiring plants growing in soil, *New Phytologist* 199(4): 1034–1044.

Chapter 6

Identifying and exploiting genes controlling root system architecture for improving drought tolerance in cereals

Eric S. Ober and Anya Heathcote, NIAB, UK

1 Introduction

Understanding the nature of the drought a farmer might encounter is important prior to devising potential solutions, whether they are agronomic or genetic, or both. Each drought scenario should be considered, including the timing and severity of the water deficit, the probabilities of drought occurrence over time, and trade-offs that might have to be balanced when there are costs associated with the expression of a trait or combination of traits that may enhance the resilience of the crop to drought (Tardieu and Tuberosa, 2010; Tardieu, 2012; Comas et al., 2013; Tardieu et al., 2018). As genomic studies increase our understanding of the complex genetic networks that regulate root growth and development (Wachsman et al., 2015; Slovak et al., 2016; Zaidi et al., 2016; Siddiqui et al., 2021), there is increasing focus on root traits as a means of improving yields for water-limited conditions (Koevoets et al., 2016; Voss-Fels et al., 2018; White, 2019; Li et al., 2021; Lynch, 2022). Yet, important questions remain about which combination of root traits confers net economic and environmental benefits to achieve an optimised and sustainable farming

http://dx.doi.org/10.19103/AS.2022.0109.10

system for drought-prone areas. The list of potential root traits to consider is extensive (Freschet et al., 2021). Ideotypes have been formulated for root system architectures (RSA) that conceivably would be advantageous under certain drought scenarios (Voss-Fels et al., 2018; Li et al., 2022; Rambla et al., 2022), but it is unlikely that such benefits would be observed universally (Lynch, 2018).

2 Root system architecture ideotypes for water-limited environments

The concept of 'designing' root systems based on fundamental biological insight and principles of environmental physics has a long history (Passioura, 1972). For instance, in regions with terminal drought, crops are reliant on stored soil moisture to complete grain filling. Excessive consumption of water too early in the season leaves too little moisture available in later stages and limits grain filling (Wasson et al., 2012). Removal of soil water by the crop can be limited by reducing transpiration via stomatal closure at high vapour pressure deficit (VPD; Messina et al., 2015), reducing water transport from roots to shoots by decreasing hydraulic conductivity of the vascular system (Richards and Passioura, 1989) and by decreasing the biomass of active roots in the soil volume (Lynch, 2018). Selection for genotypes with small diameters of open xylem elements in seminal roots was suggested as an effective way to increase axial resistance of water flow in roots and thereby conserve soil moisture (Richards and Passioura, 1989). Much less attention has been given to the role of phloem pressure unloading in maintaining root growth, which is critical for the delivery of assimilates to expanding cells (Stanfield and Bartlett, 2022; Zhou et al., 2022). Root systems that can reach deeper soil layers to access greater volumes of stored soil water at depth also appear to be beneficial (Lopes and Reynolds, 2010; Rich et al., 2015), but with little disadvantage in well-watered crops, as shown in wheat (Manschadi et al., 2008; Christopher et al., 2013; Maccaferri et al., 2008; Maccaferri et al., 2016; Li et al., 2019), maize (Hund et al., 2009; Lynch, 2013) and rice (Uga et al., 2013; Pandit et al., 2020). In sorghum, QTLs for the stay-green phenotype (*Stg* loci) under drought co-located with QTLs for root:shoot ratio and nodal root angle (Borrell et al., 2014). Nevertheless, the metabolic cost of producing deep roots and the potential yield drag need to be balanced with the benefit accrued (Li et al., 2022). In one study, there was a yield advantage to steeper, deeper roots in water-limited environments, but a yield loss in well-watered environments, compared to lines with wide, shallow roots (el Hassouni et al., 2018). Another study did not observe any cost to above-ground biomass or yield in deep-rooted lines (Severini et al., 2020). The relationships between root angle and yield were context-dependent in barley (Robinson et al., 2018) and durum wheat (Alahmad et al., 2019). As Passioura

(1983) pointed out: 'There is no point in a droughted crop investing a parcel of assimilate in its roots if the extra water thereby obtained does not allow the shoots to at least replace the assimilate so spent'. A mathematical formalisation of the concept led to the definition of 'root system efficiency', which expresses the carbon investment in root mass for water acquisition (or transpiration per unit leaf area), per unit dry root biomass (van Oosterom et al., 2016; Li et al., 2022).

The realisation of the genetic potential for deep root growth depends on soil conditions that are permissive to growth (Rich and Watt, 2013). In some regions, hardpans, bedrock, low/high pH, sodicity, salinity, high levels of aluminium or boron, and root pests and pathogens may pose a greater limitation to root proliferation than the genotype (Richards et al., 2014; Hobson et al., 2022). Furthermore, following the extraction of subsoil water by roots, it is necessary that the soil profile is recharged by precipitation or irrigation in time for the subsequent crop if there is to be any advantage gained by deeper root growth (Ahmed et al., 2018).

2.1 Genes controlling root growth angle and roots in deep soil layers

Based on simple geometry, deeper soil layers can be reached more quickly by the root system when the growth trajectory of roots is steeper, compared with a shallow root angle that would promote greater proliferation of roots in shallow soil layers (Lilley and Kirkegaard, 2011). A number of studies have discovered loci that control root angle, such as a QTL on chromosome 6A in durum wheat that has a major effect on seminal root angle (Sanguineti et al., 2007; Maccaferri et al., 2016; Alahmad et al., 2019). This QTL also collocates with the QTL cluster controlling osmotic adjustment in roots (Condorelli et al., 2022). Recent research has revealed a suite of genes that are part of the regulatory network that controls the gravitropic response that governs the angle at which roots grow away from the central root axis (Toal et al., 2018). This 'gravitropic setpoint angle' (GSA) relates to the tissue responsiveness to auxin gradients established within the elongation zone of root tips (Roychoudhry et al., 2017). This is offset by a separate antigravitropic mechanism that counteracts the positive response to the gravity stimulus, thus maintaining the growth trajectory. In Arabidopsis, genes regulating GSA are members of the *TIR1/AFB* auxin response mechanism. In barley, two mutants that show extreme gravitropic response (steep seminal root angle) are *egt1* and *egt2*, identified from a Morex TILLing population. TILLing mutants are an effective reverse-genetics resource for the functional characterisation of candidate genes. Wheat orthologues of *egt1* and *egt2* were also identified recently in both tetraploid durum and hexaploid bread wheat. The *egt1* locus is part of the antigravitropic mechanism, encoding a TUBBY-like F-box protein that regulates cell wall remodelling during growth resulting in cortical cell wall stiffening, which counteracts the cell wall loosening in epidermal cells that occurs as roots respond to gravity-induced auxin fluxes (Fusi et al., 2022). The *egt2* orthologue encodes a STERILE ALPHA MOTIF protein that regulates the positive gravitropic response (Kirschner et al., 2021). In the proposed model, root bending towards gravity is caused by an egt2-regulated increase in peroxidase activity in epidermal cells, resulting in hydroxy radical-induced wall loosening (Somssich et al., 2016) and enhanced elongation of epidermal cells on the side of the root in opposition to the gravity vector. This particular phenomenon of the establishment of root growth angle progresses from activity at the gene level in small groups of cells in the root tip to the relative distribution of root system biomass throughout the soil profile, with implications for the access to and uptake of soil water.

In the case of *egt1* and *egt2*, the mode of action is on cell wall mechanics, but the differential growth process clearly involves many processes, each of which is regulated in turn by different components of the regulatory gene network. For example, synthesis, transport and response to hormonal signals (Ubeda-Tomás et al., 2012); electrophysiological setpoints controlling solute transport; osmotic potentials that drive water fluxes for the maintenance of turgor pressure (Ober and Sharp, 2003); homeostasis of secondary messengers Ca^{2+} and reactive oxygen species (Voothuluru et al., 2020), etc. Each process represents a multitude of potential gene targets for selection and manipulation by breeding.

Inadvertent selection for root angle may have resulted from pleiotropic effects of the *VRN1* gene that controls the expression of winter/spring types in wheat and barley via a MADS box transcription factor (Voss-Fels et al., 2017). In European and Australian wheat lines, the winter allele of *VRN1* is associated with a narrow root angle, although the molecular mechanism is undetermined.

2.2 Root hairs

Root hairs are the tubular outgrowths from specialised epidermal cells that greatly expand the total surface area in contact with the soil (Vissenberg et al., 2020). Experiments comparing wildtype and mutants exhibiting defective root hairs showed that root hairs were vital for water uptake under water-limited conditions (Carminati, et al., 2017; Klamer et al., 2019; Marin et al., 2021), but perhaps less so in the absence of stress (Dodd and Diatloff, 2016).

Root hairs are also important for rhizosheath formation (the soil that adheres to the root surface), which is mediated by ABA and auxin effects on root hair elongation under soil drying (Zhang et al., 2021). Rhizodeposition, or the exudation of mucilage and other compounds by the root, also plays a role in forming the rhizosheath (McCully, 1999; Holz et al., 2018). The rhizosheath combination of soil, root hairs and mucilage forms an essential bridge of hydraulic connectivity between the root surface and the bulk soil (Carminati et al., 2017), allowing the transfer of water (Cai and Ahmed, 2022) and nutrients (Ober and Parry, 2011; Kang et al., 2022). A subset of wheat lines of the Avalon × Cadenza doubled-haploid mapping population showed significant genotypic variation for rhizosheath size and mucilage composition (Marr, 2020).

Details of the complex gene networks controlling root hair growth and development are being unravelled (Salazar-Henao et al., 2016), partly through genetic dissection and functional characterisation of mutants that show defective root hairs (Chmielewska et al., 2014). While this is a useful approach, mutant lines are not suitable for breeding; donor lines that show advantageous alleles are required for genetic gain. As there is a multitude of potential gene targets, focus on the master regulators in the control pathway is warranted. An example in wheat is the gene *TaRsl4*, which encodes a bHLH transcription factor, modulating root hair length (Han et al., 2015). Interestingly, in allopolyploid wheat, there is a genomic asymmetry in *RSL4* expression, such that the homologue on the A chromosome shows greater expression than the B or D genomes, perhaps due to differential chromatin modification. Using sequence information from the conserved regions of the gene, a survey of allelic variation across various germplasm pools should reveal haploblocks that are associated with longer root hairs. Forward genetic screens of diversity panels via genome-wide association study can reveal epistatic interactions with other loci and also identify individuals representative of long root hair haplotypes, providing donor material for introgression into elite backgrounds for evaluation within a breeding programme. This kind of haplotype-led breeding is useful for allele stacking (Brinton et al., 2020). Other potential gene targets for root hair density, which is usually correlated with length (Nestler et al., 2014), are *RSL2,* a transcription factor (Han et al., 2017)); *RHD3*, a GTPase responsible for vesicle trafficking (Shan et al., 2005); *AUX1*, an auxin transporter in rice (Giri et al., 2018) and *RTH5*, an NADPH oxidase (Nestler et al., 2014).

2.3 Allocation of biomass to roots and its spatial distribution

It would seem intuitive that greater root biomass, and therefore, greater increased soil water extraction, would confer an advantage during drought. However, the yield response depends greatly on how root biomass is distributed in space and time in relation to the phase of development of the crop and rainfall patterns. For instance, lines with smaller root biomass were beneficial in terminal drought situations, as they consumed less soil water during the pre-anthesis stage than lines with greater root biomass (Figueroa-Bustos et al., 2020). This was also reflected in a modelling exercise and subsequent tests of the model (van Oosterom et al., 2016). There is evidence that root biomass was inadvertently selected against as a result of breeding for selections for heading date (Voss-Fels et al., 2018). In European bread wheat, the major haplotype variant controlling heading date has two flanking haplotypes controlling root biomass, and selection against the late heading data allele resulted in linkage drag, reducing root biomass simultaneously. A relatively rare haplotype variant was associated with larger root biomass, which was linked to two candidate genes that play roles in root growth: expansin (catalyses wall loosening) and endo-beta-mannanase.

Other studies also show changes in RSA traits over time as a consequence of breeding and selection for yield under high input conditions and high plant population densities (compared with ancient practices). For example, examining lines representing 100 years of wheat breeding in Switzerland, there was a negative correlation between year of variety release, plant height and rooting depth, although this was observed only under well-watered conditions, not drought (Friedli et al., 2019). Similarly, the year of release was negatively associated with root system size in wheat lines from several decades of breeding (Fradgley et al., 2020). Zhu et al. (2019) concluded that 'Green Revolution' alleles for semi-dwarfing and plant yield were selected more strongly than rooting depth. It is interesting to speculate that as plant population densities increased with intensive modern agriculture, the competitive ability of older varieties was indirectly selected against, in favour of genotypes that were less sensitive to neighbours, as shown in a study comparing an old and modern line of wheat (Fang et al., 2011). At the root level, this could manifest as a greater tolerance to the co-exploration of roots from different individuals within a set soil space (Fang et al., 2013; Hecht et al., 2019), which could more rapidly deplete soil moisture reserves.

While a smaller root system size may be beneficial in terminal drought environments, in other environments with episodic periods without rainfall and intermittent drought eventually relieved by rainfall, loss of yield potential due to stress can be minimised by more efficient extraction of stored soil water. In an analysis of winter wheat crops in the UK, it was shown that the mean root length density (RLD) in soil layers 0–30 cm from the surface typically exceeded the critical value of 1 cm root length per 1 cm^3 soil volume (there was superfluous root production), whereas at 50 cm and below, RLD was less than 1 cm cm^{-3} (inadequate root biomass) (White et al., 2015). A critical RLD of 1 cm cm^{-3} was shown by modelling and empirical measurement to allow the full extraction of available soil water from a typical mineral soil (Gregory, 1989). However, where certain spatial patterns of root biomass distribution exhibit trade-offs with unstressed yield potential, and drought occurrence is unpredictable, a stochastic approach to finding the optimal balance in RSA (Tardieu et al., 2018), or a highly plastic root system (see later) is necessary.

There has been little systematic testing of root biomass traits in a uniform genetic background with sufficient numbers of lines. To achieve this, a major QTL for root biomass in wheat was discovered on chromosome 5B (Voss-Fels et al., 2017), and a KASP marker was developed for this locus (Makhoul et al., 2020). Marker-assisted backcrossing into an elite background was used in conjunction with a simple visual scoring method to select lines that diverged in seedling root biomass (Rambla et al., 2022). In combination with these lines, additional lines that contrasted in seminal root angle were also backcrossed into the same background, such that sets of introgression lines comprising four combinations were developed: high vs. low biomass combined with wide vs. narrow root angle (Fig. 1). These materials are being used in field trials to assess the potential contribution of each combination in different environments. The combination of phenotypic selection and speed breeding (Watson et al., 2017) allows the rapid development of near-isogenic lines (NILs) in breeder-relevant backgrounds to assess relative trait values.

2.4 The ability of roots to penetrate soils

From the discussion so far, it is clear that for the root system to become established within the three-dimensional soil space, root apices must extend through the soil matrix. In the field, soils are often inhospitable, and high soil strength or compaction can create a formidable barrier. Many soils also harden as they dry, presenting plants with an additional challenge during drought (Whalley et al., 2008). In fact, the extent of soil drying, caused principally by the removal of water by roots, can be used as a proxy measure for genotypic differences in yield in water-limited environments by using a simple soil penetrometer (Whalley et al., 2008).

In order to continue the exploration of the soil, roots must either penetrate soil layers, grow around areas of high soil strength until roots are able to penetrate or find cracks, fissures or 'biopores' within the soil peds. Biopores

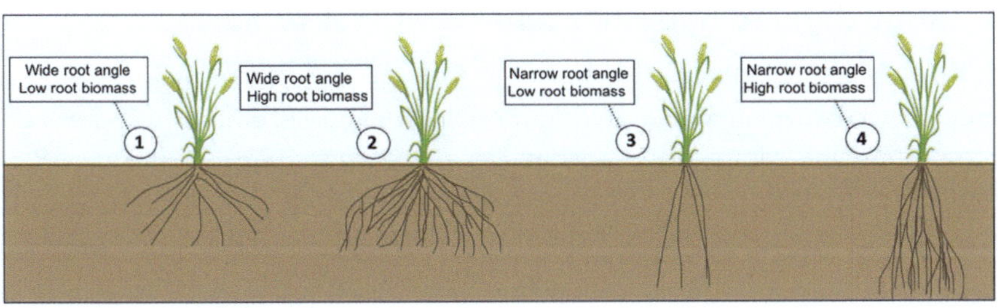

Figure 1 An example of different root ideotypes that can be generated by combining different genetic loci controlling root angle and root biomass. Used with permission from Ober et al. (2021) under the Creative Commons Licence http://creativecommons.org/licenses/by/4.0/.

are natural holes created by roots of previous crops that have subsequently decayed or by earthworms. It has been shown that most of the roots that grow to depth in soil with high penetration resistance do so via these pores (White and Kirkegaard, 2010). Genotypes with thicker roots have greater tensile strength and are often better able to penetrate hard soils, as shown in maize (Chimungu et al., 2014) and rice (Clark et al., 2008). Increased wall thickening in multiseriate cortical sclerenchyma (Schneider et al., 2021) or increased radial expansion of cortical cells is related to the accumulation of ethylene as a result of reduced gaseous diffusion through the soil around the root tip (Pandey et al., 2021). A mapping study revealed a MEI2-like RNA binding protein co-located with a major QTL peak for MCS formation (Schneider et al., 2021). Ethylene-induced deposition of lignin may also increase root stiffness. Alternatively, thin roots are better able to enter small cracks between peds than thicker roots.

Using micro-computed x-ray tomography, Atkinson et al. (2019) found that in compacted soil, 76% of roots changed direction (altered trajectory of growth), compared with 21% in loose soil. A peculiar behaviour of roots is that growth often follows a spiral and waving pattern called circumnutation; it has been suggested that this helps in obstacle avoidance (Fang et al., 2013) and increases the chances that a root tip will encounter a crack or macropore (Taylor et al., 2021). Key regulatory genes in rice, a histidine kinase (*OsHK1*) and an auxin influx transporter (*OsAUX1*), may prove useful breeding targets for problem soils. By using wax layers of variable hardness to mimic strong soil, one study found that the number of root axes that managed to penetrate a hard layer was simply related to the total number of root axes produced by a genotype (Whalley et al., 2013). The wax layer method was used to map QTLs for penetration ability, but many QTLs with small effects limited marker development (Botwright Acuña et al., 2014). Other features that assist with penetration ability are anchorage via root hairs (Bengough et al., 2016) and lubrication via mucilage exudation (Oleghe et al., 2017).

3 Root-derived hydraulic and hormonal signals

Regulation of the balance between root and shoot (R:S) growth plays an important role in maintaining yields during drought. A wild emmer introgression into durum wheat improved transpiration efficiency through alteration of R:S, and restricting transpiration to periods of the day when VPD was low (Bacher et al., 2022). Root growth can continue at water potentials (ψ_w) that completely inhibit shoot growth, indicating fundamental differences in how the tissues of above-ground and below-ground structures respond to drought (Sharp and Davies, 1979) (Fig. 2). This can be viewed as an adaptive mechanism that limits shoot biomass and leaf area, which is susceptible to transpirational water losses, and partitions more assimilate to root growth, needed for water acquisition.

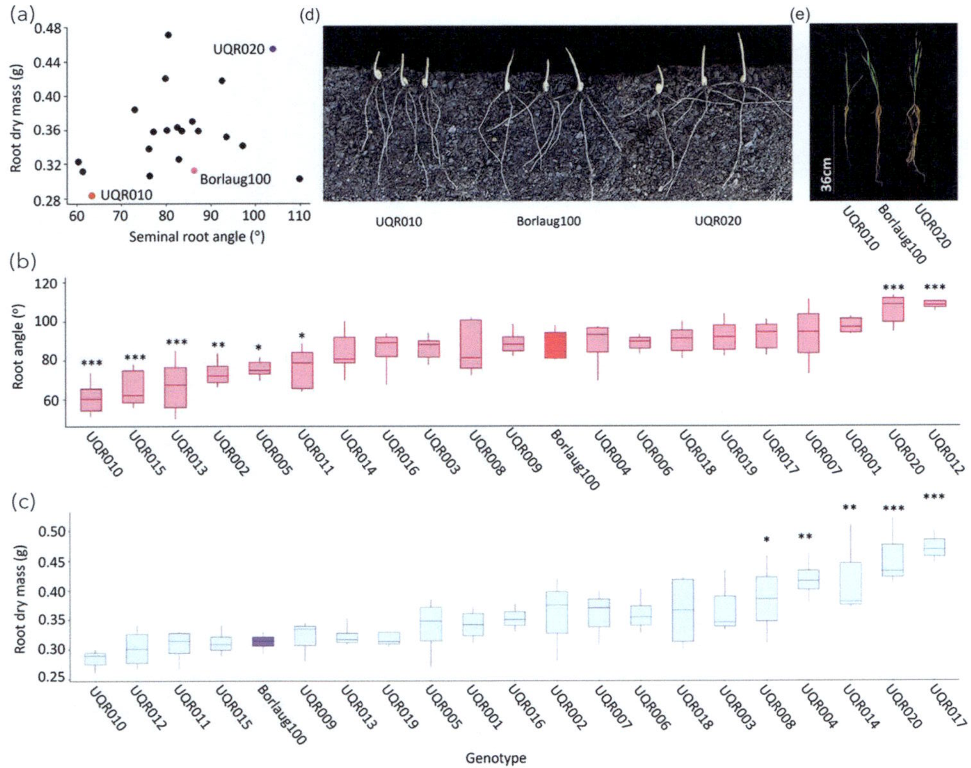

Figure 2 Phenotypes of wheat introgression lines selected for contrasting seminal root angle and root biomass. Lines are BC_2F_4:F_5 in a Borlaug100 background. (a) A plot of 20 lines diverging in root biomass and root angle. (b) Boxplots representing the seminal root angle and (c) root dry biomass for each introgression line. Genotypic means of lines were compared with Borlaug 100 using Fisher's protected LSD test (*, ** and *** indicate significance at $P \leq 0.05$, $P \leq 0.01$ and $P \leq 0.001$, respectively). (d) Comparison of the seminal root angle between Borlaug100 and the introgression lines UQR010 (narrow) and UQR020 (wide). (e) Plants typical of the root biomass phenotypes of UQR020 (high) and UQR010 (low) compared with Borlaug100. Used with permission from Rambla et al., 2022.

Using the maize primary root as a model system, studies have shown that in low ψ_w media, the maintenance of the growth of cells in the apical region of the elongation zone involves complex interactions of many factors. For instance, growth depends on the production of abscisic acid (ABA), which keeps a check on the growth-inhibitory effects of ethylene; the accumulation of solutes for osmotic adjustment (Condorelli et al., 2022); cell wall modification via ROS and wall loosening enzymes, which allows root cells to elongate at diminished levels of turgor pressure (reviewed in Ober and Sharp, 2013). A key target in maize suggested by a combination of kinematic growth analysis and root zone-specific RNAseq is oxalate oxidase, which plays a role in H_2O_2 production (Voothuluru et al., 2020; Kang et al., 2022).

Some of the earliest responses that trigger these changes occur at the cell membrane, where low ψ_w induces hyperpolarisation of the membrane potential via upregulation of H^+-ATPase activity (Ober and Sharp, 2003). Under prolonged exposure to low ψ_w, the ABA-dependent homeostatic setpoint of the membrane potential is adjusted to a more negative value, maintaining a greater driving force for cation uptake, such as K^+, important for osmotic adjustment along with organic compounds such as proline (Bhaskara et al., 2015).

3.1 Regulation of plant hydraulic conductivity

While production of ABA may be necessary for root growth maintenance at low ψ_w, ABA that enters the transpiration stream via the xylem can lead to inhibition of the growth of leaves (Saab and Sharp, 1989), decreased stomatal conductance, and changes in hydraulic conductivity (Huntenburg et al., 2022). Thus, ABA produced in roots and leaves can act as a non-hydraulic signal but can also induce hydraulic effects, with different responses in different tissues in different environments (e.g. night vs. day; low vs. high VPD; (Tardieu et al., 2010).

ABA and other signalling factors can affect the hydraulic conductivity of roots via modulation of aquaporin (AQP) activity, proteins that facilitate water transport across the cell membrane (PIP gene subfamily) and the tonoplast (TIP gene subfamily; Maurel et al., 2021). Xylem vessel properties (discussed earlier) affect axial root conductivity and suberin deposition (Calvo-Polanco et al., 2021), and AQP affect radial conductivity (Maurel and Nacry, 2020). Transpiration and hydraulic conductance are closely coupled by bi-directional signalling between roots and shoots (Huntenburg et al., 2022). In the early stages of soil drying, root AQP expression is down-regulated and water channel activity is modulated via phosphorylation/dephosphorylation to reduce hydraulic conductivity, hence conserving moisture (Javot and Maurel, 2002; Kapilan et al., 2018). A simplified summary of a large volume of research is that up-regulation of AQP activity in roots to increase root hydraulic conductivity, and a decrease in AQP activity (perhaps via expression of different tissue-specific isoforms or differential post-translational modification) in shoots to decrease transpiration rates could help maintain favourable water status of plants during drought (Shekoofa and Sinclair, 2018). In terminal drought environments, down-regulation of AQP activity could decrease hydraulic conductivity and conserve soil moisture.

3.2 Hydraulic lift in mixed cropping and agroforestry

We have seen that improving crop production in dry environments entails complexity on many levels, beyond merely the genetics of the plant, involving also the physics and biology of the soil environment. An additional level to consider is the cropping system. Compared with conventional monocrops, utilisation of cover crops is now common practice in many areas, and intercropping and agroforestry may expand beyond smallholder farming to large-scale operations. In addition, there is growing interest in varietal mixtures rather than pure stands. In such systems, roots of different species or varieties share the same soil space (Kemper et al., 2022), yet we understand little of how these interactions may benefit the overall resilience of the mixed crop during a water deficit. In the section earlier, we mention how the suberisation of roots controls radial hydraulic conductivity, limiting the loss of water to dry soil from mature root tissues. However, there is evidence that roots do lose water to the soil (Bormann, 1957) and that nearby roots from the same or different plants can utilise this water (Neumann and Cardon, 2012). An example of 'hydraulic lift' or upward redistribution was shown in an interplanting arrangement of alfalfa and maize, where deeper roots of alfalfa supplied water to the roots of maize in upper soil layers, extending the period of survival during drought (Corak et al., 1987). Mechanisms of this 'reverse flow' are not well understood (Caldwell et al., 1998), but it would be surprising if the rectification of water transport was not tightly regulated by the plant. Roots exhibit a curious tendency for non-random 'swarming' behaviour (Ciszak et al., 2012), such that neighbouring roots are often closely appressed, or clumped together, particularly in soil pores (White and Kirkegaard, 2010). In Arabidopsis, roots can apparently avoid obstacles, but grow towards roots, with a preference for the same genotype (Fang et al., 2013). While this may not be a theoretically efficient distribution of root biomass, it may assist water transfer from one part of the root system/soil profile to another.

4 Plasticity of root growth

It is well known that plant form is genetically programmed, yet the final shape and directionality of growth also depend on environmental signals, some of which trigger various 'tropisms (Darwin, 1880). So too, the underground plant

Published by Burleigh Dodds Science Publishing Limited, 2024.

form follows a pre-determined genetic pattern, but the expression of many root traits can be modulated by the heterogeneous environment of the rhizosphere (Hodge, 2009). The sensitivity of plant tissues to abiotic stresses and the manner in which this information is transduced by cellular machinery to effect changes in developmental patterns are also encoded in the genome and/or epigenome (Lloyd and Lister, 2022). Some interesting and useful examples of this plastic behaviour that impinges on drought tolerance are discussed later.

4.1 Biomass allocation patterns

In a field study that compared isogenic wheat lines that differed only in the wheat-rye 1RS translocation, the 1RS lines showed greater root biomass and yield than the background line without the translocation under well-watered conditions (Ehdaie et al., 2012). However, under water deficit, the line lacking the translocation increased total root biomass in response to stress, whereas the root biomass response to stress was negative in 1RS.1DL or neutral in 1RS.1BL and 1RS.1AL. In a year with mild stress, the positive plastic response was associated with a smaller reduction in yield loss than the 1RS lines, whereas in a year with greater stress intensity, the line with the plastic response showed a greater yield loss, presumably due to greater allocation of root biomass in shallow layers that exhausted limited water supplies, and shifted partitioning of assimilates away from shoots to roots. The region controlling plasticity was located on 1BS using homologous recombinant lines (Ehdaie et al., 2012).

In rice, QTLs associated with total root length (TRL) were not consistent across well-watered and droughted conditions, and QTLs for plasticity (the relative change in TRL across environments) did not co-locate with QTLs for TRL (Kadam et al., 2017), indicating separate levels of control. Another rice study discovered single-nucleotide polymorphisms (SNPs) related to root biomass plasticity (albeit in a small panel of lines grown in rhizoboxes), which were correlated with yield stability. The alleles were contributed by donor lines adapted to marginal environments and conferred greater root plasticity to an elite line grown in high-yield environments (Sandhu et al., 2016). To identify loci associated with root plasticity under fluctuating soil water conditions, a panel of inbred lines was created from a cross between the recurrent parent Nipponbare and a line that showed enhancement of root branching in shallow soil layers during drought (Owusu-Nketia et al., 2018). One of the NILS showed similar rooting patterns to the recurrent parent under continuously well-watered conditions, but greater lateral and nodal root length under drought, which was associated with greater stomatal conductance and grain yield.

In maize, QTLs for root trait plasticity across environments did not co-locate with QTLs associated with the traits in individual environments (Schneider et al., 2020). For example, SNPs related to lateral root length and branching density were located at different chromosomal regions for the plasticity of expression, compared with trait expression in the separate well-watered and stressed environments, except for one common SNP on chromosome 10 that co-located with a candidate gene encoding an ethylene response transcription factor.

4.2 Hydrotropism and hydropatterning

For a time it was debated whether hydrotropism (root bending) or hydropatterning (root branching) - which leads to the proliferation of roots in patches of soil with greater matric potential - was merely the net result of biomass allocation due to slow growth in drier soils, or was the result of directionality of growth and developmental patterning towards moisture (Giehl and von Wirén, 2018). Carefully designed experiments have shown that root cell elongation on different sides of the root does indeed respond to favourable gradients in hydraulic conductivity and water potential of the external medium (Bao et al., 2014), although they may not be necessary (Lind et al., 2021). Regulated by opposing auxin gradients established within the root via PIN transporters (Robbins and Dinneny, 2018), the growth trajectory responds simultaneously to the gravity and water potential vectors (Morohashi et al., 2017), and likely other signals as well (Knight, 1811). Thus, water potential gradients can cause growth to deviate from the genetically determined GSA (mentioned earlier) and follow a new trajectory, directing the placement of roots where there is greater resource availability.

The initiation and growth of lateral roots are additional plastic responses to local soil environmental conditions. Roots growing through a large soil macropore produce lateral roots only where the root is in contact with the wall of the pore and repress the formation of lateral roots where the root is in contact with the air space within the pore (Bao et al., 2014). Inhibition of lateral root formation is dependent on the action of ABA (Orman-Ligeza et al., 2018), which represses the auxin pathway factor ARF7 in Arabidopsis (Orosa-Puente et al., 2018). The absence of water stimulates the modification of the ARF7 protein by sumoylation, which then blocks the recruitment of auxin factors to the 'air side' of the root. On the 'wet side' of the root, auxin-responsive genes such as *LBD16* are free to initiate asymmetric cell division of lateral root founder cells in the pericycle (Babé et al., 2012).

Unlike most axile roots, lateral roots can exhibit determinate growth: cell production from the meristem ceases after a period of growth, defining their final length. However, the genetic programme that determines the duration of development can be modulated by external water potential. In maize, lateral root growth can continue for an additional 3 days under mild water deficits compared with well-watered plants, a response that is genotype-dependent (Dowd et al., 2020). In another study, mild water deficit had no effect on the linear frequency of lateral roots along the primary or seminal root axes but did increase the final length of first-order lateral roots (Ito et al., 2006).

4.3 Developmental instability, cryptic variation and noise

The RSA can be determined by the angle, velocity, duration and branching of root growth, which respond to both genetic and environmental signals. While plasticity can provide an adaptive or yield advantage, stability of developmental patterns - buffering the phenotype against every tiny or momentary change in the environment - is also vital, and plants manage some kind of balance. In addition to relatively predictable phenotypes due to genetics or environmental responses (plasticity), there appears to be an element of randomness, or noise in many developmental processes. This injection of stochasticity into the system also affords a level of protection against unpredictable environmental challenges at the individual and population level as a 'bet-hedging' strategy (Forde, 2009). Intriguingly, the level of developmental instability (plasticity and noise) that is tolerated appears to be under genetic control in plants (Sangster et al., 2008) and may be mediated in part by chromatin remodelling rather than changes in DNA sequence (Lloyd and Lister, 2022). Changes in phytohormone levels or sensitivity may directly affect chromatin modifiers to transduce environmental stresses to alterations in root meristem maintenance and developmental patterning (Xiao et al., 2017; Maury et al., 2019).

In the constant search for novel variation for breeding improvements, geneticists debate whether cryptic variation - revealed through alteration of chromatin remodelling that can affect epigenetic buffering of development programs - can lead to heritable changes to the epigenome lasting more than a few generations before they reset (Hauser et al., 2011; Bloomfield et al., 2014). It is unclear if the net effect of stress-induced changes to the epigenome is agronomically beneficial (Lloyd and Lister, 2022). Important questions are to what extent increased plasticity and/or developmental stability in root development are desirable in certain environments, and if these alleles or epialleles are amendable to improvement through selective breeding.

5 Root–microbe interactions

5.1 The rhizosphere microbiome

The rhizosphere microbial community and plant roots form complex interrelationships that are rich territory for scientific exploration and potentially exploitation to benefit agriculture and ecosystem services. There is evidence that arbuscular mycorrhizal associations can promote drought tolerance in wheat (Lehnert et al., 2017). Drought-induced changes in root-associated bacterial communities can have positive and negative effects (Aslam et al., 2022), partly due to the severity of the stress (Naylor and Coleman-Derr, 2017). Benefits to the plant may come from recruiting bacteria that produce growth-promoting hormones (Rolfe et al., 2019) or inhibition of ethylene production (which inhibits root growth) via ACC deaminase (Sayer et al., 2021). The microbiome is supported by carbon and nitrogen substrates exuded by the host plant roots. Metagenomic analysis of the rhizosheath soil isolated from individuals of the Avalon × Cadenza mapping population of wheat showed differences in bacterial and fungal diversity and abundances, which may be related to genotypic differences in the compositional profile of mucilage (Marr, 2020). In future, it may be possible to engineer solutions for droughted conditions by soil inoculation of selected beneficials, in combination with plant genetics that favour a microbial balance that helps protect the plant. However, some evidence suggests that modern cultivars and intensive agriculture management practices have reduced the rhizo- and endosphere species richness (Kavamura et al., 2019).

5.2 Beneficial endophytes

The root endosphere microbiome may also contribute to drought tolerance by the composition of non-mycorrhizal endophytic species hosted by the plant tissues. For instance, drought during early development resulting in severe stress enhances *Actinobacteria* in millet (Simmons et al., 2020). The endophytic fungus *Acremonium coenophialum* in tall fescue promotes tiller survival and recovery from drought (Elemi and West, 1995). A study in rice showed that the host genotype affected the endo-fungal community composition (mostly the Ascomycota subphylum Pezizomycotina), which was associated with enhanced drought tolerance in the field (Andreo-Jimenez et al., 2019). Wheat plants inoculated with Ascomycetous fungal isolates isolated from the roots of field-grown durum wheat resulted in greater seed yield compared with endophyte-free plants (Hubbard et al., 2014). However, these are complex experiments to conduct, and conclusions from controlled environments can be inconclusive (Decunta et al., 2021).

6 Role of roots in supporting yield stability across water regimes

Farmers and breeders traditionally focus on yield potential as the important metric of a variety's performance, but stability across locations and years is becoming increasingly important as climate variability increases with global warming. The need to breed varieties with greater climate resilience and therefore yield stability may increase the focus on the optimisation of RSA (Topp, 2016). Perhaps it will be revealed that some loci that associate with stability co-locate with loci controlling root traits (Mickelbart et al., 2015). Analysis of the natural allelic variation for root traits and stability will inform the haplotype diversity within pools of germplasm (Qian et al., 2017), which will help breeders choose the best lines for crossing in order to stack beneficial alleles in progeny lines (Brinton et al., 2020).

Varietal performance is usually dependent on interactions with management practices and biotic and abiotic factors in a defined location, over a specific time period (Annicchiarico, 2002). In this context of genotype × environment interactions, yield resilience or static stability can be understood as the insensitivity of a cultivar to environment influences and achieve its yield potential unperturbed (Eberhart and Russell, 1966). This definition does not, however, allow for beneficial interactions that may increase yield in favourable environments, which requires a dynamic view of stability. The concept of yield stability is not new and yet, despite its agronomic and commercial value, there is little agreement on how it should be defined, measured or implemented in practice (Lin et al., 1986; Reckling et al., 2021).

At a physiological level, biomass production and how it is allocated is thought to contribute to yield stability across a range of environments (Weiner et al., 2021). Increased allocation to root biomass in superficial soil layers is typical of cereal crops in arid and semi-arid climates (Wang et al., 2014; Palta et al., 2011; Hamblin and Tennant, 1987). While there are plausible explanations for the advantage each trait previously described would confer to yield stability, the manipulation of root biomass and root growth angle (see Section 2) may pose the most obvious solution. One hypothesis is that in temperate climates a variety with narrow root angles and increased root proliferation in the subsoil would reduce the probability of yield losses at water-limited sites (thus contributing to yield stability), but in semi-arid environments, reduced root distribution in upper soil layers may lead to more consistent yields (Yang et al., 2022; Fang et al., 2021). This suggests that genetic material with narrow root angles such as the *egt1* and *egt2* mutants (see Section 2.1) and introgression lines with high root biomass (see Section 2.3) would have greater yield stability, when compared with the wild-type or recurrent parent, respectively. Experiments to test this hypothesis are underway.

Perhaps, though, the key to realising stable yields in cereal crops across water regimes may not be held by one ideotype, but instead by a combination of root traits. Varietal mixtures (see Section 3.2) comprising contrasting but complementary RSA could more efficiently occupy the soil volume, and therefore in the context of a plant stand (rather than an individual plant) could strike an optimum balance between biomass allocation, soil water extraction and grain yield. The idea that varietal mixtures could support yield stability was introduced alongside the term itself (Eberhart and Russell, 1966), and more recently mixtures have found renewed focus in efforts to improve the sustainability of 'regenerative' farming systems.

7 Breeding for root traits to improve drought tolerance

7.1 Ask breeders what they need

There is no shortage of published papers which suggest that their trait or gene of interest will be useful for breeders. Very few of these traits or genes have been incorporated into active breeding programmes (Passioura, 2012). Jackson et al. (1996) considered this disconnect between fundamental plant science and breeding and arrived at some suggestions, which are still relevant today, particularly for root traits. To find utility in a breeding programme, fundamentally traits should have high heritability, show good genetic correlation with yield and/or sustainability metrics in target environments, be amenable and affordable to selection on a large scale, be characterised in relevant elite material in the field, be identified in useful donor lines representative of major or novel haplotypes, etc. (Richards, 2006). Breeders need no convincing that roots are important and have been overlooked in breeding. But they need to know what to select for, a way to make those selections, and evidence for the impact of those selections on performance in the field in each predicted target environment.

Research on seedling traits in controlled environments or containerised experiments can be useful but does not fully address the earlier needs, as juvenile traits do not always reflect trait expression in more mature plants in the field (Bai et al., 2019; Rich et al., 2020). Measurement of roots in the field at

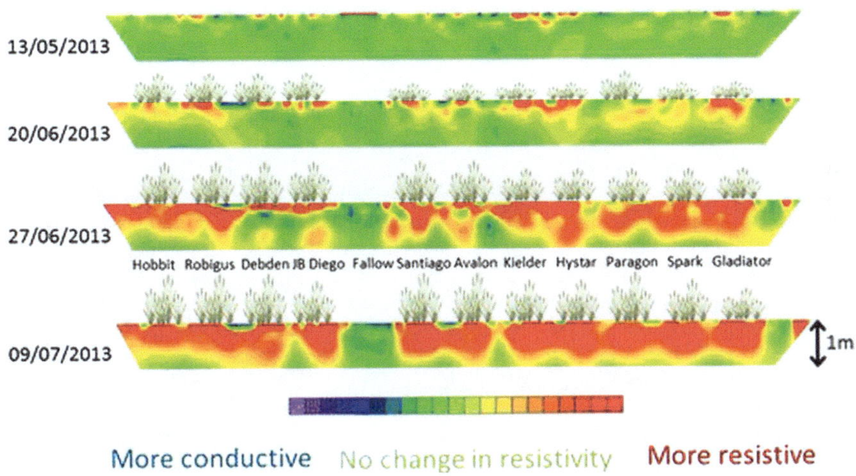

Figure 3 Progressive soil drying shown via time-lapse images created from inversion of electrical resistance tomography measurements in the soil underneath field plots of different wheat varieties. Colours indicate relative changes in soil resistivity after the baseline date 23 April 2013 (red: greater resistivity; green: no change; blue: smaller resistivity). The position of plants in each plot along the ERT transect is indicated, as well as the central bare fallow plot. Used with permission from Whalley et al., 2017.

a medium scale can be accomplished via soil coring and shallow excavations (shovelomics) but is laborious and time-consuming, limiting the throughput capacity (Tracy et al., 2020). Therefore, non-invasive proxy measures that relate to rooting behaviour have been investigated. Two such methods show some promise: inferences about soil water extraction patterns by roots can be made using canopy temperature measurement via thermal remote-sensing (Pinto and

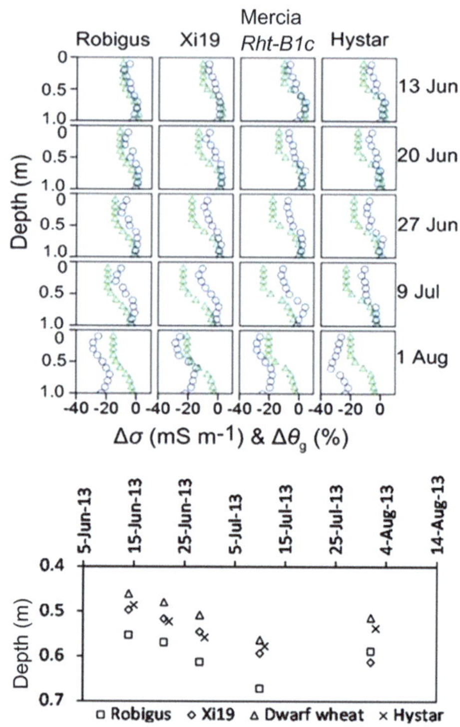

Figure 4 Top: The change in electrical conductivity (Δσ) (baseline date 14 May 2013) over time determined from electromagnetic induction (EMI) measurements (triangles) and the change in gravimetric soil moisture content ($\Delta\theta_g$) (circles), plotted against depth from the soil surface. Negative values indicate a relative decrease in soil moisture content. Different wheat genotypes are indicated at the top. Mercia *Rht-B1c* is an introgression line showing an extreme dwarf phenotype (but despite small shoot biomass, showed similar patterns of soil water extraction to standard varieties). Hystar is a hybrid wheat but did not demonstrate deeper rooting or greater water extraction than other varieties. Robigus is the only variety in this set containing the 1RS.1BL chromosomal translocation from rye. Bottom: Changes in the inflection point of the measured Δσ over time in each genotype. The inflection point relates to the soil depth where soil moisture content changes rapidly from low to high water content, roughly indicating the rooting front as the root system progresses downward. Measurements were made during a period of little rainfall and therefore plants were removing stored soil moisture over this period. Symbols are the mean of four replicate plots. Modified and used with permission from Shanahan et al. (2015) under the Creative Commons Licence http://creativecommons.org/licenses/by/4.0/.

Reynolds, 2015; Li et al., 2019) and adaptation of electromagnetic induction methods (Figs. 3 and 4) (Whalley et al., 2017; McLachlan et al., 2021).

7.2 Validation: do small phenotypic changes make a difference?

Like many quantitative traits, differences in RSA between elite lines drawn from major germplasm pools are often subtle and difficult to detect. Greater differences can be observed comparing exotic material and agronomically improved lines, but with such diverse sets, the contribution of root traits to yield, drought tolerance or effective use of water can be surmised based only on correlative evidence. Validation of trait value can be better done using NILs that differ only in the locus in question. Even though genotypic differences in trait characters may be numerically small, the impact can be significant. For instance, a deeper root system that can mine an extra 10 mm of subsoil water can translate to an additional 600 kg grain/ha, using an average water use efficiency of 60 kg/ha/mm (Kirkegaard et al., 2007). In wheat, root dry matter at 1 m depth comprises only 3% of the total root biomass, which seems almost negligible, but during a dry period, these roots can contribute 20% of the plant's evapotranspiration (Gregory et al., 1978).

Genotypic differences in root growth angle also can be numerically small, but important. For example, simplifying the root system to a cone (with the point at the root/shoot junction) and applying some geometry, the result of a 10° reduction from 55° (from vertical) to a more narrow angle of 45° is that the root tips reach a depth of 1 m 27 days earlier, assuming a constant trajectory and growth rate of 1.2 cm/day (three-dimensional root simulation models would produce a more realistic result). In an environment where crop phenological development is racing against the onset of terminal drought, quicker access to deep water stores could be vital. There is evidence that a shift in root angle can be achieved quickly through speed breeding (Richard et al., 2018) and that a more narrow RSA can produce a yield benefit in droughted environments, as discussed in the previous section.

Models suggest that a small increase in root hair length from 0.4 mm to 0.8 mm could increase P uptake by 52% in dry soils, compared with a 26% boost in well-watered soils (Wissuwa and Kant, 2021). Field trials have supported the value of root hairs in droughted conditions (e.g. Vetterlein et al., 2022), but these experiments commonly involve comparisons with mutants exhibiting defective root hair development. More research is needed to identify useful donor lines within natural variants and to test the model predictions in a near-isogenic background.

While there is pressure on crop scientists and breeders to deliver step-changes in performance characteristics of cereals to keep pace with the demand of a growing world population, we see that even small changes in the expression of certain traits can have a significant impact.

8 Future trends in research

A recent survey outlined the current research needs of the root biology community (Delory et al., 2022). Chief among those highlighted by the respondents was high-throughput phenotyping methods (Atkinson et al., 2019; Tracy et al., 2020) and a better understanding of the relationship between traits and function.

8.1 Common genetic panels for community-wide phenotyping

The creation of common panels of appropriate genetic materials in elite backgrounds that can be widely shared for research – and with publically available genotype data – might advance understanding faster than traditional correlative experiments that compare small sets of highly diverse lines that are unique to each study (Uga, 2021). Some such sets already exist, such as the Global Durum Panel, various MAGIC populations and NAM panels. Some biparental mapping populations have been used for many studies, such as the Avalon × Cadenza doubled-haploid winter wheat population, with published data ranging from seedlings to the field.

8.2 Root image analysis

Much root phenotyping is not done on roots but on images of roots. Due to efforts by many groups over a number of years, there is a selection of different software applications that automate or semi-automate root image analysis (Falk et al., 2020). Recently, these tools utilise machine learning methods to increase the speed and accuracy of feature identification and quantification (Pound et al., 2017; Wang et al., 2019; Yasrab et al., 2019; Teramoto and Uga, 2020; Bauer et al., 2021; Narisetti et al., 2021; Seethepalli et al., 2021) While some packages require a level of manual curation during the process, or familiarity with artificial intelligence methods to train software to new species or background conditions, newer methods allow non-expert users to train the software quickly using smaller numbers of annotated images (Smith et al., 2022). For analysis of older plants with more complex RSA, and the inevitable overlap and occlusion of root segments, additional care must be taken to obtain accurate phenotype data (Lobet et al., 2017).

8.3 Models for simulating root growth, development and function

There have been a number of advances in the development of models that simulate root growth and function, and these have become more accessible as commonplace computing power increases. Still, simulation models are under-utilised, and as the accuracy and functionality of these models increase, their incorporation into research should help underpin selection and breeding strategies (Lynch, 2018). For instance, functional-structural models such as OpenSimRoot (Postma et al., 2017) and CPlantBox (Zhou et al., 2020) can be used to predict interactions between RSA and nutrient and water uptake, particularly when coupled with soil hydrological models (Tron et al., 2015; Ruiz et al., 2020). Using the whole-plant APSIM maize model, Hammer et al. (2009) showed that RSA (water acquisition) had a greater effect on grain yield than canopy architecture (light interception) in high-density planting with adequate moisture availability. In particular, there was a yield advantage of steeper root angles (25° from vertical) vs. wider angles of 45° in non-stressed environments. Models that incorporate management practices at the farming system level (e.g. Cooper et al., 2021) provide a more holistic view of the potential impact of genetic manipulation of RSA (Thorup-Kristensen and Kirkegaard, 2016).

Other topics introduced above show promise for greater exploration and exploitation in the future. For instance, it is important to better understand the genetic control of RSA plasticity and how genes that control plasticity can be incorporated into improvement strategies (Suralta et al., 2018; Schneider and Lynch, 2020; Schneider et al., 2020).

8.4 Roots and the rhizosphere microbiome

An exciting area for new discovery is the interaction between host plant genetics and the rhizosphere microbiome, aided by technical advances in metagenomics. Harnessing the beneficial effects of soil microbes that can improve the resilience of crops to drought shows promise (Hong et al., 2022). However, we require a greater understanding of the nexus between RSA, root exudates, physical soil properties, soil tillage and crop management practices such as fungicide use, legacy effects in a crop rotation, etc., to provide a holistic view of how practical improvements could be made (Kawasaki et al., 2016; Marr, 2020; Herms et al., 2022). In future, the genetics of the crop and the microbiome could be (naturally or artificially) engineered (Gopal and Gupta, 2016; Yu and Hochholdinger, 2018). Evidence suggests that selective breeding for agronomic traits has indirectly affected the composition of the microbiome, so it follows that this also could be done in a targeted manner (Kavamura et al., 2019; Tkacz et al., 2020).

The research summarised in this Chapter has largely focussed on the major cereal crops, and we have not drawn on a large body of information from other species. A challenge for the future is to support root research and genetic improvement of RSA in other cereals such as millet and other crop species that would be characterised as 'orphan' or neglected. These crops receive less R&D funding but can be vital for small-scale production systems or can be further developed to diversify rotations in farming systems of industrialised countries.

9 Conclusion

The literature cited here illustrates the extensive body of information that supports improvements in roots and the rhizosphere to increase productivity in water-limited conditions. The list of QTLs and candidate genes for root traits is growing rapidly (Kulkarni et al., 2017; Klein et al., 2020; Li et al., 2021). Many of the fundamental principles have been understood for some time: RSA and physiology converge with soil hydraulics to define water acquisition and supply to the shoot, while shoot growth and transpiration converge to modulate the volume of water drawn from soil stores via atmospheric demand (Boyer, 1985). We will no doubt eventually see new varieties released that have genetically modified root systems, or varieties with novel root phenotypes derived using gene editing. These are only techniques that augment conventional breeding methods; the same fundamental questions apply in matching root traits with environments and the usual attention that must be paid to pleiotropic effects or 'off-target' events, which may be deleterious or beneficial.

After the briefest acquaintance with roots in the field, it becomes apparent that phenotypic plasticity to macro- or microenvironmental variation, coupled with developmental instability defines the phenotype as much or more than the fundamental genetic program, and that RSA is not uniform even within a constant genetic background. For those interested in selective breeding for root traits, this cannot be safely ignored. Rather, taking this into consideration, the lessons for crop improvement may be to courageously embrace the noise rather than finding clever statistical methods of partitioning it out of the way, to quantify phenotypic variation to characterised environmental cues rather than treat 'environment' as a nuisance and to employ the complexity of epigenetics rather than exploiting only the simply inherited phenotypes.

The stochastic nature of root trait expression and drought occurrence must always be taken into account when designing solutions. And to deliver those solutions, open and active dialogue between plant biologists, geneticists, breeders, agronomists and farmers must be constantly fostered.

There is no single crop improvement strategy that will fit all situations, particularly with regard to RSA ideotypes (Clarke and McCaig, 1993).

Ideotype breeding may result in highly specialised varieties for targeted environments (van der Bom et al., 2020), while the provision of a 'toolbox' of donor lines carrying a variety of alleles controlling different RSA types (Rambla et al., 2022) will allow breeders to mix and match for different environments differing in resource availability, or for wide adaptation. It is unclear how many robust molecular markers may eventually appear in the toolbox that will be implemented in a marker-assisted selection of root traits; at the moment, there are very few. It is also uncertain if genomic selection methods could be applied to RSA if the requirement is for large training sets of root phenotype data from the field. It can be argued that attention to yield stability across locations and years is as important to farmers as yield potential. Optimising root function may play an important part in reducing volatility in yields that compounds the volatility in market prices and weather that cause farm businesses to struggle.

10 Where to look for further information

The International Society of Root Research (https://www.rootresearch.org/) is a good place to find more information about current research on roots, and the ISRR hosts a conference every three years. Web portals to access software for root modelling and image analysis are available, such as https://www.quantitative-plant.org/. Although peer-reviewed journal articles are the best place to find the most recent advances in root research, there are many oft-cited books on roots, such as 'Plant Roots: The Hidden Half' and the classics that have well-worn pages: Weaver's 'Root Development of Field Crops', Böhm's 'Methods of Studying Root Systems', and Smit et al's Handbook on Root Methods, to mention a few:

- Böhm, W. (1979). *Methods of Studying Root Systems*. Ecological Studies. 33. Springer. https://doi.org/10.1007/978-3-642-67282-8.
- Eshel, A. and Beeckman, T. (Eds.). (2013). *Plant Roots: The Hidden Half*, Fourth Edition (4th ed.). CRC Press. https://doi.org/10.1201/b14550.
- Smit, A. L., Bengough, A. G., Engels, C., van Noordwijk, M., Pellerin, S. and van de Geijn, S. C. (2000). *Root Methods: A Handbook*. Plant Methods. Springer Science & Business Media.
- Weaver, J. E. (1926). *Root Development of Field Crops*. McGraw-Hill.

11 References

Ahmed, M. A., Passioura, J. and Carminati, A. (2018). Hydraulic processes in roots and the rhizosphere pertinent to increasing yield of water-limited grain crops: A critical review, *Journal of Experimental Botany* 69(13), 3255–3265. https://doi.org/10.1093/jxb/ery183.

Alahmad, S., El Hassouni, K., Bassi, F. M., Dinglasan, E., Youssef, C., Quarry, G., Aksoy, A., Mazzucotelli, E., Juhász, A., Able, J. A., Christopher, J., Voss-Fels, K. P. and Hickey, L. T. (2019). A major root architecture QTL responding to water limitation in durum wheat, *Frontiers in Plant Science* 10, 436. Available at: https://www.frontiersin.org/article/10.3389/fpls.2019.00436.

Andreo-Jimenez, B., Vandenkoornhuyse, P., Lê Van, A., Heutinck, A., Duhamel, M., Kadam, N., Jagadish, K., Ruyter-Spira, C. and Bouwmeester, H. (2019). Plant host and drought shape the root associated fungal microbiota in rice, *PeerJ* 7, e7463. https://doi.org/10.7717/peerj.7463.

Annicchiarico, P. (2002). Genotype x environment interactions: Challenges and opportunities for plant breeding and cultivar recommendation. In: *Quantitative Genetics, Genomics and Plant Breeding*. Wallingford, UK: CABI Publishing, pp. 365–384.

Aslam, M. M., Okal, E. J., Idris, A. L., Qian, Z., Xu, W., Karanja, J. K., Wani, S. H. and Yuan, W. (2022). Rhizosphere microbiomes can regulate plant drought tolerance, *Pedosphere* 32(1), 61–74. https://doi.org/10.1016/S1002-0160(21)60061-9.

Atkinson, J. A., Hawkesford, M. J., Whalley, W. R., Zhou, H. and Mooney, S. J. (2019). Soil strength influences wheat root interactions with soil macropores, *Plant, Cell and Environment* 43, 235–245. https://doi.org/10.1111/pce.13659.

Atkinson, J. A., Pound, M. P., Bennett, M. J. and Wells, D. M. (2019). Uncovering the hidden half of plants using new advances in root phenotyping, *Current Opinion in Biotechnology* 55, 1–8. Available at: http://www.sciencedirect.com/science/article/pii/S0958166918300521.

Babé, A., Lavigne, T., Séverin, J. P., Nagel, K. A., Walter, A., Chaumont, F., Batoko, H., Beeckman, T. and Draye, X. (2012). Repression of early lateral root initiation events by transient water deficit in barley and maize, *Philosophical Transactions of the Royal Society of London: Series B, Biological Sciences* 367(1595), 1534–1541. https://doi.org/10.1098/rstb.2011.0240.

Bacher, H., Sharaby, Y., Walia, H. and Peleg, Z. (2022). Modifying root-to-shoot ratio improves root water influxes in wheat under drought stress, *Journal of Experimental Botany* 73(5), 1643–1654. https://doi.org/10.1093/jxb/erab500.

Bai, C., Ge, Y., Ashton, R. W., Evans, J., Milne, A., Hawkesford, M. J., Whalley, W. R., Parry, M. A. J., Melichar, J., Feuerhelm, D., Basler, P. B. and Bartsch, M. (2019). The relationships between seedling root screens, root growth in the field and grain yield for wheat, *Plant and Soil* 440(1–2), 311–326. https://doi.org/10.1007/s11104-019-04088-9.

Bao, Y., Aggarwal, P., Robbins, N. E., Sturrock, C. J., Thompson, M. C., Tan, H. Q., Tham, C., Duan, L., Rodriguez, P. L., Vernoux, T., Mooney, S. J., Bennett, M. J. and Dinneny, J. R. (2014). Plant roots use a patterning mechanism to position lateral root branches toward available water, *Proceedings of the National Academy of Sciences of the United States of America* 111(25), 9319–9324. https://doi.org/10.1073/pnas.1400966111.

Bauer, F. M., Lärm, L., Morandage, S., Lobet, G., Vanderborght, J., Vereecken, H. and Schnepf, A. (2021). Combining deep learning and automated feature extraction to analyze minirhizotron images: Development and validation of a new pipeline, *bioRxiv*. https://doi.org/10.1101/2021.12.01.470811.

Bengough, A. G., Loades, K. and McKenzie, B. M. (2016). Root hairs aid soil penetration by anchoring the root surface to pore walls, *Journal of Experimental Botany* 67(4), 1071–1078. http://doi.org/10.1093/jxb/erv560.

Bhaskara, G. B., Yang, T. H. and Verslues, P. E. (2015). Dynamic proline metabolism: Importance and regulation in water limited environments, *Frontiers in Plant Science* 6, 484. https://doi.org/10.3389/fpls.2015.00484.

Bloomfield, J. A., Rose, T. J. and King, G. J. (2014). Sustainable harvest: Managing plasticity for resilient crops, *Plant Biotechnology Journal* 12(5), 517–533. https://doi.org/10.1111/pbi.12198.

Bormann, F. H. (1957). Moisture transfer between plants through intertwined root systems, *Plant Physiology* 32(1), 48–55. https://doi.org/10.1104/pp.32.1.48.

Borrell, A. K., Mullet, J. E., George-Jaeggli, B., van Oosterom, E. J., Hammer, G. L., Klein, P. E. and Jordan, D. R. (2014). Drought adaptation of stay-green sorghum is associated with canopy development, leaf anatomy, root growth, and water uptake, *Journal of Experimental Botany* 65(21), 6251–6263. https://doi.org/10.1093/jxb/eru232.

Botwright Acuña, T. L., Rebetzke, G. J., He, X., Maynol, E. and Wade, L. J. (2014). Mapping quantitative trait loci associated with root penetration ability of wheat in contrasting environments, *Molecular Breeding* 34(2), 631–642. https://doi.org/10.1007/s11032-014-0063-x.

Boyer, J. S. (1985). Water transport, *Annual Review of Plant Physiology* 36(1), 473–516. https://doi.org/10.1146/annurev.pp.36.060185.002353.

Brinton, J., Ramirez-Gonzalez, R. H., Simmonds, J., Wingen, L., Orford, S., Griffiths, S., 10 Wheat Genome Project, Haberer, G., Spannagl, M., Walkowiak, S., Pozniak, C. and Uauy, C. (2020). A haplotype-led approach to increase the precision of wheat breeding, *Communications Biology* 3(1), 712. https://doi.org/10.1038/s42003-020-01413-2.

Cai, G. and Ahmed, M. A. (2022). The role of root hairs in water uptake: Recent advances and future perspectives, *Journal of Experimental Botany* 73(11), 3330–3338. https://doi.org/10.1093/jxb/erac114.

Caldwell, M. M., Dawson, T. E. and Richards, J. H. (1998). Hydraulic lift: Consequences of water efflux from the roots of plants, *Oecologia* 113(2), 151–161. https://doi.org/10.1007/s004420050363.

Calvo-Polanco, M., Ribeyre, Z., Dauzat, M., Reyt, G., Hidalgo-Shrestha, C., Diehl, P., Frenger, M., Simonneau, T., Muller, B., Salt, D. E., Franke, R. B., Maurel, C. and Boursiac, Y. (2021). Physiological roles of Casparian strips and suberin in the transport of water and solutes, *New Phytologist* 232(6), 2295–2307. https://doi.org/10.1111/nph.17765.

Carminati, A., Benard, P., Ahmed, M. A. and Zarebanadkouki, M. (2017). Liquid bridges at the root-soil interface, *Plant and Soil* 417(1–2), 1–15. http://doi.org/10.1007/s11104-017-3227-8.

Carminati, A., Passioura, J. B., Zarebanadkouki, M., Ahmed, M. A., Ryan, P. R., Watt, M. and Delhaize, E. (2017). Root hairs enable high transpiration rates in drying soils, *New Phytologist* 216(3), 771–781. https://doi.org/10.1111/nph.14715.

Chimungu, J. G., Brown, K. M. and Lynch, J. P. (2014). Large root cortical cell size improves drought tolerance in maize, *Plant Physiology* 166(4), 2166–2178. Available at: http://www.ncbi.nlm.nih.gov/pmc/articles/PMC4256844/.

Chmielewska, B., Janiak, A., Karcz, J., Guzy-Wrobelska, J., Forster, B. P., Nawrot, M., Rusek, A., Smyda, P., Kedziorski, P., Maluszynski, M. and Szarejko, I. (2014). Morphological, genetic and molecular characteristics of barley root hair mutants, *Journal of Applied Genetics* 55(4), 433–447. https://doi.org/10.1007/s13353-014-0225-x.

Christopher, J., Christopher, M., Jennings, R., Jones, S., Fletcher, S., Borrell, A., Manschadi, A. M., Jordan, D., Mace, E. and Hammer, G. (2013). QTL for root angle and number in a population developed from bread wheats (*Triticum aestivum*) with contrasting adaptation to water-limited environments, *TAG: Theoretical and Applied Genetics. Theoretische und Angewandte Genetik* 126(6), 1563–1574. https://doi.org/10.1007/s00122-013-2074-0.

Ciszak, M., Comparini, D., Mazzolai, B., Baluska, F., Arecchi, F. T., Vicsek, T. and Mancuso, S. (2012). Swarming behavior in plant roots, *PLoS ONE* 7(1), e29759. https://doi.org/10.1371/journal.pone.0029759.

Clark, L. J., Price, A. H., Steele, K. A. and Whalley, W. R. (2008). Evidence from near-isogenic lines that root penetration increases with root diameter and bending stiffness in rice, *Functional Plant Biology* 35(11), 1163–1171. https://doi.org/10.1071/FP08132.

Clarke, J. M. and McCaig, T. N. (1993). Breeding for efficient root systems. In: Hayward, M. D., Bosemark, N. O. and Romagosa, T. (Eds.), *Plant Breeding: Principles and Prospects*. Dordrecht: Springer Netherlands, pp. 485–499. https://doi.org/10.1007/978-94-011-1524-7_29.

Comas, L. H., Becker, S. R., Cruz, V. M., Byrne, P. F. and Dierig, D. A. (2013). Root traits contributing to plant productivity under drought, *Frontiers in Plant Science* 4, 442. https://doi.org/10.3389/fpls.2013.00442.

Condorelli, G. E., Newcomb, M., Groli, E. L., Maccaferri, M., Forestan, C., Babaeian, E., Tuller, M., White, J. W., Ward, R., Mockler, T., Shakoor, N. and Tuberosa, R. (2022). Genome wide association study uncovers the QTLome for osmotic adjustment and related drought adaptive traits in durum wheat, *Genes* 13(2), 293. https://doi.org/10.3390/genes13020293.

Cooper, M., Voss-Fels, K. P., Messina, C. D., Tang, T. and Hammer, G. L. (2021). Tackling G × E × M interactions to close on-farm yield-gaps: Creating novel pathways for crop improvement by predicting contributions of genetics and management to crop productivity, *TAG: Theoretical and Applied Genetics. Theoretische und Angewandte Genetik* 134(6), 1625–1644. https://doi.org/10.1007/s00122-021-03812-3.

Corak, S. J., Blevins, D. G. and Pallardy, S. G. (1987). Water transfer in an alfalfa/maize association: Survival of maize during drought, *Plant Physiology* 84(3), 582–586. https://doi.org/10.1104/pp.84.3.582.

Darwin, C. R. (1880). *The Power of Movements in Plants*. London: John Murray. Available at: http://darwin-online.org.uk/.

Decunta, F. A., Pérez, L. I., Malinowski, D. P., Molina-Montenegro, M. A. and Gundel, P. E. (2021). A systematic review on the effects of Epichloë fungal endophytes on drought tolerance in cool-season grasses, *Frontiers in Plant Science* 12, 644731. Available at: https://www.frontiersin.org/articles/10.3389/fpls.2021.644731.

Delory, B. M., Hernandez-Soriano, M. C., Wacker, T. S., Dimitrova, A., Ding, Y., Greeley, L. A., Ng, J. L. P., Mesa-Marín, J., Xie, L., Zheng, C. and York, L. M. (2022). A snapshot of the root phenotyping landscape in 2021, *bioRxiv*. https://doi.org/10.1101/2022.01.28.478001.

Dodd, I. C. and Diatloff, E. (2016). Enhanced root growth of the brb (bald root barley) mutant in drying soil allows similar shoot physiological responses to soil water deficit as wild-type plants, *Functional Plant Biology* 43(2), 199–206. https://doi.org/10.1071/FP15303.

Dowd, T. G., Braun, D. M. and Sharp, R. E. (2020). Maize lateral root developmental plasticity induced by mild water stress. II: Genotype-specific spatio-temporal effects on determinate development, *Plant, Cell and Environment* 43(10), 2409–2427. https://doi.org/10.1111/pce.13840.

Eberhart, S. A. and Russell, W. A. (1966). Stability parameters for comparing varieties, *Crop Science Crop Science* 42, 423–427. https://doi.org/10.2135/cropsci1966.0011183X000600010011x.

Ehdaie, B., Layne, A. P. and Waines, J. G. (2012). Root system plasticity to drought influences grain yield in bread wheat, *Euphytica* 186(1), 219–232. https://doi.org/10.1007/s10681-011-0585-9.

El Hassouni, K., Alahmad, S., Belkadi, B., Filali-Maltouf, A., Hickey, L. T. and Bassi, F. M. (2018). Root system architecture and its association with yield under different water regimes in durum wheat, *Crop Science* 58(6), 2331–2346. https://doi.org/10.2135/cropsci2018.01.0076.

Elemi, A. A. and West, C. P. (1995). Endophyte infection effects on stomatal conductance, osmotic adjustment and drought recovery of tall fescue, *New Phytologist* 131(1), 61–67. https://doi.org/10.1111/j.1469-8137.1995.tb03055.x.

Falk, K. G., Jubery, T. Z., Mirnezami, S. V., Parmley, K. A., Sarkar, S., Singh, A., Ganapathysubramanian, B. and Singh, A. K. (2020). Computer vision and machine learning enabled soybean root phenotyping pipeline, *Plant Methods* 16(1), 5. https://doi.org/10.1186/s13007-019-0550-5.

Fang, S., Clark, R. T., Zheng, Y., Iyer-Pascuzzi, A. S., Weitz, J. S., Kochian, L. V., Edelsbrunner, H., Liao, H. and Benfey, P. N. (2013). Genotypic recognition and spatial responses by rice roots, *Proceedings of the National Academy of Sciences of the United States of America* 110(7), 2670–2675. https://doi.org/10.1073/pnas.1222821110.

Fang, Y., Liang, L., Liu, S., Xu, B., Siddique, K. H., Palta, J. A. and Chen, Y. (2021). Wheat cultivars with small root length density in the topsoil increased post-anthesis water use and grain yield in the semi-arid region on the Loess Plateau, *European Journal of Agronomy* 124, 126243. https://doi.org/10.1016/j.eja.2021.126243.

Fang, Y., Liu, L., Xu, B. and Li, F. (2011). The relationship between competitive ability and yield stability in an old and a modern winter wheat cultivar, *Plant and Soil* 347(1–2), 7–23. https://doi.org/10.1007/s11104-011-0780-4.

Figueroa-Bustos, V., Palta, J. A., Chen, Y., Stefanova, K. and Siddique, K. H. M. (2020). Wheat cultivars with contrasting root system size responded differently to terminal drought, *Frontiers in Plant Science* 11, 1285. Available at: https://www.frontiersin.org/article/10.3389/fpls.2020.01285.

Forde, B. G. (2009). Is it good noise? The role of developmental instability in the shaping of a root system, *Journal of Experimental Botany* 60(14), 3989–4002. https://doi.org/10.1093/jxb/erp265.

Fradgley, N., Evans, G., Biernaskie, J. M., Cockram, J., Marr, E. C., Oliver, A. G., Ober, E. and Jones, H. (2020). Effects of breeding history and crop management on the root architecture of wheat, *Plant and Soil* 452(1), 587–600. https://doi.org/10.1007/s11104-020-04585-2.

Freschet, G. T., Roumet, C., Comas, L. H., Weemstra, M., Bengough, A. G., Rewald, B., Bardgett, R. D., De Deyn, G. B., Johnson, D., Klimešová, J., Lukac, M., McCormack, M. L., Meier, I. C., Pagès, L., Poorter, H., Prieto, I., Wurzburger, N., Zadworny, M., Bagniewska-Zadworna, A., Blancaflor, E. B., Brunner, I., Gessler, A., Hobbie, S. E., Iversen, C. M., Mommer, L., Picon-Cochard, C., Postma, J. A., Rose, L., Ryser, P., Scherer-Lorenzen, M., Soudzilovskaia, N. A., Sun, T., Valverde-Barrantes, O. J., Weigelt, A., York, L. M. and Stokes, A. (2021). Root traits as drivers of plant and ecosystem functioning: Current understanding, pitfalls and future research needs, *New Phytologist* 232(3), 1123–1158. https://doi.org/10.1111/nph.17072.

Friedli, C. N., Abiven, S., Fossati, D. and Hund, A. (2019). Modern wheat semi-dwarfs root deep on demand: Response of rooting depth to drought in a set of Swiss era wheats covering 100 years of breeding, *Euphytica* 215(4), 85. https://doi.org/10.1007/s10681-019-2404-7.

Fusi, R., Rosignoli, S., Lou, H., Sangiorgi, G., Bovina, R., Pattem, J. K., Borkar, A. N., Lombardi, M., Forestan, C., Milner, S. G., Davis, J. L., Lale, A., Kirschner, G. K., Swarup, R., Tassinari, A., Pandey, B. K., York, L. M., Atkinson, B. S., Sturrock, C. J., Mooney, S. J., Hochholdinger, F., Tucker, M. R., Himmelbach, A., Stein, N., Mascher, M., Nagel, K. A., De Gara, L., Simmonds, J., Uauy, C., Tuberosa, R., Lynch, J. P., Yakubov, G. E., Bennett, M. J., Bhosale, R. and Salvi, S. (2022). Root angle is controlled by EGT1 in cereal crops employing an antigravitropic mechanism, *Proceedings of the National Academy of Sciences of the United States of America* 119(31), e2201350119. https://doi.org/10.1073/pnas.2201350119.

Giehl, R. F. H. and von Wirén, N. (2018). Hydropatterning-how roots test the waters, *Science* 362(6421), 1358–1359. Available at: http://science.sciencemag.org/content/362/6421/1358.abstract.

Giri, J., Bhosale, R., Huang, G., Pandey, B. K., Parker, H., Zappala, S., Yang, J., Dievart, A., Bureau, C., Ljung, K., Price, A., Rose, T., Larrieu, A., Mairhofer, S., Sturrock, C. J., White, P., Dupuy, L., Hawkesford, M., Perin, C., Liang, W., Peret, B., Hodgman, C. T., Lynch, J., Wissuwa, M., Zhang, D., Pridmore, T., Mooney, S. J., Guiderdoni, E., Swarup, R. and Bennett, M. J. (2018). Rice auxin influx carrier OsAUX1 facilitates root hair elongation in response to low external phosphate, *Nature Communications* 9(1), 1408. https://doi.org/10.1038/s41467-018-03850-4.

Gopal, M. and Gupta, A. (2016). Microbiome selection could spur next-generation plant breeding strategies, *Frontiers in Microbiology* 7, 1971. Available at: https://www.frontiersin.org/article/10.3389/fmicb.2016.01971.

Gregory, P. J. (1989). The role of root characteristics in moderating the effects of drought. In: Baker, F. W. G. (Ed.), *Drought Resistance in Cereals*, pp. 141–150. CABI Publishing.

Gregory, P. J., McGowan, M. and Biscoe, P. V. (1978). Water relations of winter wheat. 2. Soil water relations, *Journal of Agricultural Science* 91(1), 103–116. Available at: https://doi.org/doi:10.1017/S0021859600056665.

Hamblin, A. P. and Tennant, D. (1987). Root length density and water uptake in cereals and grain legumes: How well are they correlated, *Australian Journal of Agricultural Research* 38(3), 513–527.

Hammer, G. L., Dong, Z., McLean, G., Doherty, A., Messina, C., Schussler, J., Zinselmeier, C., Paszkiewicz, S. and Cooper, M. (2009). Can changes in canopy and/or root system architecture explain historical maize yield trends in the U. S. corn belt?, *Crop Science* 49(1), 299–312. https://doi.org/10.2135/cropsci2008.03.0152.

Han, H., Wang, H., Han, Y., Hu, Z., Xin, M., Peng, H., Yao, Y., Sun, Q. and Ni, Z. (2017). Altered expression of the TaRSL2 gene contributed to variation in root hair length during allopolyploid wheat evolution, *Planta* 246(5), 1019–1028. https://doi.org/10.1007/s00425-017-2735-3.

Han, Y., Xin, M., Huang, K., Xu, Y., Liu, Z., Hu, Z., Yao, Y., Peng, H., Ni, Z. and Sun, Q. (2016). Altered expression of TaRSL4 gene by genome interplay shapes root hair length in allopolyploid wheat, *New Phytologist* 209(2), 721–732. https://doi.org/10.1111/nph.13615.

Hauser, M. T., Aufsatz, W., Jonak, C. and Luschnig, C. (2011). Transgenerational epigenetic inheritance in plants, *Biochimica et Biophysica Acta* 1809(8), 459–468. https://doi.org/10.1016/j.bbagrm.2011.03.007.

Hecht, V. L., Temperton, V. M., Nagel, K. A., Rascher, U., Pude, R. and Postma, J. A. (2019). Plant density modifies root system architecture in spring barley (*Hordeum vulgare* L.) through a change in nodal root number, *Plant and Soil* 439(1–2), 179–200. https://doi.org/10.1007/s11104-018-3764-9.

Herms, C. H., Hennessy, R. C., Bak, F., Dresbøll, D. B. and Nicolaisen, M. H. (2022). Back to our roots: Exploring the role of root morphology as a mediator of beneficial plant-microbe interactions, *Environmental Microbiology* 24(8), 3264–3272. https://doi.org/10.1111/1462-2920.15926.

Hobson, D. J., Harty, M. A., Langton, D., McDonnell, K. and Tracy, S. R. (2022). The establishment of winter wheat root system architecture in field soils: The effect of soil type on root development in a temperate climate, *Soil Use and Management* 39, 198–208. https://doi.org/10.1111/sum.12795.

Hodge, A. (2009). Root decisions, *Plant, Cell and Environment* 32(6), 628–640. https://doi.org/10.1111/j.1365-3040.2008.01891.x.

Holz, M., Zarebanadkouki, M., Kaestner, A., Kuzyakov, Y. and Carminati, A. (2018). Rhizodeposition under drought is controlled by root growth rate and rhizosphere water content, *Plant and Soil* 423(1–2), 429–442. https://doi.org/10.1007/s11104-017-3522-4.

Hong, Y., Zhou, Q., Hao, Y. and Huang, A. C. (2022). Crafting the plant root metabolome for improved microbe-assisted stress resilience, *New Phytologist* 234(6), 1945–1950. https://doi.org/10.1111/nph.17908.

Hubbard, M., Germida, J. J. and Vujanovic, V. (2014). Fungal endophytes enhance wheat heat and drought tolerance in terms of grain yield and second generation seed viability, *Journal of Applied Microbiology* 116(1), 109–122. https://doi.org/10.1111/jam.12311.

Hund, A., Ruta, N. and Liedgens, M. (2009). Rooting depth and water use efficiency of tropical maize inbred lines, differing in drought tolerance, *Plant and Soil* 318(1–2), 311–325. https://doi.org/10.1007/s11104-008-9843-6.

Huntenburg, K., Puértolas, J., de Ollas, C. and Dodd, I. C. (2022). Bi-directional, long-distance hormonal signalling between roots and shoots of soil water availability, *Physiologia Plantarum* 174(3), e13697. https://doi.org/10.1111/ppl.13697.

Ito, K., Tanakamaru, K., Morita, S., Abe, J. and Inanaga, S. (2006). Lateral root development, including responses to soil drying, of maize (*Zea mays*) and wheat (*Triticum aestivum*) seminal roots, *Physiologia Plantarum* 127(2), 260–267. https://doi.org/10.1111/j.1399-3054.2006.00657.x.

Jackson, P., Robertson, M., Cooper, M. and Hammer, G. (1996). The role of physiological understanding in plant breeding; from a breeding perspective, *Field Crops Research* 49(1), 11–37.

Javot, H. and Maurel, C. (2002). The role of aquaporins in root water uptake, *Annals of Botany* 90(3), 301–313. https://doi.org/10.1093/aob/mcf199.

Kadam, N. N., Tamilselvan, A., Lawas, L. M. F., Quinones, C., Bahuguna, R. N., Thomson, M. J., Dingkuhn, M., Muthurajan, R., Struik, P. C., Yin, X. and Jagadish, S. V. K. (2017). Genetic control of plasticity in root morphology and anatomy of rice in response to water deficit, *Plant Physiology* 174(4), 2302–2315. https://doi.org/10.1104/PP.17.00500.

Kang, J., Peng, Y. and Xu, W. (2022). Crop root responses to drought stress: Molecular mechanisms, nutrient regulations, and interactions with microorganisms in the rhizosphere, *International Journal of Molecular Sciences* 23(16), 9310.

Kang, J., Voothuluru, P., Hoyos-Miernyk, E., Alexander, D., Oliver, M. J. and Sharp, R. E. (2022). Antioxidant metabolism underlies different metabolic strategies for primary root growth maintenance under water stress in cotton and maize, *Antioxidants* 11(5), 820.

Kapilan, R., Vaziri, M. and Zwiazek, J. J. (2018). Regulation of aquaporins in plants under stress, *Biological Research* 51(1), 4. https://doi.org/10.1186/s40659-018-0152-0.

Kavamura, V. N., Robinson, R. J., Hayat, R., Clark, I. M., Hughes, D., Rossmann, M., Hirsch, P. R., Mendes, R. and Mauchline, T. H. (2019). Land management and microbial seed load effect on rhizosphere and endosphere bacterial community assembly in wheat, *Frontiers in Microbiology* 10, 2625. Available at: https://www.frontiersin.org/article/10.3389/fmicb.2019.02625.

Kawasaki, A., Donn, S., Ryan, P. R., Mathesius, U., Devilla, R., Jones, A. and Watt, M. (2016). Microbiome and exudates of the root and rhizosphere of *Brachypodium distachyon*, a model for wheat, *PLoS ONE* 11(10), e0164533. https://doi.org/10.1371/journal.pone.0164533.

Kemper, R., Döring, T. F., Legner, N., Meinen, C. and Athmann, M. (2022). Root traits in cover crop mixtures of blue lupin and winter rye, *Plant and Soil* 475(1–2), 309–328. https://doi.org/10.1007/s11104-022-05366-9.

Kirkegaard, J. A., Lilley, J. M., Howe, G. N. and Graham, J. M. (2007). Impact of subsoil water use on wheat yield, *Australian Journal of Agricultural Research* 58(4), 303–315. https://doi.org/10.1071/AR06285.

Kirschner, G. K., Rosignoli, S., Guo, L., Vardanega, I., Imani, J., Altmüller, J., Milner, S. G., Balzano, R., Nagel, K. A., Pflugfelder, D., Forestan, C., Bovina, R., Koller, R., Stöcker, T. G., Mascher, M., Simmonds, J., Uauy, C., Schoof, H., Tuberosa, R., Salvi, S. and Hochholdinger, F. (2021). Enhanced gravitropism 2 encodes a STERILE alpha MOTIVE containing protein that controls root growth angle in barley and wheat, *Proceedings of the National Academy of Sciences of the United States of America* 118(35), e2101526118. https://doi.org/10.1073/pnas.2101526118.

Klamer, F., Vogel, F., Li, X., Bremer, H., Neumann, G., Neuhäuser, B., Hochholdinger, F. and Ludewig, U. (2019). Estimating the importance of maize root hairs in low phosphorus conditions and under drought, *Annals of Botany* 124(6), 961–968. https://doi.org/10.1093/aob/mcz011.

Klein, S. P., Schneider, H. M., Perkins, A. C., Brown, K. M. and Lynch, J. P. (2020). Multiple integrated root phenotypes are associated with improved drought tolerance, *Plant Physiology* 183(3), 1011–1025. https://doi.org/10.1104/pp.20.00211.

Knight, T. A. (1811). XI. On the causes which influence the direction of the growth of roots. By T. A. Knight, esq. F. R. S. In a letter to the Right Hon. Sir Joseph Banks, Bart. K. B. P. R. S, *Philosophical Transactions of the Royal Society of London* 101, 209–219. https://doi.org/10.1098/rstl.1811.0013.

Koevoets, I. T., Venema, J. H., Elzenga, J. T. and Testerink, C. (2016). Roots withstanding their environment: Exploiting root system architecture responses to abiotic stress to improve crop tolerance, *Frontiers in Plant Science* 7, 1335. Available at: https://www.frontiersin.org/article/10.3389/fpls.2016.01335.

Kulkarni, M., Soolanayakanahally, R., Ogawa, S., Uga, Y., Selvaraj, M. G. and Kagale, S. (2017). Drought response in wheat: Key genes and regulatory mechanisms controlling root system architecture and transpiration efficiency, *Frontiers in Chemistry* 5, 106. Available at: https://www.frontiersin.org/article/10.3389/fchem.2017.00106.

Lehnert, H., Serfling, A., Enders, M., Friedt, W. and Ordon, F. (2017). Genetics of mycorrhizal symbiosis in winter wheat (*Triticum aestivum*), *New Phytologist* 215(2), 779–791. https://doi.org/10.1111/nph.14595.

Li, B., Zhang, X., Morita, S., Sekiya, N., Araki, H., Gu, H., Han, J., Lu, Y. and Liu, X. (2022). Are crop deep roots always beneficial for combating drought: A review of root structure and function, regulation and phenotyping, *Agricultural Water Management* 271, 107781. https://doi.org/10.1016/j.agwat.2022.107781.

Li, C., Li, L., Reynolds, M. P., Wang, J., Chang, X., Mao, X. and Jing, R. (2021). Recognizing the hidden half in wheat: Root system attributes associated with drought tolerance, *Journal of Experimental Botany* 72(14), 5117–5133. https://doi.org/10.1093/jxb/erab124.

Li, H., Li, L., Liu, N., Chen, S., Shao, L., Sekiya, N. and Zhang, X. (2022). Root efficiency and water use regulation relating to rooting depth of winter wheat, *Agricultural Water Management* 269, 107710. https://doi.org/10.1016/j.agwat.2022.107710.

Li, X., Ingvordsen, C. H., Weiss, M., Rebetzke, G. J., Condon, A. G., James, R. A. and Richards, R. A. (2019). Deeper roots associated with cooler canopies, higher normalized difference vegetation index, and greater yield in three wheat populations grown on stored soil water, *Journal of Experimental Botany* 70(18), 4963–4974. https://doi.org/10.1093/jxb/erz232.

Lilley, J. M. and Kirkegaard, J. A. (2011). Benefits of increased soil exploration by wheat roots, *Field Crops Research* 122(2), 118–130. https://doi.org/10.1016/j.fcr.2011.03.010.

Lin, C. S., Binns, M. R. and Lefkovitch, L. P. (1986). Stability analysis: Where do we stand?, *Crop Science* 26(5), 894–900. https://doi.org/10.2135/cropsci1986.0011183X002600050012x.

Lind, K. R., Siemianowski, O., Yuan, B., Sizmur, T., VanEvery, H., Banerjee, S. and Cademartiri, L. (2021). Evidence for root adaptation to a spatially discontinuous water availability in the absence of external water potential gradients, *Proceedings of the National Academy of Sciences of the United States of America* 118(1), e2012892118. https://doi.org/10.1073/pnas.2012892118.

Lloyd, J. P. B. and Lister, R. (2022). Epigenome plasticity in plants, *Nature Reviews Genetics* 23(1), 55–68. https://doi.org/10.1038/s41576-021-00407-y.

Lobet, G., Koevoets, I. T., Noll, M., Meyer, P. E., Tocquin, P., Pagès, L. and Périlleux, C. (2017). Using a structural root system model to evaluate and improve the accuracy of root image analysis pipelines, *Frontiers in Plant Science* 8, 447. Available at: https://www.frontiersin.org/articles/10.3389/fpls.2017.00447.

Lopes, M. S. and Reynolds, M. P. (2010). Partitioning of assimilates to deeper roots is associated with cooler canopies and increased yield under drought in wheat, *Functional Plant Biology* 37(2), 147–156. https://doi.org/10.1071/FP09121.

Lynch, J. P. (2013). Steep, cheap and deep: An ideotype to optimize water and N acquisition by maize root systems, *Annals of Botany* 112(2), 347–357. http://doi.org/10.1093/aob/mcs293.

Lynch, J. P. (2018). Rightsizing root phenotypes for drought resistance, *Journal of Experimental Botany* 69(13), 3279–3292. https://doi.org/10.1093/jxb/ery048.

Lynch, J. P. (2022). Harnessing root architecture to address global challenges, *The Plant Journal: For Cell and Molecular Biology* 109(2), 415–431. https://doi.org/10.1111/tpj.15560.

Maccaferri, M., El-Feki, W., Nazemi, G., Salvi, S., Canè, M. A., Colalongo, M. C., Stefanelli, S. and Tuberosa, R. (2016). Prioritizing quantitative trait loci for root system architecture in tetraploid wheat, *Journal of Experimental Botany* 67(4), 1161–1178. http://doi.org/10.1093/jxb/erw039.

Maccaferri, M., Sanguineti, M. C., Corneti, S., Ortega, J. L., Salem, M. B., Bort, J., DeAmbrogio, E., del Moral, L. F., Demontis, A., El-Ahmed, A., Maalouf, F., Machlab, H., Martos, V., Moragues, M., Motawaj, J., Nachit, M., Nserallah, N., Ouabbou, H., Royo, C., Slama, A. and Tuberosa, R. (2008). Quantitative trait loci for grain yield and adaptation of durum wheat (*Triticum durum* Desf.) across a wide range of water availability, *Genetics* 178(1), 489–511.

Makhoul, M., Rambla, C., Voss-Fels, K. P., Hickey, L. T., Snowdon, R. J. and Obermeier, C. (2020). Overcoming polyploidy pitfalls: A user guide for effective SNP conversion into KASP markers in wheat, *TAG. Theoretical and Applied Genetics. Theoretische und Angewandte Genetik* 133(8), 2413–2430. https://doi.org/10.1007/s00122-020-03608-x.

Manschadi, A. M., Hammer, G. L., Christopher, J. T. and deVoil, P. (2008). Genotypic variation in seedling root architectural traits and implications for drought adaptation in wheat (*Triticum aestivum* L.), *Plant and Soil* 303(1–2), 115–129. https://doi.org/10.1007/s11104-007-9492-1.

Marin, M., Feeney, D. S., Brown, L. K., Naveed, M., Ruiz, S., Koebernick, N., Bengough, A. G., Hallett, P. D., Roose, T., Puértolas, J., Dodd, I. C. and George, T. S. (2021). Significance of root hairs for plant performance under contrasting field conditions and water deficit, *Annals of Botany* 128(1), 1–16. https://doi.org/10.1093/AOB/MCAA181.

Marr, E. (2020). Unlocking the genetic potential of the root system, rhizosheath mucilage and microbiome of wheat (*Triticum aestivum* L.). PhD. University of Cambridge.

Maurel, C. and Nacry, P. (2020). Root architecture and hydraulics converge for acclimation to changing water availability, *Nature Plants* 6(7), 744–749. https://doi.org/10.1038/s41477-020-0684-5.

Maurel, C., Tournaire-Roux, C., Verdoucq, L. and Santoni, V. (2021). Hormonal and environmental signaling pathways target membrane water transport, *Plant Physiology* 187(4), 2056–2070. https://doi.org/10.1093/plphys/kiab373.

Maury, S., Sow, M. D., Le Gac, A. L., Genitoni, J., Lafon-Placette, C. and Mozgova, I. (2019). Phytohormone and chromatin crosstalk: The missing link for developmental plasticity?, *Frontiers in Plant Science* 10, 395. Available at: https://www.frontiersin.org/articles/10.3389/fpls.2019.00395.

McCully, M. E. (1999). Roots in soil: Unearthing the complexities of roots and their rhizospheres, *Annual Review of Plant Physiology and Plant Molecular Biology* 50(1), 695–718. https://doi.org/10.1146/annurev.arplant.50.1.695.

McLachlan, P., Blanchy, G. and Binley, A. (2021). EMagPy: Open-source standalone software for processing, forward modeling and inversion of electromagnetic induction data, *Computers and Geosciences* 146, 104561. https://doi.org/10.1016/j.cageo.2020.104561.

Messina, C. D., Sinclair, T. R., Hammer, G. L., Curan, D., Thompson, J., Oler, Z., Gho, C. and Cooper, M. (2015). Limited-transpiration trait may increase maize drought tolerance in the US corn Belt, *Agronomy Journal* 107(6), 1978–1986. https://doi.org/10.2134/agronj15.0016.

Mickelbart, M. V., Hasegawa, P. M. and Bailey-Serres, J. (2015). Genetic mechanisms of abiotic stress tolerance that translate to crop yield stability, *Nature Reviews Genetics* 16(4), 237–251. http://doi.org/10.1038/nrg3901.

Morohashi, K., Okamoto, M., Yamazaki, C., Fujii, N., Miyazawa, Y., Kamada, M., Kasahara, H., Osada, I., Shimazu, T., Fusejima, Y., Higashibata, A., Yamazaki, T., Ishioka, N., Kobayashi, A. and Takahashi, H. (2017). Gravitropism interferes with hydrotropism via counteracting auxin dynamics in cucumber roots: Clinorotation and spaceflight experiments, *New Phytologist* 215(4), 1476–1489. https://doi.org/10.1111/nph.14689.

Narisetti, N., Henke, M., Seiler, C., Junker, A., Ostermann, J., Altmann, T. and Gladilin, E. (2021). Fully-automated root image analysis (faRIA), *Scientific Reports* 11(1), 16047. https://doi.org/10.1038/s41598-021-95480-y.

Naylor, D. and Coleman-Derr, D. (2017). Drought stress and root-associated bacterial communities, *Frontiers in Plant Science* 8, 2223. Available at: https://www.frontiersin.org/articles/10.3389/fpls.2017.02223.

Nestler, J., Liu, S., Wen, T. J., Paschold, A., Marcon, C., Tang, H. M., Li, D., Li, L., Meeley, R. B., Sakai, H., Bruce, W., Schnable, P. S. and Hochholdinger, F. (2014). Roothairless5, which functions in maize (*Zea mays* L.) root hair initiation and elongation encodes a monocot-specific NADPH oxidase, *The Plant Journal: For Cell and Molecular Biology* 79(5), 729–740. https://doi.org/10.1111/tpj.12578.

Neumann, R. B. and Cardon, Z. G. (2012). The magnitude of hydraulic redistribution by plant roots: A review and synthesis of empirical and modeling studies, *New Phytologist* 194(2), 337–352. https://doi.org/10.1111/j.1469-8137.2012.04088.x.

Ober, E. S., Alahmad, S., Cockram, J., Forestan, C., Hickey, L. T., Kant, J., Maccaferri, M., Marr, E., Milner, M., Pinto, F., Rambla, C., Reynolds, M., Salvi, S., Sciara, G., Snowdon, R. J., Thomelin, P., Tuberosa, R., Uauy, C., Voss-Fels, K. P., Wallington, E. and Watt, M. (2021). Wheat root systems as a breeding target for climate resilience, *TAG. Theoretical and Applied Genetics. Theoretische und Angewandte Genetik* 134(6), 1645–1662. https://doi.org/10.1007/s00122-021-03819-w.

Ober, E. S. and Sharp, R. E. (2003). Electrophysiological responses of maize roots to low water potentials: Relationship to growth and ABA accumulation, *Journal of Experimental Botany* 54(383), 813–824.

Ober, E. S. and Parry, M. A. J. (2011). Drought and implications for nutrition. In: Hawkesford, M. J. (Ed.), *Molecular and Physiological Basis of Nutrient Use Efficiency in Crops*. Wiley-Blackwell, pp. 431–441.

Ober, E. S. and Sharp, R. E. (2013). Maintaining root growth in drying soil: A review of progress and gaps in understanding. In: Eshel, A. (Ed.), *Plant Roots: The Hidden Half* (4th edn.). Boca Raton: CRC Press.

Oleghe, E., Naveed, M., Baggs, E. M. and Hallett, P. D. (2017). Plant exudates improve the mechanical conditions for root penetration through compacted soils, *Plant and Soil* 421(1), 19–30. https://doi.org/10.1007/s11104-017-3424-5.

Orman-Ligeza, B., Morris, E. C., Parizot, B., Lavigne, T., Babé, A., Ligeza, A., Klein, S., Sturrock, C., Xuan, W., Novák, O., Ljung, K., Fernandez, M. A., Rodriguez, P. L., Dodd, I. C., De Smet, I., Chaumont, F., Batoko, H., Périlleux, C., Lynch, J. P., Bennett, M. J., Beeckman, T. and Draye, X. (2018). The Xerobranching response represses lateral root formation when roots are not in contact with water, *Current Biology* 28(19), 3165–3173.e5. https://doi.org/10.1016/j.cub.2018.07.074.

Orosa-Puente, B., Leftley, N., von Wangenheim, D., Banda, J., Srivastava, A. K., Hill, K., Truskina, J., Bhosale, R., Morris, E., Srivastava, M., Kümpers, B., Goh, T., Fukaki, H., Vermeer, J. E. M., Vernoux, T., Dinneny, J. R., French, A. P., Bishopp, A., Sadanandom, A. and Bennett, M. J. (2018). Root branching toward water involves posttranslational modification of transcription factor ARF7, *Science* 362(6421), 1407–1410. Available at: http://science.sciencemag.org/content/362/6421/1407.abstract.

Owusu-Nketia, S., Inukai, Y., Ohashi, S., Suralta, R. R., Doi, K., Mitsuya, S., Kano-Nakata, M., Niones, J. M., Nguyen, D. T. N., Takuya, K., Makihara, D. and Yamauchi, A. (2018). Root plasticity under fluctuating soil moisture stress exhibited by backcross inbred line of a rice variety, Nipponbare carrying introgressed segments from KDML105 and detection of the associated QTLs, *Plant Production Science* 21(2), 106–122. https://doi.org/10.1080/1343943X.2018.1446759.

Palta, J. A., Chen, X., Milroy, S. P., Rebetzke, G. J., Dreccer, M. F. and Watt, M. (2011). Large root systems: Are they useful in adapting wheat to dry environments?, *Functional Plant Biology* 38(5), 347–354. https://doi.org/10.1071/FP11031.

Pandey, B. K., Huang, G., Bhosale, R., Hartman, S., Sturrock, C. J., Jose, L., Martin, O. C., Karady, M., Voesenek, L. A. C. J., Ljung, K., Lynch, J. P., Brown, K. M., Whalley, W. R., Mooney, S. J., Zhang, D. and Bennett, M. J. (2021). Plant roots sense soil compaction through restricted ethylene diffusion, *Science* 371(6526), 276–280. https://doi.org/10.1126/science.abf3013.

Pandit, E., Panda, R. K., Sahoo, A., Pani, D. R. and Pradhan, S. K. (2020). Genetic relationship and structure analysis of root growth angle for improvement of drought avoidance in early and mid-early maturing rice genotypes, *Rice Science* 27(2), 124–132. Available at: http://www.sciencedirect.com/science/article/pii/S1672630820300056.

Passioura, J. B. (1972). Effect of root geometry on yield of wheat growing on stored water, *Australian Journal of Agricultural Research* 23(5), 745–752.

Passioura, J. B. (1983). Roots and drought resistance, *AG Water Management* 7, 265–280.

Passioura, J. B. (2012). Phenotyping for drought tolerance in grain crops: When is it useful to breeders?, *Functional Plant Biology* 39(11), 851–859. https://doi.org/10.1071/FP12079.

Pinto, R. S. and Reynolds, M. P. (2015). Common genetic basis for canopy temperature depression under heat and drought stress associated with optimized root distribution in bread wheat, *TAG. Theoretical and Applied Genetics. Theoretische und Angewandte Genetik* 128(4), 575–585. https://doi.org/10.1007/s00122-015-2453-9.

Postma, J. A., Kuppe, C., Owen, M. R., Mellor, N., Griffiths, M., Bennett, M. J., Lynch, J. P. and Watt, M. (2017). OpenSimRoot: Widening the scope and application of root architectural models, *New Phytologist* 215(3), 1274–1286. https://doi.org/10.1111/nph.14641.

Pound, M. P., Fozard, S., Torres Torres, M., Forde, B. G. and French, A. P. (2017). AutoRoot: Open-source software employing a novel image analysis approach to support fully-automated plant phenotyping, *Plant Methods* 13(1), 12. https://doi.org/10.1186/s13007-017-0161-y.

Qian, L., Hickey, L. T., Stahl, A., Werner, C. R., Hayes, B., Snowdon, R. J. and Voss-Fels, K. P. (2017). Exploring and harnessing haplotype diversity to improve yield stability in crops, *Frontiers in Plant Science* 8, 1534. Available at: https://www.frontiersin.org/article/10.3389/fpls.2017.01534.

Rambla, C., Van Der Meer, S., Voss-Fels, K. P., Makhoul, M., Obermeier, C., Snowdon, R., Ober, E. S., Watt, M., Alahmad, S. and Hickey, L. T. (2022). A toolkit to rapidly modify root systems through single plant selection, *Plant Methods* 18(1), 2. https://doi.org/10.1186/s13007-021-00834-2.

Reckling, M., Ahrends, H., Chen, T., Eugster, W., Hadasch, S., Knapp, S., Laidig, F., Linstädter, A., Macholdt, J., Piepho, H., Schiffers, K. and Döring, T. F. (2021). Methods of yield stability analysis in long-term field experiments. A review, *Agronomy for Sustainable Development* 41(2), 27. https://doi.org/10.1007/s13593-021-00681-4.

Rich, S. M., Christopher, J., Richards, R. and Watt, M. (2020). Root phenotypes of young wheat plants grown in controlled environments show inconsistent correlation with mature root traits in the field, *Journal of Experimental Botany* 71(16), 4751–4762. https://doi.org/10.1093/jxb/eraa201.

Rich, S. M., Wasson, A. P., Richards, R. A., Katore, T., Prashar, R., Chowdhary, R., Saxena, D. C., Mamrutha, H. M., Zwart, A., Misra, S. C. and Prasad, S. S. (2015). Wheats developed for high yield on stored soil moisture have deep vigorous root systems, *Plant Methods* 43(2), 173–188. https://doi.org/10.1071/FP15182.

Rich, S. M. and Watt, M. (2013). Soil conditions and cereal root system architecture: Review and considerations for linking Darwin and Weaver, *Journal of Experimental Botany* 64(5), 1193–1208. https://doi.org/10.1093/jxb/ert043.

Richard, C., Christopher, J., Chenu, K., Borrell, A., Christopher, M. and Hickey, L. (2018). Selection in early generations to shift allele frequency for seminal root angle in wheat, *Plant Genome* 11(2), 1–12. https://doi.org/10.3835/plantgenome2017.08.0071.

Richards, R. A. (2006). Physiological traits used in the breeding of new cultivars for water-scarce environments, *Agricultural Water Management* 80(1–3), 197–211.

Richards, R. A., Hunt, J. R., Kirkegaard, J. A. and Passioura, J. B. (2014). Yield improvement and adaptation of wheat to water-limited environments in Australia: A case study, *Crop and Pasture Science* 65(7), 676–689. https://doi.org/10.1071/CP13426.

Richards, R. A. and Passioura, J. B. (1989). A breeding program to reduce the diameter of the major xylem vessel in the seminal roots of wheat and its effect on grain yield in rain-fed environments, *Australian Journal of Agricultural Research* 40(5), 943–950. https://doi.org/10.1071/AR9890943.

Robbins, N. E. and Dinneny, J. R. (2018). Growth is required for perception of water availability to pattern root branches in plants, *Proceedings of the National Academy of Sciences of the United States of America* 115(4), E822–E831. https://doi.org/10.1073/pnas.1710709115.

Robinson, H., Kelly, A., Fox, G., Franckowiak, J., Borrell, A. and Hickey, L. (2018). Root architectural traits and yield: Exploring the relationship in barley breeding trials, *Euphytica* 214(9), 151. https://doi.org/10.1007/s10681-018-2219-y.

Rolfe, S. A., Griffiths, J. and Ton, J. (2019). Crying out for help with root exudates: Adaptive mechanisms by which stressed plants assemble health-promoting soil microbiomes, *Current Opinion in Microbiology* 49, 73–82. https://doi.org/10.1016/j.mib.2019.10.003.

Roychoudhry, S., Kieffer, M., Del Bianco, M., Liao, C. Y., Weijers, D. and Kepinski, S. (2017). The developmental and environmental regulation of gravitropic setpoint angle in Arabidopsis and bean, *Scientific Reports* 7, 42664. http://doi.org/10.1038/srep42664.

Ruiz, S., Koebernick, N., Duncan, S., Fletcher, D. M., Scotson, C., Boghi, A., Marin, M., Bengough, A. G., George, T. S., Brown, L. K., Hallett, P. D. and Roose, T. (2020). Significance of root hairs at the field scale – Modelling root water and phosphorus uptake under different field conditions, *Plant and Soil* 447(1), 281–304. https://doi.org/10.1007/s11104-019-04308-2.

Saab, I. N. and Sharp, R. E. (1989). Non-hydraulic signals from maize roots in drying soil: Inhibition of leaf elongation but not stomatal conductance, *Planta* 179(4), 466–474. https://doi.org/10.1007/BF00397586.

Salazar-Henao, J. E., Vélez-Bermúdez, I. C. and Schmidt, W. (2016). The regulation and plasticity of root hair patterning and morphogenesis, *Development* 143(11), 1848–1858. https://doi.org/10.1242/dev.132845.

Sandhu, N., Raman, K. A., Torres, R. O., Audebert, A., Dardou, A., Kumar, A. and Henry, A. (2016). Rice root architectural plasticity traits and genetic regions for adaptability to variable cultivation and stress conditions, *Plant Physiology* 171(4), 2562–2576. Available at: http://www.plantphysiol.org/content/171/4/2562.abstract.

Sangster, T. A., Salathia, N., Undurraga, S., Milo, R., Schellenberg, K., Lindquist, S. and Queitsch, C. (2008). HSP90 affects the expression of genetic variation and developmental stability in quantitative traits, *Proceedings of the National Academy of Sciences of the United States of America* 105(8), 2963–2968. https://doi.org/10.1073/pnas.0712200105.

Sanguineti, M. C., Li, S., Maccaferri, M., Corneti, S., Rotondo, F., Chiari, T. and Tuberosa, R. (2007). Dissection of genetic variability for root architecture in elite durum wheat germplasm, *Annals of Applied Biology* 151(3), 291–305.

Sayer, E. J., Crawford, J. A., Edgerley, J., Askew, A. P., Hahn, C. Z., Whitlock, R. and Dodd, I. C. (2021). Adaptation to chronic drought modifies soil microbial community responses to phytohormones, *Communications Biology* 4(1), 516. https://doi.org/10.1038/s42003-021-02037-w.

Schneider, H. M., Klein, S. P., Hanlon, M. T., Kaeppler, S., Brown, K. M. and Lynch, J. P. (2020). Genetic control of root anatomical plasticity in maize, *The Plant Genome* 13(1), e20003. https://doi.org/10.1002/tpg2.20003.

Schneider, H. M., Klein, S. P., Hanlon, M. T., Nord, E. A., Kaeppler, S., Brown, K. M., Warry, A., Bhosale, R. and Lynch, J. P. (2020). Genetic control of root architectural plasticity in maize, *Journal of Experimental Botany* 71(10), 3185–3197. https://doi.org/10.1093/jxb/eraa084.

Schneider, H. M. and Lynch, J. P. (2020). Should root plasticity be a crop breeding target?, *Frontiers in Plant Science* 11, 546. Available at: https://www.frontiersin.org/article/10.3389/fpls.2020.00546.

Schneider, H. M., Strock, C. F., Hanlon, M. T., Vanhees, D. J., Perkins, A. C., Ajmera, I. B., Sidhu, J. S., Mooney, S. J., Brown, K. M. and Lynch, J. P. (2021). Multiseriate cortical sclerenchyma enhance root penetration in compacted soils, *Proceedings of the National Academy of Sciences of the United States of America* 118(6), e2012087118. https://doi.org/10.1073/pnas.2012087118.

Seethepalli, A., Dhakal, K., Griffiths, M., Guo, H., Freschet, G. T. and York, L. M. (2021). RhizoVision explorer: Open-source software for root image analysis and measurement standardization, *AoB Plants* 13(6), plab056. https://doi.org/10.1093/aobpla/plab056.

Severini, A. D., Wasson, A. P., Evans, J. R., Richards, R. A. and Watt, M. (2020). Root phenotypes at maturity in diverse wheat and triticale genotypes grown in three field experiments: Relationships to shoot selection, biomass, grain yield, flowering time, and environment, *Field Crops Research* 255, 107870. https://doi.org/10.1016/j.fcr.2020.107870.

Shan, L., Zhao, S.-Y. and Xia, G.-M. (2005). Cloning of the full-length cDNA of the wheat involved in salt stress: Root hair defective 3 gene (RHD3), *Journal of Integrative Plant Biology* 47(7), 881–891. https://doi.org/10.1111/j.1744-7909.2005.00090.x.

Shanahan, P. W., Binley, A., Whalley, W. R. and Watts, C. W. (2015). The use of electromagnetic induction to monitor changes in soil moisture profiles beneath different wheat genotypes, *Soil Science Society of America Journal Madison, WI* 79(2), 459–466. https://doi.org/10.2136/sssaj2014.09.0360.

Sharp, R. E. and Davies, W. J. (1979). Solute regulation and growth by roots and shoots of water-stressed maize plants, *Planta* 147(1), 43–49.

Shekoofa, A. and Sinclair, T. R. (2018). Aquaporin activity to improve crop drought tolerance, *Cells* 7(9), 123. https://doi.org/10.3390/cells7090123.

Siddiqui, M. N., Léon, J., Naz, A. A. and Ballvora, A. (2021). Genetics and genomics of root system variation in adaptation to drought stress in cereal crops, *Journal of Experimental Botany* 72(4), 1007–1019. https://doi.org/10.1093/jxb/eraa487.

Simmons, T., Styer, A. B., Pierroz, G., Gonçalves, A. P., Pasricha, R., Hazra, A. B., Bubner, P. and Coleman-Derr, D. (2020). Drought drives spatial variation in the millet root microbiome, *Frontiers in Plant Science* 11, 599. Available at: https://www.frontiersin.org/articles/10.3389/fpls.2020.00599.

Slovak, R., Ogura, T., Satbhai, S. B., Ristova, D. and Busch, W. (2016). Genetic control of root growth: From genes to networks, *Annals of Botany* 117(1), 9–24. https://doi.org/10.1093/aob/mcv160.

Smith, A. G., Han, E., Petersen, J., Olsen, N. A. F., Giese, C., Athmann, M., Dresbøll, D. B. and Thorup-Kristensen, K. (2022). RootPainter: Deep learning segmentation of biological images with corrective annotation, *New Phytologist* 236(2), 774–791. https://doi.org/10.1111/nph.18387.

Somssich, M., Khan, G. A. and Persson, S. (2016). Cell wall heterogeneity in root development of Arabidopsis, *Frontiers in Plant Science* 7, 1242. Available at: https://www.frontiersin.org/articles/10.3389/fpls.2016.01242.

Stanfield, R. C. and Bartlett, M. K. (2022). Coordination between phloem loading and structure maintains carbon transport under drought, *Frontiers in Plant Science* 13, 787837. Available at: https://www.frontiersin.org/articles/10.3389/fpls.2022.787837.

Suralta, R. R., Kano-Nakata, M., Niones, J. M., Inukai, Y., Kameoka, E., Tran, T. T., Menge, D., Mitsuya, S. and Yamauchi, A. (2018). Root plasticity for maintenance of productivity under abiotic stressed soil environments in rice: Progress and prospects, *Field Crops Research* 220, 57–66. Available at: http://www.sciencedirect.com/science/article/pii/S037842901630209X.

Tanner, C. B. and Sinclair, T. R. (1983). Efficient water use in crop production: Research or re-search? In: Asa, C. S. S. A. and Books, S. S. S. A. (Eds.), *Limitations to Efficient Water Use in Crop Production*, pp. 1–27. https://doi.org/10.2134/1983.limitationstoefficientwateruse.c1.

Tardieu, F. (2012). Any trait or trait-related allele can confer drought tolerance: Just design the right drought scenario, *Journal of Experimental Botany* 63(1), 25–31. https://doi.org/10.1093/jxb/err269.

Tardieu, F., Parent, B. and Simonneau, T. (2010). Control of leaf growth by abscisic acid: Hydraulic or non-hydraulic processes?, *Plant, Cell and Environment* 33(4), 636–647. https://doi.org/10.1111/j.1365-3040.2009.02091.x.

Tardieu, F., Simonneau, T. and Muller, B. (2018). The physiological basis of drought tolerance in crop plants: A scenario-dependent probabilistic approach, *Annual Review of Plant Biology* 69(1), 733–759. https://doi.org/10.1146/annurev-arplant-042817-040218.

Tardieu, F. and Tuberosa, R. (2010). Dissection and modelling of abiotic stress tolerance in plants, *Current Opinion in Plant Biology* 13(2), 206–212.

Taylor, I., Lehner, K., McCaskey, E., Nirmal, N., Ozkan-Aydin, Y., Murray-Cooper, M., Jain, R., Hawkes, E. W., Ronald, P. C., Goldman, D. I. and Benfey, P. N. (2021). Mechanism and function of root circumnutation, *Proceedings of the National Academy of Sciences of the United States of America* 118(8), e2018940118. https://doi.org/10.1073/pnas.2018940118.

Teramoto, S. and Uga, Y. (2020). A deep learning-based phenotypic analysis of rice root distribution from field images, *Plant Phenomics* 2020, 3194308. https://doi.org/10.34133/2020/3194308.

Thorup-Kristensen, K. and Kirkegaard, J. (2016). Root system-based limits to agricultural productivity and efficiency: The farming systems context, *Annals of Botany* 118(4), 573–592. https://doi.org/10.1093/aob/mcw122.

Tkacz, A., Pini, F., Turner, T. R., Bestion, E., Simmonds, J., Howell, P., Greenland, A., Cheema, J., Emms, D. M., Uauy, C. and Poole, P. S. (2020). Agricultural Selection of Wheat Has Been Shaped by Plant-Microbe Interactions, *Frontiers in Microbiology* 11, 132. https://doi.org/10.3389/fmicb.2020.00132.

Toal, T. W., Ron, M., Gibson, D., Kajala, K., Splitt, B., Johnson, L. S., Miller, N. D., Slovak, R., Gaudinier, A., Patel, R. and De Lucas, M. (2018). Regulation of root angle and gravitropism, *G3: Genes|Genomes|Genetics* 8(12), 3841. Available at: http://www.g3journal.org/content/8/12/3841.abstract.

Topp, C. N. (2016). Hope in change: The role of root plasticity in crop yield stability, *Plant Physiology* 172(1), 5–6. Available at: http://www.plantphysiol.org/content/172/1/5.abstract.

Tracy, S. R., Nagel, K. A., Postma, J. A., Fassbender, H., Wasson, A. and Watt, M. (2020). Crop improvement from phenotyping roots: Highlights reveal expanding opportunities, *Trends in Plant Science* 25(1), 105–118. https://doi.org/10.1016/j.tplants.2019.10.015.

Tron, S., Bodner, G., Laio, F., Ridolfi, L. and Leitner, D. (2015). Can diversity in root architecture explain plant water use efficiency? A modeling study, *Ecological Modelling* 312, 200–210. https://doi.org/10.1016/j.ecolmodel.2015.05.028.

Ubeda-Tomás, S., Beemster, G. T. S. and Bennett, M. J. (2012). Hormonal regulation of root growth: Integrating local activities into global behaviour, *Trends in Plant Science* 17(6), 326–331. https://doi.org/10.1016/j.tplants.2012.02.002.

Uga, Y. (2021). Challenges to design-oriented breeding of root system architecture adapted to climate change, *Breeding Science* 71(1), 3–12. https://doi.org/10.1270/jsbbs.20118.

Uga, Y., Sugimoto, K., Ogawa, S., Rane, J., Ishitani, M., Hara, N., Kitomi, Y., Inukai, Y., Ono, K., Kanno, N., Inoue, H., Takehisa, H., Motoyama, R., Nagamura, Y., Wu, J., Matsumoto, T., Takai, T., Okuno, K. and Yano, M. (2013). Control of root system architecture by deeper rooting 1 increases rice yield under drought conditions, *Nature Genetics* 45(9), 1097–1102. http://doi.org/10.1038/ng.2725.

van der Bom, F. J. T., Williams, A. and Bell, M. J. (2020). Root architecture for improved resource capture: Trade-offs in complex environments, *Journal of Experimental Botany* 71(19), 5752–5763. https://doi.org/10.1093/jxb/eraa324.

van Oosterom, E. J., Yang, Z., Zhang, F., Deifel, K. S., Cooper, M., Messina, C. D. and Hammer, G. L. (2016). Hybrid variation for root system efficiency in maize: Potential links to drought adaptation, *Functional Plant Biology* 43(6), 502-511. https://doi.org/10.1071/FP15308.

Vetterlein, D., Phalempin, M., Lippold, E., Schlüter, S., Schreiter, S., Ahmed, M. A., Carminati, A., Duddek, P., Jorda, H., Bienert, G. P., Bienert, M. D., Tarkka, M., Ganther, M., Oburger, E., Santangeli, M., Javaux, M. and Vanderborght, J. (2022). Root hairs matter at field scale for maize shoot growth and nutrient uptake, but root trait plasticity is primarily triggered by texture and drought, *Plant and Soil* 478(1-2), 119-141. https://doi.org/10.1007/s11104-022-05434-0.

Vissenberg, K., Claeijs, N., Balcerowicz, D. and Schoenaers, S. (2020). Hormonal regulation of root hair growth and responses to the environment in Arabidopsis, *Journal of Experimental Botany* 71(8), 2412-2427. https://doi.org/10.1093/jxb/eraa048.

Voothuluru, P., Mäkelä, P., Zhu, J., Yamaguchi, M., Cho, I. J., Oliver, M. J., Simmonds, J. and Sharp, R. E. (2020). Apoplastic hydrogen peroxide in the growth zone of the maize primary root. Increased levels differentially modulate root elongation under well-watered and water-stressed conditions, *Frontiers in Plant Science* 11, 392. Available at: https://www.frontiersin.org/article/10.3389/fpls.2020.00392.

Voss-Fels, K. P., Robinson, H., Mudge, S. R., Richard, C., Newman, S., Wittkop, B., Stahl, A., Friedt, W., Frisch, M., Gabur, I., Miller-Cooper, A., Campbell, B. C., Kelly, A., Fox, G., Christopher, J., Christopher, M., Chenu, K., Franckowiak, J., Mace, E. S., Borrell, A. K., Eagles, H., Jordan, D. R., Botella, J. R., Hammer, G., Godwin, I. D., Trevaskis, B., Snowdon, R. J. and Hickey, L. T. (2018). Vernalization1 modulates root system architecture in wheat and barley, *Molecular Plant* 11(1), 226-229. http://doi.org/10.1016/j.molp.2017.10.005.

Voss-Fels, K. P., Snowdon, R. J. and Hickey, L. T. (2018). Designer roots for future crops, *Trends in Plant Science* 23(11), 957-960. https://doi.org/10.1016/j.tplants.2018.08.004.

Wachsman, G., Sparks, E. E. and Benfey, P. N. (2015). Genes and networks regulating root anatomy and architecture, *New Phytologist* 208(1), 26-38. https://doi.org/10.1111/nph.13469.

Wang, C., Liu, W., Li, Q., Ma, D., Lu, H., Feng, W., Xie, Y., Zhu, Y. and Guo, T. (2014) Effects of different irrigation and nitrogen regimes on root growth and its correlation with above-ground plant parts in high-yielding wheat under field conditions, *Field Crops Research* 165, 138-149. https://doi.org/10.1016/j.fcr.2014.04.011.

Wang, T., Rostamza, M., Song, Z., Wang, L., McNickle, G., Iyer-Pascuzzi, A. S., Qiu, Z. and Jin, J. (2019). SegRoot: A high throughput segmentation method for root image analysis, *Computers and Electronics in Agriculture* 162, 845-854. https://doi.org/10.1016/j.compag.2019.05.017.

Wasson, A. P., Richards, R. A., Chatrath, R., Misra, S. C., Prasad, S. V., Rebetzke, G. J., Kirkegaard, J. A., Christopher, J. and Watt, M. (2012). Traits and selection strategies to improve root systems and water uptake in water-limited wheat crops, *Journal of Experimental Botany* 63(9), 3485-3498. https://doi.org/10.1093/jxb/ers111.

Watson, A., Ghosh, S., Williams, M. J., Cuddy, W. S., Simmonds, J., Rey, M. D., Asyraf Md Hatta, M., Hinchliffe, A., Steed, A., Reynolds, D. and Adamski, N. M. (2017). Speed breeding is a powerful tool to accelerate crop research and breeding, Nature Plants 4, 23-29. https://doi.org/10.1038/s41477-017-0083-8.

Weiner, J., Du, Y. L., Zhao, Y. M. and Li, F. M. (2021). Allometry and yield stability of cereals, *Frontiers in Plant Science* 12, 681490. Available at: https://www.frontiersin.org/articles/10.3389/fpls.2021.681490.

Whalley, W. R., Binley, A., Watts, C. W., Shanahan, P., Dodd, I. C., Ober, E. S., Ashton, R. W., Webster, C. P., White, R. P. and Hawkesford, M. J. (2017). Methods to estimate changes in soil water for phenotyping root activity in the field, *Plant and Soil* 415(1), 407–422. https://doi.org/10.1007/s11104-016-3161-1.

Whalley, W. R., Dodd, I. C., Watts, C. W., Webster, C. P., Phillips, A. L., Andralojc, J., White, R. P., Davies, W. J. and Parry, M. A. J. (2013). Genotypic variation in the ability of wheat roots to penetrate wax layers, *Plant and Soil* 364(1–2), 171–179. https://doi.org/10.1007/s11104-012-1342-0.

Whalley, W. R., Watts, C. W., Gregory, A. S., Mooney, S. J., Clark, L. J. and Whitmore, A. P. (2008). The effect of soil strength on the yield of wheat, *Plant and Soil* 306(1–2), 237–247. https://doi.org/10.1007/s11104-008-9577-5.

White, C. A., Sylvester-Bradley, R. and Berry, P. M. (2015). Root length densities of UK wheat and oilseed rape crops with implications for water capture and yield, *Journal of Experimental Botany* 66(8), 2293–2303. https://doi.org/10.1093/jxb/erv077.

White, P. J. (2019). Root traits benefitting crop production in environments with limited water and nutrient availability, *Annals of Botany* 124(6), 883–890. https://doi.org/10.1093/aob/mcz162.

White, R. G. and Kirkegaard, J. A. (2010). The distribution and abundance of wheat roots in a dense, structured subsoil - Implications for water uptake, *Plant, Cell and Environment* 33(2), 133–148. https://doi.org/10.1111/j.1365-3040.2009.02059.x.

Wissuwa, M. and Kant, J. (2021). Does half a millimetre matter? Root hairs for yield stability. A commentary on "Significance of root hairs for plant performance under contrasting field conditions and water deficit", *Annals of Botany* 128(1), iii–iiv. https://doi.org/10.1093/AOB/MCAB027.

Xiao, J., Jin, R. and Wagner, D. (2017). Developmental transitions: Integrating environmental cues with hormonal signaling in the chromatin landscape in plants, *Genome Biology* 18(1), 88. https://doi.org/10.1186/s13059-017-1228-9.

Yang, W., Chang, F., Ma, D., Wang, S. and Yin, L. (2022). Subsoil water use to attain stable high yields of winter wheat in drylands Loess Plateau of China, *European Journal of Agronomy* 139, 126558. https://doi.org/10.1016/j.eja.2022.126558.

Yasrab, R., Atkinson, J. A., Wells, D. M., French, A. P., Pridmore, T. P. and Pound, M. P. (2019). RootNav 2.0: Deep learning for automatic navigation of complex plant root architectures, *GigaScience* 8(11), giz123. https://doi.org/10.1093/gigascience/giz123.

Yu, P. and Hochholdinger, F. (2018). The role of host genetic signatures on root-microbe interactions in the rhizosphere and endosphere, *Frontiers in Plant Science* 9, 1896. Available at: https://www.frontiersin.org/article/10.3389/fpls.2018.01896.

Zaidi, P. H., Seetharam, K., Krishna, G., Krishnamurthy, L., Gajanan, S., Babu, R., Zerka, M., Vinayan, M. T. and Vivek, B. S. (2016). Genomic regions associated with root traits under drought stress in tropical maize (*Zea mays* L.), *PLoS ONE* 11(10), e0164340. http://doi.org/10.1371/journal.pone.0164340.

Zhang, Y., Xu, F., Ding, Y., Du, H., Zhang, Q., Dang, X., Cao, Y., Dodd, I. C. and Xu, W. (2021). Abscisic acid mediates barley rhizosheath formation under mild soil drying by promoting root hair growth and auxin response, *Plant, Cell and Environment* 44(6), 1935–1945. https://doi.org/10.1111/pce.14036.

Zhou, X.-R., Schnepf, A., Vanderborght, J., Leitner, D., Lacointe, A., Vereecken, H. and Lobet, G. (2020). CPlantBox, a whole-plant modelling framework for the simulation of water- and carbon-related processes, *In Silico Plants* 2(1), diaa001. https://doi.org/10.1093/insilicoplants/diaa001.

Zhou, X.-R., Schnepf, A., Vanderborght, J., Leitner, D., Vereecken, H. and Lobet, G. (2022). Phloem anatomy restricts root system architecture development: Theoretical clues from in silico experiments, *bioRxiv*. https://doi.org/10.1101/2022.10.04.510862.

Zhu, Y. H., Weiner, J., Yu, M. X. and Li, F. M. (2019). Evolutionary agroecology: Trends in root architecture during wheat breeding, *Evolutionary Applications* 12(4), 733–743. https://doi.org/10.1111/eva.12749.

Chapter 7

Identifying and exploiting photosynthetic genes in improving drought resistance in cereals

Weiguo Chen, Daizhen Sun and Runzhi Li, Shanxi Agricultural University, China; and Ruilian Jing, Chinese Academy of Agricultural Sciences, China

1 Introduction

Photosynthesis is a fundamental biological process required for the survival of green plants and is thereby essential for human and other life on earth. In this redox (oxidation-reduction) process, plants assimilate atmospheric CO_2 and

http://dx.doi.org/10.19103/AS.2022.0109.11

use light energy to produce oxygen and glucose that are directly or indirectly utilized by consumers (from the microscopic scale upwards) at various trophic levels in our ecosystem. The chemical reaction of oxygenic photosynthesis is formulated as follows (Miglani et al., 2021):

$$6CO_2 + 12H_2O + \text{light energy} \rightarrow C_6H_{12}O_6 + 6O_2 + 6H_2O$$

Two sets of partial reactions occur in the photosynthesis process, namely light reactions and dark reactions (Fig. 1) (Ashraf and Harris, 2013). In light reactions, light energy is converted into adenosine triphosphate (ATP) and nicotinamide adenine dinucleotide phosphate (NADPH), along with discharged oxygen. In dark reactions, CO_2 is reduced (via the addition of H atoms) to carbohydrate utilizing ATP and NADPH generated by light reactions (Lawlor, 2001; Gu et al., 2019; Li et al., 2021). Plants have two major pathways for CO_2 fixation (C_3 and C_4 pathways) in which the initial fixation of CO_2 forms 3-carbon acids and 4-carbon acids, respectively. Plants can therefore be divided into three types - C_3, C_4, or C_3-C_4 intermediate plants - depending on the spatial distribution of these two pathways within leaf tissues or as Crassulacean acid metabolism (CAM) plants with a temporal distribution, i.e. it is C_3 in light and C_4 in the dark (Doubnerová and Ryšlavá 2011; Freschi and Mercier, 2012). C_3 plants grow well in cool and wet climates where light intensity is low. Hot and dry weather with high light intensity is required for C_4 plants. Generally, C_4 plants have higher water-use efficiency (WUE) and higher photosynthetic efficiency as they possess an additional carbon fixation pathway and anatomical characteristics to restrict photorespiration, which makes them well suited to arid, high light intensity, or hot climatic conditions. Additionally, CAM plants can efficiently retain metabolic energy and water in severe climatic conditions by closing their stomata throughout the day (Ashraf and Harris, 2013; Shi et al., 2021).

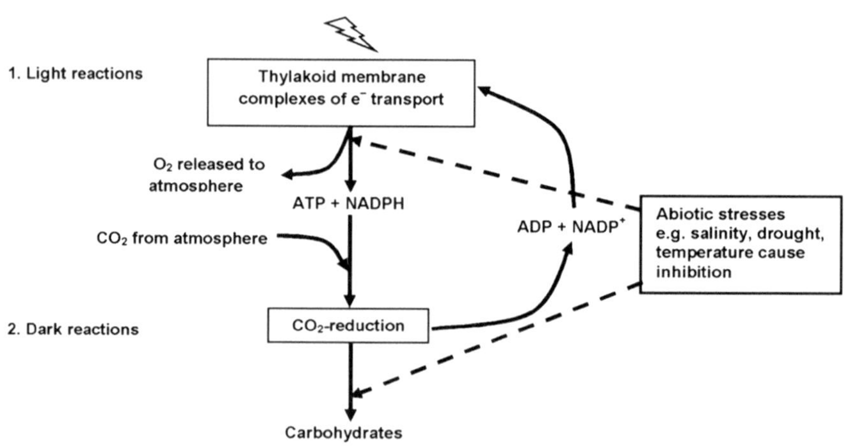

Figure 1 Light and dark reactions of photosynthesis. Source: Ashraf and Harris (2013).

Water shortage is a key restriction that hinders plant growth and development by interfering with critical processes in plants. Drought stress causes physiological, biochemical, and molecular changes in plants, affecting the photosynthesis, respiration, ion and nutrient translocation processes, and growth regulator activities (Zargar et al., 2017). Drought stress severely affects plant photosynthesis (Flexas et al., 2012), resulting in altered intracellular state, lower electron transfer rates, photosynthetic phosphorylation uncoupling, increased hydrolysis, and reduced chlorophyll biosynthesis. In addition, reactive oxygen species accumulation in the chloroplast under drought stress can induce oxidative damage and inhibit photosynthesis (Gollan et al., 2015; Exposito-Rodriguez et al., 2017).

Crops can cope with stress through a variety of mechanisms, including osmotic adjustment (OA), protein accumulation, and antioxidant defense systems (Cui et al., 2016; Gürel et al., 2016). Depending on their carbon fixation mechanisms (C_3, C_4, and CAM), plants' ability to adapt to various abiotic stimuli differs from species to species (Rao and Chaitanya, 2016). These regulatory pathways communicate with one another to establish a drought-resistant network that allows plants to sustain photosynthesis in the presence of drought, mitigating its negative effects on yield (Redillas et al., 2012; Hu and Xiong, 2014; He et al., 2016; Shi et al., 2018). Plants' ability to retain photosynthesis in the presence of drought stress is a crucial determinant of drought tolerance (Ma et al., 2016; Blum and Tuberosa, 2018; Shi et al., 2018).

The main objective of this chapter is to review the photosynthetic response of plants to drought stress and to identify and exploit the genes related to photosynthesis to improve drought resistance in cereal crops. We will focus on the following issues:

- Indicators for photosynthetic traits;
- Factors affecting photosynthetic efficiency;
- Impact of drought stress on photosynthetic assimilation efficiency;
- Photosynthetic metabolic pathways of photosynthesis associated with drought tolerance in crops;
- Contribution of photosynthetic rate maintenance to crop drought tolerance under drought stress;
- Engineering C_4 traits into C_3 plants to improve crop drought tolerance;
- Drought tolerance genes related to photosynthesis;
- QTL mapping for photosynthetic traits;
- Effects of SG phenotype on crop yield under drought conditions; and
- Introducing the SG trait into plants to enhance photosynthesis under drought stress.

The chapter also includes a case study on the enhancement of sorghum tolerance to post-flowering drought stress using SG QTL.

2 Indicators for photosynthetic traits

Photosynthetic characteristics have proven to be highly sensitive to different environmental stresses, which makes them a suitable indicator for studying the impact of drought stress on plants (Massacci et al., 2008). Stomatal closure restricts CO_2 diffusion to chloroplasts, which is the earliest response of the leaf to drought stress (Hura et al., 2007). Leaf gas exchange should be measured to provide a foundation for estimating the stomatal limitation of photosynthetic CO_2 assimilation. Leaf gas exchange parameters include stomatal conductance (g_s), internal concentration of CO_2 (C_i), net photosynthesis (P_N), and transpiration rate (E).

The use of chlorophyll fluorescence characteristics is a highly effective strategy to identify variations in the photosynthetic system under stressful conditions (Hura et al., 2007). Stomatal closure limits CO_2 assimilation, which may cause an imbalance between photochemical activity at photosystem II (PSII) and electron supplies for photosynthesis, resulting in overexcitation and subsequent photoinhibition degradation to PSII. Thus, photosynthetic machinery damage may happen as a result of this, causing non-stomatal constraint to the process (Souza et al., 2004). Alterations in chlorophyll production, functional and structural fluctuations in chloroplasts, and disturbances to the assimilation products' accumulation and distribution are some of the non-stomatal processes that occur during severe soil dryness (Medrano et al., 2002). Fluorescence parameter variations can be used to investigate non-stomatal photosynthetic constraints as well as drought-related defensive mechanisms (Hura et al., 2007).

Chlorophyll fluorescence parameters such as F_v/F_m ratio, the ratio of chlorophyll fluorescence (ChlF) decrease, F_0-quenching coefficient (q_0), relative electron transport rate, photochemical (q_P) and non-photochemical (q_N) quenching of variable chlorophyll fluorescence yield, non-photochemical quenching of ChlF (NPQ), and $\Delta\Phi_F/\Phi_{Fm}$ ratio, have all been researched in the last few decades and successively introduced into photosynthesis research. All other chlorophyll fluorescence parameters can be determined by using the quartet of parameters, i.e. the maximum quantum yield of *PS2* photochemistry (Φ_{P0}), the photochemical quenching of variable ChlF (q_P), the NPQ of variable ChlF (q_N), and the relative change of minimum ChlF (q_0). The parameters, Φ_{P0}, q_P, q_N, q_0, Φ_2, and NPQ, are the most accessible chlorophyll fluorescence characteristics that can provide meaningful insight into the photosynthetic processes of chloroplasts (Roháček, 2002).

In addition to chlorophyll fluorescence parameters and gas exchange parameters, chlorophyll content, canopy temperature, and SG can also be used as indicators to evaluate the photosynthetic capacity of crop plants. For example, the amount of chlorophyll in the plant has been proposed as a valid metric for determining the metabolic balance between photosynthesis and yield performance (Araus et al., 1998; Liu et al., 2015). The difference in canopy temperature across genotypes can be utilized as an indication of transpiration and stomatal conductance variations (Jackson et al., 1981; Jones et al., 2009; Kaler et al., 2018). Crop genotypes with a functional SG character, such as delayed leaf senescence, can retain photosynthetic capability and a favorable supply of assimilates to grain for extended periods, resulting in higher grain yields (GYs) (Gong et al., 2005; Christopher et al., 2008; Chen et al., 2010; Yang et al., 2016).

3 Factors affecting photosynthetic efficiency

Photosynthetic efficiency is a key determinant of crop productivity. The constraints affecting photosynthetic efficiency include photorespiration, photoprotective heat dissipation, and plant canopy structure (Ahmad et al., 2020; Miglani et al., 2021). Plant carbon fixation is limited by photorespiration, in which CO_2 is released and cellular energy is used to recycle the RuBisCO oxygenation product, 2-phosphoglycolate, back into the Calvin cycle. CO_2 release and energy dissipation in this process hinder agricultural production (Trudeau et al., 2018; Miglani et al., 2021). To avoid damage to photosynthetic machinery caused by high light intensities in photosystems, NPQ directs surplus excitations toward heat dissipation rather than the photochemical process (Zhu et al., 2008, 2010; Miglani et al., 2021). When there is no additional light, photoprotective heat dissipation does not continue to drain energy from regular photosynthesis. In addition, plant canopy structure must be taken into account because photoassimilates generated at the whole plant level are required for crop productivity. In any case, rather than separate processes such as single-leaf photosynthetic capability, integrated canopy photosynthesis is very important in determining ultimate output. For cereal crops, canopy photosynthesis is reliant not only on the photosynthesis of individual leaves but also on non-laminar components such as the ears and panicles (Murchie et al., 2018; Sanchez-Bragado et al., 2020). All these traits related to photosynthesis should be considered in genetic modifications to crop photosynthesis efficiency.

4 How does drought stress affect photosynthetic assimilation efficiency?

Drought stress is one of the most important abiotic stress factors affecting crop yield. Global warming and decreased underground water availability reduce soil moisture and atmospheric humidity, leading to increased evapotranspiration that promotes global drought. Drought directly influences plant productivity and growth (Muhammad et al., 2021; Piveta et al., 2021) and plants generally need to make morphological and physiological adjustments to cope with drought stress, such as stunting, OA, stomata closure (to decrease transpiration), and photosynthesis reduction (Blum, 2011).

In general, under mild and moderate water stress, the main reason for a decrease in photosynthesis is stomatal constraint. To protect water loss from the leaf, the stomata has to close, blocking CO_2 from entering the leaf cells and thus inhibiting photosynthesis. In addition, photosynthesis is also affected by various non-stomatal mechanisms. Stomatal closure disrupts CO_2 intake, which alters enzyme activity, causes membrane disruption, and reduces ATP synthesis and ribulose-1,5-bisphosphate regeneration. Consequently, both RuBisCO activity and photosynthetic processes are reduced (Fig. 2) (Muhammad et al., 2021). In addition, water loss in leaves reduces chlorophyll biosynthesis and accelerates its degradation, leading to decreased leaf chlorophyll content. The chlorophyll content within a certain range also directly affects leaf photosynthetic capacity. Photosynthesis, the most vital physiological process, is greatly influenced by drought stress, which ultimately reduces plant growth and productivity (Blum, 2011; Jia et al., 2020b). Therefore, there is an urgent need to explore photosynthetic

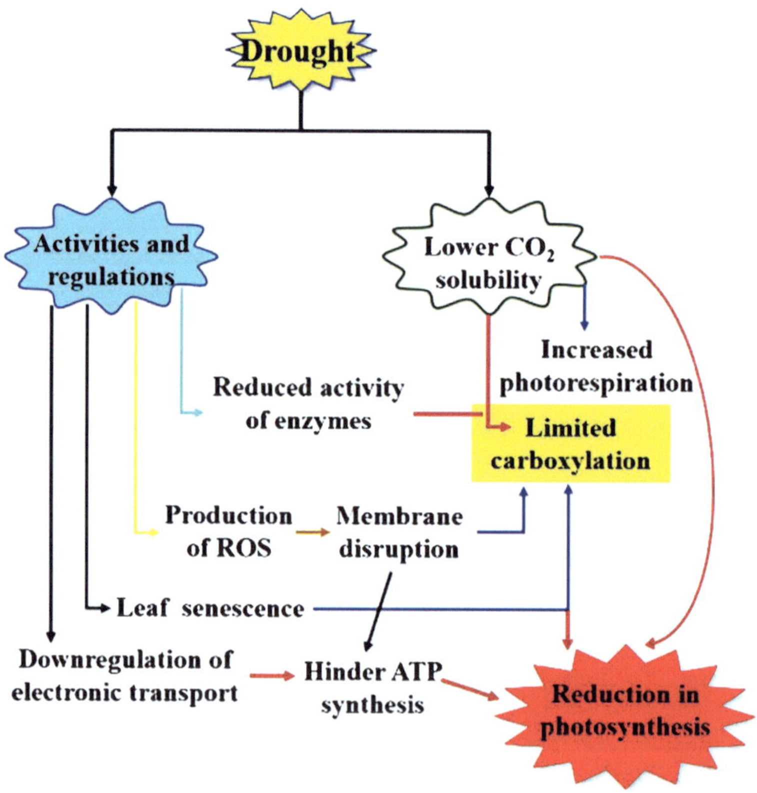

Figure 2 Schematic representation of the photosynthesis performance under drought. Source: Part of the figure from Muhammad et al. (2021).

responses during water stress, which will help predict photosynthesis constraints under drought stress (Flexas et al., 2012).

5 Photosynthetic metabolic pathways of photosynthesis associated with drought tolerance in crops

Numerous photosynthesis metabolic pathways linked with drought stress tolerance in cereal crops have already been revealed via comparative physiological and omics research approaches. Drought stress has a serious impact on chloroplasts, specifically the thylakoid membrane–PSII reaction centers (Tai et al., 2011; Zhao et al., 2016; Zenda et al., 2020). A number of genes have a crucial function in plant photosynthesis machinery, consisting of light-harvesting complexes (LHCs) and PSII electron transport systems. For example, the proteins encoded by *lhcb5-1* and *lhcb5-2* act as peripheral antenna systems, allowing more effective light energy absorption. lhch5-1 is also engaged in the NPQ and cysteine production processes within the cell (Melkozernov and Blankenship, 2006; Murata et al., 2012; Zenda et al., 2018; Zenda et al., 2020). Previously, photosynthesis pathways were found to be substantially enriched in response to drought stress in maize, and chlorophyll a/b binding proteins were upregulated in an abscisic acid (ABA)-dependent manner in pearl millet (Zhao et al., 2016; Dudhate et al., 2018; Zenda et al., 2018; Zenda et al., 2020). This photosynthesis-related mechanism is essential for balancing light assembly and preventing PSII photoinhibition (Zhao et al., 2016; Zenda et al., 2020).

Although significant progress has been made to understand the nature of events occurring in drought-stress plants, an integrated picture of the metabolic regulation still requires clarification. Therefore, multilevel analysis is needed to integrate physiological, transcriptome, proteome, and metabolome data. Development of new experimental and computational approaches will allow a more detailed view of photosynthetic metabolic pathways in response to drought stress (Pinheiro and Chaves, 2011).

6 Does maintaining photosynthetic rate under drought stress contribute to crop drought tolerance?

Photosynthesis is a crucial mechanism that underpins agricultural plant yield and stress sensitivity (Ahmad et al., 2020). Drought stress reduces cereal crop yields depending on the severity and length of the stress period as well as photosynthetic damage due to decreased leaf growth, poorer photosynthesis machinery, and early leaf senescence (Aslam et al., 2015; Sehgal et al., 2018; Zenda et al., 2020). Principally, the cumulative rate of photosynthesis during the growing season determines the GY, which is governed by the following factors (Passioura, 1977):

$$GY = LI \times RUE \times HI$$

where LI = light interception by the crop; RUE = radiation-use efficiency; and HI = the harvest index (ratio of harvestable versus total aerial biomass) (Fig. 3).

Over the last few decades, significant advances in both HI and LI have been made, with increases in HI following the Green Revolution and the introduction of dwarfing (Rht) genes (Gale et al., 1985; Calderini et al., 1995; Simkin et al., 2020). Current efforts are focusing on RUE, which largely relates to photosynthesis and the conversion of light energy into fixed carbon (Reynolds et al., 2009; Parry et al., 2011; Simkin et al., 2020).

Productivity is determined by photosynthates and their mobilization from stored reserves. Increasing or maintaining the rate of photosynthesis offers substantial stress tolerance to crops and minimizes yield losses (Gupta and Thind, 2015; Miglani et al., 2021). The C_4 pathway of photosynthesis is a mechanism to reduce water loss from transpiration by partial stomatal closure in hot dry conditions so that C_4 plants can make more efficient use of water for growth than C_3 plants (Furbank and Taylor, 1995; Ku et al., 1996; Bandyopadhyay et al., 2007). In addition, C_4 plants (e.g. maize and sorghum) have a CO_2-concentrating mechanism that works by progressively prefixing CO_2 into C_4-carbon molecules through phosphoenolpyruvate carboxylase (PEPC) inside mesophyll cells. The decarboxylation of the bound carbon takes place in gas-tight bundle sheath cells. Such mechanisms increase the intercellular CO_2 concentration (C_i) surrounding the Calvin cycle's RuBisCO

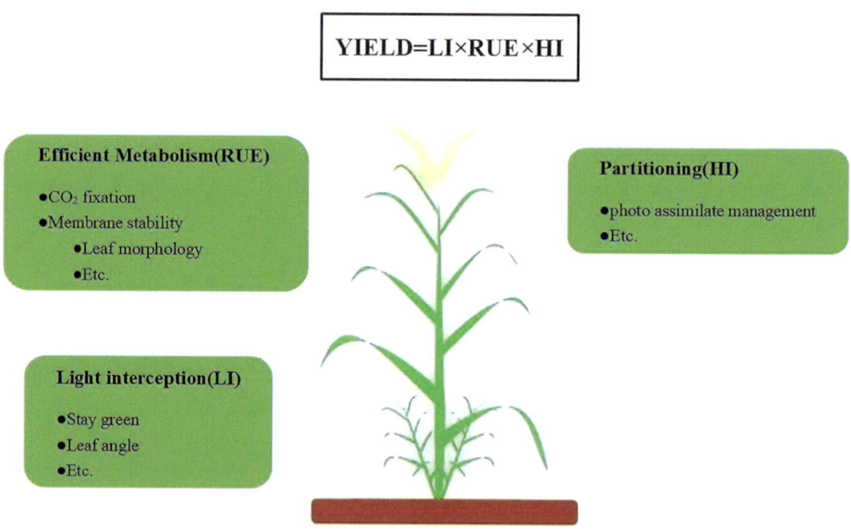

Figure 3 Grain yield improvement model developed by Passioura (1977). LI refers to light interception by the crop; RUE refers to radiation use efficiency, and HI refers to harvest index. Source: Morales et al. (2020).

enzyme. Accordingly, this mode allows C_4 plants to compete in the case of C_i shortage produced by high temperatures, high light intensities, and drought (Häusler et al., 2002; Edwards et al., 2004; Shen et al., 2015). However, wheat, rice, soybean, and potato (the key food crops) are C_3 plants. They absorb atmospheric CO_2 directly, bypassing the metabolic pump. These crop yields are restricted by environmental factors such as drought stress (Raines, 2006; Bandyopadhyay et al., 2007). Integrating C_4 traits into C_3 plants such as rice could therefore improve crop sensitivity to water scarcity (Häusler et al., 2002; Shen et al., 2015).

Recent studies have shown that changes in the key enzymes in the Calvin-Benson cycle (CBC), electron transport, photorespiration, and the kinetics of NPQ can improve photosynthesis in leaf tissue, which have great potential to increase yields of crops cultivated in a greenhouse as well as in the field (Simkin et al., 2020). Transient expression of the *PEPC* gene significantly increases the photosynthetic capacity of plants under stress conditions and photo-oxidation tolerance in rice and maize (Ku et al., 2000; Jiao et al., 2002; Zhou et al., 2011). In addition, *PEPC* overexpression also improves drought tolerance in transgenic rice by refixing internal CO_2 and reducing oxidative damage to the photosynthetic system (Shen et al., 2015). Wang et al. (2018) demonstrated that transgenic maize overexpressing *ZmNF-YB16* exhibited significant drought resistance by increasing the photosynthetic rate, thereby increasing yield under drought conditions. Under soil drought conditions, C_4 photosynthesis enzyme (PPDK and PCK) overexpression increased GYs and overexpression of *HvCBF4* improved the plants' tolerance to abiotic stresses by increasing F_v/F_m without restricting plant growth in rice (Oh et al., 2007; Gu et al., 2013). Under drought stress, ferrochelatase gene (*HvFC1* and *HvFC2*) overexpressing barley plants had better photosynthetic rates and less photo-oxidative damage (Nagahatenna et al., 2020). Jia et al. (2020a) also found that drought tolerance and photosynthesis were improved in transgenic inbred maize lines overexpressing *ZmVPP1* at the seedling stage. Overall, abiotic stress tolerance in crops can be enhanced by modifying the leaf metabolome to sustain the photosynthetic rate in adverse environmental conditions.

7 Can engineering of C_4 traits into C_3 plants improve crop drought tolerance?

Yield potentiality is a product of plant biomass and harvest index (HI), which can be achieved when a crop is cultivated with adequate water and nutrient inputs as well as no pest-disease infections (Evans, 1993). HI has been optimized by breeding in most cereals. Currently, HI values for most crops are almost at their theoretical maximum (Foulkes et al., 2007; Long et al., 2015; Carmo-Silva et al., 2017; Miglani et al., 2021; Roell et al., 2021; Yang et al., 2021), although the efficiency of energy conversion into biomass remains lower, with most crop species achieving just one-third of their theoretical maximum (Carmo-Silva et al., 2017). There is still plenty of room for improvement in energy conversion efficiency (Reynolds and Tuberosa, 2008; Zhu et al., 2008; Ort et al., 2011; Carmo-Silva et al., 2017). In support of this, crop photosynthesis has been greatly improved by approaches such as free-air CO_2 enrichment and increasing activities of photosynthetic enzymes in the CBC. Such photosynthesis improvement leads to higher agricultural yields (Bernacchi et al., 2006; Long et al., 2006a,b; Carmo-Silva et al., 2017; López-Calcagno et al., 2020), so enhancing photosynthesis is widely accepted as a critical method of advancing crop yield to meet growing global food demand (Parry et al., 2011; Long et al., 2015; Wu et al., 2019).

Under drought stress, C_4 plants perform much better than C_3 plants in terms of minimizing water loss, reducing photorespiration, and enhancing photosynthetic efficiency. Breeding C_3 plants to possess C_4 pathways has huge potential for world food security because many important food crops only have C_3 photosynthesis pathways (Edwards and Walker, 1983; Basu et al., 2016). Increasing efforts have been focused on engineering C_4 traits into C_3 plants, either by overexpressing the key C_3 enzymes or by transferring C_4 enzymes to C_3 plants with the aim of improving photosynthetic performance and crop yield (Zhou et al., 2011; Ashraf and Harris, 2013; Gu et al., 2013; Shen et al., 2015; Chatterjee et al., 2021; Cui, 2021; Ermakova et al., 2021; Jiang et al., 2021; Kopriva and Weber, 2021; Roell et al., 2021). Transgenic rice plants overexpressing the *C_4-PEPC* gene also displayed increased photosynthetic efficiency and tolerance to photo-oxidation, reduced photorespiration, increased drought and low nitrogen tolerance, and a significant increase in GY (Zhou et al., 2011; Gu et al., 2013; Chen et al., 2014; Lian et al., 2014; Ding et al., 2015; Huo et al., 2017; Tang et al., 2018; He et al., 2020). Drought-induced oxidative stress in the photosynthetic system exhibited site-dependent tendencies, with PSI downregulation being the most important cause. *PEPC* overexpression largely improved the drought tolerance of transgenic rice by refixing internal CO_2 and reducing oxidative damage to the photosynthetic system (Shen et al., 2015). In addition, Gu et al. (2013) found that rice plants overexpressing C_4 photosynthesis enzymes (PPDK and PCK) had greater GYs than WT plants, particularly when the soil was dry. The transgenic rice lines displayed a better yield performance and drought tolerance, which is associated with a number of improved traits including greater photosynthetic rate in leaves; higher levels of leaf water content and chlorophyll and nitrogen content; increased transpiration efficiency, PEPC, and carbonic anhydrase (CA) activities in leaves; higher root oxidation stability; and a stronger active oxygen scavenging system.

As described previously, developing transgenic lines of C_3 plants overexpressing C_4 photosynthetic enzymes is a viable strategy for improving C_3 plants' photosynthetic capacity and drought tolerance. However, the effectiveness of such strategy is also determined by other factors, including the localization of introduced C_4 enzymes within leaf tissues, and the coordination of introduced C_4 enzymes with existing CO_2 fixing pathways or other allied pathways in C_3 plants. Although the overexpression of a single C_4 enzyme can alter photosynthetic metabolism in C_3 plants, photosynthesis does not increase substantially in most cases. Furthermore, the special leaf anatomical structure in C_4 plants has not been transferred into C_3 plants, which might be the major reason behind failed attempts to genetically modify C_3 photosynthetic capability to reach the same level in C_4 plants. Alternatively, efforts could be undertaken to increase the effectiveness of CO_2-concentrating mechanisms in C_3 plants by genetically manipulating the activity of important enzymes (Ashraf and Harris, 2013).

Recently, Ermakova et al. (2021) developed a single construct containing five enzymes necessary for forming a minimum NADP-ME C_4 cycle. The five genes were successfully transformed into suitable cell types with the proper subcellular compartment in rice. *In vivo* incorporation of CO_2 into C_4 acids (C_4 carboxylation) was discovered using $^{13}CO_2$ labeling and metabolomics, but no subsequent decarboxylation was observed. This indicates that the C_4 cycle partially functions in the transgenic rice lines. Such C_4 carboxylation exhibited a modest physiological impact in rice, with very slight changes in photosynthetic light induction and no significant side effect on the global leaf transcriptome. In these transgenic rice lines, the five C_4 enzyme genes exhibited a cell-specific expression pattern that was mostly acceptable for two-cell C_4 photosynthesis. The transgenic rice plants displayed high levels of photosynthesis, which is compatible with the occurrence of C_4 carboxylation *in vivo*. These findings imply that rice may be engineered to assemble the entire C_4 metabolic pathway.

8 Drought tolerance genes related to photosynthesis

Drought is a major abiotic stress that seriously affects crop growth, development, and yield. Photosynthesis is a major primary process affected by water constraints. Therefore, enhancing photosynthesis under drought stress is essential to improve drought resistance and ensure food security. Genetic engineering has the unique advantage of achieving precise and efficient improvement of specific traits using a wide range of gene sources, showing a broad application prospect. To date, researchers have identified many drought-responsive genes related to photosynthesis from different crops, and a number of them found in cereals are listed in Table 1.

Table 1 A list of genes related to drought tolerance and photosynthesis that have been functionally validated in cereal crops

No.	Gene name	Donor species	Experimental host	Experimental findings	References
1	*FC1, FC2*	Barley	Barley	Under drought stress, photosynthetic rates have improved and photo-oxidative damage has decreased.	Nagahatenna et al. (2020)
2	*NF-YB1*	Arabidopsis	Maize	Photosynthesis efficiency is increased.	Yang et al. (2010); Nelson et al. (2007)
3	*Nicotiana protein kinase (NPK1)*	Tobacco	Maize	Keeping the photosynthetic mechanism from being damaged by dehydration.	Shou et al. (2004)
4	*SbER2et*	Sorghum	Maize	A higher WUE and a greater Pn rate.	Li et al. (2019)
5	Light harvest complex-related genes (*LHCA1, LHCB, LHCA3, LHCA2*)	Maize	Maize	In the PSII, balancing the light capture.	Zenda et al. (2018); Min et al. (2016)
6	*NF_YB16*	Maize	Maize	Under control and drought-stress conditions, photosynthesis and antioxidant enzyme activity were higher, membrane stability was improved, and electrolyte leakage was reduced.	Wang, et al. (2018)
7	*NF-YB2*	Maize	Maize	Improved drought resistance based on responses to a range of stress-related indicators, such as chlorophyll content, stomatal conductance, leaf temperature, decreased wilting, and photosynthetic maintenance.	Nelson et al. (2007)
8	*CspB*	*Escherichia coli*	Maize	Improved growth, chlorophyll content, and photosynthetic rates under water-deficit stress test.	Castiglioni et al. (2008)
9	*VPP1*	Maize	Maize	At the seedling stage, transgenic inbred lines showed improved drought tolerance and photosynthesis.	Jia et al. (2020a)

10	*TPP1*	Rice	Maize	Increased kernel set and harvest index resulted in yield increases of tolerance and photosynthesis, chlorophyll content, stomatal conductance, and leaf temperature.	Nuccio et al. (2015)
11	*ABF3*	Arabidopsis	Rice	Increased F_v/F_m value.	Oh et al. (2005)
12	*HARDY*	Arabidopsis	Rice	WUE and photosynthetic efficiency have both improved.	Karaba et al. (2007)
13	*CspA, CspB*	*Escherichia coli*	Rice	Increased yield and resistance to salt and drought stresses, stay-green phenotype.	Guddimalli et al. (2021)
14	*PEPC*	Maize	Rice	Drought tolerance has improved.	Zhou et al. (2011); Shen et al. (2015)
15	*PPDK*	Maize	Rice	Increased photosynthesis and drought tolerance.	Gu et al. (2013)
16	*CBF4*	Barley	Rice	The ratio of F_v/F_m was roughly 30% increased.	Oh et al. (2007)
17	*HRD*	Arabidopsis	Rice	Enhanced photosynthetic assimilation and reduced transpiration to enhance water usage efficiency or the ratio of biomass produced to water used.	Karaba et al. (2007)
18	*TPSP fusion* gene	*Escherichia coli*	Rice	Under salt, drought, and low-temperature stress conditions, plants have more sustained growth, less photo-oxidative damage, and a more favorable mineral balance.	Garg et al. (2002)

9 Quantitative trait loci mapping for photosynthetic traits

The process of photosynthesis is critical for crop growth and development. GY is linked to photosynthetic capacity during the reproductive period (Liu et al., 2015). According to recent studies, there is genetic heterogeneity in photosynthesis across existing germplasm, and yield progress in cereals via breeding has been linked to improved photosynthesis (Fischer and Edmeades, 2010; Gu et al., 2012; Kamal et al., 2019; Morales et al., 2020).

Many QTLs have been identified using molecular genetics techniques to reduce the impact of drought on plant photosynthesis. These QTLs either enhance the yield of the plant under stress or boost the expression of characteristics or genes that have a predisposition to endure drought conditions. The QTLs for photosynthetic properties that have been identified for crop improvement are listed in Table 2.

The increasing number of studies identifying QTLs for photosynthetic attributes in drought-stressed crops will help researchers better understand the genomic plant loci that affect trait variation. It is worthy of note that several photosynthetic characteristics boost grain productivity during drought (Borrell et al., 2000a, b; Kebede et al., 2001; Kassahun et al., 2010; Rama Reddy et al., 2014), and SG phenotypes are often correlated with GY traits (Verma et al., 2004; Yoo et al., 2007; Bogard et al., 2011; Fu et al., 2011; Wehner et al., 2015; Yang et al., 2017; Hassan et al., 2021). Under drought conditions, wheat production has been linked to chlorophyll, normalized difference vegetation index (NDVI), and canopy temperature, which implies that their genetic origins are linked to the same genomic loci (Pinto et al., 2010). The co-location of QTL for photosynthetic characteristics and QTL for yield is supported by various experimental findings. In wheat, QTLs for leaf senescence, transpiration rate, stomatal conductance, WUE, chlorophyll, NDVI, canopy temperature, net photosynthetic rate, intercellular CO_2 concentration, and stomatal characteristics were shown to be co-located with yield QTLs (Peleg et al., 2009; Zhang et al., 2009; Pinto et al., 2010; Bogard et al., 2011; Sukumaran et al., 2015; Osipova et al., 2016; Shahinnia et al., 2016; Zhu et al., 2016; Xu et al., 2017; Hassan et al., 2021). Similarly, various QTLs for photosynthetic traits were co-located with QTLs for yield in other crops (Ishimaru et al., 2001; Zhao et al., 2008; Zheng et al., 2009; Fu et al., 2011; Rama Reddy et al., 2014; Yin et al., 2015; Arriagada et al., 2017). These studies suggest that pyramiding the QTLs for photosynthetic characteristics and yield traits by marker-assisted selection breeding will contribute to the genetic improvement of crop yield.

Table 2 List of selected examples of QTLs identified for photosynthetic traits in major cereals

Quantitative trait loci	Mapping population	Parents	Genotyping markers	Environment	Chromosome	Crop	References
12 QTLs for IPAR	IL	Caeserea 26-24 and Harrington	SNP	Field	4H, 6H	Barley	Arriagada et al. (2017)
9 QTLs for loss of green color	DH	Vlamingh, Buloke, VB9524 and ND11231*12	SNP and SSR	Field	1H, 2H, 4H, 5H, 6H, 7H	Barley	Emebiri (2013)
10 QTLs for stay-green	DH	ND24260 and Flagship	DArT	Glasshouse	3H, 4H, 5H, 6H, 7H	Barley	Gous et al. (2016)
35 QTLs for photosynthetic efficiency	RIL	Maresi and CamB	SSR, SNP	Growth chamber	2H, 3H, 5H, 6H, 7H	Barley	Gudys et al. (2018)
27 QTLs for F_o, F_v, F_m, F_v/F_m, and chlorophyll content	RIL	Arta and *Hordeum spontaneum* 41-1	SSR and AFLP	Greenhouse	1H, 2H, 4H, 5H, 6H, 7H	Barley	Guo et al. (2008)
2 QTLs for chlorophyll fluorescence and 4 QTLs for SPAD	IL	Scarlett and ISR42-8	SNP	Greenhouse	1H, 4H, 6H	Barley	Honsdorf et al. (2014)
37 QTLs for FLL, FLW, RWC, chlorophyll content, and F_v/F_m	148 barley cultivar		AFLP and SSR	Field	1H, 2H, 3H, 4H, 5H, 6H, 7H	Barley	Jabbari et al. (2019)
207 marker-trait associations for physiological traits	148 barley cultivars		AFLP and SSR	Field	1H, 2H, 3H, 4H, 5H, 6H, 7H	Barley	Jabbari et al. (2021)
38 QTLs for physiological and morphological traits	DH	Huaai 11 and Huadamai 6	SSR	Rain shelter	1H, 2H, 3H, 4H, 6H, 7H	Barley	Liu et al. (2015)

(*Continued*)

Table 2 (*Continued*)

Quantitative trait loci	Mapping population	Parents	Genotyping markers	Environment	Chromosome	Crop	References
22 QTLs for chlorophyll fluorescence parameters	RIL	Badia and Kavir	SSR, ISSR, IRAP, CAAT, Scot, iPBS	Greenhouse	1, 3, 4, 6, 7	Barley	Makhtoum et al. (2021)
17 QTLs for RWC and F_v/F_m	RCSL	Hordeum spontaneum and Harrington	SNP	Field	1H, 3H, 4H, 7H	Barley	Mora et al. (2016)
6 QTLs for NDVI and leaf chlorophyll content	DH	Commander, Fleet, and WI4304	SNP	Field	1H, 3H, 4H	Barley	Obsa et al. (2016)
2 marker-trait associations for SPAD	223 barley accessions		DArT, SNP, and SSR	Field	5, 6	Barley	Varshney et al. (2012)
6 SNP significantly associated with leaf senescence genes	156 winter barley genotypes		SNP	Greenhouse	1H, 2H, 3H, 6H, 7H	Barley	Wehner et al. (2016)
6 marker-trait associations for SPAD	156 winter barley genotypes		SNP	Greenhouse	2H, 6H, 7H	Barley	Wehner et al. (2015)
58 marker-trait associations for physiological parameter	109 lines		DArT and SNP	Growth chambers	2H, 3H, 4H, 5H, 6H	Barley	Wójcik-Jagła et al. (2018)
85 marker-trait associations for canopy temperature	299 cultivars and advanced lines		SNP	Field	1A, 2A, 5A, 6A, 7A, 2B, 3B, 6B, 2D, 3D, 4D, 6D	Wheat	Awad (2015)
6 QTLs for flag leaf senescence	F_2	Variant-2 and Cham-6	RAPD, ISSR, SSR	Greenhouse	2D	Wheat	Barakat et al. (2013)
28 QTLs for leaf senescence	DH	Toisondor and CF9107	DArT, SSR, SNP	Field	2A, 1B, 5B, 6B, 7B, 2D, 4D, 7D	Wheat	Bogard et al. (2011)

Tackx4 gene	RIL	Jing 411 and Hongmangchun 21	EST	Field	3A	Wheat	Chang et al. (2015)
160 QTLs for chlorophyll fluorescence, SPAD, Chl_{a+b}, and carotenoid contents	DH	Chinese Spring and SQ2	RFLP, AFLP, SSR, DArT	Open-sided greenhouse	All except 7B	Wheat	Czyczyło-Mysza et al. (2013)
20 QTLs for chlorophyll fluorescence parameters	DH	Chinese Spring and SQ1	RFLP, AFLP, SSR	Rainout shelter	2A, 4A, 1B, 4B, 6B, 2D, 4D, 7D	Wheat	Czyczyło-Mysza et al. (2011)
63 QTLs for canopy temperature depression, 43 for chlorophyll content, 7 for fluorescence indices	RIL	Jennah Khetifa and Cham-1	RFLP, SSR, AFLP, EST	Field	1A, 2A, 3A, 4A, 5A, 6A, 7A, 1B, 2B, 3B, 4B, 5B, 6B, 7B	Durum wheat	Diab et al. (2008)
26 QTLs for vegetative indices	DH	Yangmai16 and Zhongmai895	SNP	Field	3A, 5A, 7A, 1B, 2B, 4B, 5B, 7B, 3D, 4D, 5D, 6D, 7D	Wheat	Hassan et al. (2021)
16 QTLs for total chlorophyll content and parameters of chlorophyll fluorescence kinetics	DH	Opata and SH223	SSR	Field andgreenhouse	4A, 7A, 1B, 2B, 3B, 5B, 7B, 7D	Wheat	Ilyas et al. (2014)
47 marker-trait associations for leaf pigment composition, hydration, and model-based prediction of overall photosynthetic capacity	149 bread wheat spring types		SNP	Field	2A, 3A, 5A, 6A, 7A, 1B, 2B, 3B, 5B, 6B, 7B, 1D, 2D, 3D, 6D, 7D	Wheat	Joynson et al. (2021)

(Continued)

Table 2 (*Continued*)

Quantitative trait loci	Mapping population	Parents	Genotyping markers	Environment	Chromosome	Crop	References
3 QTLs for stay-green	RIL	Chirya 3 and Sonalika	SSR	Field	1A, 3B, 7D	Wheat	Kumar et al. (2010)
17 QTLs for F_v/F_m, chlorophyll content, and leaf temperature	RIL	C306 and HUW206	SSR	Greenhouses and growth chambers	1A, 3A, 6A, 2B, 3B, 4B, 5B, 1D, 4D	Wheat	Kumar et al. (2012)
43 QTLs for chlorophyll content and chlorophyll fluorescence	DH	Hanxuan 10 and Lumai 14	SSR and AFLP	Growth chamber	2A, 6A, 7A, 1B, 2B, 3B, 6B, 7B, 1D, 3D, 5D	Wheat	Li et al. (2012)
3 QTLs for F_v/F_m	DH	Huapei 3 and Yumai 57	SSR, EST, ISSR	Field	5A, 6A, 6D	Wheat	Liang et al., (2010)
1 QTL for P_n	F_2	Chakwal-86 and 6544-6	SSR	Plastic tub	2A	Wheat	Malik and Malik (2015)
1 QTL for photosynthetic rate	F_2	Chakawal-50 and 9436	ISSR	Field	3A	Wheat	Maqsood et al. (2017)
8 QTLs for canopy temperature	RIL	Len and Halberd	SNP	Field	1A, 6A, 2B, 3B, 7B	Wheat	Mohammed et al. (2021)
9 QTLs for stay-green	RIL	Reeder (PI 613586) and Conan (PI 607549)	SSR, DArT	Field, greenhouse	1B, 2D, 3A, 3B, 4A, 5B, 7B	Wheat	Naruoka et al. (2012)
53 QTLs for key photosynthesis traits and shoot biomass	IL	CS and Syn	SSR	Climatic chamber	1D, 2D, 4D, 5D, 6D, 7D	Wheat	Osipova et al. (2016)
11 QTLs for chlorophyll content	RIL	Durum wheat (cv. Langdon) and wild emmer (acc# G18-16)	SSR, DArT	Insect-proof screenhouse	1A, 4A, 5A, 6A, 7A, 1B, 2B, 5B	Tetraploid wheat	Peleg et al. (2009)
46 QTLs for canopy temperature, NDVI, chlorophyll content	RIL	Seri M82 and Babax	SSR, DArT, AFLP	Field	1B, 2B, 3B, 4A, 5A, 7A, 7B	Wheat	Pinto et al. (2010)

5 QTLs for Y(II), ETR and chlorophyllcontent	SCRDH	Saratovskaya 29 and Yanetzkis Probat	SNP	Climatic chamber, greenhouse, and field	2A	Wheat	Pshenichnikova et al. (2021)
21 QTLs for stay-green	RIL	CN18 and T1208	SNP	Field	2A, 3A, 4A, 6A, 1B, 2B, 5B, 6B, 7B, 2D, 3D, 5D, 7D	Wheat	Ren et al. (2022)
50 QTLs for stomatal traits and yield	DH	RAC875 and Kukri	SSR, DArT, and SNP	Field and glasshouse	1A, 3A, 4A, 5A, 6A, 7A, 1B, 2B, 3B, 4B, 5B, 7B, 2D, 3D, 5D, 6D, 7D	Wheat	Shahinnia et al. (2016)
71 QTLs for chlorophyll content and NDVI	DH	Hanxuan 10 and Lumai 14	SSR and AFLP	Field	1A, 2A, 3A, 4A, 5A, 6A, 7A, 1B, 2B, 3B, 4B, 6B, 7B, 1D, 2D, 3D, 4D, 6D	Wheat	Shi et al. (2017)
21 marker-trait Associations for key photosynthetic traits	200 wheat cultivars		SNP	Field, rainout shelter	2A, 3A, 4A, 5A, 7A, 1B, 2B, 4B, 6B, 2D	Wheat	Siddiqui et al. (2021)
7 marker-trait associations for chlorophyll index, canopy temperature, and NDVI	287 spring wheat lines		SNP	Field	6A, 1B, 3B, 5B, 6B, 4D	Wheat	Sukumaran et al. (2015)

(Continued)

Table 2 (*Continued*)

Quantitative trait loci	Mapping population	Parents	Genotyping markers	Environment	Chromosome	Crop	References
22 additive QTLs and 5 pairs of epistatic QTLs for chlorophyll and fluorescence parameters	DH	Huapei 3 and Yumai 57	SSR and EST-SSR	Field	2A, 3A, 4A, 5A, 6A, 7A, 1B, 2B, 3B, 4B, 7B, 2D, 3D, 5D, 6D	Wheat	Tian et al. (2015)
4 QTLs for stay-green	DH	Beaver and Soissons	AFLP and SSR	Field	2B, 2D	Wheat	Verma et al. (2004)
18 QTLs for G_s, P_n, and T_r	DH	Hanxuan 10 and Lumai 14	SSR and AFLP	Field	3A, 4A, 6A, 2B, 3B, 5B, 7B, 1D, 2D, 6D	Wheat	Wang et al. (2015)
225 QTLs for yield and physiological traits	RIL	Chuan 35050 and Shannong 483	SSR, DArT, EST-SSR	Field	1A, 2A, 3A, 4A, 5A, 6A, 7A, 1B, 2B, 3B, 4B, 5B, 6B, 7B , 1D, 2D, 3D, 4D, 5D, 6D, 7D	Wheat	Xu et al. (2017)
29 QTLs for FLL, FLW, FLA, and chlorophyll content	IL	Lumai 14, Jing 411, and Shaanhan 8675	SSR	Field	1A, 2A, 3A, 4A, 6A, 1B, 3B, 4B, 2D, 3D, 5D	Wheat	Yan et al. (2020)
18 QTLs for chlorophyll content and parameters of chlorophyll fluorescence kinetics	DH	Hanxuan 10 and Lumai 14	SSR and AFLP	Field	1A, 5A, 6A, 7A, 1B, 3B, 4D, 7D	Wheat	Yang et al. (2007)

65 additive QTLs and 42 pairs of epistatic QTLs for the greenness of flag leaf	RIL	Longjian 19 and Q9086	SSR	Field	1A, 2A, 3A, 4A, 6A, 7A, 1B, 2B, 3B, 4B, 5B, 6B, 7B, 1D, 2D, 3D, 4D, 5D, 6D, 7D	Wheat	Yang et al. (2016)
16 additive QTLs and 11 pairs of epistatic QTLs for chlorophyll fluorescence traits	RIL	W-7984 and Opata85	RFLP	Growth chamber	1A, 3A, 4A, 5A, 7A, 1B, 2B, 4B, 5B, 1D, 2D, 3D, 4D, 5D, 6D, 7D	Wheat	Zhang et al. (2010)
8 additive QTLs and 3 pairs of epistatic QTLs for chlorophyll content	DH	Huapei 3 and Yumai 57	SSR	Field	2A, 4A, 5A, 7A, 1B, 2D, 5D	Wheat	Zhang et al. (2009)
Tabas1-B1 gene	RIL	Jing411 and Hongmangchun21	SSR	Field	2B	Wheat	Zhu et al. (2016)
1 QTL for canopy temperature	DH	CT9993-5-10-1-M and IR62266-42-6-2	RFLP, SSR, and AFLP	Field	2	Rice	Babu et al. (2003)
6 QTLs for stay-green	RIL	Suweon490, Andabyeo, and SNU-SG1	SSR	Field	2, 4, 5, 6, 9, 12	Rice	Fu et al. (2011)
7 QTLs for canopy temperature difference	BC_1F_6	Koshihikari and Takanari	SSR	Field	3, 9, 10, 11, 12	Rice	Fukuda et al. (2018)
29 QTLs for photosynthetic traits and drought sensitivity	IL	Shennong265 and Haogelao	SSR	Field	1, 2, 3, 4, 6, 7, 8, 9, 11	Rice	Gu et al. (2012)
25 QTLs for physiological traits	BC_1	Nipponbare and Kasalath	RFLP	Field	1, 2, 3, 4, 5, 6, 8, 9, 11, 12	Rice	Ishimaru et al. (2001)
60 additive QTLs and 52 epistatic QTL pairs for six chlorophyll traits	DH and BC	Zhenshan 97 and Wuyujing 2	SSR	Field	1, 2, 3, 4, 6, 7, 8, 9, 11, 12	Rice	Jiang et al. (2014)

(*Continued*)

Table 2 *(Continued)*

Quantitative trait loci	Mapping population	Parents	Genotyping markers	Environment	Chromosome	Crop	References
46 main-effect QTLs and 50 digenic interactions for stay-green traits	DH	Zhenshan 97 and Wuyujing 2	SSR	Field	All	Rice	Jiang et al. (2004)
2 QTLs for chlorophyll content	RIL	Azucena and Bala	RFLP, SSR, and AFLP	Field	6	Rice	Lafitte et al. (2004)
5 QTLs for canopy temperature	RIL	IR20 and Nootripathu	SSR	Field	2, 4	Rice	Michael Gomez et al. (2010)
2 QTLs for resting stomatal conductance	F_2	Azucena and Bala	RFLP	Glasshouse	3, 12	Rice	Price et al. (1997)
3 QTLs for canopy temperature	RIL	IR20 and Nootripathu	SSR	Field	4, 6, 7	Rice	Prince et al. (2015)
143 QTLs for chlorophyll fluorescence parameters	154 genotypes		SNP	Growth room	All	Rice	Quero et al. (2021)
1 QTL for leaf temperature	Inter-varietal BILs	Swarna and WAB 450	SSR	Poly house	2	Rice	Sangodele et al. (2014)
41 QTLs for the morpho-physiological traits	RIL	IR64 and Hawara Bunar	SNP	Greenhouse	1, 3, 6, 8, 9, 12	Rice	Satrio et al. (2021)
49 QTLs associated with delayed leaf senescence traits	IL	Swarna and Moroberekan	SNP	Field	1, 2, 3, 5, 6, 7, 10	Rice	Singh et al. (2020)
8 QTLs for SPAD, G_s, and senescence	NF-RIL	Norungan, IR64 and IR50	SSR	Field	1, 4, 8, 9, 11	Rice	Subashri et al. (2009)
8 QTLs for photosynthesis and related physiological characters	DH	ZYQ8 and JX17	RFLP and SSR	Field	1, 3, 4, 6, 7, 8	Rice	Teng et al. (2004)

69 QTLs for PSII chlorophyll fluorescence parameters	283 accessions		SNP	Field	1, 2, 3, 4, 5, 6, 7, 8, 9, 10, 11, 12	Rice	Vilas et al. (2020)
11 QTLs for chlorophyll content	DH	Zhaiyeqing 8 and Jingxi 17	RFLP, SSR, and AFLP	Field	2, 3, 5, 6, 9, 11	Rice	Yang et al. (2008)
28 QTLs for stay-green traits	F_2 and RIL	SNU-SG1 and M23	SSR and STS	Field	1, 3, 4, 5, 6, 7, 9, 12	Rice	Yoo et al. (2007)
6 QTLs for the degree of greenness and 14 QTLs for stay-green-related traits	RIL	Zhenshan 97 and IRAT 109	SSR	Field	1, 2, 3, 4, 5, 7, 8, 9	Rice	Yue et al. (2006)
40 QTLs for morpho-physiological traits and grain yield	IL	Teqing and Lemont	SSR	Field	All	Rice	Zhao et al. (2008)
8 meta-QTLs for stay-green	RIL, $F_{2:3}$	CML444, MALAWI, CML440, CML504, CML441	SSR and SNP	Field	1, 3, 4, 5, 8, 10	Maize	Almeida (2012)
80 QTLs for stay-green	RIL, $F_{2:3}$, $F_{2:3}$	CML444, MALAWI, CML440, CML504, CML441	SNP and SSR	Field	1, 2, 3, 4, 5, 6, 7, 8, 9, 10	Maize	Almeida et al. (2014)
8 QTLs for stay-green	Topcross and F_4 progeny	B73 and MO17	RFLP	Field	1, 2, 6, 8, 9	Maize	Beavis et al. (1994)
17 QTLs for stay-green	BC	L-14-04B and L-08-05F	SSR	Field	1, 2, 3, 4, 6, 9	Maize	Belícuas et al. (2014)
170 SNPs significantly associated with visual senescence, chlorophyll, PSII quenching, chlorophyll fluorescence	MAGIC		SNP	Field	ALL	Maize	Caicedo et al. (2021)

(Continued)

Table 2 (*Continued*)

Quantitative trait loci	Mapping population	Parents	Genotyping markers	Environment	Chromosome	Crop	References
1 QTL for stay-green	RIL	CG60 and CG102	SSR and SNP	Field	8	Maize	Khanal et al. (2015)
12 QTLs for G_s, F_m′, and chlorophyll concentration	F_2	F-2 and Polj17	RFLP	Greenhouse	1, 2, 3, 5, 6, 7, 8, 10	Maize	Lebreton et al. (1995)
4 QTLs for leaf temperature differences	RIL	Zong3 and 87-1	SSR	Greenhouse	2, 9, 10	Maize	Liu et al. (2011)
32 QTLs for leaf chlorophyll, plant senescence, and electric root capacitance	RIL	CML444 and SC-Malawi	SSR	Field	1, 2, 4, 5, 6, 10	Maize	Messmer et al. (2011)
10 main-effect QTLs and 11 couples of interaction QTLs for net CO_2 uptake, transpiration, and G_s	RIL	F-2 and MBS847	RFLP	Greenhouse	1, 2, 3, 4, 5, 6, 7, 8, 9		Pelleschi et al. (2006)
5 QTLs for leaf surface area	$F_{2:3}$	DTP79 and B73	RFLP	Field	3, 9	Maize	Rahman et al. (2011)
61 QTLs for the grain yield and nine stay-green traits	RIL	M35-1 and B35	SSR	Field	1, 2, 3, 4, 5, 6, 7, 9, 10	Maize	Rama Reddy et al. (2014)
7 QTLs for leaf senescence	F_4	H16 and K64	SSR	Field	2, 4, 6, 7, 8	Maize	Ribaut et al. (2004)
10 QTLs for chlorophyll a fluorescence parameters	RIL	B73 and Mo17	RFLP and SSR	Field	1, 5, 7, and 8	Maize	Šimić et al. (2014)

5 eQTLs for *ZmRCAβ* expression	RIL	RA and M53	SSR, SNP	Field	2, 4	Maize	Sun et al. (2017)
21 overdominant QTLs and 7 non-overdominant QTLs for stay-green	BC_1 $F_{2:3}$	CML491, DTPWC9F104 and LPSC7F64	SNP	Field	1, 2, 4, 5, 6, 8, 10	Maize	Trachsel et al. (2016)
22 QTLs for stay-green	F_2	A150-3-2 and Mo17	SSR	Field	1, 2, 3, 4, 5, 6, 9	Maize	Wang et al. (2012)
23 QTLs for chlorophyll content, photosystem II photochemical efficiency, and stay-green area	RIL	Zheng58 and B73	SSR	Field	1, 2, 3, 4, 5, 6, 8, 9, 10	Maize	Yang et al. (2017)
12 QTLs for chlorophyll fluorescence parameters	RIL	RA and M53	SNP and SSR	Field	1, 5, 9	Maize	Yin et al. (2015)
41 SNPs significantly associated with fast chlorophyll fluorescence parameters	404 inbred lines		SNP	Field	1, 2, 4, 5, 6, 7, 9	Maize	Yin et al. (2016)
54 QTLs for photosynthetic performance parameters	F_4	Chang7-2,TS141, Langhuang	SSR	Field	All	Maize	Zhao and Zhong (2021)
14 QTLs for stay-green	F_2	Q319 × Mo17	SSR	Field	1, 2, 3, 5, 6, 8, 9	Maize	Zheng et al. (2009)
16 QTLs for canopy temperature depression	RIL	Tx642 and Tx7000	DG	Field and glasshouse	3, 4, 6, 8, 10	Sorghum	Awika et al. (2017)

(Continued)

Table 2 (*Continued*)

Quantitative trait loci	Mapping population	Parents	Genotyping markers	Environment	Chromosome	Crop	References
7 QTLs for stay-green	RIL	B35 and TX430	RFLP	Field	A, B, D, G, I, J	Sorghum	Crasta et al. (1999)
41 QTLs for stay-green	RIL	IS9830, E36-1, and N13	AFLP, RFLP, SSR, and RAPD	Field	A, B, C, D, E, G, H	Sorghum	Haussmann et al. (2002)
9 QTLs for stay-green	RIL	SC56 and Tx7000	RFLP	Field	A, B, C, D, E, F, G, J	Sorghum	Kebede et al. (2001)
33 QTLs for stay-green	$F_{2:4}$	RSG04008-6 and J2614-11	SNP	Field	10	Sorghum	Kiranmayee et al. (2020)
181 SNP-trait correlations for photosynthesis and chlorophyll fluorescence traits	324 accessions		SNP	Growth chamber	1, 2, 3, 4, 5, 6, 7, 8, 10	Sorghum	Ortiz and Salas-Fernandez (2021)
5 QTLs for stay-green and 3 QTLs for chlorophyll content	RIL	B35 and TX7000	RFLP, SSR, and RAPD	Field	A, D, E, J	Sorghum	Subudhi et al. (2000)
5 QTLs for stay-green	RIL	QL39 and QL41	RFLP and SSR	Field	A, B, C, G, I	Sorghum	Tao et al. (2000)
6 QTLs for stay-green	RIL	B35 and TX7078	RAPD and RFLP	Field	B, F, G, H, I	Sorghum	Tuinstra et al. (1997)
7 QTLs for stay-green and chlorophyll content	RIL	B35 and Tx7000	RFLP	Field	A, D, J	Sorghum	Xu et al. (2000)

10 Does the stay-green phenotype lead to increased crop yield under drought conditions?

The leaves of healthy crop plants remain green due to high levels of chlorophyll. During plant senescence or under adverse environmental conditions, leaves will fade, and consequently photosynthesis and transpiration are inhibited, affecting the growth and development of the plant. Drought generally causes crop leaf yellowing and accelerates plant senescence and death, despite crop senescence phenotypes varying by plant species and variety. During grain filling, leaves in various genotypes exhibiting significant drought tolerance remain green and maintain vigorous photosynthesis under drought stress. This property is known as SG.

For a given crop, high yield depends on the following factors: (i) radiation uptake throughout the whole growth period of the crop; (ii) RUE (also termed photosynthetic efficiency); and (iii) HI (Fig. 1). To date, most studies to improve crop yield through photosynthesis have focused on increasing the RUE of specific organs, mostly leaves, by increasing photosynthesis rates per unit organ area. The underlying assumption is that any enhancement in photosynthesis rate per unit leaf area would result in increased biomass and yield. In comparison, considerably less time has been spent on boosting the canopy's photosynthesis. Changes in canopy architecture, such as more erect leaves and the SG phenotype, can promote canopy photosynthesis (Fig. 4) (Araus et al., 2021). It is therefore necessary to pay more attention to the SG phenotype. In particular, functional SG traits under drought conditions are important to increase crop production.

Plants produce carbohydrates and absorb nitrogen in their vegetative organs (source). These nutrients are remobilized into newly growing tissues or reproductive organs (sink) throughout the developmental process (Gregersen et al., 2013). Higher GYs can be achieved by increasing source strength in cereal crops. Transfer from the carbon (C) capture to the nitrogen (N) remobilization phase of foliar growth has been linked to the functional SG characteristic. In functional stay-green plants, the C-N transition point is delayed, or the transition occurs on time but subsequent yellowing and N remobilization occur slowly (Lee and Tollenaar, 2007; Zhang et al., 2019). SG therefore appears to be the most important photosynthetic feature linked to increased production potential in plants under drought stress (Peingao, 2013; Borrell et al., 2014; Graziani et al., 2014; Jaegglia et al., 2017; Kamal et al., 2018; Kamal et al., 2019; Zhang et al., 2019). Overexpression of RNA-binding bacterial chaperones in rice results in a SG phenotype that has enhanced yield and tolerance against salt and drought stress (Guddimalli et al., 2021). The SG feature has proven to be beneficial in enhancing drought tolerance to help mitigate the effects of an ever-worsening environment for agriculture and declining freshwater supply (Sekhon et al., 2019).

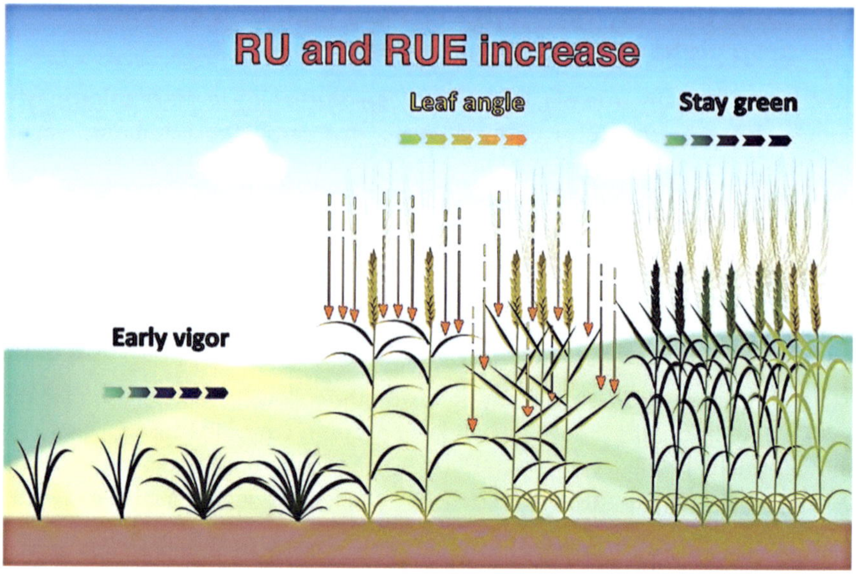

Figure 4 Accumulated radiation uptake (RU) of the crop: organization levels of the plant involved and some traits that are amenable to plant breeding, which may modify RU, early vigor, and stay-green. RUE refers to radiation use efficiency. Source: Araus et al. (2021).

Functional SG is also concerned with the water status of plants, especially OA. Dwarf nature, high rates of transpiration, and reduced rates of leaf senescence during drought stress are characteristic of landraces with greater OA, and they are more likely to exhibit SG phenotypes. By preventing leaf rolling and tissue death, OA boosts agricultural production (Blum, 1996) and the efficient leaf photosynthetic area increases with decreased leaf rolling and senescence (Nadew et al., 2021). Indeed, a lack of water inhibits development prior to photosynthesis, causing sugars to be released in an attempt to rectify the osmotic balance. On the other hand, OA has long been known to support photosynthesis by maintaining turgor when stress increases. As a result, OA can be associated with growth inhibition while also protecting photosynthesis (Blum, 2017). For instance, OA was assessed and compared with yield under stress in different genotypes of 12 crops, namely barley, wheat, maize, sorghum, chickpea, pea, pigeon pea, soybean, canola, mustard, castor bean, and sunflower. The findings suggest that OA is a key adaptive feature in supporting agricultural production under water stress (Blum, 2017).

11 Introducing the stay-green trait into plants to enhance photosynthesis under drought stress

'SG' is an essential feature for the genetic improvement of various crops, allowing plants to maintain active photosynthetic levels in their leaves even under stress conditions (Abdelrahman et al., 2017). There are several ways to introduce the SG trait into plants to boost photosynthesis: (i) knockout of chlorophyll degradation genes; (ii) manipulation of endogenous cytokinin and senescence genes; (iii) overexpressing genes responsible for chlorophyll biosynthesis; and (iv) developing varieties/cultivars with delayed senescence (Ambler et al., 1992; Kassahun et al., 2010; Kusaba et al., 2013; Luche et al., 2015; Chen et al., 2018; Miglani et al., 2021).

12 Case study: enhancement of sorghum tolerance to post-flowering drought stress using stay-green quantitative trait loci

The SG gene allows plants to maintain photosynthetic activity in their leaves, which improves grain filling even in stressful situations (Rosenow, 1983; Gregersen et al., 2006; Peingao, 2013; Borrell et al., 2014; Jaegglia et al., 2017; Kamal et al., 2018; Kamal et al., 2019; Zhang et al., 2019). Genes or QTLs controlling SG can be used to enhance drought tolerance in drought-sensitive crops. A series of QTLs related to SG phenotypes have been identified in crop plants. The QTLs for SG of major cereal crops and their specific functional definitions are listed in Table 2.

Decreased leaf senescence (SG) has been shown to increase sorghum tolerance to post-flowering drought stress (Harris et al., 2007; Kassahun et al., 2010), and several sorghum genotypes - BTx642, ET36-1, M35, SC56, and K19 - have been studied for their SG phenotypes. The BTx642 genotype, formerly known as B35, has been the most prevalent source of the SG characteristic, and many markers associated with the SG QTLs have been discovered in B35 (Harris et al., 2007; Kassahun et al., 2010; Ouedraogo et al., 2017). Six SG QTLs for post-flowering drought tolerance have been found, four of which (designated as Stg1, Stg2, Stg3, and Stg4) are significant major QTLs and two (StgA and StgB) are minor QTLs for the SG phenotype (Ouedraogo et al., 2017; Kamal et al., 2018). Stg1 and Stg2 have been located on chromosome 3 of sorghum and are associated with around 20% and 30% of the phenotypic variation, respectively. Stg3 and Stg4 have been found on chromosomes 2 and 5 and account for 16% and 10% of phenotypic variation, respectively (Sanchez et al., 2002; Harris et al., 2007; Ouedraogo et al., 2017). These QTLs have been found in a range of environments and from different genetic backgrounds (Ouedraogo et al., 2017). Kassahun et al. (2010) first introduced four stable SG QTLs (StgB, Stg1, Stg3, and Stg4) partially (BC_2F_3/BC_1F_4 generations) to an early senescence variety R16 from its donor parent B35 in both well water-supplied and drought-stress conditions at the post-flowering stage and successfully introgressed into the recurrent parent R16's genetic background. Before and during leaf senescence, the majority of the QTL introgression lines displayed greater leaf chlorophyll content (soil and plant analyzer development (SPAD) reading) and had a slower rate of loss of green leaf area (drought-stress condition at post-flowering), delaying the onset of leaf senescence (supplementally irrigated environment). In two of the three post-flowering stress contexts, delayed leaf senescence (relative percent green leaf area in the stress/irrigated conditions) was linked to increased grain production. Therefore, the research provides valuable evidence that marker-assisted backcrossing of SG QTLs from B35 into an acclimated yet early senescence background can improve sorghum tolerance to post-flowering water deficit. Mwamahonje et al. (2021) employed marker-assisted backcross to improve the SG features of Tanzanian farmer-preferred sorghum cultivars. The findings suggested increased tolerance to drought in locally adapted sorghum cultivars in Africa. In maize, SG maize genotypes also had high tolerance to drought stresses compared to non-SG genotypes. Several studies have found that maize breeding efforts aimed at increasing grain production improved the SG during the selection process (Kamal et al., 2019).

Overall, SG plants had longer grain-filling time, and, as a result, a higher yield was obtained compared to non-SG plants. Functional SG breeding has helped enhance agricultural yields, especially when paired with other beneficial yield traits. Thus, in conjunction with yield adaptation and stabilization, the SG phenotype has been identified as a preferable attribute for economically bred cereal selection to overcome existing yield decline (Kamal et al., 2019).

13 Conclusion and future trends in research

Increasing agricultural production is the most effective approach to address the need to double global food production by 2050 (Tilman et al., 2011; Ray et al., 2013; Long et al., 2015; South et al., 2018; Wang et al., 2020). Photosynthesis, together with respiration, is the major mechanism underlying plant growth and agricultural production (Gago et al., 2019), while crop biomass and GYs are mostly influenced by photosynthetic efficiency (Heyneke and Fernie, 2018). Therefore, the second Green Revolution required to ensure mankind's food security in the next few decades will focus on increasing crop photosynthetic efficiency (Long et al., 2006a; Zhu et al., 2008; Von Caemmerer et al., 2012; Long et al., 2015; Wang et al., 2020). Plants with improved photosynthesis exhibit several advantages such as better adaptability to changing environmental circumstances, greater abiotic stress tolerance, delayed senescence, and more biomass production (Miglani et al., 2021). To further improve yield through photosynthesis, more efforts should be invested in crops' SG traits in addition to the rate of photosynthesis per unit organ area. SG crops have higher green leaf areas and photosynthetic activity under stress conditions such as drought and water shortages, which also prolongs nutrient and water uptake (Nehe et al., 2021). A high photosynthetic rate of shoots secures high root activity by supplying a sufficient amount of photosynthates to the roots. Conversely, high root activity maintains a high photosynthetic rate by supplying a sufficient amount of nutrients to shoots (Fu et al., 2009).

The growing body of research has shown that a significant increase in photosynthetic capacity and/or efficiency is required to enhance agricultural production, backed up by evidence that increases in photosynthesis have been positively linked to crop production capacity in the past. Therefore, crop breeding programs should continue to focus on increasing plant CO_2 assimilation efficiency, together with improving WUE and nutrient-use efficiency under multiple stress conditions (Morales et al., 2020). To take advantage of cutting-edge technologies to sustainably enhance agricultural production, future research will invest in the following aspects: (i) rapidly obtaining a large amount of plant phenotypic information through high-throughput plant phenotyping platforms to identify photosynthetic genes for drought resistance; (ii) continuing to explore the excellent allelic variants of photosynthetic genes to improve crop photosynthesis and drought resistance through molecular marker-assisted breeding; (iii) continuing to use genetic engineering technologies (including genome editing technologies) to improve photosynthetic pathways of crops and drought resistance; and (iv) using '-omics' analyses (such as genomics, transcriptomics, proteomics, metabolomics, ionomics, and hormonomics) to identify novel targets and networks underlying the photosynthesis machinery, leading to new ways of increasing photosynthesis efficiency, crop productivity, and quality (Baslam et al., 2020).

Further research into plants with delayed senescence (SG) traits is also required, as this scenario keeps the plant photosynthetic active for longer, resulting in a prolonged period of grain filling (Morales et al., 2020). It is also important to focus on crop improvement by modifying site-specific photosynthesis, an ability of the same plant to display distinct photosynthetic characteristics at different locations (Dehigaspitiya et al., 2019). Genetic improvement in crop photosynthesis is particularly important for overall production of essential C_3 food crops such as rice and wheat (Dehigaspitiya and Seneweera, 2021). Speculatively, to increase crop productivity through enhanced photosynthesis, multidimensional strategies need to be developed, including addressing photosynthesis at various levels, from the molecular level to the entire canopy and ears in cereals. Additionally, root system architecture and its interaction with the rhizosphere should also be considered, due to the key role of roots in acquiring water and nutrients and regulating ABA signaling to the shoot (Ober et al., 2021). An interdisciplinary research strategy is urgently required for photosynthesis and crop yield improvements (Araus et al., 2021).

14 Where to look for further information

The following books help readers understand the basics of plant stress physiology, plant responses to drought stress, molecular breeding for crop abiotic stress tolerance, and plant breeding for drought environments.

1. *Plant Stress Physiology* (2021). Edited by Akbar Hossain. United Kingdom, IntechOpen.
2. The Response of Photosynthesis to Soil Water Stress (2012). Flexas, J., Gallé, A., Galmés, J., Ribas-Carbo, M. and Medrano, H. In: Aroca, R. (eds) *Plant Responses to Drought Stress*. Springer, Berlin, Heidelberg.
3. *Molecular Breeding for Rice Abiotic Stress Tolerance and Nutritional Quality* (2021). Edited by Mohammad Anwar Hossain, Lutful Hassan, Khandakar Md. Iftekharuddaula, Arvind Kumar and Robert Henry.
4. *Plant breeding for water-limited environments* (2010) by Abraham Blum. Springer Science & Business Media.

To find out what is being done or has been done to increase photosynthetic efficiency in recent years, you can further visit the following projects and websites:

1. The C_4 rice project, https://C4rice.com/.
2. The '3to4: Converting C_3 to C_4 photosynthesis for sustainable agriculture' project, https://cordis.europa.eu/project/id/289582.
3. Realizing increased photosynthetic efficiency (RIPE), http://ripe.illinois.edu/.
4. ARC Centre of Excellence for Translational Photosynthesis, https://photosynthesis.org.au/.

For phenotypic data on plants under drought conditions, please visit the following website:

1. The Planteome project, http://planteome.org/; https://github.com/Planteome/.
2. European Plant Phenotyping Network, (EPPN2020), https://eppn2020.plant-phenotyping.eu/.
3. www.plantstress.com.
4. The leaf senescence database (LSD) 3.0, https://bigd.big.ac.cn/lsd/.

15 References

Abdelrahman, M., El-Sayed, M., Jogaiah, S., Burritt, D. J. and Tran, L. P. (2017). The "STAY-GREEN" trait and phytohormone signaling networks in plants under heat stress, *Plant Cell Rep.* 36(7), 1009–1025.

Ahmad, N., Zaidi, S. S. and Mansoor, S. (2020). Alternative routes to improving photosynthesis in field crops, *Trends Plant Sci.* 25(10), 958–960.

Almeida, G. D., Nair, S., Borém, A., Cairns, J., Trachsel, S., Ribaut, J. M., Bänziger, M., Prasanna, B. M., Crossa, J. and Babu, R. (2014). Molecular mapping across three populations reveals a QTL hotspot region on chromosome 3 for secondary traits associated with drought tolerance in tropical maize, *Mol. Breed.* 34(2), 701–715.

Almeida, G. D. D. (2012). QTL mapping reveals constitutive and adaptive genomic regions for drought tolerance in tropical maize (*Zea mays* L.), PhD Thesis, Universidade Federal de Viçosa, Brasil.

Ambler, J. R., Morgan, P. W. and Jordan, W. R. (1992). Amounts of zeatin and zeatin riboside in xylem sap of senscent and nonsenescent sorghum, *Crop Sci.* 32(2), 411–419.

Araus, J. L., Amaro, T., Voltas, J., Nakkoul, H. and Nachit, M. M. (1998). Chlorophyll fluorescence as a selection criterion for grain yield in durum wheat under Mediterranean conditions, *Field Crops Res.* 55(3), 209–223.

Araus, J. L., Sanchez-Bragado, R. and Vicente, R. (2021). Improving crop yield and resilience through optimization of photosynthesis: Panacea or pipe dream?, *J. Exp. Bot.* 72(11), 3936–3955.

Arriagada, O., Mora, F., Quitral, Y. and Del Pozo, A. (2017). Identification of QTL underlying agronomic, morphological and physiological traits in barley under rainfed conditions using SNP markers, *Acta Sci. Agron.* 39(3), 321–329.

Ashraf, M. and Harris, P. J. C. (2013). Photosynthesis under stressful environments: An overview, *Photosynthetica* 51(2), 163–190.

Aslam, M., Maqbool, M. A. and Cengiz, R. (2015). Drought stress in maize (*Zea mays* L.): Effects, resistance mechanisms, global achievements and biological strategies for improvement. In: *Springer Briefs in Agriculture*, Springer, Cham.

Awad, W. (2015). Genome-wide association study and drought tolerance evaluation of a winter wheat association mapping panel, Doctoral Dissertation, Colorado State University.

Awika, H. O., Hays, D. B., Mullet, J. E., Rooney, W. L. and Weers, B. D. (2017). QTL mapping and loci dissection for leaf epicuticular wax load and canopy temperature depression and their association with QTL for staygreen in Sorghum bicolor under stress, *Euphytica* 213(9), 1–22.

Babu, R. C., Nguyen, B. D., Chamarerk, V., Shanmugasundaram, P., Chezhian, P., Jeyaprakash, P., Ganesh, S. K., Palchamy, A., Sadasivam, S., Sarkarung, S., Wade, L. J. and Nguyen, H. T. (2003). Genetic analysis of drought resistance in rice by molecular markers: Association between secondary traits and field performance, *Crop Sci.* 43(4), 1457–1469.

Bandyopadhyay, A., Datta, K., Zhang, J., Yang, W., Raychaudhuri, S., Miyao, M. and Datta, S. K. (2007). Enhanced photosynthesis rate in genetically engineered indica rice expressing PEPC gene cloned from maize, *Plant Sci.* 172(6), 1204–1209.

Barakat, M. N., Wahba, L. E. and Milad, S. I. (2013). Molecular mapping of QTLs for wheat flag leaf senescence under water-stress, *Biol. Plant.* 57(1), 79–84.

Baslam, M., Mitsui, T., Hodges, M., Priesack, E., Herritt, M. T., Aranjuelo, I. and Sanz-Sáez, Á. (2020). Photosynthesis in a changing global climate: Scaling up and scaling down in crops, *Front. Plant Sci.* 11, 882.

Basu, S., Ramegowda, V., Kumar, A. and Pereira, A. (2016). Plant adaptation to drought stress, *F1000Research* 5, Rev-1554.

Beavis, W. D., Smith, O. S., Grant, D. and Fincher, R. (1994). Identification of quantitative trait loci using a small sample of topcrossed and F_4 progeny from maize, *Crop Sci.* 34(4), 882–896.

Belícuas, P. R., Aguiar, A. M., Bento, D. A. V., Câmara, T. M. M. and de Souza Junior, C. L. (2014). Inheritance of the stay-green trait in tropical maize, *Euphytica* 198(2), 163–173.

Bernacchi, C. J., Leakey, A. D., Heady, L. E., Morgan, P. B., Dohleman, F. G., McGrath, J. M., Gillespie, K. M., Wittig, V. E., Rogers, A., Long, S. P. and Ort, D. R. (2006). Hourly and seasonal variation in photosynthesis and stomatal conductance of soybean grown at future CO_2 and ozone concentrations for 3 years under fully open-air field conditions, *Plant Cell Environ.* 29(11), 2077–2090.

Blum, A. (1996). Crop responses to drought and the interpretation of adaptation, *Plant Growth Regul.* 20(2), 135–148.

Blum, A. (2011). *Plant Breeding for Water-Limited Environments*, Springer Science & Business Media, New York, NY.

Blum, A. (2017). Osmotic adjustment is a prime drought stress adaptive engine in support of plant production, *Plant Cell Environ.* 40(1), 4–10.

Blum, A. and Tuberosa, R. (2018). Dehydration survival of crop plants and its measurement, *J. Exp. Bot.* 69(5), 975–981.

Bogard, M., Jourdan, M., Allard, V., Martre, P., Perretant, M. R., Ravel, C., Heumez, E., Orford, S., Snape, J., Griffiths, S., Gaju, O., Foulkes, J. and Le Gouis, J. (2011). Anthesis date mainly explained correlations between post-anthesis leaf senescence, grain yield, and grain protein concentration in a winter wheat population segregating for flowering time QTLs, *J. Exp. Bot.* 62(10), 3621–3636.

Borrell, A. K., Hammer, G. L. and Douglas, A. C. L. (2000a). Does maintaining green leaf area in sorghum improve yield under drought? I. Leaf growth and senescence, *Crop Sci.* 40(4), 1026–1037.

Borrell, A. K., Hammer, G. L. and Henzell, R. G. (2000b). Does maintaining green leaf area in sorghum improve yield under drought? II. Dry matter production and yield, *Crop Sci.* 40(4), 1037–1048.

Borrell, A. K., Mullet, J. E., George-Jaeggli, B., van Oosterom, E. J., Hammer, G. L., Klein, P. E. and Jordan, D. R. (2014). Drought adaptation of stay-green sorghum is associated with canopy development, leaf anatomy, root growth, and water uptake, *J. Exp. Bot.* 65(21), 6251–6263.

Caicedo, M., Munaiz, E. D., Malvar, R. A., Jiménez, J. C. and Ordas, B. (2021). Precision mapping of a maize MAGIC population identified a candidate gene for the senescence-associated physiological traits, *Front. Genet.* 12, 716821.

Calderini, D. F., Dreccer, M. F. and Slafer, G. A. (1995). Genetic improvement in wheat yield and associated traits. A re-examination of previous results and the latest trends, *Plant Breed.* 114(2), 108–112.

Carmo-Silva, E., Andralojc, P. J., Scales, J. C., Driever, S. M., Mead, A., Lawson, T., Raines, C. A. and Parry, M. A. J. (2017). Phenotyping of field-grown wheat in the UK highlights contribution of light response of photosynthesis and flag leaf longevity to grain yield, *J. Exp. Bot.* 68(13), 3473–3486.

Castiglioni, P., Warner, D., Bensen, R. J., Anstrom, D. C., Harrison, J., Stoecker, M., Abad, M., Kumar, G., Salvador, S., D'Ordine, R., Navarro, S., Back, S., Fernandes, M., Targolli, J., Dasgupta, S., Bonin, C., Luethy, M. H. and Heard, J. E. (2008). Bacterial RNA chaperones confer abiotic stress tolerance in plants and improved grain yield in maize under water-limited conditions, *Plant Physiol.* 147(2), 446–455.

Chang, C., Lu, J., Zhang, H. P., Ma, C. X. and Sun, G. (2015). Copy number variation of cytokinin oxidase gene *Tackx4* associated with grain weight and chlorophyll content of flag leaf in common wheat, *PLoS ONE* 10(12), e0145970.

Chatterjee, J., Coe, R. A., Acebron, K., Thakur, V., Yennamalli, R. M., Danila, F., Lin, H. C., Balahadia, C. P., Bagunu, E., Padhma, P. P. O. S., Bala, S., Yin, X., Rizal, G., Dionora, J., Furbank, R. T., von Caemmerer, S. and Quick, W. P. (2021). Identification of a low CO_2 responsive mutant from chemical mutagenesis of Setaria viridis shows that reduced carbonic anhydrase severely limits C_4 photosynthesis, *J. Exp. Bot.* 72(8), 3122–3136.

Chen, G., Wu, C., He, L., Qiu, Z., Zhang, S., Zhang, Y., Guo, L., Zeng, D., Hu, J., Ren, D., Qian, Q. and Zhu, L. (2018). Knocking out the gene *RLS1* induces hypersensitivity to oxidative stress and premature leaf senescence in rice, *Int. J. Mol. Sci.* 19(10), 2853.

Chen, J., Liang, Y., Hu, X., Wang, X., Tan, F., Zhang, H., Ren, Z. and Luo, P. (2010). Physiological characterization of 'stay green' wheat cultivars during the grain filling stage under field growing conditions, *Acta Physiol. Plant.* 32(5), 875–882.

Chen, P., Li, X., Huo, K., Wei, X., Dai, C. and Lv, C. (2014). Promotion of photosynthesis in transgenic rice over-expressing of maize C_4 phosphoenolpyruvate carboxylase gene by nitric oxide donors, *J. Plant Physiol.* 171(6), 458–466.

Christopher, J. T., Manschadi, A. M., Hammer, G. L. and Borrell, A. K. (2008). Developmental and physiological traits associated with high yield and stay-green phenotype in wheat, *Aust. J. Agric. Res.* 59(4), 354–364.

Crasta, O. R., Xu, W. W., Rosenow, D. T., Mullet, J. and Nguyen, H. T. (1999). Mapping of post-flowering drought resistance traits in grain sorghum: Association between QTLs influencing premature senescence and maturity, *Mol. Gen. Genet.* 262(3), 579–588.

Cui, H. (2021). Challenges and approaches to crop improvement through C_3-to-C_4 engineering, *Front. Plant Sci.* 12, 715391.

Cui, Y., Wang, M., Zhou, H., Li, M., Huang, L., Yin, X., Zhao, G., Lin, F., Xia, X. and Xu, G. (2016). OsSGL, a novel DUF1645 domain-containing protein, confers enhanced drought tolerance in transgenic rice and *Arabidopsis*, *Front. Plant Sci.* 7, 2001.

Czyczyło-Mysza, I., Marcińska, I., Skrzypek, E., Chrupek, M., Grzesiak, S., Hura, T., Stojałowski, S., Myśków, B., Milczarski, P. and Quarrie, S. (2011). Mapping QTLs for yield components and chlorophyll a fluorescence parameters in wheat under three levels of water availability, *Plant Genet. Res.* 9(2), 291–295.

Czyczyło-Mysza, I., Tyrka, M., Marcińska, I., Skrzypek, E., Karbarz, M., Dziurka, M., Hura, T., Dziurka, K. and Quarrie, S. A. (2013). Quantitative trait loci for leaf chlorophyll fluorescence parameters, chlorophyll and carotenoid contents in relation to biomass and yield in bread wheat and their chromosome deletion bin assignments, *Mol. Breed.* 32(1), 189–210.

Dehigaspitiya, P., Milham, P., Ash, G. J., Arun-Chinnappa, K., Gamage, D., Martin, A., Nagasaka, S. and Seneweera, S. (2019). Exploring natural variation of photosynthesis in a site-specific manner: Evolution, progress, and prospects, *Planta* 250(4), 1033–1050.

Dehigaspitiya, P. and Seneweera, S. (2021). Manipulation of photosynthesis to increase rice yield potential. In: Hossain, M. A., Hassan, L., Ifterkharuddaula, K. M., Kumar, A. and Henry, R. (Eds.), *Molecular Breeding for Rice Abiotic Stress Tolerance and Nutritional Quality*, John Wiley & Sons, Ltd, pp. 274–286.

Diab, A. A., Kantety, R. V., Ozturk, N. Z., Benscher, D., Nachit, M. M. and Sorrells, M. E. (2008). Drought-inducible genes and differentially expressed sequence tags associated with components of drought tolerance in durum wheat, *Sci. Res. Essay* 3, 9-26.

Ding, Z. S., Sun, X. F., Huang, S. H., Zhou, B. Y. and Zhao, M. (2015). Response of photosynthesis to short-term drought stress in rice seedlings overexpressing C_4 phosphoenolpyruvate carboxylase from maize and millet, *Photosynthetica* 53(4), 481–488.

Doubnerová, V. and Ryšlavá, H. (2011). What can enzymes of C_4 photosynthesis do for C_3 plants under stress?, *Plant Sci.* 180(4), 575–583.

Dudhate, A., Shinde, H., Tsugama, D., Liu, S. and Takano, T. (2018). Transcriptomic analysis reveals the differentially expressed genes and pathways involved in drought tolerance in pearl millet [*Pennisetum glaucum* (L.) R. Br], *PLoS ONE* 13(4), e0195908.

Edwards, G. and Walker, D. (1983). *C_3, C_4: Mechanisms, and Cellular and Environmental Regulation, of Photosynthesis*, University of California Press, CA.

Edwards, G. E., Franceschi, V. R. and Voznesenskaya, E. V. (2004). Single-cell C_4 photosynthesis versus the dual-cell (Kranz) paradigm, *Annu. Rev. Plant Biol.* 55, 173–196.

Emebiri, L. C. (2013). QTL dissection of the loss of green colour during post-anthesis grain maturation in two-rowed barley, *Theor. Appl. Genet.* 126(7), 1873–1884.

Ermakova, M., Arrivault, S., Giuliani, R., Danila, F., Alonso-Cantabrana, H., Vlad, D., Ishihara, H., Feil, R., Guenther, M., Borghi, G. L., Covshoff, S., Ludwig, M., Cousins, A. B., Langdale, J. A., Kelly, S., Lunn, J. E., Stitt, M., von Caemmerer, S. and Furbank, R. T. (2021). Installation of C_4 photosynthetic pathway enzymes in rice using a single construct, *Plant Biotechnol. J.* 19(3), 575–588.

Evans, L. T. (1993). *Crop Evolution, Adaptation and Yield*, Cambridge University Press, Cambridge, UK.

Exposito-Rodriguez, M., Laissue, P. P., Yvon-Durocher, G., Smirnoff, N. and Mullineaux, P. M. (2017). Photosynthesis-dependent H_2O_2 transfer from chloroplasts to nuclei provides a high-light signalling mechanism, *Nat. Commun.* 8(1), 49.

Fischer, R. A. T. and Edmeades, G. O. (2010). Breeding and cereal yield progress, *Crop Sci.* 50, S85.

Flexas, J., Gallé, A., Galmés, J., Ribas-Carbo, M. and Medrano, H. (2012). The response of photosynthesis to soil water stress. In: Aroca, R. (Ed.), *Plant Responses to Drought Stress*, Springer, Berlin, Heidelberg, pp. 129–144.

Foulkes, M. J., Snape, J. W., Shearman, V. J., Reynolds, M. P., Gaju, O. and Sylvester-bradley, R. (2007). Paper presented at international workshop on increasing wheat yield potential, Cimmyt, Obregon, Mexico, 20–24 March 2006 genetic progress in yield potential in wheat: Recent advances and future prospects, *J. Agric. Sci.* 145(1), 17–29.

Freschi, L. and Mercier, H. (2012). Connecting environmental stimuli and crassulacean acid metabolism expression: Phytohormones and other signaling molecules, *Prog. Bot.* 73, 231–255.

Fu, J. D., Yan, Y. F., Kim, M. Y., Lee, S. H. and Lee, B. W. (2011). Population-specific quantitative trait loci mapping for functional stay-green trait in rice (*Oryza sativa* L.), *Genome* 54(3), 235–243.

Fu, J. D., Yan, Y. F. and Lee, B. W. (2009). Physiological characteristics of a functional stay-green rice "SNU-SG1" during grain-filling period, *J. Crop Sci. Biotechnol.* 12(1), 47–52.

Fukuda, A., Kondo, K., Ikka, T., Takai, T., Tanabata, T. and Yamamoto, T. (2018). A novel QTL associated with rice canopy temperature difference affects stomatal conductance and leaf photosynthesis, *Breed. Sci.* 68(3), 305–315.

Furbank, R. T. and Taylor, W. C. (1995). Regulation of photosynthesis in C_3 and C_4 plants: A molecular approach, *Plant Cell* 7(7), 797–807.

Gago, J., Carriquí, M., Nadal, M., Clemente-Moreno, M. J., Coopman, R. E., Fernie, A. R. and Flexas, J. (2019). Photosynthesis optimized across land plant phylogeny, *Trends Plant Sci.* 24(10), 947–958.

Gale, M. D., Youssefian, S. and Russell, G. E. (1985). Dwarfing genes in wheat. In: Russell, G. E. (Ed.), *Progress in Plant Breeding*, Butterworth-Heinemann, London, UK, pp. 1–35.

Garg, A. K., Kim, J. K., Owens, T. G., Ranwala, A. P., Choi, Y. D., Kochian, L. V. and Wu, R. J. (2002). Trehalose accumulation in rice plants confers high tolerance levels to different abiotic stresses, *Proc. Natl. Acad. Sci. U. S. A.* 99(25), 15898–15903.

Gollan, P. J., Tikkanen, M. and Aro, E. M. (2015). Photosynthetic light reactions: Integral to chloroplast retrograde signalling, *Curr. Opin. Plant Biol.* 27, 180–191.

Gong, Y.-H., Zhang, J., Gao, J.-F., Lu, J.-Y. and Wang, J.-R. (2005). Slow export of photoassimilate from stay-green leaves during late grain-filling stage in hybrid winter wheat (*Triticum aestivum* L.), *J. Agron. Crop Sci.* 191(4), 292–299.

Gous, P. W., Hickey, L., Christopher, J. T., Franckowiak, J. and Fox, G. P. (2016). Discovery of QTL for stay-green and heat-stress in barley (*Hordeum vulgare*) grown under simulated abiotic stress conditions, *Euphytica* 207(2), 305–317.

Graziani, M., Maccaferri, M., Royo, C., Salvatorelli, F. and Tuberosa, R. (2014). QTL dissection of yield components and morpho-physiological traits in a durum wheat elite population tested in contrasting thermo-pluviometric conditions, *Crop Pasture Sci.* 65(1), 80–95.

Gregersen, N., Bross, P., Vang, S. and Christensen, J. H. (2006). Protein misfolding and human disease, *Annu. Rev. Genomics Hum. Genet.* 7, 103–124.

Gregersen, P. L., Culetic, A., Boschian, L. and Krupinska, K. (2013). Plantsenescence and crop productivity, *Plant Mol. Biol.* 82(6), 603–622.

Gu, J., Yin, X., Struik, P. C., Stomph, T. J. and Wang, H. (2012). Using chromosome introgression lines to map quantitative trait loci for photosynthesis parameters in rice (*Oryza sativa* L.) leaves under drought and well-watered field conditions, *J. Exp. Bot.* 63(1), 455–469.

Gu, J. F., Qiu, M. and Yang, J. C. (2013). Enhanced tolerance to drought in transgenic rice plants overexpressing C_4 photosynthesis enzymes, *Crop J.* 1(2), 105–114.

Gu, L., Han, J., Wood, J. D., Chang, C. Y. and Sun, Y. (2019). Sun-induced Chl fluorescence and its importance for biophysical modeling of photosynthesis based on light reactions, *New Phytol.* 223(3), 1179–1191.

Guddimalli, R., Somanaboina, A. K., Palle, S. R., Edupuganti, S., Kummari, D., Palakolanu, S. R., Naravula, J., Gandra, J., Qureshi, I. A., Marka, N., Polavarapu, R. and Kavi Kishor, P. B. (2021). Overexpression of RNA-binding bacterial chaperones in rice leads to stay-green phenotype, improved yield and tolerance to salt and drought stresses, *Physiol. Plant.* 173(4), 1351–1368. DOI: 10.1111/ppl.13369.

Gudys, K., Guzy-Wrobelska, J., Janiak, A., Dziurka, M. A., Ostrowska, A., Hura, K., Jurczyk, B., Żmuda, K., Grzybkowska, D., Śróbka, J., Urban, W., Biesaga-Koscielniak, J., Filek, M., Koscielniak, J., Mikołajczak, K., Ogrodowicz, P., Krystkowiak, K., Kuczyńska, A., Krajewski, P. and Szarejko, I. (2018). Prioritization of candidate genes in QTL regions for physiological and biochemical traits underlying drought response in barley (*Hordeum vulgare* L.), *Front. Plant Sci.* 9, 769.

Guo, P., Baum, M., Varshney, R. K., Graner, A., Grando, S. and Ceccarelli, S. (2008). QTLs for chlorophyll and chlorophyll fluorescence parameters in barley under post-flowering drought, *Euphytica* 163(2), 203–214.

Gupta, N. and Thind, S. (2015). Improving photosynthetic performance of bread wheat under field drought stress by foliar applied glycine betaine, *J. Agr. Sci. Tech.-Iran.* 17(1), 75–86.

Gürel, F., Öztürk, Z. N., Uçarlı, C. and Rosellini, D. (2016). Barley genes as tools to confer abiotic stress tolerance in crops, *Front. Plant Sci.* 7, 1137.

Harris, K., Subudhi, P. K., Borrell, A., Jordan, D., Rosenow, D., Nguyen, H., Klein, P., Klein, R. and Mullet, J. (2007). Sorghum stay-green QTL individually reduce post-flowering drought-induced leaf senescence, *J. Exp. Bot.* 58(2), 327–338.

Hassan, M. A., Yang, M., Rasheed, A., Tian, X., Reynolds, M., Xia, X., Xiao, Y. and He, Z. (2021). Quantifying senescence in bread wheat using multispectral imaging from an unmanned aerial vehicle and QTL mapping, *Plant Physiol.* 187(4), 2623–2636.

Häusler, R. E., Hirsch, H. J., Kreuzaler, F. and Peterhänsel, C. (2002). Overexpression of C_4-cycle enzymes in transgenic C_3 plants: A biotechnological approach to improve C_3-photosynthesis, *J. Exp. Bot.* 53(369), 591–607.

Haussmann, B. I., Mahalakshmi, V., Reddy, B. V., Seetharama, N., Hash, C. T. and Geiger, H. H. (2002). QTL mapping of stay-green in two sorghum recombinant inbred populations, *Theor. Appl. Genet.* 106(1), 133–142.

He, G. H., Xu, J. Y., Wang, Y. X., Liu, J. M., Li, P. S., Chen, M., Ma, Y. Z. and Xu, Z. S. (2016). Drought-responsive WRKY transcription factor genes *TaWRKY1* and *TaWRKY33* from wheat confer drought and/or heat resistance in Arabidopsis, *BMC Plant Biol.* 16(1), 1–16.

He, Y. F., Xie, Y. F., Li, X. and Yang, J. (2020). Drought tolerance of transgenic rice overexpressing maize *C_4-PEPC* gene related to increased anthocyanin synthesis regulated by sucrose and calcium, *Biol. Plant.* 64, 136–149.

Heyneke, E. and Fernie, A. R. (2018). Metabolic regulation of photosynthesis, *Biochem. Soc. Trans.* 46(2), 321–328.

Honsdorf, N., March, T. J., Hecht, A., Eglinton, J. and Pillen, K. (2014). Evaluation of juvenile drought stress tolerance and genotyping by sequencing with wild barley introgression lines, *Mol. Breed.* 34(3), 1475–1495.

Hu, H. and Xiong, L. (2014). Genetic engineering and breeding of drought-resistant crops, *Annu. Rev. Plant Biol.* 65, 715–741.

Huo, K., Li, X., He, Y. F., Wei, X. D., Lu, W., Zhao, C. F. and Wang, C. L. (2017). Exogenous ATP enhance signal response of suspension cells of transgenic rice (*Oryza sativa* L.) expressing maize *C4-pepc* encoded phosphoenolpyruvate carboxylase under PEG treatment, *Plant Growth Regul.* 82(1), 55–67.

Hura, T., Hura, K., Grzesiak, M. and Rzepka, A. (2007). Effect of long-term drought stress on leaf gas exchange and fluorescence parameters in C_3 and C_4 plants, *Acta Physiol. Plant.* 29(2), 103–113.

Ilyas, M., Ilyas, N., Arshad, M., Kazi, A. G., Kazi, A. M. and Waheed, A. (2014). QTL mapping of wheat doubled haploids for chlorophyll content and chlorophyll fluorescence kinetics under drought stress imposed at an thesis stage, *Pak. J. Bot.* 46(5), 1889–1897.

Ishimaru, K., Yano, M., Aoki, N., Ono, K., Hirose, T., Lin, S. Y., Monna, L., Sasaki, T. and Ohsugi, R. (2001). Toward the mapping of physiological and agronomic characters on a rice function map: QTL analysis and comparison between QTLs and expressed sequence tags, *Theor. Appl. Genet.* 102(6-7), 793–800.

Jabbari, M., Fakheri, B. A., Aghnoum, R., Darvishzadeh, R., Mahdi Nezhad, N., Ataei, R., Koochakpour, Z. and Razi, M. (2021). Association analysis of physiological traits in spring barley (*Hordeum vulgare* L.) under water-deficit conditions, *Food Sci. Nutr.* 9(3), 1761–1779.

Jabbari, M., Fakheri, B. A., Aghnoum, R., Nezhad, N. M., Ataei, R. and Koochakpour, Z. (2019). Association mapping of morphological and physiological traits of flag leaf related to drought tolerance in barley, *J. Neotrop. Agri.* 6(2), 7–18.

Jackson, R. D., Idso, S. B., Reginato, R. J. and Pinter, P. J. (1981). Canopy temperature as a crop water stress indicator, *Water Resour. Res.* 17(4), 1133–1138.

Jaegglia, B. G., Mortlockb, M. Y. and Borrell, A. K. (2017). Bigger is not always better: Reducing leaf area helps stay-green sorghum use soil water more slowly, *Environ. Exp. Bot.* 138, 119–129.

Jia, T., Jing-jing, L., Wang, L., Cao, Y., Ma, J., Wang, H., Zhang, D. and Li, H. (2020a). Evaluation of drought tolerance in *ZmVPP1*-overexpressing transgenic inbred maize lines and their hybrids, *J. Integr. Agr.* 19(9), 2177–2187.

Jia, Y., Xiao, W., Ye, Y., Wang, X., Liu, X., Wang, G., Li, G. and Wang, Y. (2020b). Response of photosynthetic performance to drought duration and re-watering in maize, *Agronomy* 10(4), 533.

Jiang, D., Wang, F., Zhang, H., Gao, W., Tong, X., Lv, C. and Chen, G. (2021). Formation of proto-Kranz in C_3 rice induced by spike-stalk injection method, *Int. J. Mol. Sci.* 22(9), 4305.

Jiang, G., Zeng, J. and He, Y. (2014). Analysis of quantitative trait loci affecting chlorophyll content of rice leaves in a double haploid population and two backcross populations, *Gene* 536(2), 287–295.

Jiang, G. H., He, Y. Q., Xu, C. G., Li, X. H. and Zhang, Q. (2004). The genetic basis of stay-green in rice analyzed in a population of doubled haploid lines derived from an indica by japonica cross, *Theor. Appl. Genet.* 108(4), 688–698.

Jiao, D., Huang, X., Li, X., Chi, W., Kuang, T., Zhang, Q., Ku, M. S. and Cho, D. (2002). Photosynthetic characteristics and tolerance to photo-oxidation of transgenic rice expressing C_4 photosynthesis enzymes, *Photosynth. Res.* 72(1), 85–93.

Jones, H. G., Serraj, R., Loveys, B. R., Xiong, L., Wheaton, A. and Price, A. H. (2009). Thermal infrared imaging of crop canopies for the remote diagnosis and quantification of plant responses to water stress in the field, *Funct. Plant Biol.* 36(11), 978–989.

Joynson, R., Molero, G., Coombes, B., Gardiner, L. J., Rivera-Amado, C., Piñera-Chávez, F. J., Evans, J. R., Furbank, R. T., Reynolds, M. P. and Hall, A. (2021). Uncovering candidate genes involved in photosynthetic capacity using unexplored genetic variation in Spring Wheat, *Plant Biotechnol. J.* 19(8), 1537–1552.

Kaler, A. S., Ray, J. D., Schapaugh, W. T., Asebedo, A. R., King, C. A., Gbur, E. E. and Purcell, L. C. (2018). Association mapping identifies loci for canopy temperature under drought in diverse soybean genotypes, *Euphytica* 214(8), 1–18.

Kamal, N. M., Gorafi, Y. S. A., Abdelrahman, M., Abdellatef, E. and Tsujimoto, H. (2019). Stay-green trait: A prospective approach for yield potential, and drought and heat stress adaptation in globally important cereals, *Int. J. Mol. Sci.* 20(23), 5837.

Kamal, N. M., Gorafi, Y. S. A., Tsujimoto, H. and Ghanim, A. M. A. (2018). Stay-green QTLs response in adaptation to post-flowering drought depends on the drought severity, *BioMed Res. Int.* 2018, 7082095.

Karaba, A., Dixit, S., Greco, R., Aharoni, A., Trijatmiko, K. R., Marsch-Martinez, N., Krishnan, A., Nataraja, K. N., Udayakumar, M. and Pereira, A. (2007). Improvement of water use efficiency in rice by expression of *HARDY*, an Arabidopsis drought and salt tolerance gene, *Proc. Natl. Acad. Sci. U. S. A.* 104(39), 15270–15275.

Kassahun, B., Bidinger, F. R., Hash, C. T. and Kuruvinashetti, M. S. (2010). Stay-green expression in early generation sorghum [*Sorghum bicolor* (L.) Moench] QTL introgression lines, *Euphytica* 172(3), 351–362.

Kebede, H., Subudhi, P. K., Rosenow, D. T. and Nguyen, H. T. (2001). Quantitative trait loci influencing drought tolerance in grain sorghum (*Sorghum bicolor* L. Moench), *Theor. Appl. Genet.* 103(2–3), 266–276.

Khanal, R., Navabi, A. and Lukens, L. (2015). Linkage map construction and quantitative trait loci (QTL) mapping using intermated vs. selfed recombinant inbred maize line (*Zea mays* L.), *Can. J. Plant Sci.* 95(6), 1133–1144.

Kiranmayee, K. N. S. U., Hash, C. T., Sivasubramani, S., Ramu, P., Amindala, B. P., Rathore, A., Kishor, P. B. K., Gupta, R. and Deshpande, S. P. (2020). Fine-mapping of sorghum stay-green QTL on chromosome10 revealed genes associated with delayed senescence, *Genes* 11(9), 1026.

Kopriva, S. and Weber, A. P. M. (2021). Genetic encoding of complex traits, *J. Exp. Bot.* 72(1), 1–3.

Ku, M. S., Kano-Murakami, Y. and Matsuoka, M. (1996). Evolution and expression of C_4 photosynthesis genes, *Plant Physiol.* 111(4), 949–957.

Ku, M. S. B., Ranade, U., Hsu, T.-P., Cho, D., Li, X., Jiao, D. M., Ehleringer, J., Miyao, M. and Matsuoka, M. (2000). Photosynthetic performance of transgenic rice plants overexpressing maize C4 photosynthesis enzymes. In: Sheehy, J. E., Mitchell, P. L. and Hardy, B. (Eds.), *Redesigning Rice Photosynthesis to Increase Yield*. Proceedings of the Workshop on the Quest to Reduce Hunger: Redesigning Rice Photosynthesis, 30 November–3 December 1999, Los Bafios. *Philippines: International Rice Research Institute and Amsterdam (The Netherlands), Studies in Plant Science*, Elsevier, pp. 193–204.

Kumar, S., Sehgal, S. K., Kumar, U., Prasad, P. V. V., Joshi, A. K. and Gill, B. S. (2012). Genomic characterization of drought tolerance-related traits in spring wheat, *Euphytica* 186(1), 265–276.

Kumar, U., Joshi, A. K., Kumari, M., Paliwal, R., Kumar, S. and Röder, M. S. (2010). Identification of QTLs for stay green trait in wheat (*Triticum aestivum* L.) in the 'Chirya 3'×'Sonalika'population', *Euphytica* 174(3), 437–445.

Kusaba, M., Tanaka, A. and Tanaka, R. (2013). Stay-green plants: What do they tell us about the molecular mechanism of leaf senescence, *Photosynth. Res.* 117(1–3), 221–234.

Lafitte, H. R., Price, A. H. and Courtois, B. (2004). Yield response to water deficit in an upland rice mapping population: Associations among traits and genetic markers, *Theor. Appl. Genet.* 109(6), 1237–1246.

Lawlor, D. W. (2001). *Photosynthesis* (3rd edn.), Bios Scientific Publishers Ltd, Oxford, UK.

Lebreton, C., Lazić-Jančić, V., Steed, A., Pekić, S. and Quarrie, S. A. (1995). Identification of QTL for drought responses in maize and their use in testing causal relationships between traits, *J. Exp. Bot.* 46(7), 853–865.

Lee, E. A. and Tollenaar, M. (2007). Physiological basis of successful breeding strategies for maize grain yield, *Crop Sci.* 47, S-202.

Li, H., Han, X., Liu, X., Zhou, M., Ren, W., Zhao, B., Ju, C., Liu, Y. and Zhao, J. (2019). A leucine-rich repeat-receptor-like kinase gene *SbER2-1* from sorghum (*Sorghum bicolor* L.) confers drought tolerance in maize, *BMC Genomics* 20(1), 737.

Li, H., Lin, F., Wang, G., Jing, R., Zheng, Q., Li, B. and Li, Z. (2012). Quantitative trait loci mapping of dark-induced senescence in winter wheat (*Triticum aestivum*), *J. Integr. Plant Biol.* 54(1), 33-44.

Li, Y., Hui, D., Sun, Y., Wang, Y., Wu, Z., Wang, C. and Zhao, J. (2021). Boosting thermo-photocatalytic CO_2 conversion activity by using photosynthesis-inspired electron-proton-transfer mediators, *Nat. Commun.* 12(1), 123.

Lian, L., Wang, X., Zhu, Y., He, W., Cai, Q., Xie, H., Zhang, M. and Zhang, J. (2014). Physiological and photosynthetic characteristics of indica Hang2 expressing the sugarcane *PEPC* gene, *Mol. Biol. Rep.* 41(4), 2189–2197.

Liang, Y., Zhang, K., Zhao, L., Liu, B., Meng, Q., Tian, J. and Zhao, S. (2010). Identification of chromosome regions conferring dry matter accumulation and photosynthesis in wheat (*Triticum aestivum* L.), *Euphytica* 171(1), 145–156.

Liu, L., Sun, G., Ren, X., Li, C. and Sun, D. (2015). Identification of QTL underlying physiological and morphological traits of flag leaf in barley, *BMC Genet.* 16(1), 29.

Liu, Y., Subhash, C., Yan, J., Song, C., Zhao, J. and Li, J. (2011). Maize leaf temperature responses to drought: Thermal imaging and quantitative trait loci (QTL) mapping, *Environ. Exp. Bot.* 71(2), 158–165.

Long, S. P., Ainsworth, E. A., Leakey, A. D., Nösberger, J. and Ort, D. R. (2006a). Food for thought: Lower-than-expected crop yield stimulation with rising CO_2 concentrations, *Science* 312(5782), 1918–1921.

Long, S. P., Zhu, X. G., Naidu, S. L. and Ort, D. R. (2006b). Can improvement in photosynthesis increase crop yields?, *Plant Cell Environ.* 29(3), 315–330.

Long, S. P., Marshall-Colon, A. and Zhu, X. G. (2015). Meeting the global food demand of the future by engineering crop photosynthesis and yield potential, *Cell* 161(1), 56–66.

López-Calcagno, P. E., Brown, K. L., Simkin, A. J., Fisk, S. J., Vialet-Chabrand, S., Lawson, T. and Raines, C. A. (2020). Stimulating photosynthetic processes increases productivity and water-use efficiency in the field, *Nat. Plants* 6(8), 1054–1063.

Luche, H. D. S., Silva, J. A. G. D., Maia, L. C. D. and Oliveira, A. C. D. (2015). Stay-green: A potentiality in plant breeding, *Cienc. Rural* 45(10), 1755–1760.

Ma, X., Xia, H., Liu, Y., Wei, H., Zheng, X., Song, C., Chen, L., Liu, H. and Luo, L. (2016). Transcriptomic and metabolomic studies disclose key metabolism pathways contributing to well-maintained photosynthesis under the drought and the consequent drought-tolerance in rice, *Front. Plant Sci.* 7, 1886.

Makhtoum, S., Sabouri, H., Gholizadeh, A., et al. (2021). Quantitative genes controlling chlorophyll fluorescence attributes in barley (*Hordeum vulgare* L.), *J. Genet. Resour.* 7(1), 72-86.

Malik, S. and Malik, T. A. (2015). Genetic mapping of potential QTLs associated with drought tolerance in wheat, *J. Anim. Plant Sci.* 25(4), 1032–1040.

Maqsood, R. H., Amjid, M. W., Saleem, M. A., Shabbir, G. and Khaliq, I. (2017). Identification of genomic regions conferring drought tolerance in bread wheat using ISSR markers, *Pak. J. Bot.* 49(5), 1821–1827.

Massacci, A., Nabiev, S. M., Pietrosanti, L., Nematov, S. K., Chernikova, T. N., Thor, K. and Leipner, J. (2008). Response of the photosynthetic apparatus of cotton (*Gossypium hirsutum*) to the onset of drought stress under field conditions studied by gas-exchange analysis and chlorophyll fluorescence imaging, *Plant Physiol. Biochem.* 46(2), 189–195.

Medrano, H., Escalona, J. M., Bota, J., Gulías, J. and Flexas, J. (2002). Regulation of photosynthesis of C_3 plants in response to progressive drought: Stomatal conductance as a reference parameter, *Ann. Bot.* 89(7), 895–905.

Melkozernov, A. N. and Blankenship, R. E. (2006). Photosynthetic functions of chlorophylls. In: Grimm, B., Porra, R. J., Rüdiger, W. and Scheer, H. (Eds.), *Chlorophylls and Bacteriochlorophylls*, Springer, Dordrecht, pp. 397–412.

Messmer, R., Fracheboud, Y., Bänziger, M., Stamp, P. and Ribaut, J. (2011). Drought stress and tropical maize: QTLs for leaf greenness, plant senescence, and root capacitance, *Field Crops Res.* 124(1), 93–103.

Michael Gomez, S., Manikanda Boopathi, N., Satheesh Kumar, S., Ramasubramanian, T., Chengsong, Z., Jeyaprakash, P., Senthil, A. and Chandra Babu, R. (2010). Molecular mapping and location of QTLs for drought-resistance traits in indica rice (*Oryza sativa* L.) lines adapted to target environments, *Acta Physiol. Plant.* 32(2), 355–364.

Miglani, G. S., Kaur, R., Sharma, P. and Gupta, N. (2021). Leveraging photosynthetic efficiency toward improving crop yields, *J. Crop Improv.* 35(3), 361–402.

Min, H., Chen, C., Wei, S., Shang, X., Sun, M., Xia, R., Liu, X., Hao, D., Chen, H. and Xie, Q. (2016). Identification of drought tolerant mechanisms in maize seedlings based on transcriptome analysis of recombination inbred lines, *Front. Plant Sci.* 7, 1080.

Mohammed, S., Huggins, T., Mason, E., Beecher, F., Chick, C., Sengodon, P., Paudel, A., Ibrahim, A., Tilley, M. and Hays, D. (2021). Mapping the genetic loci regulating leaf epicuticular wax, canopy temperature, and drought susceptibility index in *Triticum aestivum*, *Crop Sci.* 61(4), 2294–2305.

Mora, F., Quitral, Y. A., Matus, I., Russell, J., Waugh, R. and Del Pozo, A. (2016). SNP-based QTL mapping of 15 complex traits in barley under rain-fed and well-watered conditions by a mixed modeling approach, *Front. Plant Sci.* 7, 909.

Morales, F., Ancín, M., Fakhet, D., González-Torralba, J., Gámez, A. L., Seminario, A., Soba, D., Ben Mariem, S., Garriga, M. and Aranjuelo, I. (2020). Photosynthetic metabolism under stressful growth conditions as a bases for crop breeding and yield improvement, *Plants (Basel)* 9(1), 88.

Muhammad, I., Shalmani, A., Ali, M., et al. (2021). Mechanisms regulating the dynamics of photosynthesis under abiotic stresses, *Front. Plant Sci.* 11, 2310.

Murata, N., Allakhverdiev, S. I. and Nishiyama, Y. (2012). The mechanism of photoinhibition in vivo: Re-evaluation of the roles of catalase, α-tocopherol, non-photochemical quenching, and electron transport, *Biochim. Biophys. Acta* 1817(8), 1127–1133.

Murchie, E. H., Kefauver, S., Araus, J. L., Muller, O., Rascher, U., Flood, P. J. and Lawson, T. (2018). Measuring the dynamic photosynthome, *Ann. Bot.* 122(2), 207–220.

Mwamahonje, A., Eleblu, J. S. Y., Ofori, K., Feyissa, T., Deshpande, S., Garcia-Oliveira, A. L., Bohar, R., Kigoni, M. and Tongoona, P. (2021). Introgression of QTLs for drought tolerance into farmers' preferred sorghum varieties, *Agriculture* 11(9), 883.

Nadew, D., Bejiga, T. and Teressa, T. (2021). Breeding of sorghum crop for resistance and tolerance to drought, *Int. Res. Plant Crop Sci.* 6(2), 188–195.

Nagahatenna, D. S. K., Parent, B., Edwards, E. J., Langridge, P. and Whitford, R. (2020). Barley plants overexpressing *Ferrochelatases* (*HvFC1* and *HvFC2*) show improved photosynthetic rates and have reduced photo-oxidative damage under drought stress than non-transgenic controls, *Agronomy* 10(9), 1351.

Naruoka, Y., Sherman, J. D., Lanning, S. P., Blake, N. K., Martin, J. M. and Talbert, L. E. (2012). Genetic analysis of green leaf duration in spring wheat, *Crop Sci.* 52(1), 99–109.

Nehe, A. S., Foulkes, M. J., Ozturk, I., Rasheed, A., York, L., Kefauver, S. C., Ozdemir, F. and Morgounov, A. (2021). Root and canopy traits and adaptability genes explain drought tolerance responses in winter wheat, *PLoS ONE* 16(4), e0242472.

Nelson, D. E., Repetti, P. P., Adams, T. R., Creelman, R. A., Wu, J., Warner, D. C., Anstrom, D. C., Bensen, R. J., Castiglioni, P. P., Donnarummo, M. G., Hinchey, B. S., Kumimoto, R. W., Maszle, D. R., Canales, R. D., Krolikowski, K. A., Dotson, S. B., Gutterson, N., Ratcliffe, O. J. and Heard, J. E. (2007). Plant nuclear factor Y (NF-Y) B subunits confer drought tolerance and lead to improved corn yields on water-limited acres, *Proc. Natl. Acad. Sci. U. S. A.* 104(42), 16450–16455.

Nuccio, M. L., Wu, J., Mowers, R., Zhou, H. P., Meghji, M., Primavesi, L. F., Paul, M. J., Chen, X., Gao, Y., Haque, E., Basu, S. S. and Lagrimini, L. M. (2015). Expression of trehalose-6-phosphate phosphatase in maize ears improves yield in well-watered and drought conditions, *Nat. Biotechnol.* 33(8), 862–869.

Obsa, B. T., Eglinton, J., Coventry, S., March, T., Langridge, P. and Fleury, D. (2016). Genetic analysis of developmental and adaptive traits in three doubled haploid populations of barley (*Hordeum vulgare* L.), *Theor. Appl. Genet.* 129(6), 1139–1151.

Oh, S. J., Kwon, C. W., Choi, D. W., Song, S. I. and Kim, J. K. (2007). Expression of barley HvCBF4 enhances tolerance to abiotic stress in transgenic rice, *Plant Biotechnol. J.* 5(5), 646–656.

Oh, S. J., Song, S. I., Kim, Y. S., Jang, H. J., Kim, S. Y., Kim, M., Kim, Y. K., Nahm, B. H. and Kim, J. K. (2005). Arabidopsis CBF3/DREB1A and ABF3 in transgenic rice increased tolerance to abiotic stress without stunting growth, *Plant Physiol.* 138(1), 341–351.

Ort, D. R., Melis, A. and Melis, A. (2011). Optimizing antenna size to maximize photosynthetic efficiency, *Plant Physiol.* 155(1), 79–85.

Ortiz, D. and Salas-Fernandez, M. G. (2021). Dissecting the genetic control of natural variation in sorghum photosynthetic response to drought stress, *J. Exp. Bot.* 73, 3251. DOI: 10.1093/jxb/erab502.

Osipova, S., Permyakov, A., Permyakova, M., Pshenichnikova, T., Verkhoturov, V., Rudikovsky, A., Rudikovskaya, E., Shishparenok, A., Doroshkov, A. and Börner, A. (2016). Regions of the bread wheat D genome associated with variation in key photosynthesis traits and shoot biomass under both well watered and water deficient conditions, *J. Appl. Genet.* 57(2), 151–163.

Ouedraogo, N., Sanou, J., Traore, H., Gracen, V., Tongoona, P. and Danquah, E. Y. (2017). Genetic diversity among sorghum landraces and polymorphism assessment of local improved varieties for stay-green trait, *Int. J. Bio. Chem. Sci.* 11(1), 1–14.

Parry, M. A., Reynolds, M., Salvucci, M. E., Raines, C., Andralojc, P. J., Zhu, X. G., Price, G. D., Condon, A. G. and Furbank, R. T. (2011). Raising yield potential of wheat. II. Increasing photosynthetic capacity and efficiency, *J. Exp. Bot.* 62(2), 453–467.

Passioura, J. B. (1977). Grain yield, harvest index, and water use of wheat, *J. Aust. Inst. Agric. Sci.* 43, 117–120.

Peingao, L. (2013). Structural and biochemical mechanism responsible for the stay-green phenotype in common wheat, *Chi. Sci. Bull.* 51, 2595–2603.

Peleg, Z. V. I., Fahima, T., Krugman, T., Abbo, S., Yakir, D., Korol, A. B. and Saranga, Y. (2009). Genomic dissection of drought resistance in durum wheat× wild emmer wheat recombinant inbreed line population, *Plant Cell Environ.* 32(7), 758–779.

Pelleschi, S., Leonardi, A., Rocher, J.-P., Cornic, G., de Vienne, D., Thévenot, C. and Prioul, J.-L. (2006). Analysis of the relationships between growth, photosynthesis and carbohydrate metabolism using quantitative trait loci (QTLs) in young maize plants subjected to water deprivation, *Mol. Breed.* 17(1), 21–39.

Pinheiro, C. and Chaves, M. M. (2011). Photosynthesis and drought: Can we make metabolic connections from available data?, *J. Exp. Bot.* 62(3), 869–882.

Pinto, R. S., Reynolds, M. P., Mathews, K. L., McIntyre, C. L., Olivares-Villegas, J. J. and Chapman, S. C. (2010). Heat and drought adaptive QTL in a wheat population designed to minimize confounding agronomic effects, *Theor. Appl. Genet.* 121(6), 1001–1021.

Piveta, L. B., Roma-Burgos, N., Noldin, J. A., Viana, V. E., Oliveira, Cd., Lamego, F. P. and Avila, L. Ad. (2021). Molecular and physiological responses of rice and weedy rice to heat and drought stress, *Agriculture* 11(1), 9.

Price, A. H., Young, E. M. and Tomos, A. D. (1997). Quantitative trait loci associated with stomatal conductance, leaf rolling and heading date mapped in upland rice (*Oryza sativa*), *New Phytol.* 137(1), 83–91.

Prince, S. J., Beena, R., Gomez, S. M., Senthivel, S. and Babu, R. C. (2015). Mapping consistent rice (*Oryza sativa* L.) yield QTLs under drought stress in target rainfed environments, *Rice (N Y)* 8(1), 53.

Pshenichnikova, T. A., Osipova, S. V., Smirnova, O. G., Leonova, I. N., Permyakova, M. D., Permyakov, A. V., Rudikovskaya, E. G., Konstantinov, D. K., Verkhoturov, V. V., Lohwasser, U. and Börner, A. (2021). Regions of chromosome 2a of bread wheat (*Triticum aestivum* L.) Associated with variation in physiological and agronomical traits under contrasting water regimes, *Plants (Basel)* 10(5), 1023.

Quero, G., Bonnecarrère, V., Simondi, S., Santos, J., Fernández, S., Gutierrez, L., Garaycochea, S. and Borsani, O. (2021). Genetic architecture of photosynthesis energy partitioning as revealed by a genome-wide association approach, *Photosynth. Res.* 150(1–3), 97–115.

Rahman, H., Pekic, S., Lazic-Jancic, V., Quarrie, S. A., Shah, S. M., Pervez, A. and Shah, M. M. (2011). Molecular mapping of quantitative trait loci for drought tolerance in maize plants, *Genet. Mol. Res.* 10(2), 889–901.

Raines, C. A. (2006). Transgenic approaches to manipulate the environmental responses of the C_3 carbon fixation cycle, *Plant Cell Environ.* 29(3), 331–339.

Rama Reddy, N. R., Ragimasalawada, M., Sabbavarapu, M. M., Nadoor, S. and Patil, J. V. (2014). Detection and validation of stay-green QTL in post-rainy sorghum involving widely adapted cultivar, M35-1 and a popular stay-green genotype B35, *BMC Genomics* 15(1), 909.

Rao, D. E. and Chaitanya, K. V. (2016). Photosynthesis and antioxidative defense mechanisms in deciphering drought stress tolerance of crop plants, *Biol. Plant.* 60(2), 201–218.

Ray, D. K., Mueller, N. D., West, P. C. and Foley, J. A. (2013). Yield trends are insufficient to double global crop production by 2050, *PLoS ONE* 8(6), e66428.

Redillas, M. C., Jeong, J. S., Kim, Y. S., Jung, H., Bang, S. W., Choi, Y. D., Ha, S. H., Reuzeau, C. and Kim, J. K. (2012). The overexpression of *OsNAC9* alters the root architecture of rice plants enhancing drought resistance and grain yield under field conditions, *Plant Biotechnol. J.* 10(7), 792–805.

Ren, T., Fan, T., Chen, S., Chen, Y., Ou, X., Jiang, Q., Peng, W., Ren, Z., Tan, F., Luo, P. and Li, Z. (2022). Identification and validation of quantitative trait loci for the functional

stay green trait in common wheat (*Triticum aestivum* L.) via high-density SNP-based genotyping, *Theor. Appl. Genet.* 135(4), 1429–1441.
Reynolds, M., Foulkes, M. J., Slafer, G. A., Berry, P., Parry, M. A., Snape, J. W. and Angus, W. J. (2009). Raising yield potential in wheat, *J. Exp. Bot.* 60(7), 1899–1918.
Reynolds, M. and Tuberosa, R. (2008). Translational research impacting on crop productivity in drought-prone environments, *Curr. Opin. Plant Biol.* 11(2), 171–179.
Ribaut, J., Edmeades, G., Hoisington, D., Setter, T. and Banziger, M. (2004). Genetic dissection of drought tolerance in maize: A case study. In: Ribaut, J. M. (Ed.), *Physiology and Biotechnology Integration for Plant Breeding*, Marcel Dekker, New York, pp. 571–609.
Roell, M. S., von Borzykowski, L. S., Westhoff, P., Plett, A., Paczia, N., Claus, P., Urte, S., Erb, T. J. and Weber, A. P. M. (2021). A synthetic C_4 shuttle via the β-hydroxyaspartate cycle in C_3 plants, *Proc. Natl. Acad. Sci. U. S. A.* 118(21), e2022307118.
Roháček, K. (2002). Chlorophyll fluorescence parameters: The definitions, photosynthetic meaning, and mutual relationships, *Photosynthetica* 40(1), 13–29.
Rosenow, D. T. (1983). Breeding for resistance to root and stalk rots in Texas. In: *Sorghum Root and Stalk Rots, A Critical Review*, ICRISTAT, Patancheru, AP, India, pp. 209–217.
Sanchez, A. C., Subudhi, P. K., Rosenow, D. T. and Nguyen, H. T. (2002). Mapping QTLs associated with drought resistance in sorghum (*Sorghum bicolor* L. Moench), *Plant Mol. Biol.* 48(5–6), 713–726.
Sanchez-Bragado, R., Vicente, R., Molero, G., Serret, M. D., Maydup, M. L. and Araus, J. L. (2020). New avenues for increasing yield and stability in C_3 cereals: Exploring ear photosynthesis, *Curr. Opin. Plant Biol.* 56, 223–234.
Sangodele, E. A., Hanchinal, R. R., Hanamaratti, N. G., Shenoy, V. and Kumar, V. M. (2014). Analysis of drought tolerant QTL linked to physiological and productivity component traits under water-stress and non-stress in rice (*Oryza sativa* L.), *Int. J. Curr. Res. Acad. Rev.* 2(5), 108–113.
Satrio, R. D., Fendiyanto, M. H., Supena, E. D. J., Suharsono, S. and Miftahudin, M. (2021). Genome-wide SNP discovery, linkage mapping, and analysis of QTL for morpho-physiological traits in rice during vegetative stage under drought stress, *Physiol. Mol. Biol. Plants* 27(11), 2635–2650.
Sehgal, A., Sita, K., Siddique, K. H. M., Kumar, R., Bhogireddy, S., Varshney, R. K., HanumanthaRao, B., Nair, R. M., Prasad, P. V. V. and Nayyar, H. (2018). Drought or/and heat-stress effects on seed filling in food crops: Impacts on functional biochemistry, seed yields, and nutritional quality, *Front. Plant Sci.* 9, 1705.
Sekhon, R. S., Saski, C., Kumar, R., Flinn, B. S., Luo, F., Beissinger, T. M., Ackerman, A. J., Breitzman, M. W., Bridges, W. C., de Leon, N. and Kaeppler, S. M. (2019). Integrated genome-scale analysis identifies novel genes and networks underlying senescence in maize, *Plant Cell* 31(9), 1968–1989.
Shahinnia, F., Le Roy, J., Laborde, B., Sznajder, B., Kalambettu, P., Mahjourimajd, S., Tilbrook, J. and Fleury, D. (2016). Genetic association of stomatal traits and yield in wheat grown in low rainfall environments, *BMC Plant Biol.* 16(1), 150.
Shen, W. J., Chen, G. X., Xu, J. G., Jiang, Y., Liu, L., Gao, Z. P., Ma, J., Chen, X., Chen, T. H. and Lv, C. F. (2015). Overexpression of maize phosphoenolpyruvate carboxylase improves drought tolerance in rice by stabilization the function and structure of thylakoid membrane, *Photosynthetica* 53(3), 436–446.
Shi, S., Azam, F. I., Li, H., Chang, X., Li, B. and Jing, R. (2017). Mapping QTL for stay-green and agronomic traits in wheat under diverse water regimes, *Euphytica* 213(11), 1–19.

Shi, W., Cheng, J., Wen, X., Wang, J., Shi, G., Yao, J., Hou, L., Sun, Q., Xiang, P., Yuan, X., Dong, S., Guo, P. and Guo, J. (2018). Transcriptomic studies reveal a key metabolic pathway contributing to a well-maintained photosynthetic system under drought stress in foxtail millet (*Setaria italica* L.), *PeerJ* 6, e4752.

Shi, Y., Zhang, X., Chang, X., Yan, M., Zhao, H., Qin, Y. and Wang, H. (2021). Integrated analysis of DNA methylome and transcriptome reveals epigenetic regulation of CAM photosynthesis in pineapple, *BMC Plant Biol.* 21(1), 19.

Shou, H., Bordallo, P. and Wang, K. (2004). Expression of the *Nicotiana* protein kinase (NPK1) enhanced drought tolerance in transgenic maize, *J. Exp. Bot.* 55(399), 1013–1019.

Siddiqui, M. N., Teferi, T. J., Ambaw, A. M., Gabi, M. T., Koua, P., Léon, J. and Ballvora, A. (2021). New drought-adaptive loci underlying candidate genes on wheat chromosome 4b with improved photosynthesis and yield responses, *Physiol. Plant.* 173(4), 2166–2180.

Šimić, D., Lepeduš, H., Jurković, V., Antunović, J. and Cesar, V. (2014). Quantitative genetic analysis of chlorophyll a fluorescence parameters in maize in the field environments, *J. Integr. Plant Biol.* 56(7), 695–708.

Simkin, A. J., Faralli, M., Ramamoorthy, S. and Lawson, T. (2020). Photosynthesis in non-foliar tissues: Implications for yield, *Plant J.* 101(4), 1001–1015.

Singh, U. M., Sinha, P., Dixit, S., Abbai, R., Venkateshwarlu, C., Chitikineni, A., Singh, V. K., Varshney, R. K. and Kumar, A. (2020). Unraveling candidate genomic regions responsible for delayed leaf senescence in rice, *PLoS ONE* 15(10), e0240591.

South, P. F., Cavanagh, A. P., Lopez-Calcagno, P. E., Raines, C. A. and Ort, D. R. (2018). Optimizing photorespiration for improved crop productivity, *J. Integr. Plant Biol.* 60(12), 1217–1230.

Souza, R. P., Machado, E. C., Silva, J. A. B., Lagôa, A. M. M. A. and Silveira, J. A. G. (2004). Photosynthetic gas exchange, chlorophyll fluorescence and some associated metabolic changes in cowpea (*Vigna unguiculata*) during water stress and recovery, *Environ. Exp. Bot.* 51(1), 45–56.

Subashri, M., Robin, S., Vinod, K. K., Rajeswari, S., Mohanasundaram, K. and Raveendran, T. S. (2009). Trait identification and QTL validation for reproductive stage drought resistance in rice using selective genotyping of near flowering RILs, *Euphytica* 166(2), 291–305.

Subudhi, P. K., Rosenow, D. T. and Nguyen, H. T. (2000). Quantitative trait loci for the stay green trait in sorghum (*Sorghum bicolor* L. Moench): Consistency across genetic backgrounds and environments, *Theor. Appl. Genet.* 101(5–6), 733–741.

Sukumaran, S., Dreisigacker, S., Lopes, M., Chavez, P. and Reynolds, M. P. (2015). Genome-wide association study for grain yield and related traits in an elite spring wheat population grown in temperate irrigated environments, *Theor. Appl. Genet.* 128(2), 353–363.

Sun, Q., Zhang, Y., Chen, B., Jia, B., Zhang, Z. L., Cui, M., Kan, X., Shi, H. B., Deng, D. X. and Yin, Z. T. (2017). Expression quantitative trait loci analysis of the RuBisCO Activase gene in maize, *Photosynthetica* 55(2), 329–337.

Tai, F. J., Yuan, Z. L., Wu, X. L., Zhao, P. F., Hu, X. L. and Wang, W. (2011). Identification of membrane proteins in maize leaves, altered in expression under drought stress through polyethylene glycol treatment, *Plant OMICS, J.* 4(5), 250–256.

Tang, Y. T., Li, X., Lu, W., Wei, X., Zhang, Q., Lv, C. and Song, N. (2018). Transgenic rice over-expressing maize C_4 phosphoenolpyruvate carboxylase gene tolerant to low nitrogen stress by enhancing photorespiration, *Plant Physiol. Biochem.* 130, 577–588.

Tao, Y. Z., Henzell, R. G., Jordan, D. R., Butler, D. G., Kelly, A. M. and McIntyre, C. L. (2000). Identification of genomic regions associated with stay green in sorghum by testing RILs in multiple environments, *Theor. Appl. Genet.* 100(8), 1225–1232.

Teng, S., Qian, Q., Zeng, D., Kunihiro, Y., Fujimoto, K., Huang, D. and Zhu, L. (2004). QTL analysis of leaf photosynthetic rate and related physiological traits in rice (*Oryza sativa* L.), *Euphytica* 135(1), 1–7.

Tian, J., Deng, Z., Zhang, K., Yu, H., Jiang, X. and Li, C. (2015). Genetic analysis of main physiological and morphological traits. In: *Genetic Analyses of Wheat and Molecular Marker-Assisted Breeding*, Springer, Dordrecht, pp. 351–443.

Tilman, D., Balzer, C., Hill, J. and Befort, B. L. (2011). Global food demand and the sustainable intensification of agriculture, *Proc. Natl. Acad. Sci. U. S. A.* 108(50), 20260–20264.

Trachsel, S., Sun, D., SanVicente, F. M., Zheng, H., Atlin, G. N., Suarez, E. A., Babu, R. and Zhang, X. (2016). Identification of QTL for early vigor and stay-green conferring tolerance to drought in two connected advanced backcross populations in tropical maize (*Zea mays* L.), *PLoS ONE* 11(3), e0149636.

Trudeau, D. L., Edlich-Muth, C., Zarzycki, J., Scheffen, M., Goldsmith, M., Khersonsky, O., Avizemer, Z., Fleishman, S. J., Cotton, C. A. R., Erb, T. J., Tawfik, D. S. and Bar-Even, A. (2018). Design and in vitro realization of carbon-conserving photorespiration, *Proc. Natl. Acad. Sci. U. S. A.* 115(49), E11455–E11464.

Tuinstra, M. R., Grote, E. M., Goldsbrough, P. B. and Ejeta, G. (1997). Genetic analysis of post-flowering drought tolerance and components of grain development in *Sorghum bicolor* (L.) Moench, *Mol. Breed.* 3(6), 439–448.

Varshney, R. K., Paulo, M. J., Grando, S., van Eeuwijk, F. A., Keizer, L. C. P., Guo, P., Ceccarelli, S., Kilian, A., Baum, M. and Graner, A. (2012). Genome wide association analyses for drought tolerance related traits in barley (*Hordeum vulgare* L.), *Field Crops Res.* 126, 171–180.

Verma, V., Foulkes, M. J., Worland, A. J., Sylvester-Bradley, R., Caligari, P. D. S. and Snape, J. W. (2004). Mapping quantitative trait loci for flag leaf senescence as a yield determinant in winter wheat under optimal and drought-stressed environments, *Euphytica* 135(3), 255–263.

Vilas, J. M., Burgos, E., Puig, M. L., Colazo, J., Livore, A., Ruiz, O. A., Carrari, F., Rodriguez, A. A. and Maiale, S. J. (2020). Genetic structure of photosystem II functionality in rice unraveled by GWAS analysis, *bioRxiv*.

Von Caemmerer, S., Quick, W. P. and Furbank, R. T. (2012). The development of C_4 rice: Current progress and future challenges, *Science* 336(6089), 1671–1672.

Wang, A. Y., Li, Y. and Zhang, C. Q. (2012). QTL mapping for stay-green in maize (*Zea mays*), *Can. J. Plant Sci.* 92(2), 249–256.

Wang, B., Li, Z., Ran, Q., Li, P., Peng, Z. and Zhang, J. (2018). *ZmNF-YB16* overexpression improves drought resistance and yield by enhancing photosynthesis and the antioxidant capacity of maize plants, *Front. Plant Sci.* 9, 709.

Wang, L. M., Shen, B. R., Li, B. D., Zhang, C. L., Lin, M., Tong, P. P., Cui, L. L., Zhang, Z. S. and Peng, X. X. (2020). A synthetic photorespiratory shortcut enhances photosynthesis to boost biomass and grain yield in rice, *Mol. Plant* 13(12), 1802–1815.

Wang, S. G., Jia, S. S., Sun, D. Z., Wang, H. Y., Dong, F. F., Ma, H. X., Jing, R. L. and Ma, G. (2015). Genetic basis of traits related to stomatal conductance in wheat cultivars in response to drought stress, *Photosynthetica* 53(2), 299–305.

Wehner, G., Balko, C., Humbeck, K., Zyprian, E. and Ordon, F. (2016). Expression profiling of genes involved in drought stress and leaf senescence in juvenile barley, *BMC Plant Biol.* 16(1), 3.

Wehner, G., Balko, C., Enders, M., Humbeck, K. and Ordon, F. (2015). Identification of genomic regions involved in tolerance to drought stress and drought stress induced leaf senescence in juvenile barley, *BMC Plant Biol.* 15(1), 125.

Wójcik-Jagła, M., Fiust, A., Kościelniak, J. and Rapacz, M. (2018). Association mapping of drought tolerance-related traits in barley to complement a traditional biparental QTL mapping study, *Theor. Appl. Genet.* 131(1), 167–181.

Wu, A., Hammer, G. L., Doherty, A., von Caemmerer, S. and Farquhar, G. D. (2019). Quantifying impacts of enhancing photosynthesis on crop yield, *Nat. Plants* 5(4), 380–388.

Xu, W., Subudhi, P. K., Crasta, O. R., Rosenow, D. T., Mullet, J. E. and Nguyen, H. T. (2000). Molecular mapping of QTLs conferring stay-green in grain sorghum (*Sorghum bicolor* L. Moench), *Genome* 43(3), 461–469.

Xu, Y. F., Li, S. S., Li, L. H., Ma, F., Fu, X., Shi, Z., Xu, H., Ma, P. and An, D. (2017). QTL mapping for yield and photosynthetic related traits under different water regimes in wheat, *Mol. Breed.* 37(3), 1–18.

Yan, X., Wang, S., Yang, B., Zhang, W., Cao, Y., Shi, Y., Sun, D. and Jing, R. (2020). QTL mapping for flag leaf-related traits and genetic effect of QFLW-6A on flag leaf width using two related introgression line populations in wheat, *PLoS ONE* 15(3), e0229912.

Yang, D., Li, M., Liu, Y., Chang, L., Cheng, H., Chen, J. and Chai, S. (2016). Identification of quantitative trait loci and water environmental interactions for developmental behaviors of leaf greenness in wheat, *Front. Plant Sci.* 7, 273.

Yang, D. L., Jing, R. L., Chang, X. P. and Li, W. (2007). Quantitative trait loci mapping for chlorophyll fluorescence and associated traits in wheat (*Triticum aestivum*), *J. Integr. Plant Biol.* 49(5), 646–654.

Yang, G., Tu, S., Li, S., Feng, L., Kong, J., Li, H. and Li, Y. (2008). Analysis of quantitative trait loci underlying the traits related to chlorophyll content of the flag leaf in rice, *Front. Biol. China* 3(4), 443–448.

Yang, H., Mo, P., Chen, Y., Chen, R., Wei, T., Xie, W., Xiang, X., Huang, X., Zheng, T. and Fan, G. (2021). Genetic progress in grain yield radiation and nitrogen use efficiency of dryland winter wheat in Southwest China since 1965: Progress and prospect for improvements, *Crop Sci.* 61(6), 4255–4272.

Yang, S., Vanderbeld, B., Wan, J. and Huang, Y. (2010). Narrowing down the targets: Towards successful genetic engineering of drought-tolerant crops, *Mol. Plant* 3(3), 469–490.

Yang, Z., Li, X., Zhang, N., Wang, X., Zhang, Y., Ding, Y., Kuai, B. and Huang, X. (2017). Mapping and validation of the quantitative trait loci for leaf stay-green-associated parameters in maize, *Plant Breed.* 136(2), 188–196.

Yin, Z., Qin, Q., Kan, X., Chen, Y., Cheng, Q. and Deng, D. (2016). Genome-wide association analysis of fast chlorophyll fluorescence parameters in maize, *Agric. Biotechnol.* 5(6), 4.

Yin, Z., Qin, Q., Wu, F., Zhang, J., Chen, T., Sun, Q., Zhang, Y., Wang, H. and Deng, D. (2015). Quantitative trait locus mapping of chlorophyll a fluorescence parameters using a recombinant inbred line population in maize, *Euphytica* 205(1), 25–35.

Yoo, S. C., Cho, S. H., Zhang, H., Paik, H. C., Lee, C. H., Li, J., Yoo, J. H., Lee, B. W., Koh, H. J., Seo, H. S. and Paek, N. C. (2007). Quantitative trait loci associated with functional stay-green SNU-SG1 in rice, *Mol. Cells* 24(1), 83–94.

Yue, B., Xue, W. Y., Luo, L. J. and Xing, Y. Z. (2006). QTL analysis for flag leaf characteristics and their relationships with yield and yield traits in rice, *Yi Chuan Xue Bao* 33(9), 824–832.

Zargar, S. M., Gupta, N., Nazir, M., Mahajan, R., Malik, F. A., Sofi, N. R., Shikari, A. B. and Salgotra, R. K. (2017). Impact of drought on photosynthesis: Molecular perspective, *Plant Gene* 11, 154–159.

Zenda, T., Liu, S. and Duan, H. (2020). Adapting cereal grain crops to drought stress: 2020 and beyond. In: Fahad, S., Saud, S., Chen, Y., Wu, C. and Wang, D. (Eds.), *Abiotic Stress in Plants*, IntechOpen, London, UK. DOI: 10.5772/intechopen.93845. Available at: https://www.intechopen.com/chapters/73281.

Zenda, T., Liu, S., Wang, X., Jin, H., Liu, G. and Duan, H. (2018). Comparative proteomic and physiological analyses of two divergent maize inbred lines provide more insights into drought stress tolerance mechanisms, *Int. J. Mol. Sci.* 19(10), 3225.

Zhang, J., Fengler, K. A., Van Hemert, J. L., Gupta, R., Mongar, N., Sun, J., Allen, W. B., Wang, Y., Weers, B., Mo, H., Lafitte, R., Hou, Z., Bryant, A., Ibraheem, F., Arp, J., Swaminathan, K., Moose, S. P., Li, B. and Shen, B. (2019). Identification and characterization of a novel stay-green QTL that increases yield in maize, *Plant Biotechnol. J.* 17(12), 2272–2285.

Zhang, K., Zhang, Y., Chen, G. and Tian, J. (2009). Genetic analysis of grain yield and leaf chlorophyll content in common wheat, *Cereal Res. Commun.* 37(4), 499–511.

Zhang, Z. B., Xu, P., Jia, J. Z. and Zhou, R. H. (2010). Quantitative trait loci for leaf chlorophyll fluorescence traits in wheat, *Aust. J. Crop Sci.* 4(8), 571-579.

Zhao, F., Zhang, D., Zhao, Y., Wang, W., Yang, H., Tai, F., Li, C. and Hu, X. (2016). The difference of physiological and proteomic changes in maize leaves adaptation to drought, heat, and combined both stresses, *Front. Plant Sci.* 7, 1471.

Zhao, X. Q., Xu, J. L., Zhao, M., Lafitte, R., Zhu, L., Fu, B., Gao, Y. and Li, Z. (2008). QTLs affecting morph-physiological traits related to drought tolerance detected in overlapping introgression lines of rice (*Oryza sativa* L.), *Plant Sci.* 174(6), 618-625.

Zhao, X. Q. and Zhong, Y. (2021). Genetic dissection of the photosynthetic parameters of maize (*Zea mays* L.) in drought-stressed and well-watered environments. *Russ. J. Plant Physiol.* 68(6), 1125–1134.

Zheng, H. J., Wu, A. Z., Zheng, C. C., Wang, Y. F., Cai, R., Shen, X. F., Xu, R. R., Liu, P., Kong, L. J. and Dong, S. T. (2009). QTL mapping of maize (*Zea mays*) stay-green traits and their relationship to yield, *Plant Breed.* 128(1), 54–62.

Zhou, B. Y., Ding, Z. S. and Zhao, M. (2011). Alleviation of drought stress inhibition on photosynthesis by overexpression of PEPC in rice, *Acta Agron. Sin.* 37(1), 112-118.

Zhu, X. F., Zhang, H. P., Hu, M. J., Wu, Z., Jiang, H., Cao, J., Xia, X., Ma, C. and Chang, C. (2016). Cloning and characterization of *Tabas1-B1* gene associated with flag leaf chlorophyll content and thousand-grain weight and development of a gene-specific marker in wheat, *Mol. Breed.* 36(10), 1–12.

Zhu, X. G., Long, S. P. and Ort, D. R. (2008). What is the maximum efficiency with which photosynthesis can convert solar energy into biomass?, *Curr. Opin. Biotechnol.* 19(2), 153–159.

Zhu, X. G., Long, S. P. and Ort, D. R. (2010). Improving photosynthetic efficiency for greater yield, *Annu. Rev. Plant Biol.* 61, 235–261.

Chapter 8

Genomic selection, gene editing and genetic engineering for drought tolerance in cereals

Rodomiro Ortiz, Swedish University of Agricultural Sciences (SLU), Sweden

1 Introduction

Asymmetry in crop research led to knowledge gaps, thus losing derived technology options for improving productivity and resilience in agro-ecosystems (Reynolds et al., 2021). In this regard, omics, genetic engineering and bioinformatics offer means to both search for desired traits and genes controlling them as well as for assisting breeding to develop new cultivars for the target population of environments (TPEs), where each crop thrives. Drought and heat are among the main abiotic factors affecting the crop life cycle at critical development time points, and are often associated (Mwadzingeni et al., 2016). The long-term plant breeding gains for adapting crops to such stress-prone environments relate to the accumulation of favourable, small effects of various genes providing both yield gains and stability under stress (Snowdon et al., 2021). Understanding both the physiology and genetics behind target traits for stress adaptation facilitates the desired genetic modification as well as to identify the most appropriate germplasm source(s), breeding approach, methods and tools. Furthermore, the genotype-by-environment interaction (GEI) challenges any breeding approach because it underpins crop productivity (Cooper et al., 2021). Hence, plant breeding must consider GEI in any strategy for deploying sustainable productivity increases through the release of new cultivars in the TPEs. Indeed, developing improved cultivars and delivering genetic gains therein to farmers are key elements for increasing crop productivity. Superb

http://dx.doi.org/10.19103/AS.2022.0109.12

plant breeding programmes achieve genetic gain rates above 1.5% per year (CGIAR Excellence in Breeding Platform, 2020), while poor programs produce rates below 1% (Gary Atlin, BMGF, Seattle, WA, USA, personal communication).

Plant breeding programs are cyclical schemes aiming to gradually improve the populations from which new cultivars are selected. The germplasm enhancement of genetic resources to sustain genetic gains in plant breeding include various steps, namely: (1) defining plant breeding objectives with end users, (2) identify a useful trait in breeding populations or gene banks, (3) managing genetic variation of useful characteristics based on genetics and knowledge derived from omics research, particularly cloning, (4) leveraging genes in a usable form(s) for later use in crossing blocks, (5) using genome editing or a transgenic approach if target traits are not available in the gene bank or breeding population, (6) utilizing DNA markers to monitor chromosomal changes and as a selection aid in conjunction with phenomics for accurate phenotyping, (7) multi-site trials plus e-phenotyping with drones in the target population(s) and over years for subsequent release of cultivars (which may involve participatory methods) and (8) providing genetically improved seeds that incorporate genetic gains for the benefit of farmers, consumers and other end users. Genomic estimated breeding values (GEBV) for selection (also known for short as genomic selection), gene editing for inducing variants and genetic engineering for transgenic breeding are the most recent methods of crop improvement. They belong to the third and fourth phase of plant breeding as described by Ramstein et al. (2019). As indicated recently by Varshney et al. (2021a), genomic tools along with knowledge ensuing from plant physiology facilitates both high-throughput finding of key drought adaptive traits and developing marker-aided crossbreeding methods for designing stress-resilient cultivars under a changing climate.

This chapter provides an overview on advances on new breeding methods for improving cereal crops grown under drought stress. Further up-to-date information on drought stress and its impact is available at: https://plantstress.com/drought/. This web site –initiated in 2000 by the late drought physiologist Abraham Blum – provides details on selection methods under stress.

2 Genomic estimated breeding values for selection

Genomic-aided breeding (GAB) led to exploiting the potential of modern genome resources for characterizing and using allelic variation available in breeding programs and gene banks for both population improvement and cultivar development (Varshney et al., 2021b). GAB allows cost-effectively and timely accumulating favourable alleles and purging deleterious alleles and in the breeding populations. Modern breeding methods provide means for assessing the genetic worth of plant genetic resources for their further use in crop improvement. One of such methods is genomic selection, which also enables both speeding the selection of outstanding gene bank accessions or breeding lines, as well as shortening the breeding cycles (Crossa et al., 2017). Furthermore, combining genomic selection with high-throughput phenotyping using hyperspectral image technology may lead to develop new cultivars for a TPEs under drought stress. In this regard, Moreira et al. (2020) provide an update on high-throughput phenotyping and statistical genomic methods that are available for improving characteristics that are expressed continuously through the life cycle of a plant, including the response to stressors.

In genomic selection, the allele effects (α) are estimated in a training model population (TP) that will be used in all subsequent steps (Ventorim Ferrao et al., 2017). In the breeding population (BP), these effects are used to estimate the predicted genomic values (GEBVs), which are correlated with the actual phenotypic values. This correlation value is called predictive accuracy (ρ) and is an important indicator of efficiency (Abera Desta and Ortiz, 2014). The BP is the genomic prediction target. Prediction based on genomic information is made and genotypes are selected in the early stages (seedlings), using the effects of alleles as selection criteria. It seems that genomic prediction in plant breeding works well when using simple models that are as good as complex (Abera Desta and Ortiz, 2014). Likewise, a few hundreds of dense DNA markers (such as single-nucleotide polymorphism, SNPs) and a training population of a couple of hundreds may suffice to estimate GEBVs. Furthermore, the TP should be related to the BP, the best option being to use full siblings rather than mega-panels of unrelated elite germplasm (Ventorim Ferrao et al., 2017). Last but not least, the analysis of quantitative trait loci (QTL) and the use of the related gene-linked SNPs or other DNA markers (used sometimes as breeding aid for indirect selection) may increase ρ significantly. Voss-Fels et al. (2019) indicate that GEBVs - based on additive genetic effects - accelerate the delivery of genetic gains, which may be further improved by adding non-additive genetic effects such as epistasis that facilitate the access to other source of genetic variation.

2.1 Maize

Progress in grain yield under drought stress ensued from selection using multi-site trials for component traits (Ortiz, 2013 and references therein). For example, 4.1% yearly genetic gains for grain yield were estimated after breeding extra-early maturity maize hybrids targeting West African drought-prone sites (Badu-Apraku et al., 2021). Although, crossbreeding led to releasing promising hybrids outyielding by up to 26% the best available hybrids, their grain yield repeatability was low in drought-prone sub-Saharan African environments (Gedil and Menkir, 2019). Crossbreeding with the aid of DNA markers offers option for significantly and predictably improving drought tolerance in new maize cultivars. As stated recently by Gedil and Menkir (2019) 'genomic tools enable genetic dissections of complex traits and promote an understanding of the physiological basis of key agronomic and nutritional quality traits', including maize performance in drought-prone environments benefited from the use of DNA markers. Research demonstrated that backcrossing could be useful for introgressing a few target QTL into elite germplasm, but it does not seem effective if many QTL with small effects are involved. QTL are also germplasm specific (Ortiz et al., 2007 and references therein). Erikssson et al. (2018 and references therein) indicated that the average gains resulting from marker aided in African drought-prone sites (229 kg ha^{1} $year^{1}$) and under irrigation (389.3 kg ha^{1} $year^{1}$) were three- to seven-fold higher, respectively, than those from phenotypic selection alone.

Beyene et al. (2015) estimated genetic gains for grain yield ensuing from genomic selection after using eight bi-parental maize populations under drought stress. Single tester crossing from a complementary heterotic group was pursued with each population. The testcrossing-derived offsprings were thereafter included in field yield trials under both drought and irrigation across African sites. They were also genotyped using 191 to 286 SNPs. Selection of top ten performing offspring in C_0 was based on a selection phenotypic index, and they were intermated within each population to obtain C_1. GEBVs were used to select in C_1 and C_2. Pedigree selection was also undertaken to advance the best C_0 lines. The average genetic gain resulting from genomic selection per cycle across the eight populations was 86 kg ha^{-1}. The C_3-derived hybrids had a significant grain yield than those hybrids derived from C_0. Their grain yield was 7.3% (176 kg ha^{-1}) above that of hybrids resulting from pedigree selection. This research article (Beyene et al., 2015) was among the first to demonstrate and clearly described the effectiveness of genomic selection in tropical maize under drought.

Two biparental maize breeding populations resulting from crossing two elite Asian-bred yellow grain lines with an African drought-tolerant white grain inbred were included in a study to determine the feasibility of genomic selection in South Asia (Vivek et al., 2017). C_1 included the offspring after recombining the top 10% of the $F_{2:3}$ according to the testcrossing results. Thereafter, there were two C_2s, namely one C_2 derived from recombining C_1 plants showing superior phenotypes *per se* (C_2[PerSe-PS]), and the other C_2 ensued after crossing a second set of C_1 plants according to their GEBVs (C_2[TC-GS]). C_2[TC-GS] exhibited 10–20% superiority over C_2[PerSe-PS], and their top-crossing derived offspring had 4–43% more grain yield that C_2[PerSe-PS] under drought, thus showing that genomic selections allows achieving rapid genetic gains in such stress-prone environments. Nepolean et al. (2018) suggest that genomic selection coupled with accelerated line breeding (e.g. doubled haploids or speed breeding) will further increase genetic gains due to reducing the time for inbreeding. Speed breeding also shortens the time for cultivar development, release and marketing (Samantara et al., 2022; Wanga et al., 2021; Watson et al., 2018), as well as it allows exploiting genebank accessions and library of mutants for a rapid gene discovery and use in crop improvement (Li et al., 2018).

After testing the GEBVs of 240 subtropical maize lines under drought stress, Shikha et al. (2017) concluded that Bayes B – which uses both shrinkage and variable selection methods – was the best due to their ρ. They were able to further select 77 SNPs (out of 1053) related to ten drought-responsive transcription factors that were associated with characteristics such as stomatal closure, root development, hormonal signalling and photosynthesis.

Genomic selection improved grain yield of two multi-parent yellow synthetic population of tropical maize in South Asia under both drought and waterlogging (Das et al., 2021). The use of multi-parent populations with trait donors for target stress and diverse elite bred germplasm appears to be an effective breeding method for simultaneous tolerance to different water stress-prone environments without losing grain yield when any of them occur. Das et al. (2020) estimated realized genetic gains of 110 kg ha−1 year−1 and 135 kg ha^{-1} year^{-1} on the two multi-parent yellow synthetic populations. Their results further show that two cycles of genomic selection did not lead to a yield penalty neither to losing diversity.

Although genomic selection was able to improve grain yield of early maturity maize biparental populations simultaneously in both *Striga*-infested sites (498 kg ha^{-1} cycle^{1} or 16.9% cycle^{1}) and as well as under optimal environments in Nigeria and Zimbabwe (522 kg ha^{1} cycle^{1} or 12.6% cycle^{1}), Badu-Apraku et al. (2019) indicated that it did not increase grain yield under drought stress (5.1 kg ha^{1} cycle^{1} or 0.22% cycle^{1}). Instead, they noted delayed flowering, poor anthesis-silking synchrony during flowering and increased ear height.

2.2 Rice

Drought tolerance is regarded as a quantitative trait in rice because it includes varying phenotypes according to the plant development stage. About 250–350 QTL related to various shoot and root responses under drought are known in rice (Oladosu et al., 2019; Sahebi et al., 2018), but their use through marker-aided breeding is yet to prove being cost-effective for developing new cultivars. QTL with a large and consistent effect on grain yield under drought were found and further use for pyramiding them in popular rice mega-cultivars or for developing stress-tolerant breeding lines (Mallikarjuna Swamy and Kumar, 2013; Yadav et al., 2021), thus validating DNA marker-aided breeding for incorporating such traits to improve grain yield of rice under drought. Bhandari et al. (2019) found that drought tolerance was up to 32% better predicted by multi-environment models when using selection of trait-specific markers and multi-environment models for rice breeding and cross-validation scenarios. Reasonable prediction was achieved with about 3000 SNPs.

2.3 Wheat

Drought affects significantly both the growth and productivity of wheat. Introducing favourable alleles for stress tolerance through crossbreeding with the aid of DNA markers may increase grain yield of wheat when grown under drought (Mohammadi, 2018). There are some characteristics that improve grain yield of wheat in drought-prone environments, e.g. root (Alemu et al., 2021; Ober et al., 2021) and canopy traits (Rutkoski et al., 2016). Recently, Nehe et al. (2021) indicated that the screening for drought tolerance in wheat should include assessing canopy green area at anthesis, root number per shoot, and adaptability markers such as *TaCwi.4A* associated with high yield under stress, and *Dreb1* that increases root surface area and root volume. Notably, a recent GWAS study conducted under drought on durum wheat showed that *Dreb1* co-located with a major QTL controlling root growth angle that has been demonstrated to play a relevant role in maintaining water access in deep soils during wheat terminal drought (Condorelli et al., 2022).

Landraces and resynthesized wheat (resulting from crossing crop wild relatives) are interesting genetic resources for improving drought tolerance in wheat (Khadka et al., 2020; Mwadzingeni et al., 2017). A multi-disciplinary approach including the physiology of drought-tolerance traits as proxies along with the genetic resources bearing those traits and genomics tools to aid on selection provide means for developing wheat germplasm for specific drought-prone sites (Fleury et al., 2010; Salvi and Tuberosa, 2015; Tuberosa and Salvi, 2006). As noted previously, tolerance to drought in wheat involves various traits and several genes with small effects controlling them. Hence, mapping such genes and incorporating them into elite germplasm seems to be challenging, even for marker-aided selection (Khan et al., 2016), which calls for the use of GEBVs for selection under drought stress.

Bassi et al. (2016) gave some guidelines on genomic selection in wheat, particularly about genotyping, train population design, selection intensity, inbreeding, generation of selection and recurrent schemes involving either half- or full-sibs. They found that F_3 and F_4 recurrent selection schemes had 0.19 and 0.11 genetic gains per year, respectively. The relatively low-cost genotyping-by-sequencing (GBS) provides a suitable platform for genomic selection in wheat after finding that the GEBV for selection accuracy for various agronomic traits (including grain yield) had an improvement of 0.1–0.2 as compared to the then available DNA marker platform for wheat (Polland et al., 2012). High-throughput phenotyping (HTP) along with GEBV for selection accelerate breeding gains when developing climate-resilient wheat (Juliana et al., 2019). Multi-variate prediction of grain yield may be based on the green normalized difference vegetation index (NVDI). Integrating HTP with GEBV allows both increasing population size for drought screening as well as evaluating large unphenotyped nursery sets for stress resilience of wheat germplasm within years.

Prediction accuracy ranged from low (−0.32) to moderate (0.52) in trials under drought and irrigation of a panel of 286 cultivars (87) and landraces (199), which were genotyped with 9047 SNPs using GBS (Shabannejad et al., 2021). GEBV for selection accuracy for all traits increased after including 93 significant ($P < 0.001$) DNA markers resulting from genome-wide association genetics research. Shabannejad et al. (2021) indicated that prediction accuracy is affected by linkage disequilibrium extent, trait genetic architecture, population genetic diversity and genomic selection method. Furthermore, genomic prediction accuracy increased after including in the model physiological traits and effects of both GEI or physiology-by-environment interactions (Guo et al., 2020). The physiological traits were canopy temperature (CT), SPAD chlorophyll content, membrane thermostability, senescence rate, stay green and NDVI. Similarly, Rutkoski et al. (2016) found that CT and NDVI improve prediction accuracy for grain yield under stress in wheat. These secondary traits increased genomic prediction accuracy by 70%.

2.4 Other cereals

Recently, Yabe and Iwata (2020) gave an overview on the few attempts of implementing genomic selection in "minor" cereals (oat, pearl millet, rye, sorghum) and in buckwheat but none of the cited journal articles refers to breeding for drought-prone environments. Previously, Serba and Yadav (2016) indicated that genomic selection tools were limited for breeding pearl millet targeting drought-prone sites although they anticipated that it may speed up genetic gains. In this regard, Srivastava et al. (2020) revealed that GEBV for selection revamped in pearl millet due to advances in GBS, sequencing, tuneable GBS and whole genome resequencing. Indeed, Varshney et al. (2017) used resequencing data to undertake genomic selection to predict grain yield for test crosses under four environments, which included early and late stress. They estimated a ρ of 0.6 for grain yield across environments, which suggests that GEBV for selection may significantly increase genetic gains per year. Genomic selection considering additive and dominance effects in the model was also useful for predicting hybrid performance or to identify breeding lines never used before for producing hybrids (Varshney et al., 2017), thereby accelerating further genetic gains from selection.

Hunt et al. (2018) demonstrated that genomic prediction models based on DNA marker relationships had higher selection accuracy than those based on pedigree relationships in sorghum. Nevertheless, the inclusion of pedigree information together with DNA marker information in the model improves prediction accuracy. They further indicated that breeding lines not included in trials may be predicted accurately based on DNA markers alone. Accuracy declines as the relationship between TP and the predicted breeding line decreases. GEI remains as a limiting factor for implementing GEBV for selection in sorghum. A low correlation among sites leads to a low ability for predicting genotype performance across sites.

Drought reduces significantly grain yield (30–70%) in sorghum (Batista et al., 2019). Modelling GEI increases ρ and reduces genomic prediction bias for broad sorghum germplasm adaptation across different sites, e.g. under drought and non-stress environments (Velazco et al., 2020). Best linear unbiased prediction models combining DNA markers and pedigrees (K-BLUPs) improve genomic prediction for specific adaptation to either drought or non-stress sites. Velazco et al. (2020) concluded that improving sorghum adaptability to varying environments may be feasible when using genomic prediction models that include GEI and locus-specific genetic variances. Moreover, Bayes A and Bayes ridge regression (BRR) were the best models according to both their ρ and computation time for predicting several traits (including grain yield) for sweet sorghum under irrigation and water stress in Haiti (Doval, 2019). Multi-trait models may improve further genomic prediction for sorghum grain yield and stay green – being both traits important for broad adaptation – across sites, which include drought-prone environments (Velazco et al., 2019). These multi-trait modelling allows both predicting crop productivity and drought adaptability in sorghum.

3 Genetic engineering

There are several known genes that affect crop adaptation to drought-prone environments (Hu and Xiong, 2014). Most of these genes were found when doing research in plants grown in small pots or at seedling stage in greenhouses, but only a few were included for testing in field trials, e.g. *ZmNF-YB2* in maize or *SNAC1*, *OsNAC9*, *OsNAC10* and *OsLEA3-1* in rice (He et al., 2020 and references therein). Martignago et al. (2020) provide a recent overview, based on research in thale grass (*Arabidopsis thaliana*), about traits and related genes that may be used for engineering drought tolerance considering specific plant tissue responses.

Advances in gene editing allows engineering-desired trait variation for developing cultivars suitable for drought-prone sites (Varshney et al., 2021a). Clustered regularly interspaced short palindromic repeat and their associated proteins (CRISPR–Cas) is a powerful and advanced gene editing tool that shows a great potential for precision plant breeding (Zhang et al., 2018). It recognizes specific sequences in the plant genome and cut them, thus acting as a molecular scissor in target sites. The follow-up repair mechanisms bring defined alterations at the specific cutting site. Very recently, a new gene editing tool named retron library recombineering (RLR) was proposed (Schubert et al., 2021). RLR allows constructing a library of millions of elements to be built using as input natural DNA or random variation, thus providing means for screening at once an entire mutant pool.

Joshi et al. (2020) lists some target genes for CRISPR–Cas gene editing aiming drought tolerance in maize and rice. Modifying transcription factors (TFs), which are regulatory genes, may also lead to drought-tolerant crops (Ahmed et al., 2020). TFs increase stress signals controlling various downstream genes simultaneously. Very recently, gene editing was also pursued to modify the binding site of the enzyme trehalase, which catalyzes the conversion of trehalose into glucose (Nuñez-Muñoz et al., 2021). The sugar trehalose, which consists of two glucose molecules, plays a role as osmoprotector against various stress types including drought in plants such as *Selaginella lepidophylla*, *S. tamariscina*, *Myrothamnus flabellifolia*, *Sporobolus stapfianus*, *Fagus sylvatica* and *Botrychium lunaria*. These plants accumulate remarkably high levels of trehalose, which helps them to keep their cell viability by using different mechanisms. Decreasing trehalase activity through either decreasing the enzyme amount or its activity allows accumulating trehalose, which may allow getting plant germplasm with tolerance to drought-prone sites. Nuñez-Muñoz et al. (2021) were able to obtain thale grass plants bearing variants of the trehalase substrate binding-domain that had tolerance to drought. Although, an interesting result with a model plant system in small pots, such findings should be taken with great caution because, as noted by Blum (2014, 2018), there has been a limited impact of genetic engineering on global dryland agriculture because the few transgenic cultivar releases for drought-prone environments.

3.1 Maize

Transgenic maize with enhanced adaptation to drought-stress environments became available in the early decades of this century (Liu and Qin, 2021). Overexpressing *ZmNF-YB2* or *ZmASR1*, and incorporating either a gene that encodes rice's trehalose-6-phosphate phosphatase or constitutive bacterial RNA chaperones such as CspA from *Escherichia coli* and CspB from *Bacillus subtilis* improved grain yield of maize under drought. There are, however, other factors to be taken into account for deploying transgenic maize for drought-prone environments (Cattivelli et al., 2008), e.g. transgene suitability as well as both its dosage effect and stress tolerance level, or any related grain yield penalty, and, last but not least, end-user acceptability of transgenic maize as food source.

The former Monsanto (which was acquired few years ago by Bayer AG and merged with its division Bayer CropScience) released in 2013 DroughtGard™ hybrids that derived from a transgenic maize (MON 87460) based on the bacterial cold shock protein (*CspB*) gene, which allows the plant to act more quickly against drought by slowing its growth and conserving water to make it available for key function after the onset of drought stress (Ortiz et al., 2014 and references therein). Before the release of DroughtGard™ hybrids there were 250 large-scale on farm trials on about 4000 ha in the US Great Plains. DroughtGard™ hybrids had higher yield than other hybrids with enhanced WUE and may save about 2.5 mm of water inputs per hectare. As indicated by Stecker (2012), this transgenic maize provides a means for managing risks and buffering grain yield loss in drought-prone sites.

ARGOS is defined as an auxin-regulated gene involved in organ size. ARGOS8 is a negative regulator of responses to ethylene, modulates ethylene signal transduction and improves grain yield under drought when overexpressed in transgenic maize (Tenorio Berrío et al., 2022). Shi et al. (2017) used CRISPR-Cas9 gene editing to get ARGOS8 variants in maize. These variants exhibited high levels of ARGOS8 transcripts as compared to the wild type that were noticed throughout testing tissues after using the native maize promoter GOS2. ARGOS8 variants increased, on average, grain yield by 326 kg ha[1] under drought stress at flowering and without showing any grain yield penalty under irrigation.

3.2 Rice

Research on drought-tolerant transgenic rice has given priority to study the overexpression of different genes such as oat's *ARGININE DECARBOXYLASE* (*ADC*) gene, rice's *ANNEXIN* (*OsANN3*) and *DROUGHT AND SALT STRESS RESPONSE-1* (*OsDSSR1*) genes, *late embryogenesis-abundant* (*LEA*) genes such as *OsLEA3-1*, receptor like kinases (*OsESG1*, OsNADK), hormonal receptors (*OsPYL10*, *OsIAA6*, *OsPIN3t*) or genes related to regulatory proteins (*OsSDIR1*); plus altering the expression of rice genes *OsSGL*, *OsGRXS17*, *OsPM1* or OsSCE3 (Khan et al., 2020 and references therein). The drought-tolerant transgenic rice cultivar 'Pusa Sugandhi 2' uses the individual overexpression of *OsRab16A* and *AtDREB1A* genes involved in ABA-dependent and ABA-independent pathways, respectively (Ganguly et al., 2020). The transgenic plants had reduced root and shoot length inhibition, leaf relative water content increase, enhanced catalase (CAT) activity to reduce oxidative damage, both chlorophyll and proline content augmented, plus relatively high filled grain, spikelet fertility and grain yield under drought. Hence, changing gene expression and metabolic pathways through genetic engineering provide means for developing rice germplasm for drought-prone sites.

Mutations in abscisic acid (ABA) receptor genes improve growth and grain yield in rice. For example, enhanced plant growth and improved grain yield was noticed in rice after mutating genes encoding for ABA receptors pyrabactin resistance 1-like 1 (PYL1), PYL4, and PYL6 (Miao et al., 2018). It seems that these variants evolved for regulating plant growth and may also be useful for improving sustainably rice productivity. In this regard, gene editing with CRISPR-Cas9 allows producing mutants that may show tolerance to stressful environments (Romero and Gática-Arias, 2019), e.g. *osera1*-mutant lines displaying enhanced sensitivity to ABA and an enhanced reaction to drought stress based on stomatal regulation (Ogata et al., 2020), or a *drought and salt tolerant* (*dst*) mutant with broad leaf width and reduced stomatal density that allows increased water retention under drought stress (Kumar et al., 2020). These encouraging results highlight the advantages of using CRISPR-Cas9 gene editing to better understand traits relevant for adapting to stressful environments as well as for developing rice cultivars for drought-prone sites.

Yu et al. (2021) proposed recently producing polyploid rice through gene editing to address climate change. There are accumulating evidences correlating polyploidization with environmental change or stress (Van de Peer et al, 2017). Indeed, polyploidy may bring extra chromosomes sets bearing genes that may allow adapting to stressful environments (Wang, 2021). The *de novo* domestication of rice involves first identifying through screening the wild crop relatives as germplasm source(s). Thereafter, sequencing their genome and developing efficient gene editing tools. The next steps will be undertaking gene editing and crossing variants for developing the new crop. Following this approach, Yu et al. (2021) were able to breed six important agronomic traits in the wild amphidiploid rice species *Oryza alta*. The traits bred were seed shattering loss, reduced awn length, increased grain length, decreased plant height, thick stem diameter, and both long and short flowering times.

3.3 Wheat

Modifications resulting from genetic engineering in wheat include gain-of-function phenotypes due to, among others, gene overexpression, loss-of-function phenotypes owing to expressing antisense RNAs and manipulating gene structure or expression though CRISPR-Cas9 gene editing (Borisjuk et al., 2019 and references therein). Khan et al. (2019) list examples of transgenic wheat germplasm showing tolerance to drought. These refer from targeting osmoprotectants, metabolites and protective genes, transporter genes, changing metabolism, transcription factors, post-translational modification or protein kinase. For example, the Centro Internacional de Mejoramiento de Maíz y Trigo (CIMMYT, México) developed a transgenic wheat using dehydration-responsive element binding (*DREB*) transcription factors from thale grass (Pellegrineschi et al., 2004). The transgenic wheat plants had delayed wilting and leaf bleaching while the wild type died after 15 days of drought. This tolerance to drought of the AtDREB1A transgenic plant appears to be the result of increasing relative water content, chlorophyll, sugar and proline contents (Noor et al., 2018). In this regard, transgenic wheat expressing maize phosphoenolpyruvate carboxylase (PEPC) exhibited improved drought tolerance due to overexpressing proteins related to photosynthesis and protein metabolism (Qin et al., 2016). However, as noted by Khan et al. (2019), there are limitations because genetically engineering tolerance to drought in wheat remains complex and requires knowledge regarding altering both structure and function of target genes. Its polyploidy and large genome as well as its low transformation efficiency depending on the genotype along with phenotyping, remain as issues affecting the successful development of a transgenic wheat cultivar for drought-prone sites.

CRISPR/Cas9-based gene editing provides means for breeding wheat aiming stress-prone sites (Kumar et al., 2019). This gene editing system was also used by Liang et al. (2018) to edit a wheat homolog of *TaCer9* (*ECERIFERUM9*) to increase both drought tolerance and water use efficiency. Likewise, Kim et al. (2017) pursued gene editing in wheat protoplasts with CRISPR/Cas9 on two abiotic stress-responsive genes, namely, *TaDREB2* (*dehydration responsive element binding protein 2*) and *TaERF3* (*ethylene responsive factor 3*). Nonetheless, most of the research regarding gene editing for tolerant wheat for drought-prone environments are in early stages and additional work is required on genotyping and characterizing the target mutations. Wheat breeding through gene editing may benefit by targeting homologs already known from research in other cereals, e.g. on root architecture (Kirschner et al., 2021; Ober et al., 2021) or transpiration efficiency. Gene editing has been greatly facilitated by the use of a fusion protein that merged wheat GROWTH-REGULATING FACTOR 4 (GRF4) and its cofactor GRF-INTERACTING FACTOR 1 (GIF1) (Debernardi et al., 2020).

The most promising transgenic wheat ensued from using *HaHB4*, which is a water deficit-induced transcription factor that modulates several hundred genes (Watson et al., 2020). It was found in sunflower and belongs to the homeodomain-leucine zipper I family. A transgenic line bearing a mutated *HaHB4* had significantly 6% more grain yield and 9.4 larger water use efficiently than the control in replicated multi-site testing over 9 years in Argentina (González et al., 2019). This country approved in October 2021 this transgenic wheat – whose drought tolerance does not relate to early stomatal closure – for growth and consumption, thereby becoming the first in the world to be able to market a GM-wheat cultivar (Sheridan, 2021). Recent advances in transgenic wheat for drought-prone environments also include the insertion a barley's dehydrin responsive *HVA1* that belongs to the *Late embryogenesis abundance* (*LEA*) gene family (Habib et al., 2022) and of a cold shock gene from *Escherichia coli* (Rimpan et al., 2022). *HVA1* keeps cell membrane stability under drought, while transgenic wheat bearing the cold shock protein closes faster the stomata than the wild type.

4 Conclusion

Plant genetic resources remain as the main raw material for mining allelic variation associated with target traits for stressful environments. The use of genomic prediction to estimate breeding values for selection decreases selection time while increasing its intensity, and enhances efficiency for improving low heritability traits. Hence, it leads to elevating genetic gains while breeding cereals crops in drought-prone sites. Nonetheless, high-throughput phenotyping for drought screening remains key for accurate and fast genetic gains when pursuing such a knowledge-intensive, genomic-led, fast plant breeding approach. New methods and protocols to edit genes became available recently for various cereals that feed the world (Ansari et al., 2020; Hisano et al., 2021; Matres et al., 2021; Rajput et al., 2021). Promoting alleles through gene editing may further increase selection response by bringing more diversity and accuracy.

5 Where to look for further information

Water is a limiting factor in agriculture, and it seems to be worsening due to ongoing anthropocene-led climate change. Advances in the use of genomic tools and genetic engineering for improving crop performance under drought stress facilitate cereal breeding elsewhere. The knowledge and understanding of mechanisms providing tolerance to drought contributes to genetic gains under drought. A systemic approach integrating agronomy, genetics, genomics, and physiology along with the thoughtful use of genetics resources, phenomics. and bioinformatics allows matching genotypes to the target population of drought-prone environments. The CGIAR along with national agricultural research systems and academics worldwide are also identifying genes for further use in breeding drought tolerance in cereals through marker-aided breeding, transgenics, and gene editing. A knowledge-intensive approach will deliver germplasm-enhanced water productivity under drought, thus contributing to food and nutrition security under global warming.

6 References

Abera Desta, Z. A. and Ortiz, R. (2014). Genomic selection: genome-wide prediction in plant improvement, Trends Plant Sci. 19(9), 592–601.

Ahmed, R. F., Irfan, M., Shakir, H. A., Khan, M. and Chen, L. (2020). Engineering drought tolerance in plants by modification of transcription and signalling factors, Biotechnol. Biotechnol. Equip. 34(1), 781–789. DOI: 10.1080/13102818.2020.1805359.

Alemu, A., Feyissa, T., Maccaferri, M., Sciara, G., Tuberosa, R., Ammar, K., Badebo, A., Acevedo, M., Letta, T. and Abeyo, B. (2021). Genome-wide association analysis unveils novel QTLs for seminal root system architecture traits in Ethiopian durum wheat, BMC Genomics 22(1), 20. DOI: 10.1186/s12864-020-07320-4.

Ansari, W. A., Chandanshive, S. U., Bhatt, V., Nadaf, A. B., Vats, S., Katara, J. L., Sonah, H. and Deshmukh, R. (2020). Genome editing in cereals: approaches, applications and challenges, Int. J. Mol. Sci. 21(11), 4040. DOI: 10.3390/ijms21114040.

Badu-Apraku, B., Obesesan, O., Abiodun, A., Obeng-Bio, E. (2021), Genetic gains from selection for drought tolerance during three breeding periods in extra-early maturing maize hybrids under drought and rainfed environments, Agron. 11, 831. DOI: 10.3390/agronomy11050831.

Badu-Apraku, B., Talabi, A. O., Fakorede, M. A. B., Fasanmade, Y., Gedil, M., Magorokosho, C. and Asiedu, R. (2019). Yield gains and associated changes in an early yellow bi-parental maize population following genomic selection for Striga resistance and drought tolerance, BMC Plant Biol. 19(1), 129. DOI: 10.1186/s12870-019-1740-z.

Bassi, F. M., Bentley, A. R., Charmet, G., Ortiz, R. and Crossa, J. (2016). Breeding schemes for the implementation of genomic selection in wheat (*Triticum* spp.), Plant Sci. 242, 23–36. DOI: 10.1016/j.plantsci.2015.08.021.

Batista, P. S. C., Carvalho, A. J., Portugal, A. F., Bastos, E. A., Cardoso, M. J., Torres, L. G., Júlio, M. P. M. and de Menezes, C. B. (2019). Research Article Selection of sorghum for drought tolerance in a semiarid environment, Genet. Mol. Res. 18(1), gmr18194. DOI: 10.4238/gmr18194.

Bhandari, A., Bartholomé, J., Cao-Hamadoun, T. V., Kumari, N., Frouin, J., Kumar, A. and Ahmadi, N. (2019). Selection of trait-specific markers and multi-environment models improve genomic predictive ability in rice, PLoS ONE 14(5), e0208871. DOI: 10.1371/journal.pone.0208871.

Beyene, Y., Semang, K., Mugo, S., Tarekegne, A., Babu, R., Meisel, B., Sehabiague, P., Makumbi, D., Magorokosho, C., Oikeh, S., Gakunga, J., Vargas, M., Olsen, M., Prasanna, B. M., Banziger, M. and Crossa, J. (2015). Genetic gains in grain yield through genomic selection in eight bi-parental maize populations under drought stress, Crop Sci. 55(1), 154–163.

Blum, A. (2014). Genomics for drought resistance-getting down to earth, Funct. Plant Biol. 41(11), 1191–1198. DOI: 10.1071/FP14018.

Blum, A. (2018). Plant Breeding for Stress Environments. CRC Press, Boca Raton, FL.

Borisjuk, N., Kishchecko, O., Eliby, S., Schramm, C., Anderson, P., Jatayev, S., Kurishbayev, A. and Shavrukov, Y. (2019). Genetic modification for wheat improvement: from transgenesis to genome editing, BioMed Res. Intl. 2019, 6216304. DOI: 10.1155/2019/6216304.

Cattivelli, L., Rizza, F., Badeck, F.-W., Mazzucotelli, E., Mastrangelo, A. M., Francia, E., Marè, C., Tondelli, A. and Stanca, A. M. (2008). Drought tolerance improvement in crop plants: an integrated view from breeding to genomics, Field Crops Res. 105(1–2), 1–14. DOI: 10.1016/j.fcr.2007.07.004.

CGIAR Excellence in Breeding Platform (2020). National Breeding Programs Prepped to Measure – and Boost – Genetic Gains. Available at: https://excellenceinbreeding.org

/blog/national-breeding-programs-prepped-measure---and-boost---genetic-gains Accessed on 9th May 2021.

Condorelli, G. E., Newcomb, M., Groli, E. L., Maccaferri, M., Forestan, C., Babaeian, E., Tuller, M., White, J. W., Ward, R., Mockler, T., Shakoor, N. and Tuberosa, R. (2022). Genome Wide association study uncovers the QTLome for osmotic adjustment and related drought adaptive traits in durum wheat, Genes 13(2), 293. DOI: 10.3390/genes13020293.

Cooper, M., Voss-Fels, K. P., Messina, C. D., Tang, T. and Hammer, G. L. (2021). Tackling G × E × M interactions to close on-farm yield-gaps: creating novel pathways for crop improvement by predicting contributions of genetics and management to crop productivity, Theor. Appl. Genet. 134(6), 1625–1644. DOI: 10.1007/s00122-021-03812-3.

Crossa, J., Pérez Rodríguez, P., Cuevas, J., Montesinos-López, O., Jarquín, D., de Los Campos, G., Burgueño, J., González-Camacho, J. M., Pérez-Elizalde, S., Beyene, Y., Dreisigacker, S., Singh, R., Zhang, X., Gowda, M., Roorkiwal, M., Rutkoski, J. and Varshney, R. K. (2017). Genomic selection in plant breeding: methods, models, and perspectives, Trends Plant Sci. 22(11), 961–975.

Das, R. R., Vinayan, M. T., Patel, M. B., Phagna, R. K., Singh, S. B., Shahi, J. P., Sarma, A., Barua, N. S., Babu, R., Seetharam, K., Burgueño, J. A. and Zaidi, P. H. (2020). Genetic gains with rapid-cycle genomic selection for combined drought and waterlogging tolerance in tropical maize (*Zea mays* L.), Plant Genome 13(3), e20035. DOI: 10.1002/tpg2.20035.

Das, R. R., Vinayan, M. T., Seetharam, K., Patel, M., Phagna, R. K., Singh, S. B., Shahi, J. P., Sarma, A., Barua, N. S., Babu, R. and Zaidi, P. H. (2021). Genetic gains with genomic versus phenotypic selection for drought and waterlogging tolerance in tropical maize (*Zea mays* L.), Crop J. 9(6), 1438–1448. DOI: 10.1016/j.cj.2021.03.012.

Debernardi, J. M., Tricoli, D. M., Ercoli, M. F., Hayta, S., Ronald, P., Palatnik, J. F. and Dubcovsky, J. (2020). A GRF-GIF chimeric protein improves the regeneration efficiency of transgenic plants, Nat. Biotechnol. 38(11), 1274–1279. DOI: 10.1038/s41587-020-0703-0.

Doval, M. (2019). Genomic prediction of sweet sorghum agronomic performance under drought and irrigated environments in Haiti. MSc Thesis. University of Florida, Gainesville, FL. 84 pp.

Eriksson, D., Akoroda, M., Azmach, G., Labuschagne, M., Mahungu, N. and Ortiz, R. (2018). Measuring the impact of plant breeding on sub-Saharan African staple crops, Outlook Agric. 47(3), 163–180.

Fleury, D., Jefferies, S., Kuchel, H. and Langridge, P. (2010). Genetic and genomic tools to improve drought tolerance in wheat, J. Exp. Bot. 61(12), 3211–3222.

Ganguly, M., Roychoudhury, A., Sengupta, D. N., Datta, S. K. and Datta, K. (2020). Independent overexpression of OsRab16A and AtDREB1A exhibit enhanced drought tolerance in transgenic aromatic rice variety Pusa Sugandhi 2. J. Plant Biochem. Biotechnol. 29, 503–517. DOI: 10.1007/s13562-020-00565-w.

Gedil, M. and Menkir, A. (2019). An integrated molecular and conventional breeding scheme for enhancing genetic gain in maize in Africa, Front. Plant Sci. 10, 1430. DOI: 10.3389/fpls.2019.01430.

González, F. G., Capella, M., Ribichich, K. F., Curín, F., Giacomelli, J. I., Ayala, F., Watson, G., Otegui, M. E. and Chan, R. L. (2019). Field-grown transgenic wheat expressing the sunflower gene HaHB4 significantly outyields the wild type, J. Exp. Bot. 70(5), 1669–1681. DOI: 10.1093/jxb/erz037.

Guo, J., Pradhan, S., Shahi, D., Khan, J., Mcbreen, J., Bai, G., Murphy, J. P. and Babar, M. A. (2020). Increased prediction accuracy using combined genomic information and physiological traits in a soft wheat panel evaluated in multi-environments, Sci. Rep. 10(1), 7023. DOI: 10.1038/s41598-020-63919-3.

Habib, I., Shahzad, K., Rauf, M., Ahmad, M., Alsamadany, H., Fahad, S. and Saeed, N. A. (2022). Dehydrin responsive HVA1 driven inducible gene expression enhanced salt and drought tolerance in wheat, Plant Physiol. Biochem. 180, 124-133. DOI: 10.1016/j.plaphy.2022.03.035.

He, C., Du, Y., Fu, J., Zeng, E., Park, S., White, F., Zheng, J. and Liu, S. (2020). Early drought-responsive genes are variable and relevant to drought tolerance, G3 (Bethesda) 10(5), 1657-1670. DOI: 10.1534/g3.120.401199.

Hisano, H., Abe, F., Hoffie, R. E. and Kumlehn, J. (2021). Targeted genome modifications in cereal crops, Breed. Sci. 71(4), 405-416. DOI: 10.1270/jsbbs.21019.

Hu, H. and Xiong, L. (2014). Genetic engineering and breeding of drought-resistant crops, Annu. Rev. Plant Biol. 65, 715-741.

Hunt, C. H., van Eeuwijk, F. A., Mace, E. S., Hayes, B. J. and Jordan, D. R. (2018). Development of genomic prediction in sorghum, Crop Sci. 58(2), 690-700. DOI: 10.2135/cropsci2017.08.0469.

Joshi, R. K., Bharat, S. S. and Mishra, R. (2020). Engineering drought tolerance in plants through CRISPR-Cas genome editing, 3 Biotech 10(9), 400. DOI: 10.1007/s13205-020-02390-3.

Juliana, P., Montesinos-López, O. A., Crossa, J., Mondal, S., González Pérez, L., Poland, J., Huerta-Espino, J., Crespo-Herrera, L., Govindan, V., Dreisigacker, S., Shrestha, S., Pérez-Rodríguez, P., Pinto Espinosa, F. and Singh, R. P. (2019). Integrating genomic-enabled prediction and high-throughput phenotyping in breeding for climate-resilient bread wheat, Theor. Appl. Genet. 132(1), 177-194. DOI: 10.1007/s00122-018-3206-3.

Khadka, K., Raizada, M. N. and Navabi, A. (2020). Recent progress in germplasm evaluation and gene mapping to enable breeding of drought-tolerant wheat, Front. Plant Sci. 11, 1149. DOI: 10.3389/fpls.2020.01149.

Khan, A., Sovero, V. and Gemenet, D. (2016). Genome-assisted breeding for drought resistance, Curr. Genomics 17(4), 330-342.

Khan, M. I. R., Palakolanu, S. R., Chopra, P., Rajurkar, A. B., Gupta, R., Iqbal, N. and Maheshwari, C. (2020). Improving drought tolerance in rice: ensuring food security through multi-dimensional approaches, Physiol. Plant. 172(2), 645-668 DOI: 10.1111/ppl.13223.

Khan, S., Anwar, S., Yu, S., Sun, M., Yang, Z. and Gao, Z. Q. (2019). Development of drought-tolerant transgenic wheat: achievements and limitations, Intl. J. Mol. Sci. 20(13), 3350. DOI: 10.3390/ijms20133350.

Kim, D., Alptekin, B. and Budak, H. (2017). CRISPR/Cas9 genome editing in wheat, *Funct. Integr. Genomics* 18(1), 31-41. DOI: 10.1007/s1014.

Kirschner, G. K., Rosignoli, S., Guo, L., Vardanega, I., Imani, J., Altmüller, J., Milner, S. G., Balzano, R., Nagel, K. A., Pflugfelder, D., Forestan, C., Bovina, R., Koller, R., Stöcker, T. G., Mascher, M., Simmonds, J., Uauy, C., Schoof, H., Tuberosa, R., Salvi, S. and Hochholdinger, F. (2021). ENHANCED GRAVITROPISM 2 encodes a STERILE ALPHA MOTIF-containing protein that controls root growth angle in barley and wheat, Proc. Natl. Acad. Sci. U. S. A. 118(35), e2101526118. DOI: 10.1073/pnas.2101526118.

Kumar, R., Kaur, A., Pandey, A., Mamrutha, H. M. and Singh, G. P. (2019). CRISPR-based genome editing in wheat: a comprehensive review and future prospects, Mol. Biol. Rep. 46(3), 3557-3569. DOI: 10.1007/s11033-019-04761-3.

Kumar, V. S. S., Verma, R. K., Yadav, S. K., Yadav, P., Watts, A., Rao, M. V. and Chinnusamy, V. (2020). CRISPR-Cas9 mediated genome editing of drought and salt tolerance *OsDST* gene in indica mega rice cultivar MTU1010, Physiol. Mol. Biol. Plants 26, 1099-1110. DOI: 10.1007/s12298-020-00819-w.

Li, H., Rasheed, A., Hickey, L. T. and He, Z. (2018). Fast-forwarding genetic gain, Trends Plant Sci. 23(3), 184-186. DOI: 10.1016/j.tplants.2018.01.007.

Liang, Z., Chen, K., Yan, Y., Zhang, Y. and Gao, C. (2018). Genotyping genome-edited mutations in plants using CRISPR ribonucleoprotein complexes, Plant Biotechnol. J. 16(12), 2053-2062.

Liu, S. and Qin, F. (2021). Genetic dissection of maize drought tolerance for trait improvement, Mol. Breed. 41(2), 8. DOI: 10.1007/s11032-020-01194-w.

Mallikarjuna Swamy, B. P. and Kumar, A. (2013). Genomics-based precision breeding approaches to improve drought tolerance in rice, Biotechnol. Adv. 31(8), 1308-1318.

Martignago, D., Rico-Medina, A., Blasco-Escámez, D., Fontanet-Manzaneque, J. B. and Caño-Delgado, A. I. (2020). Drought resistance by engineering plant tissue-specific responses, Front. Plant Sci. 10, 1676. DOI: 10.3389/fpls.2019.01676.

Matres, J. M., Hilscher, J., Datta, A., Armario-Nájera, V., Baysal, C., He, W., Huang, X., Zhu, C., Valizadeh-Kamran, R., Trijatmiko, K. R., Capell, T., Christou, P., Stoger, E. and Slamet-Loedin, I. H. (2021). Genome editing in cereal crops: an overview, Transgen. Res. 30(4), 461-498. DOI: 10.1007/s11248-021-00259-6.

Miao, C., Xiao, L., Hua, K., Zou, C., Zhao, Y., Bressan, R. A. and Zhu, J. K. (2018). Mutations in a subfamily of abscisic acid receptor genes promote rice growth and productivity, Proc. Natl. Acad. Sci. U. S. A. 115(23), 6058-6063. DOI: 10.1073/pnas.1804774115.

Mwadzingeni, L., Figlan, S., Shimelis, H., Mondal, S. and Tsilo, T. J. (2017). Genetic resources and breeding methodologies for improving drought tolerance in wheat, J. Crop Improv. 31(5), 648-672, DOI: 10.1080/15427528.2017.1345816.

Mwadzingeni, L., Shimeli, H., Dube, E., Laing, M. D. and Tsilo, T. J. (2016). Breeding wheat for drought tolerance: progress and technologies, J. Integr. Agric. 15(5), 935-943.

Mohammadi, R. (2018). Breeding for increased drought tolerance in wheat: a review, Crop Pasture Sci. 69(3), 223-241. DOI: 10.1071/CP17387.

Moreira, F. F., Oliveira, H. R., Volenec, J. J., Rainey, K. M. and Brito, L. F. (2020). Integrating high-throughput phenotyping and statistical genomic methods to genetically improve longitudinal traits in crops, Front. Plant Sci. 11, 681. DOI: 10.3389/fpls.2020.00681.

Nehe, A. S., Foulkes, M. J., Ozturk, I., Rasheed, A., York, L., Kefauver, S. C., Ozdemir, F. and Morgounov, A. (2021). Root and canopy traits and adaptability genes explain drought tolerance responses in winter wheat, PLoS ONE 16(4), e0242472. DOI: 10.1371/journal.pone.0242472.

Nepolean, T., Kaul, J., Mukri, G. and Mittal, S. (2018). Genomics-enabled next-generation breeding approaches for developing system-specific drought tolerant hybrids in maize, Front. Plant Sci. 9, 361. DOI: 10.3389/fpls.2018.00361.

Noor, S., Ali, S., Rahman, H., Farhatullah, F. and Ali, G. M. (2018). Comparative study of transgenic (DREB1A) and non-transgenic wheat lines on relative water content, sugar, proline and chlorophyll under drought and salt stresses, Sarhad J. Agric. 34(4), 986-993.

Nuñez-Muñoz, L., Vargas-Hernández, B., Hinojosa-Moya, J., Ruiz-Medrano, R. and Xoconostle-Cázares, B. (2021). Plant drought tolerance provided through genome editing of the trehalase gene, Plant Signal. Behav. 16(4), 1877005. DOI: 10.1080/15592324.2021.1877005.

Ober, E. S., Lahmad, S., Cockram, J., et al. (2021). Wheat root systems as a breeding target for climate resilience, Theor. Appl. Genet. 134, 1645–1662. DOI: 10.1007/s00122-021-03819-w.

Ogata, T., Ishizaki, T., Fujita, M., Forestan, C., Hickey, L. T., Kant, J., Maccaferri, M., Marr, E., Milner, M., Pinto, F., Rambla, C., Reynolds, M., Salvi, S., Sciara, G., Snowdon, R. J., Thomelin, P., Tuberosa, R., Uauy, C., Voss-Fels, K. P., Wallington, E. and Watt, M. (2020). CRISPR/Cas9-targeted mutagenesis of OsERA1 confers enhanced responses to abscisic acid and drought stress and increased primary root growth under nonstressed conditions in rice, PLoS ONE 15, e0243376. DOI: /10.1371/journal.pone.0243376.

Oladosu, Y., Raffi, M. Y., Samuel, C., Fatai, A., Magaji, U., Kareem, I., Kamarudin, Z. S., Muhammad, I. and Kolapo, K. (2019). Drought resistance in rice from conventional to molecular a review, Int. J. Mol. Sci. 20(14), 3519. DOI: 10.3390/ijms20143519.

Ortiz, R. (2013). Drought tolerance. In: Kole, C. (Ed) Genomics and Breeding for Climate-Resilient Crops (vol. 2) Target Traits. Springer, Dordrecht, The Netherlands, pp. 203–223.

Ortiz, R., Iwanaga, M., Reynolds, M., Wu, H. and Crouch, J. H. (2007). Overview on crop genetic engineering for drought-prone environments, J. Semi-Arid Trop, Agric. Res. 4(1), Available at: http://ejournal.icrisat.org/SpecialProject/sp3.pdf.

Ortiz, R., Jarvis, A., Fox, P., Aggarwal, P. K. and Campbell, B. M. (2014). Plant genetic engineering, climate change and food security, CGIAR Research Program Climate Change, Agriculture and Food Security, Copenhagen, Denmark. CCAFS Working Paper 66, 25 pp.

Pellegrineschi, A., Reynolds, M., Pacheco, M., Brito, R. M., Almeraya, R., Yamaguchi-Shinozaki, K. and Hoisington, D. (2004). Stress-induced expression in wheat of the *Arabidopsis thaliana DREB1A* gene delays water stress symptoms under greenhouse conditions, Genome 47(3), 493–500.

Polland, J., Endelman, J., Dawson, J., Rutkoski, J., Wu, S., Manes, Y., Dreisigacker, S., Crossa, J., Sánchez-Villeda, H., Sorrells, M. and Jannink, J. (2012). Genomic selection in wheat breeding using genotyping-by-sequencing, Plant Genome 5(3), 103–113. DOI: 10.3835/plantgenome2012.06.0006.

Qin, N., Xu, W., Hu, L., Li, Y., Wang, H., Qi, X., Fang, Y. and Hua, X. (2016). Drought tolerance and proteomics studies of transgenic wheat containing the maize C4 phosphoenolpyruvate carboxylase (PEPC) gene, Protoplasma 253(6), 1503–1512.

Rajput, M., Choudhary, K., Kumar, M., Vivekanand, V., Chawade, A., Ortiz, R. and Pareek, N. (2021). RNA interference and CRISPR/Cas gene editing for crop improvement: paradigm shift towards sustainable agriculture, Plants (Basel) 10(9), 1914. DOI: 10.3390/plants10091914.

Ramstein, G. P., Jensen, S. E. and Buckler, E. S. (2019). Breaking the curse of dimensionality to identify causal variants in Breeding 4, Theor. Appl. Genet. 132(3), 559-567. DOI: 10.1007/s00122-018-3267-3.

Reynolds, M. P., Atkin, O. K., Bennett, M., Cooper, M., Dodd, I. C., Foulkes, M. J., Frohberg, C., Hammer, G., Henderson, I. R., Huang, B., Korzun, V., McCouch, S. R., Messina, C. D., Pogson, B. J., Slafer, G. A., Taylor, N. L. and Wittich, P. E. (2021). Addressing research bottlenecks to crop productivity, Trends Plant Sci. 26(6), 607–630. DOI: 10.1016/j.tplants.2021.03.011.

Rimpan, T. R., Dwany, N., Meyners, G. J., Ayustin, Y. D. and Irawati, W. (2022). Insertion of cold shock gene from *Escherichia coli* in wheat plants to produce drought tolerant properties, Biotropika, J. Trop. Biol. 10, 25–32. DOI: 10.21776/ub.biotropika.2022.010.01.03.

Romero, F. M. and Gática-Arias, A. (2019). CRISPR Cas9 development and application in rice breeding, Rice Sci. 26(5), 265–281.

Rutkoski, J., Poland, J., Mondal, S., Autrique, E., Pérez, L. G., Crossa, J., Reynolds, M. and Singh, R. (2016). Canopy temperature and vegetation indices from high-throughput phenotyping improve accuracy of pedigree and genomic selection for grain yield in wheat, G3 (Bethesda) 6(9), 2799–2808. DOI: 10.1534/g3.116.032888.

Sahebi, M., Hanafi, M. M., Raffi, M. Y., Mahmud, T. M. M., Azizi, P., Osman, M., Abiri, R., Taheri, S., Kalhori, N., Shabanimofrad, M., Miah, G. and Atabaki, N. (2018). Improvement of drought tolerance in rice genetics: genomic tools and the WRKY gene family, *BioMed Res. Intl.* 2018, 3158474. DOI: 10.1155/2018/3158474.

Salvi, S. and Tuberosa, R. (2015). The crop QTLome comes of age, Curr. Opin. Biotechnol. 32, 179–185. DOI: 10.1016/j.copbio.2015.01.001.

Samantara, K., Bohra, A., Mohapatra, S. R., Prihatini, R., Asibe, F., Singh, L., Reyes, V. P., Tiwari, A., Maurya, A. K., Croser, J. S., Wani, S. H., Siddique, K. H. M. and Varshney, R. K. (2022). Breeding more crops in less time: a perspective on speed breeding, Biology 11(2), 275. DOI: 10.3390/biology11020275.

Schubert, M. G., Goodman, D. B., Wannier, T. M., Kaur, D., Farzadfard, F., Lu, T. K., Shipman, S. L. and Church, G. M. (2021). High-throughput functional variant screens via in vivo production of single-stranded DNA, Proc. Natl. Acad. Sci. U. S. A. 118(18), e2018181118. DOI: 10.1073/pnas.2018181118.

Serba, D. D. and Yadav, R. S. (2016). Genomic tools in pearl millet breeding for drought tolerance: status and prospects, Front. Plant Sci. 7, 1724. DOI: 10.3389/fpls.2016.01724.

Shabannejad, M., Bihamta, M. R., Majidi-Hervan, E., Alipour, H. and Ebrahimi, A. (2021). A classic approach for determining genomic prediction accuracy under terminal drought stress and well-watered conditions in wheat landraces and cultivars, PLoS ONE 16(3), e0247824. DOI: 10.1371/journal.pone.0247824.

Sheridan, C. (2021). Argentina first to market with drought-resistant GM wheat, Nat. Biotechnol. 39(6), 652. DOI: 10.1038/s41587-021-00963-y.

Shi, J., Gao, H., Wang, H., Lafitte, H. R., Archibald, R. L., Yang, M., Hakimi, S. M., Mo, H. and Habben, J. E. (2017). ARGOS8 variants generated by CRISPR-Cas9 improve maize grain yield under field drought stress conditions, Plant Biotechnol. J. 15(2), 207–216. DOI: 10.1111/pbi.12603.

Shikha, M., Kanika, A., Rao, A. R., Mallikarjuna, M. G., Gupta, H. S. and Nepolean, T. (2017). Genomic selection for drought tolerance using genome-wide SNPs in maize, Front. Plant Sci. 8, 550. DOI: 10.3389/fpls.2017.00550.

Snowdon, R. J., Wittkop, B., Chen, T. W. and Stahl, A. (2021). Crop adaptation to climate change as a consequence of long-term breeding, Theor. Appl. Genet. 134(6), 1613–1623. DOI: 10.1007/s00122-020-03729-3.

Srivastava, R. K., Singh, R. B., Pujarula, V. L., Bollam, S., Pusuluri, M., Chellapilla, T. S., Yadav, R. S. and Gupta, R. (2020). Genome-wide association studies and genomic selection

in pearl millet: advances and prospects, Front. Genet. 10, 1389. DOI: 10.3389/fgene.2019.01389.
Stecker, T. (2012). Drought-tolerant corn efforts show positive early results. Available at: https://www.scientificamerican.com/article/drought-tolerant-corn-trials-show-positive-early-results/ Accessed on 9th May 2022.
Tenorio Berrío, R., Nelissen, H., Inzé, D. and Dubois, M. (2022). Increasing yield on dry fields: molecular pathways with growing potential, Plant J. 109(2), 323 -341. DOI: 10.1111/tpj.15550.
Tuberosa, R. and Salvi, S. (2006). Genomics-based approaches to improve drought tolerance of crops, Trends Plant Sci. 11(8), 405–412. DOI: 10.1016/j.tplants.2006.06.003.
Van de Peer, Y., Mizrachi, E. and Marchal, K. (2017). The evolutionary significance of polyploidy, Nat. Rev. Genet. 18(7), 411–424. DOI: 10.1038/nrg.2017.26,
Varshney, R. K., Barmukh, R., Roorkiwal, M., et al. (2021a). Breeding custom-designed crops for improved drought adaptation, Adv. Genet. 2, e202100017. DOI: 10.1002/ggn2.202100017.
Varshney, R. K., Bohra, A., Yu, J., Graner, A., Zhang, Q. and Sorrells, M. E. (2021b). Designing future crops: genomics-assisted breeding comes of age, Trends Plant Sci. 26(6), 631–649. DOI: 10.1016/j.tplants.2021.03.010.
Varshney, R. K., Shi, C., Thudi, M., Mariac, C., Wallace, J., Qi, P., Zhang, H., Zhao, Y., Wang, X., Rathore, A., Srivastava, R. K., Chitikineni, A., Fan, G., Bajaj, P., Punnuri, S., Gupta, S. K., Wang, H., Jiang, Y., Couderc, M., Katta, M. A. V. S. K., Paudel, D. R., Mungra, K. D., Chen, W., Harris-Shultz, K. R., Garg, V., Desai, N., Doddamani, D., Kane, N. A., Conner, J. A., Ghatak, A., Chaturvedi, P., Subramaniam, S., Yadav, O. P., Berthouly-Salazar, C., Hamidou, F., Wang, J., Liang, X., Clotault, J., Upadhyaya, H. D., Cubry, P., Rhoné, B., Gueye, M. C., Sunkar, R., Dupuy, C., Sparvoli, F., Cheng, S., Mahala, R. S., Singh, B., Yadav, R. S., Lyons, E., Datta, S. K., Hash, C. T., Devos, K. M., Buckler, E., Bennetzen, J. L., Paterson, A. H., Ozias-Akins, P., Grando, S., Wang, J., Mohapatra, T., Weckwerth, W., Reif, J. C., Liu, X., Vigouroux, Y. and Xu, X. (2017). Pearl millet genome sequence provides a resource to improve agronomic traits in arid environments, Nat. Biotechnol. 35(10), 969–976. DOI: 10.1038/nbt.3943.
Velazco, J. G., Jordan, D. R., Hunt, C. H., Mace, E. S. and Eeuwijk, F. A. (2020). Genomic prediction for broad and specific adaptation in sorghum accommodating differential variances of SNP effects, Crop Sci. 60(5), 2328–2342. DOI: 10.1002/csc2.20221.
Velazco, J. G., Jordan, D. R., Mace, E. S., Hunt, C. H., Malosetti, M. and van Eeuwijk, F. A. (2019). Genomic prediction of grain yield and drought-adaptation capacity in sorghum is enhanced by multi-trait analysis, Front. Plant Sci. 10, 997. DOI: 10.3389/fpls.2019.00997.
Ventorim Ferrao, L. F., Ortiz, R. and Franco García, A. (2017). Genomic selection: state of the art. In: Campos, H. and Caligari, P. (Eds) Genetic Improvement of Tropical Species. Springer International Publishing, New York, pp. 19–54.
Vivek, B. S., Krishna, G. K., Vengadessan, V., Babu, R., Zaidi, P. H., Kha, L. Q., Mandal, S. S., Grudloyma, P., Takalkar, S., Krothapalli, K., Singh, I. S., Ocampo, E. T. M., Xingming, F., Burgueño, J., Azrai, M., Singh, R. P. and Crossa, J. (2017). Use of genomic estimated breeding values results in rapid genetic gains for drought tolerance in maize, Plant Genome 10(1), 1–8. DOI: 10.3835/plantgenome2016.07.0070.

Voss-Fels, K. P., Cooper, M. and Hayes, B. J. (2019). Accelerating crop genetic gains with genomic selection, Theor. Appl. Genet. 132(3), 669–686. DOI: 10.1007/s00122-018-3270-8.

Wang, D. R. (2021). Sowing the seeds of multi-genome rice, Nature 591(7851), 537–538.

Wanga, M. A., Shimelis, H., Mashilo, J. and Laing, M. D. (2021). Opportunities and challenges of speed breeding: a review, Plant Breed. 140(2), 185–194. DOI: 10.1111/pbr.12909.

Watson, A., Ghosh, S., Williams, M. J., Cuddy, W. S., Simmonds, J., Rey, M. D., Asyraf Md Hatta, M., Hinchliffe, A., Steed, A., Reynolds, D., Adamski, N. M., Breakspear, A., Korolev, A., Rayner, T., Dixon, L. E., Riaz, A., Martin, W., Ryan, M., Edwards, D., Batley, J., Raman, H., Carter, J., Rogers, C., Domoney, C., Moore, G., Harwood, W., Nicholson, P., Dieters, M. J., DeLacy, I. H., Zhou, J., Uauy, C., Boden, S. A., Park, R. F., Wulff, B. B. H. and Hickey, L. T. (2018). Speed breeding is a powerful tool to accelerate crop research and breeding, Nat. Plants 4(1), 23–29. DOI: 10.1038/s41477-017-0083-8.

Watson, G., Pautroni, L., Simonsini, S., et al. (2020). HB4® White Paper version 2.0, Rosario. Santa Fe, Argentina. Available at: https://s26.q4cdn.com/783252186/files/doc_presentations/2020/08/White-Paper-Aug-2020.pdf Accessed on 9th May 2022.

Yabe, S. and Iwata, H. (2020). Genomics-assisted breeding in minor and pseudo-cereals, Breed. Sci. 70(1), 19–31. DOI: 10.1270/jsbbs.19100.

Yadav, S., Sandhu, N., Dixit, S., Singh, V. K., Catolos, M., Mazumder, R. R., Rahman, M. A. and Kumar, A. (2021). Genomics-assisted breeding for successful development of multiple-stress-tolerant, climate-smart rice for southern and southeastern Asia, Plant Genome 14(1), e20074. DOI: 10.1002/tpg2.20074.

Yu, H., Lin, T., Meng, X., Du, H., Zhang, J., Liu, G., Chen, M., Jing, Y., Kou, L., Li, X., Gao, Q., Liang, Y., Liu, X., Fan, Z., Liang, Y., Cheng, Z., Chen, M., Tian, Z., Wang, Y., Chu, C., Zuo, J., Wan, J., Qian, Q., Han, B., Zuccolo, A., Wing, R. A., Gao, C., Liang, C. and Li, J. (2021). Road map for domesticating multi-genome rice using gene editing, Cell 184(5), 1156–1170.e14. DOI: 10.1016/j.cell.2021.01.013.

Zhang, Y., Massel, K., Godwin, I. D. and Gao, C. (2018). Applications and potential of genome editing in crop improvement, Genome Biol. 19(1), 210. DOI: 10.1186/s13059-018-1586-y.

Chapter 9

Identifying genes for yield-related traits under drought stress conditions in durum wheat*

Ilaria Marcotuli and Agata Gadaleta, University of Bari Aldo Moro, Italy; Osvin Arriagada, Samantha Reveco and Andrés R. Schwember, Pontificia Universidad Católica de Chile, Chile; Marco Maccaferri, Matteo Campana and Roberto Tuberosa, University of Bologna, Italy; Christian Alfaro, Instituto de Investigaciones Agropecuarias (INIA), Chile; and Iván Matus, Instituto de Investigaciones Agropecuarias (INIA), Chile

1 Introduction

Durum wheat (*Triticum durum* Desf.; $2n = 4x = 28$, AABB) is the 10th most important crop in the world with a cultivated area of 16 million ha and production of 40 million tons in 2017 (Dahl, 2017). It is the most important cereal in the Mediterranean region since it is deeply connected to culinary traditions in this area (Martínez-Moreno et al., 2020). Durum wheat is primarily used for making pasta and other semolina-based products, such as freekeh, couscous, bourghul, and unleavened breads, which are widely consumed in many countries (Sharma

* This chapter is an updated version of the following article: Arriagada, O., Gadaleta, A., Marcotuli, I., Maccaferri, M., Campana, M., Reveco, S., Alfaro, C., Matus, I. and Schwember, A. R. (2022). A comprehensive meta-QTL analysis for yield-related traits of durum wheat (*Triticum turgidum* L. *var. durum*) grown under different water regimes. *Frontiers in Plant Science* 13 Art. no. 984269. doi: 10.3389/fpls.2022.984269. The article is © the authors but is an open access article published under the terms of a Creative Commons Attribution (CC-BY) License. The use of material from the article is permitted, provided the original authors/copyright holders are acknowledged and the original publication is cited.

http://dx.doi.org/10.19103/AS.2022.0109.13

et al., 2019). The main global producers of durum wheat are Spain, France, Italy, and Greece in southern Europe, Morocco, Algeria, and Tunisia in northern Africa, Turkey, and Syria in southwest Asia, Canada, United States, and Mexico in North America (Martínez-Moreno et al., 2020), and Argentina and Chile in South America (Colasuonno et al., 2019).

Durum wheat is commonly grown in arid and semi-arid regions under rainfed conditions characterized by limited and irregular rainfall (Ayed et al., 2021). Water scarcity combined with high temperatures during the grain filling period significantly affects the quality and the yields of durum wheat, causing grain-yield losses of up to 50% (Dettori et al., 2017; Soriano et al., 2021). According to FAO projections, agricultural production needs to increase by about 50% by 2050 to meet the rising global demand for food (Food and Agriculture Organization of the United Nations [FAO], 2018). In this context, the development of new durum wheat high-yielding cultivars which are tolerant to abiotic stresses is a priority, particularly in the context of global warming (Shew et al., 2020).

Improving understanding of the genetic basis of grain yield and response to drought stress is an important pre-requisite for improvement of durum wheat genotypes (Arriagada et al., 2020). The identification of quantitative trait loci (QTL) associated with molecular markers is essential for understanding the genetic basis of important traits, and an effective method for improving selection efficiency in breeding programs (Soriano et al., 2021). Hundreds of QTL, using both linkage analysis and genome wide association studies (GWAS), have been mapped across the durum wheat genome, as summarized in previous studies of grain quality (Colasuonno et al., 2019; Maccaferri et al., 2019; Marcotuli et al., 2020) and grain-yield traits (Maccaferri et al., 2019; Arriagada et al., 2020). Despite these significant advances in identifying the genetic basis for different traits related to quality and yield, only a very small number of these QTL and associated markers have been utilized in breeding programs (Cobb et al., 2019). This is because most of those QTL have minor effects and their expression is significantly affected by the environment (E), the genetic (G) background, and GxE interactions (Zheng et al., 2021). Understanding adaptation to abiotic stress conditions is particularly challenging due to the quantitative genetic basis of the molecular mechanisms adopted by plants to respond to stresses (Reynolds et al., 2005).

2 Quantitative trait loci related to yield in durum wheat

Given that the grain-yield components have a quantitative inheritance, and are therefore significantly affected by the environment (Nehe et al., 2019), the development of high-yielding varieties must incorporate QTL related to tolerating water scarcity without significantly affecting growth and yield.

Numerous studies have been carried out to identify loci associated with the main yield components under irrigated and rainfed conditions in bread wheat (Gupta et al., 2020), and to a lesser extent in durum wheat (Maccaferri et al., 2019; Arriagada et al., 2020).

Grain-yield components and their interactions determine wheat yield (Li et al., 2019). Among grain-yield components, grain weight is has been addressed most in QTL studies of durum wheat, particularly 1000-grain weight (TGW). This matches meta-QTL (MQTL) analyses carried out in bread wheat, in which grains number per spike (GNPS) and TGW are the most evaluated traits under different environmental conditions (Zhang et al., 2010; Gupta et al., 2020). This is because the main approach to augmenting crop yield has been to increase the number and the weight of grains. TGW is the most important limiting factor affecting wheat yield (Liang et al., 2017). The weight of the grain is the last component of the yield to be formed, is highly dependent on the speed and duration of the grain filling period (Takai et al., 2005), and is greatly affected by the environment (Li et al., 2019). Exploring the genetic basis of TGW and its related traits can be seen as an effective approach to increase wheat yields (Würschum et al., 2018).

According to the distribution of QTL across the durum wheat genome, the chromosomes with the highest number of QTL were 2A (76), 2B (71), and 3B (71), whilst chromosome 6A was the one with the lowest number of QTL (21). These chromosomes consistently contain the greatest number of QTL for root-related traits (Soriano and Alvaro, 2019), quality-related traits, as well as traits relating to abiotic and biotic stress in durum wheat (Soriano et al., 2021).

3 Using meta-quantitative trait loci analysis to improve the identification of quantitative trait loci

MQTL analysis has been widely used to collect data about QTL from different studies with populations of different sizes and evaluated under different environmental conditions in order to identify stable QTL in plant genomes (Shariatipour et al., 2021). This method makes it possible to identify genome regions that have pleotropic effects implicated in trait variation which reduce confidence intervals of QTL. MQTL analysis is a useful approach to combine data sets and for creating consensus positions for QTL detected in individual studies. MQTL analysis makes it possible to dissect the genetic architecture of complex traits, provide a higher mapping resolution, and allow the identification of putative molecular markers useful for marker-assisted selection (MAS) (Veyrieras et al., 2007).

MQTL analyses have been carried out on crops such as rice (Islam et al., 2019), maize (Martinez et al., 2016), and barley (Zhang et al., 2017). In wheat, the highest number of MQTL analyses have been performed in common wheat

(Quraishi et al., 2011; Tyagi et al., 2015; Kumar et al., 2020; Zheng et al., 2021). There have been few MQTL studies of durum wheat (Soriano et al., 2017, 2021).

A recent MQTL study on yield under different water regimens in durum wheat (Arriagada et al., 2022) detected fewer QTL under irrigated conditions than under rainfed conditions, because most durum wheat is sown under rainfed conditions. An extensive analysis of QTL for yield-related traits under different water regimes has been conducted using 25 studies with QTL for yield components published from 2003 to 2021 based on biparental populations. The studies comprise 26 different populations with 45 parental lines. The QTL identified in each study were classified as follows:

1 QTL under rainfed conditions which correspond to QTL identified under rainfed, water-limited, and drought conditions; and
2 QTL under irrigated conditions that correspond to QTL identified under well-water, optimal, and irrigated conditions.

A total of 19 traits associated with the two main yield components (grain weight and grains number per unit area) were selected. For the MQTL analysis, only studies that included the following information were considered:

1 Type and size of the mapping population;
2 Position of QTL (peak position and/or confidence intervals);
3 LOD (logarithm of the odds) score for each QTL; and
4 Percentage of phenotypic variance explained for each QTL (PVE or r^2).

QTL that did not meet these criteria were discarded. Each QTL was treated as an independent QTL, even if some were detected in multiple environments or genetic backgrounds.

To establish the QTL positions detected in the different studies, the original QTL data were projected onto the consensus map following the homothetic approach described by Chardon et al. (2004). The MQTL analysis was conducted with the projected QTL on the consensus map using BioMercator V4.2 software (Arcade et al., 2004). Two different approaches were performed to MQTL analysis according to the number of QTL per chromosome:

- When the number of QTL per chromosome was 10 or lower, the approach of Goffinet and Gerber (2000) was used.
- When the number of QTL per chromosome was higher than 10, the approach of Veyrieras et al. (2007) was used.

The best MQTL model was chosen according to Akaike Information Criterion (AIC), corrected Akaike Information Criterion (AICc) and modified AIC with factor 3 (AIC3), Bayesian Information Criterion (BIC) and Average Weight of Evidence (AWE) criteria. The best QTL model was selected when values of the model selection criteria were the lowest in at least three of the five models (Soriano and Alvaro, 2019).

4 Identifying quantitative trait loci related to yield and drought-stress response

A total of 724 QTL distributed throughout all 14 chromosomes (genomes A and B) were collected for the 19 yield-related traits selected. Four hundred 68 QTL were reported under rainfed conditions, and 256 QTL under irrigated conditions. In general, the number of QTL per chromosome ranged from 21 on chromosome 6A to 76 on chromosomes 2A, with an average of 51 QTL per chromosome.

Looking at key yield components, 53.72% of the QTL (389) were reported for traits related to grains number per area, and 46.27% of the QTL (335) for grain weight. The trait with the highest number of reported QTL was 1000-grain weight (TGW; 204), whereas the trait with the least reported QTL was grain weight per spike (GWPS; 9), with both traits associated with the grain weight. Confidence intervals (CI) ranged from 0 to 145 cM, with an average of 24.3 cM. The 19.19% of the QTL had a CI lower than 10 cM, and about half (48.20%) had a CI lower than 20 cM. The proportion of phenotypic variance explained (PVE) for each QTL ranged from 0.007 to 0.83, with an average of 0.138. Most of the QTL (608) had a PVE lower than 0.20.

A total of 590 out of the 724 collected QTL were projected on the consensus genetic map (Table 1). 196 QTL were projected under irrigated conditions, and 394 QTL under rainfed conditions. The remaining QTL (134) were not projected because they lacked common markers between the original and the consensus maps, and/or the QTL showed low PVE causing a large CI (Soriano and Alvaro, 2019).

Under irrigated conditions, the chromosomes 7B (26) and 4A (6) had the highest and lowest number of projected QTL, respectively, with an average of 14 QTL per chromosome. The trait with the highest number of projected QTL was TGW (66 QTL:); 49.19% of the projected QTL correspond to grain weight while grains number per area had a total of 99 projected QTL.

Under rainfed conditions, the number of QTL per chromosome ranged from 10 on chromosome 6A to 44 on chromosome 1B, with an average of 28 QTL per chromosome. The trait with the highest number of QTL was TGW (114), while those with the lowest number of QTL were GWPS (1) and SL (3). Half of the projected QTL (50%) under rainfed conditions correspond to grain weight.

Table 1 List of the QTL projected throughout all 14 chromosomes (genomes A and B) of durum wheat. QTL were collected for 19 traits related to the main yield components

Trait	Chr	LOD	R2	Peak CI (cM)	Left CI (cM)	Right_CI (cM)	Condition	Reference
GYPS	1A	9.5	0.28	12.2	8.3	16.0	Rainfed	Marcotuli et al., 2017
TGW	1A	2.8	0.03	17.5	14.4	20.6	Rainfed	Mangini et al., 2021
TGW	1A	3.0	0.07	17.5	14.4	20.6	Rainfed	Mangini et al., 2021
GNPS	1A	2.7	0.13	31.4	18.0	44.8	Rainfed	Roncallo et al., 2017
SLNS	1A	8.8	0.29	36.4	30.4	42.4	Rainfed	Roncallo et al., 2017
SLNS	1A	7.3	0.23	39.6	31.9	47.3	Rainfed	Roncallo et al., 2017
TGW	1A	4.6	0.05	44.0	26.7	61.3	Irrigated	Faris et al., 2014
SLNS	1A	3.3	0.12	46.1	32.0	60.2	Rainfed	Roncallo et al., 2017
HI	1A	3.3	0.11	47.5	32.1	62.9	Rainfed	Roncallo et al., 2017
GWPS	1A	9.0	0.06	48.0	35.3	60.7	Irrigated	Faris et al., 2014
TW	1A	3.6	0.12	91.7	83.6	99.8	Irrigated	Patil et al., 2013
TGW	1A	5.2	0.09	95.9	88.3	103.5	Rainfed	Graziani et al., 2014
TGW	1A	5.3	0.06	96.9	86.5	107.3	Rainfed	Graziani et al., 2014
TGW	1A	5.1	0.06	101.6	90.7	112.5	Rainfed	Graziani et al., 2014
TGW	1A	4.4	0.07	101.6	91.7	111.5	Irrigated	Graziani et al., 2014
GNM	1A	3.0	0.04	101.6	84.8	118.4	Irrigated	Graziani et al., 2014
GNM	1A	3.6	0.05	103.1	90.0	116.2	Rainfed	Graziani et al., 2014
TGW	1A	4.4	0.04	104.1	87.8	120.5	Rainfed	Graziani et al., 2014
TGW	1A	3.4	0.05	105.0	91.7	118.4	Rainfed	Graziani et al., 2014
TGW	1A	5.2	0.06	106.1	95.4	116.9	Rainfed	Graziani et al., 2014
GNM	1A	3.6	0.08	111.1	103.1	119.1	Irrigated	Graziani et al., 2014
SNM	1A	5.3	0.08	111.3	93.6	129.0	Rainfed	Zaïm et al., 2020
TGW	1A	5.4	0.03	113.2	102.1	114.2	Rainfed	Fatiukha et al., 2020
TGW	1A	5.4	0.05	113.2	102.1	114.2	Rainfed	Fatiukha et al., 2020
TGW	1A	5.4	0.06	113.2	102.1	114.2	Rainfed	Fatiukha et al., 2020
TGW	1A	5.4	0.03	113.2	102.1	114.2	Irrigated	Fatiukha et al., 2020
TGW	1A	5.4	0.02	113.2	102.1	114.2	Irrigated	Fatiukha et al., 2020
TGW	1A	3.6	0.11	121.1	115.0	127.3	Rainfed	Graziani et al., 2014
GW	1B	9.6	0.28	4.5	2.0	7.0	Rainfed	Desiderio et al., 2019
GNPS	1B	2.9	0.13	12.0	0.0	25.3	Irrigated	Giunta et al., 2018
GY	1B	3.3	0.10	19.3	1.6	37.0	Rainfed	Roncallo et al., 2017
GNPS	1B	4.2	0.19	23.0	13.8	32.2	Irrigated	Giunta et al., 2018
SLNS	1B	7.1	0.06	28.0	14.9	41.1	Irrigated	Faris et al., 2014
TGW	1B	4.0	0.08	34.4	19.8	49.0	Rainfed	Zaïm et al., 2020
TGW	1B	4.0	0.08	34.4	16.0	52.8	Rainfed	Zaïm et al., 2020
TGW	1B	3.8	0.14	39.6	25.7	53.4	Rainfed	Blanco et al., 2012
TGW	1B	5.7	0.21	39.6	25.7	53.4	Rainfed	Blanco et al., 2012
GNPS	1B	3.4	0.13	39.6	25.7	53.4	Rainfed	Blanco et al., 2012
GYPS	1B	8.7	0.30	39.6	25.7	53.4	Rainfed	Blanco et al., 2012
TGW	1B	4.1	0.15	39.6	25.7	53.4	Rainfed	Blanco et al., 2012
TGW	1B	3.8	0.13	39.6	26.0	53.2	Rainfed	Roncallo et al., 2017

(Continued)

Table 1 *(Continued)*

Trait	Chr	LOD	R2	Peak CI (cM)	Left CI (cM)	Right_CI (cM)	Condition	Reference
TGW	1B	2.9	0.06	42.5	30.5	47.7	Rainfed	Fatiukha et al., 2020
TGW	1B	2.9	0.03	42.5	30.5	47.7	Irrigated	Fatiukha et al., 2020
GY	1B	5.1	0.13	42.7	29.6	55.8	Rainfed	Roncallo et al., 2017
TGW	1B	10.9	0.21	43.0	37.3	48.7	Irrigated	Avni et al., 2018
GL	1B	4.6	0.07	46.0	29.1	62.9	Rainfed	Russo et al., 2014
SLNS	1B	4.0	0.13	47.1	33.1	61.1	Rainfed	Roncallo et al., 2017
GW	1B	8.1	0.09	47.8	47.1	48.4	Rainfed	Mangini et al., 2021
GL	1B	3.9	0.08	47.8	47.1	48.4	Rainfed	Mangini et al., 2021
GL	1B	8.1	0.09	47.8	47.1	48.4	Rainfed	Mangini et al., 2021
GW	1B	3.9	0.08	47.8	47.1	48.4	Rainfed	Mangini et al., 2021
GW	1B	4.8	0.13	47.8	47.1	48.4	Rainfed	Mangini et al., 2021
GL	1B	4.8	0.13	47.8	47.1	48.4	Rainfed	Mangini et al., 2021
GNP	1B	2.9	0.12	48.1	33.9	62.3	Rainfed	Roncallo et al., 2017
TGW	1B	4.5	0.03	53.1	29.7	76.5	Rainfed	Graziani et al., 2014
GNM	1B	3.4	0.04	53.1	37.5	68.7	Rainfed	Graziani et al., 2014
GY	1B	3.1	0.14	53.6	40.9	66.3	Rainfed	Roncallo et al., 2017
GNM	1B	3.6	0.04	54.6	36.4	72.8	Rainfed	Graziani et al., 2014
TGW	1B	5.7	0.06	56.1	45.2	67.0	Rainfed	Graziani et al., 2014
HI	1B	5.5	0.05	58.0	39.6	76.4	Rainfed	Peleg et al., 2009
HI	1B	5.5	0.04	58.0	39.6	76.4	Irrigated	Peleg et al., 2009
TGW	1B	7.6	0.22	66.5	63.4	69.6	Rainfed	Desiderio et al., 2019
TGW	1B	2.2	0.06	73.4	70.8	81.8	Irrigated	Fatiukha et al., 2020
GNP	1B	7.0	0.32	74.0	57.0	91.0	Rainfed	Peng et al., 2003
GY	1B	7.0	0.27	76.0	63.0	89.0	Rainfed	Peng et al., 2003
SNP	1B	7.2	0.27	76.0	68.0	84.0	Rainfed	Peng et al., 2003
HI	1B	2.2	0.07	93.1	72.7	113.5	Irrigated	Rehman Arif et al., 2020
GNM	1B	3.2	0.08	95.0	86.3	103.7	Rainfed	Graziani et al., 2014
GY	1B	2.1	0.06	95.6	71.8	119.4	Irrigated	Rehman Arif et al., 2020
TGW	1B	2.1	0.08	102.6	84.7	120.5	Irrigated	Rehman Arif et al., 2020
GNPS	1B	2.1	0.06	104.6	80.8	128.4	Irrigated	Rehman Arif et al., 2020
SLNS	1B	4.3	0.47	119.6	115.9	123.3	Rainfed	Roncallo et al., 2017
GNSL	1B	2.7	0.12	123.0	99.0	147.0	Rainfed	Peng et al., 2003
TGW	1B	6.1	0.16	124.4	115.5	133.3	Rainfed	Zaïm et al., 2020
GNPS	1B	3.7	0.17	125.0	107.0	143.0	Rainfed	Peng et al., 2003
SLNS	1B	2.7	0.13	125.0	99.0	151.0	Rainfed	Peng et al., 2003
GY	1B	7.0	0.27	132.0	115.0	149.0	Rainfed	Peng et al., 2003
GNP	1B	7.0	0.32	136.0	121.0	151.0	Rainfed	Peng et al., 2003
TGW	1B	3.9	0.19	140.6	131.4	149.8	Rainfed	Roncallo et al., 2017
GY	1B	3.2	0.10	141.1	123.4	158.8	Rainfed	Roncallo et al., 2017
SNP	1B	7.2	0.27	142.0	123.0	161.0	Rainfed	Peng et al., 2003
GNPS	1B	2.5	0.11	143.8	130.8	156.8	Irrigated	Rehman Arif et al., 2020
GNPS	1B	2.8	0.08	154.6	136.7	172.5	Irrigated	Rehman Arif et al., 2020

(Continued)

Table 1 (*Continued*)

Trait	Chr	LOD	R2	Peak CI (cM)	Left CI (cM)	Right_CI (cM)	Condition	Reference
TGW	1B	2.3	0.09	221.1	205.2	237.0	Rainfed	Rehman Arif et al., 2020
TGW	1B	2.2	0.06	230.3	206.5	254.1	Rainfed	Rehman Arif et al., 2020
TGW	2A	2.1	0.08	17.5	0.0	35.4	irrigated	Rehman Arif et al., 2020
GY	2A	2.7	0.09	26.0	10.1	41.9	Rainfed	Rehman Arif et al., 2020
TGW	2A	2.7	0.04	27.0	1.3	52.7	Rainfed	Maccaferri et al., 2016
GP	2A	9.3	0.20	32.4	29.0	35.8	Rainfed	Desiderio et al., 2019
GY	2A	36.5	0.68	35.0	34.4	35.6	Rainfed	Mangini et al., 2021
GL	2A	12.7	0.15	36.0	35.6	36.4	Rainfed	Mangini et al., 2021
TGW	2A	19.6	0.27	36.0	35.6	36.4	Rainfed	Mangini et al., 2021
TGW	2A	3.5	0.05	36.0	35.6	36.4	Rainfed	Mangini et al., 2021
GW	2A	26.1	0.42	36.0	35.6	36.4	Rainfed	Mangini et al., 2021
GNPS	2A	13.9	0.43	41.5	21.0	61.9	Rainfed	Blanco et al., 2012
TGW	2A	5.1	0.19	41.5	21.0	61.9	Rainfed	Blanco et al., 2012
GYPS	2A	6.3	0.23	41.5	21.0	61.9	Rainfed	Blanco et al., 2012
GNPS	2A	5.5	0.20	41.5	21.0	61.9	Rainfed	Blanco et al., 2012
GYPS	2A	4.8	0.18	41.5	21.0	61.9	Rainfed	Blanco et al., 2012
GNM	2A	5.3	0.08	43.9	35.9	51.9	Rainfed	Graziani et al., 2014
GNM	2A	3.6	0.04	43.9	25.7	62.1	irrigated	Graziani et al., 2014
TGW	2A	6.6	0.04	44.7	41.6	52.4	Rainfed	Fatiukha et al., 2020
TGW	2A	6.6	0.13	44.7	41.6	52.4	Rainfed	Fatiukha et al., 2020
TGW	2A	6.6	0.01	44.7	41.6	52.4	Rainfed	Fatiukha et al., 2020
TGW	2A	6.6	0.02	44.7	41.6	52.4	irrigated	Fatiukha et al., 2020
TGW	2A	6.6	0.02	44.7	41.6	52.4	irrigated	Fatiukha et al., 2020
HI	2A	4.3	0.23	48.8	41.3	56.3	Rainfed	Roncallo et al., 2017
SNP	2A	3.6	0.13	55.7	42.3	69.1	Rainfed	Roncallo et al., 2017
HI	2A	5.5	0.05	58.4	52.0	64.8	Rainfed	Peleg et al., 2009
HI	2A	5.5	0.00	58.4	52.0	64.8	irrigated	Peleg et al., 2009
TGW	2A	3.2	0.14	60.9	48.1	73.7	Rainfed	Roncallo et al., 2017
SW	2A	2.2	0.06	76.4	52.6	100.2	Rainfed	Rehman Arif et al., 2020
SW	2A	3.4	0.10	76.4	62.1	90.7	irrigated	Rehman Arif et al., 2020
SL	2A	6.9	0.21	76.4	69.6	83.2	irrigated	Rehman Arif et al., 2020
GNSL	2A	3.7	0.11	79.0	62.3	95.7	Rainfed	Roncallo et al., 2017
TGW	2A	19.3	0.26	88.6	87.5	89.6	Rainfed	Mangini et al., 2021
TGW	2A	12.2	0.33	88.6	87.5	89.6	Rainfed	Mangini et al., 2021
TGW	2A	19.4	0.36	88.6	87.5	89.6	Rainfed	Mangini et al., 2021
GL	2A	28.2	0.54	88.6	87.5	89.6	Rainfed	Mangini et al., 2021
GL	2A	29.8	0.49	88.6	87.5	89.6	Rainfed	Mangini et al., 2021
GL	2A	25.5	0.39	88.6	87.5	89.6	Rainfed	Mangini et al., 2021
GW	2A	3.5	0.09	90.5	87.5	93.5	Rainfed	Mangini et al., 2021
GW	2A	7.0	0.16	90.5	87.5	93.5	Rainfed	Mangini et al., 2021
GW	2A	4.0	0.13	94.9	84.2	105.6	irrigated	Nagel et al., 2014
TGW	2A	6.8	0.05	116.2	110.9	118.5	Rainfed	Fatiukha et al., 2020
TGW	2A	6.8	0.10	116.2	110.9	118.5	irrigated	Fatiukha et al., 2020

(*Continued*)

Table 1 (*Continued*)

Trait	Chr	LOD	R2	Peak CI (cM)	Left CI (cM)	Right_CI (cM)	Condition	Reference
GY	2A	2.5	0.08	124.6	106.7	142.5	Rainfed	Rehman Arif et al., 2020
GY	2A	2.4	0.07	125.2	104.8	145.6	irrigated	Rehman Arif et al., 2020
SNP	2A	2.2	0.11	156.0	126.0	186.0	Rainfed	Peng et al., 2003
SNP	2A	3.1	0.12	162.5	147.9	177.1	Rainfed	Roncallo et al., 2017
GA	2A	5.3	0.14	177.6	172.5	182.7	Rainfed	Desiderio et al., 2019
GNPS	2A	9.3	0.38	188.0	151.0	225.0	Rainfed	Peng et al., 2003
SLNS	2A	7.6	0.35	188.2	177.7	198.7	Rainfed	Thanh et al., 2013
SNP	2A	3.5	0.17	188.2	166.5	209.9	Rainfed	Thanh et al., 2013
GY	2A	7.9	0.32	195.0	150.0	240.0	Rainfed	Peng et al., 2003
GNP	2A	7.0	0.28	201.0	166.0	236.0	Rainfed	Peng et al., 2003
SNP	2A	2.3	0.07	209.0	188.0	230.0	Rainfed	Peng et al., 2003
GYPS	2B	3.0	0.10	25.0	19.9	30.0	Rainfed	Marcotuli et al., 2017
SNM	2B	2.4	0.09	4.1	0.0	42.7	Rainfed	Golabadi et al., 2011
HI	2B	2.3	0.07	6.1	0.0	58.0	Rainfed	Golabadi et al., 2011
GNPS	2B	2.1	0.07	10.1	0.0	57.5	Rainfed	Golabadi et al., 2011
GNPS	2B	2.5	0.13	19.1	0.0	46.7	Rainfed	Golabadi et al., 2011
HI	2B	2.3	0.11	27.1	0.0	57.7	Rainfed	Golabadi et al., 2011
GYPS	2B	3.2	0.12	36.1	30.2	41.9	Rainfed	Blanco et al., 2012
GL	2B	3.5	0.07	29.0	12.3	45.7	Rainfed	Russo et al., 2014
HI	2B	2.3	0.08	37.1	0.0	81.2	Rainfed	Golabadi et al., 2011
TGW	2B	4.1	0.24	213.5	193.4	233.5	Rainfed	Patil et al., 2013
GWPS	2B	3.7	0.17	213.5	193.4	233.5	Rainfed	Patil et al., 2013
HI	2B	5.4	0.18	17.3	7.8	26.8	Rainfed	Roncallo et al., 2017
GNSL	2B	4.0	0.18	16.0	6.2	25.8	Rainfed	Roncallo et al., 2017
SLNS	2B	8.4	0.32	11.5	6.0	17.0	Rainfed	Roncallo et al., 2017
SLNS	2B	7.1	0.22	17.3	9.3	25.3	Rainfed	Roncallo et al., 2017
GY	2B	7.1	0.23	18.8	11.2	26.4	Rainfed	Roncallo et al., 2017
HI	2B	19.4	0.13	48.8	39.6	58.0	Rainfed	Peleg et al., 2009
GY	2B	3.7	0.13	27.2	13.3	41.1	Rainfed	Roncallo et al., 2017
GY	2B	5.0	0.13	69.3	60.9	77.7	Rainfed	Zaïm et al., 2020
SL	2B	4.2	0.16	33.2	24.3	42.1	Rainfed	Rehman Arif et al., 2020
GY	2B	11.9	0.13	69.4	48.4	90.4	Rainfed	Peleg et al., 2009
GA	2B	11.8	0.34	47.3	45.3	49.3	Rainfed	Desiderio et al., 2019
TGW	2B	3.9	0.10	46.8	39.9	53.7	Rainfed	Desiderio et al., 2019
GNPS	2B	5.8	0.21	93.8	87.4	100.2	Rainfed	Blanco et al., 2012
GP	2B	4.6	0.09	61.9	54.6	69.2	Rainfed	Desiderio et al., 2019
SLNS	2B	4.7	0.15	107.4	82.1	132.7	Rainfed	Thanh et al., 2013
GL	2B	13.5	0.09	78.5	71.2	85.8	Rainfed	Desiderio et al., 2019
GL	2B	6.9	0.08	78.5	70.3	86.7	Rainfed	Desiderio et al., 2019
GY	2B	4.4	0.06	107.0	84.1	129.9	Rainfed	Zaïm et al., 2020
GY	2B	4.7	0.06	103.4	87.8	127.5	Rainfed	Milner et al., 2016
GA	2B	5.4	0.11	57.5	51.3	63.7	Rainfed	Desiderio et al., 2019

(*Continued*)

Table 1 (*Continued*)

Trait	Chr	LOD	R2	Peak CI (cM)	Left CI (cM)	Right_CI (cM)	Condition	Reference
GL	2B	3.6	0.07	57.5	47.8	67.2	Rainfed	Desiderio et al., 2019
GP	2B	6.3	0.13	57.5	52.2	62.8	Rainfed	Desiderio et al., 2019
TGW	2B	4.7	0.09	123.7	103.3	144.1	Rainfed	Zaïm et al., 2020
GY	2B	4.3	0.05	135.8	112.4	159.2	Rainfed	Zaïm et al., 2020
GY	2B	11.9	0.13	129.6	111.5	147.7	Rainfed	Peleg et al., 2009
HI	2B	19.4	0.13	117.5	111.7	123.3	Rainfed	Peleg et al., 2009
TGW	2B	13.2	0.11	146.3	140.4	152.2	Rainfed	Graziani et al., 2014
TGW	2B	11.5	0.10	145.3	138.7	151.9	Rainfed	Graziani et al., 2014
TGW	2B	6.4	0.07	142.7	133.2	152.2	Rainfed	Graziani et al., 2014
TGW	2B	10.4	0.20	137.7	134.5	141.0	Rainfed	Graziani et al., 2014
GP	2B	12.9	0.32	174.2	172.1	176.3	Rainfed	Desiderio et al., 2019
GNPS	2B	4.2	0.19	32.0	22.7	41.3	Irrigated	Giunta et al., 2018
SLNS	2B	2.8	0.13	18.0	4.3	31.7	Irrigated	Giunta et al., 2018
GNSL	2B	2.8	0.13	20.0	6.4	33.6	Irrigated	Giunta et al., 2018
GNPS	2B	6.2	0.26	42.0	35.3	48.7	Irrigated	Giunta et al., 2018
SNP	2B	3.2	0.14	44.0	32.1	55.9	Irrigated	Giunta et al., 2018
TGW	2B	3.9	0.17	213.5	193.4	233.5	Irrigated	Patil et al., 2013
TGW	2B	3.3	0.20	213.5	193.4	233.5	Irrigated	Patil et al., 2013
HI	2B	19.4	0.18	48.8	39.6	58.0	Irrigated	Peleg et al., 2009
GY	2B	11.9	0.16	69.4	48.4	90.4	Irrigated	Peleg et al., 2009
TGW	2B	11.9	0.08	114.0	104.2	123.8	Irrigated	Faris et al., 2014
GWPS	2B	14.0	0.17	114.0	109.1	118.9	Irrigated	Faris et al., 2014
GNPS	2B	17.7	0.15	114.0	108.6	119.4	Irrigated	Faris et al., 2014
GNPS	2B	2.8	0.11	57.8	48.4	67.1	Irrigated	Patil et al., 2013
HI	2B	2.4	0.08	108.8	90.9	126.7	Irrigated	Rehman Arif et al., 2020
GY	2B	11.9	0.16	129.6	111.5	147.7	Irrigated	Peleg et al., 2009
HI	2B	19.4	0.18	117.5	111.7	123.3	Irrigated	Peleg et al., 2009
GNM	2B	3.0	0.03	147.3	124.0	170.7	Irrigated	Graziani et al., 2014
GNM	2B	6.0	0.09	153.5	146.3	160.7	Irrigated	Graziani et al., 2014
GNM	2B	12.9	0.12	146.3	140.7	152.0	Irrigated	Graziani et al., 2014
TGW	2B	16.2	0.14	146.3	141.6	151.1	Irrigated	Graziani et al., 2014
TGW	2B	11.0	0.11	146.3	140.5	152.2	Irrigated	Graziani et al., 2014
TGW	2B	16.3	0.15	147.3	142.9	151.7	Irrigated	Graziani et al., 2014
TGW	3A	5.3	0.06	21.2	20.8	21.6	Rainfed	Mangini et al., 2021
HI	3A	3.3	0.11	29.5	14.0	45.0	Rainfed	Roncallo et al., 2017
GL	3A	4.5	0.12	32.0	21.8	42.2	Rainfed	Russo et al., 2014
SNM	3A	4.0	0.17	38.5	28.4	48.6	Rainfed	Roncallo et al., 2017
GNSL	3A	3.0	0.09	51.9	32.4	71.4	Rainfed	Roncallo et al., 2017
GY	3A	2.4	0.07	52.1	31.7	72.5	Irrigated	Rehman Arif et al., 2020
GNPS	3A	3.5	0.16	52.4	41.6	63.2	Rainfed	Roncallo et al., 2017
GY	3A	5.4	0.15	52.9	41.2	64.6	Rainfed	Roncallo et al., 2017
TGW	3A	6.3	0.05	53.8	52.2	60.2	Rainfed	Fatiukha et al., 2020

(*Continued*)

Table 1 *(Continued)*

Trait	Chr	LOD	R2	Peak CI (cM)	Left CI (cM)	Right_CI (cM)	Condition	Reference
TGW	3A	6.3	0.04	53.8	52.2	60.2	Rainfed	Fatiukha et al., 2020
TGW	3A	6.3	0.03	53.8	52.2	60.2	Rainfed	Fatiukha et al., 2020
TGW	3A	6.3	0.04	53.8	52.2	60.2	Irrigated	Fatiukha et al., 2020
TGW	3A	6.3	0.01	53.8	52.2	60.2	Irrigated	Fatiukha et al., 2020
HI	3A	6.5	0.30	55.9	50.0	61.8	Rainfed	Roncallo et al., 2017
GNPS	3A	2.5	0.10	56.1	41.8	70.4	Irrigated	Rehman Arif et al., 2020
GNPS	3A	3.9	0.16	57.4	46.6	68.2	Rainfed	Roncallo et al., 2017
SL	3A	2.1	0.13	61.1	50.1	72.1	Irrigated	Rehman Arif et al., 2020
GNSL	3A	2.6	0.09	61.4	41.3	81.5	Rainfed	Roncallo et al., 2017
TGW	3A	5.6	0.11	65.9	55.2	76.6	Irrigated	Avni et al., 2018
GY	3A	6.9	0.18	66.4	56.7	76.1	Rainfed	Roncallo et al., 2017
GY	3A	3.0	0.09	67.0	59.6	74.4	Irrigated	Maccaferri et al., 2008
GNPS	3A	2.4	0.10	67.9	50.5	85.3	Rainfed	Roncallo et al., 2017
GNPS	3A	3.6	0.10	68.5	54.2	82.8	Rainfed	Rehman Arif et al., 2020
SW	3A	2.6	0.08	68.5	50.6	86.4	Rainfed	Rehman Arif et al., 2020
SW	3A	3.0	0.12	72.5	60.6	84.4	Irrigated	Rehman Arif et al., 2020
TGW	3A	5.0	0.18	91.8	89.8	93.8	Rainfed	Blanco et al., 2012
GNPS	3A	2.4	0.08	96.0	64.0	128.0	Rainfed	Peng et al., 2003
SW	3A	2.6	0.08	97.8	79.9	115.7	Irrigated	Rehman Arif et al., 2020
GY	3A	3.1	0.11	98.0	75.0	121.0	Rainfed	Peng et al., 2003
GNP	3A	2.4	0.08	99.0	63.0	135.0	Rainfed	Peng et al., 2003
GNSL	3A	2.9	0.10	101.0	75.0	127.0	Rainfed	Peng et al., 2003
TGW	3A	6.1	0.06	102.9	84.8	121.0	Irrigated	Tzarfati et al., 2014
TGW	3A	3.9	0.10	111.0	101.4	120.6	Rainfed	Maccaferri et al., 2016
SNP	3A	4.0	0.16	130.7	119.5	141.9	Rainfed	Roncallo et al., 2017
GP	3A	3.5	0.07	134.2	124.5	143.9	Rainfed	Desiderio et al., 2019
TGW	3A	10.4	0.20	137.7	134.5	141.0	Rainfed	Graziani et al., 2014
GW	3A	3.1	0.09	175.1	167.5	182.7	Rainfed	Desiderio et al., 2019
TGW	3B	5.7	0.21	1.9	0.0	3.8	Rainfed	Blanco et al., 2012
GNM	3B	5.5	0.07	4.0	0.0	13.5	Irrigated	Graziani et al., 2014
TGW	3B	5.6	0.11	5.5	0.0	11.3	Rainfed	Graziani et al., 2014
TGW	3B	5.6	0.15	5.6	0.0	13.4	Rainfed	Russo et al., 2014
TGW	3B	5.4	0.06	6.0	0.0	17.3	Rainfed	Graziani et al., 2014
GYPS	3B	5.1	0.19	6.2	0.0	12.4	Rainfed	Blanco et al., 2012
GYPS	3B	9.3	0.32	6.8	0.0	13.6	Rainfed	Blanco et al., 2012
GYPS	3B	4.7	0.18	6.8	0.0	13.6	Rainfed	Blanco et al., 2012
TGW	3B	16.5	0.17	7.0	3.2	10.7	Rainfed	Graziani et al., 2014
GNM	3B	9.0	0.13	7.0	1.9	12.0	Irrigated	Graziani et al., 2014
TGW	3B	10.4	0.13	7.0	1.9	12.0	Irrigated	Graziani et al., 2014
TGW	3B	5.7	0.15	7.8	0.0	19.3	Rainfed	Roncallo et al., 2017
TGW	3B	10.6	0.11	8.0	1.7	14.2	Irrigated	Graziani et al., 2014
GNPS	3B	2.0	0.06	8.3	0.0	32.1	Irrigated	Rehman Arif et al., 2020

(Continued)

Table 1 (*Continued*)

Trait	Chr	LOD	R2	Peak CI (cM)	Left CI (cM)	Right_CI (cM)	Condition	Reference
GNPS	3B	3.4	0.13	8.9	0.0	19.3	Rainfed	Blanco et al., 2012
TGW	3B	14.2	0.13	9.0	4.1	13.9	Irrigated	Graziani et al., 2014
GNM	3B	8.8	0.10	9.0	2.2	15.7	Irrigated	Graziani et al., 2014
TGW	3B	4.5	0.06	11.0	0.6	21.4	Rainfed	Graziani et al., 2014
GNM	3B	3.9	0.06	11.1	0.0	23.0	Rainfed	Graziani et al., 2014
GYPS	3B	6.7	0.24	12.4	5.3	19.5	Rainfed	Blanco et al., 2012
TGW	3B	3.4	0.13	12.4	5.3	19.5	Rainfed	Blanco et al., 2012
GYPS	3B	5.2	0.19	12.4	5.3	19.5	Rainfed	Blanco et al., 2012
GNSL	3B	4.8	0.17	12.8	2.5	23.1	Rainfed	Roncallo et al., 2017
GNPS	3B	6.0	0.20	13.0	4.4	21.6	Rainfed	Roncallo et al., 2017
GY	3B	10.7	0.38	13.0	8.4	17.6	Rainfed	Roncallo et al., 2017
GNP	3B	4.5	0.17	13.0	2.4	23.6	Rainfed	Roncallo et al., 2017
GNPS	3B	5.7	0.21	13.6	7.1	20.2	Rainfed	Blanco et al., 2012
SW	3B	2.8	0.08	13.9	0.0	31.8	Rainfed	Rehman Arif et al., 2020
GNPS	3B	3.6	0.10	13.9	0.0	28.2	Rainfed	Rehman Arif et al., 2020
TGW	3B	4.7	0.07	15.0	1.2	28.8	Rainfed	Maccaferri et al., 2016
GYPS	3B	4.1	0.13	24.6	20.6	28.5	Rainfed	Marcotuli et al., 2017
TGW	3B	2.2	0.11	26.0	0.0	58.2	Rainfed	Golabadi et al., 2011
GY	3B	4.9	0.24	27.5	20.2	34.8	Rainfed	Roncallo et al., 2017
GY	3B	3.8	0.05	32.9	5.1	60.7	Rainfed	Zaïm et al., 2020
GY	3B	7.5	0.09	32.9	20.9	44.9	Rainfed	Zaïm et al., 2020
GNP	3B	3.1	0.15	37.3	25.7	48.9	Rainfed	Roncallo et al., 2017
TGW	3B	4.2	0.05	39.3	27.1	51.4	Rainfed	Graziani et al., 2014
GNM	3B	3.7	0.07	41.3	31.5	51.0	Rainfed	Graziani et al., 2014
GY	3B	6.2	0.19	43.8	34.7	52.9	Rainfed	Zaïm et al., 2020
TGW	3B	4.5	0.09	49.1	37.0	61.2	Rainfed	Zaïm et al., 2020
TGW	3B	4.5	0.09	49.1	40.2	58.0	Rainfed	Zaïm et al., 2020
HI	3B	5.8	0.07	60.9	17.8	104.0	Rainfed	Peleg et al., 2009
HI	3B	5.8	0.03	60.9	17.8	104.0	Irrigated	Peleg et al., 2009
SLNS	3B	4.7	0.06	62.0	48.9	75.1	Irrigated	Faris et al., 2014
SNM	3B	5.3	0.13	66.5	53.2	79.8	Rainfed	Zaïm et al., 2020
TGW	3B	2.5	0.45	86.1	78.3	93.9	Rainfed	Golabadi et al., 2011
HI	3B	2.9	0.09	89.8	73.9	105.7	Irrigated	Rehman Arif et al., 2020
SNM	3B	2.3	0.49	100.1	93.0	107.2	Irrigated	Golabadi et al., 2011
TGW	3B	2.2	0.11	114.0	83.3	144.7	Irrigated	Golabadi et al., 2011
TGW	3B	4.3	0.05	116.0	100.3	131.7	Irrigated	Faris et al., 2014
TGW	3B	5.8	0.18	130.1	126.2	134.0	Rainfed	Desiderio et al., 2019
TGW	3B	5.5	0.09	169.5	155.5	183.5	Irrigated	Avni et al., 2018
TGW	3B	3.6	0.12	179.4	164.7	194.1	Rainfed	Roncallo et al., 2017
GNSL	3B	4.4	0.13	179.4	166.0	192.8	Rainfed	Roncallo et al., 2017
GNPS	3B	2.6	0.08	182.8	164.9	200.7	Irrigated	Rehman Arif et al., 2020
SW	3B	2.3	0.12	194.8	182.9	206.7	Irrigated	Rehman Arif et al., 2020

(*Continued*)

Table 1 (*Continued*)

Trait	Chr	LOD	R2	Peak CI (cM)	Left CI (cM)	Right_CI (cM)	Condition	Reference
GY	3B	3.6	0.11	205.7	192.7	218.7	Irrigated	Rehman Arif et al., 2020
GY	3B	2.2	0.06	205.7	181.9	229.5	Irrigated	Rehman Arif et al., 2020
HI	3B	2.4	0.08	205.7	187.8	223.6	Irrigated	Rehman Arif et al., 2020
HI	3B	2.8	0.09	205.7	189.8	221.6	Irrigated	Rehman Arif et al., 2020
GL	4A	7.9	0.10	23.2	16.2	30.2	Rainfed	Desiderio et al., 2019
GL	4A	12.6	0.09	23.2	15.2	31.2	Rainfed	Desiderio et al., 2019
GY	4A	2.7	0.06	28.1	0.3	55.9	Rainfed	Roncallo et al., 2017
GNSL	4A	3.7	0.11	44.2	28.3	60.1	Rainfed	Roncallo et al., 2017
GY	4A	4.9	0.14	44.2	31.9	56.5	Rainfed	Roncallo et al., 2017
SNP	4A	3.7	0.13	44.2	30.7	57.7	Rainfed	Roncallo et al., 2017
GNP	4A	4.5	0.17	44.2	33.6	54.8	Rainfed	Roncallo et al., 2017
HI	4A	4.0	0.14	45.7	33.5	57.9	Rainfed	Roncallo et al., 2017
SLNS	4A	4.2	0.15	49.8	37.8	61.8	Rainfed	Roncallo et al., 2017
GNPS	4A	3.3	0.12	49.8	35.4	64.2	Rainfed	Roncallo et al., 2017
SLNS	4A	2.6	0.09	55.3	35.4	75.2	Rainfed	Roncallo et al., 2017
GNPS	4A	3.8	0.17	56.0	45.9	66.1	Irrigated	Giunta et al., 2018
GNPS	4A	3.2	0.12	57.1	46.0	68.2	Rainfed	Blanco et al., 2012
TGW	4A	2.9	0.05	59.2	51.5	62.6	Rainfed	Fatiukha et al., 2020
TGW	4A	2.9	0.03	59.2	51.5	62.6	Irrigated	Fatiukha et al., 2020
HI	4A	4.3	0.02	71.4	54.5	88.3	Rainfed	Peleg et al., 2009
HI	4A	4.3	0.05	71.4	54.5	88.3	Irrigated	Peleg et al., 2009
GY	4A	6.6	0.04	77.8	61.1	94.5	Rainfed	Peleg et al., 2009
GY	4A	6.6	0.08	77.8	61.1	94.5	Irrigated	Peleg et al., 2009
SL	4A	6.5	0.09	78.0	69.0	87.0	Irrigated	Faris et al., 2014
SNP	4A	2.8	0.13	90.0	76.3	103.7	Irrigated	Giunta et al., 2018
GY	4A	4.0	0.11	105.4	89.8	121.0	Rainfed	Roncallo et al., 2017
GL	4A	30.0	0.30	154.5	152.2	156.8	Rainfed	Desiderio et al., 2019
GL	4A	9.9	0.13	154.5	149.2	159.8	Rainfed	Desiderio et al., 2019
GNM	4B	3.9	0.05	0.0	0.0	12.6	Rainfed	Graziani et al., 2014
SNP	4B	2.6	0.12	2.0	0.0	16.3	Irrigated	Giunta et al., 2018
TW	4B	2.8	0.09	16.7	0.0	33.4	Irrigated	Patil et al., 2013
TGW	4B	4.2	0.20	16.7	0.0	33.4	Irrigated	Patil et al., 2013
TGW	4B	3.5	0.13	16.7	0.0	33.4	Irrigated	Patil et al., 2013
GNSL	4B	2.7	0.12	18.0	4.2	31.8	Irrigated	Giunta et al., 2018
GNPS	4B	3.0	0.12	21.7	10.1	32.7	Rainfed	Blanco et al., 2012
TGW	4B	5.4	0.20	28.0	14.8	41.2	Rainfed	Russo et al., 2014
GW	4B	6.9	0.16	28.0	20.6	35.4	Rainfed	Russo et al., 2014
GNPS	4B	4.9	0.20	35.4	26.5	44.3	Rainfed	Roncallo et al., 2017
TGW	4B	4.0	0.11	35.7	22.7	48.7	Irrigated	Rehman Arif et al., 2020
TGW	4B	2.7	0.09	35.7	19.8	51.6	Irrigated	Rehman Arif et al., 2020
SLNS	4B	2.8	0.12	35.9	21.8	50.0	Rainfed	Roncallo et al., 2017
HI	4B	2.9	0.10	36.9	19.0	54.8	Rainfed	Roncallo et al., 2017

(*Continued*)

Table 1 (*Continued*)

Trait	Chr	LOD	R2	Peak CI (cM)	Left CI (cM)	Right_CI (cM)	Condition	Reference
SNM	4B	5.0	0.19	40.4	31.2	49.6	Rainfed	Roncallo et al., 2017
TGW	4B	4.2	0.12	42.5	30.6	54.4	Irrigated	Rehman Arif et al., 2020
SNP	4B	4.9	0.21	44.0	35.9	52.1	Irrigated	Giunta et al., 2018
TGW	4B	4.4	0.15	44.6	33.4	55.8	Irrigated	Patil et al., 2013
GNSL	4B	3.8	0.16	44.9	34.1	55.7	Rainfed	Roncallo et al., 2017
GNP	4B	3.4	0.15	45.4	33.5	57.3	Rainfed	Roncallo et al., 2017
TGW	4B	15.7	0.09	52.1	50.1	54.0	Rainfed	Fatiukha et al., 2020
TGW	4B	15.7	0.13	52.1	50.1	54.0	Rainfed	Fatiukha et al., 2020
TGW	4B	15.7	0.03	52.1	50.1	54.0	Rainfed	Fatiukha et al., 2020
TGW	4B	15.7	0.06	52.1	50.1	54.0	Irrigated	Fatiukha et al., 2020
TGW	4B	15.7	0.03	52.1	50.1	54.0	Irrigated	Fatiukha et al., 2020
SL	4B	7.0	0.08	54.0	43.4	64.6	Irrigated	Faris et al., 2014
TGW	4B	9.4	0.13	59.0	51.8	66.2	Rainfed	Maccaferri et al., 2016
TGW	4B	9.3	0.06	60.4	40.6	80.2	Irrigated	Avni et al., 2018
TGW	4B	3.7	0.13	60.4	51.3	69.5	Irrigated	Avni et al., 2018
TGW	4B	7.9	0.18	60.4	53.8	67.0	Irrigated	Avni et al., 2018
GY	4B	5.7	0.06	60.5	43.3	77.7	Rainfed	Peleg et al., 2009
GY	4B	5.7	0.07	60.5	43.3	77.7	Irrigated	Peleg et al., 2009
GW	4B	3.0	0.08	68.2	66.0	70.4	Rainfed	Mangini et al., 2021
TGW	4B	5.3	0.19	74.2	67.2	81.2	Rainfed	Blanco et al., 2012
GY	4B	3.8	0.18	92.8	82.8	102.8	Rainfed	Roncallo et al., 2017
GNM	4B	4.0	0.10	97.6	90.8	104.3	Rainfed	Graziani et al., 2014
TGW	4B	3.6	0.12	97.6	93.4	101.8	Rainfed	Patil et al., 2013
TGW	4B	4.4	0.04	101.6	86.0	117.1	Irrigated	Graziani et al., 2014
TGW	4B	3.4	0.04	102.5	87.6	117.4	Rainfed	Graziani et al., 2014
GNM	4B	3.2	0.05	102.5	89.1	115.9	Rainfed	Graziani et al., 2014
GNM	4B	5.1	0.05	102.5	89.1	115.9	Irrigated	Graziani et al., 2014
TGW	4B	5.1	0.09	103.5	96.2	110.9	Rainfed	Graziani et al., 2014
GNM	4B	9.7	0.15	105.5	101.2	109.8	Irrigated	Graziani et al., 2014
GNM	4B	6.3	0.11	106.5	100.3	112.7	Rainfed	Graziani et al., 2014
SL	4B	3.9	0.13	109.9	102.3	117.5	Rainfed	Patil et al., 2013
GL	4B	7.0	0.09	110.7	102.8	118.6	Rainfed	Desiderio et al., 2019
GL	4B	15.5	0.11	110.7	104.6	116.8	Rainfed	Desiderio et al., 2019
GY	4B	3.8	0.11	120.8	118.0	123.6	Irrigated	Patil et al., 2013
HI	4B	2.5	0.08	130.6	112.7	148.5	Rainfed	Rehman Arif et al., 2020
SLNS	4B	3.6	0.19	143.7	123.6	163.7	Irrigated	Patil et al., 2013
SW	5A	2.9	0.09	0.0	0.0	15.9	Rainfed	Rehman Arif et al., 2020
GNPS	5A	2.8	0.08	1.0	0.0	18.9	Rainfed	Rehman Arif et al., 2020
GNM	5A	3.0	0.06	4.0	0.0	14.2	Rainfed	Graziani et al., 2014
HI	5A	5.7	0.05	33.8	13.7	53.9	Rainfed	Peleg et al., 2009
HI	5A	5.7	0.04	33.8	13.7	53.9	Irrigated	Peleg et al., 2009
GY	5A	5.7	0.03	38.0	24.1	51.9	Rainfed	Peleg et al., 2009

(*Continued*)

Table 1 (*Continued*)

Trait	Chr	LOD	R2	Peak CI (cM)	Left CI (cM)	Right_CI (cM)	Condition	Reference
GY	5A	5.7	0.09	38.0	24.1	51.9	Irrigated	Peleg et al., 2009
GL	5A	3.8	0.05	42.1	36.8	47.3	Rainfed	Mangini et al., 2021
GNPS	5A	5.5	0.13	49.9	44.9	54.9	Rainfed	Graziani et al., 2014
GW	5A	5.8	0.15	53.6	49.1	58.1	Rainfed	Desiderio et al., 2019
SLNS	5A	3.1	0.09	54.9	36.3	73.5	Rainfed	Roncallo et al., 2017
SLNS	5A	3.0	0.33	67.8	62.4	73.2	Rainfed	Roncallo et al., 2017
GNSL	5A	8.7	0.39	69.0	51.0	87.0	Rainfed	Peng et al., 2003
GNPS	5A	10.4	0.40	75.0	53.0	97.0	Rainfed	Peng et al., 2003
TGW	5A	16.7	0.15	78.9	71.8	86.0	Irrigated	Tzarfati et al., 2014
GNP	5A	10.5	0.52	81.0	59.0	103.0	Rainfed	Peng et al., 2003
GY	5A	10.0	0.50	82.0	56.0	108.0	Rainfed	Peng et al., 2003
SLNS	5A	6.7	0.33	84.0	50.0	118.0	Rainfed	Peng et al., 2003
TGW	5A	3.2	0.12	124.7	112.8	136.6	Irrigated	Rehman Arif et al., 2020
TGW	5A	2.0	0.07	126.7	106.3	147.1	Rainfed	Rehman Arif et al., 2020
TGW	5A	13.2	0.04	128.9	127.4	131.8	Rainfed	Fatiukha et al., 2020
TGW	5A	13.2	0.03	128.9	127.4	131.8	Rainfed	Fatiukha et al., 2020
TGW	5A	13.2	0.04	128.9	127.4	131.8	Rainfed	Fatiukha et al., 2020
TGW	5A	13.2	0.10	128.9	127.4	131.8	Irrigated	Fatiukha et al., 2020
TGW	5A	13.2	0.08	128.9	127.4	131.8	Irrigated	Fatiukha et al., 2020
SNP	5A	2.3	0.08	148.0	88.0	208.0	Rainfed	Peng et al., 2003
GY	5A	2.1	0.07	158.8	138.4	179.2	Irrigated	Rehman Arif et al., 2020
TGW	5A	5.6	0.10	158.9	146.4	171.4	Irrigated	Avni et al., 2018
GNP	5A	10.5	0.52	160.0	142.0	178.0	Rainfed	Peng et al., 2003
SL	5A	26.5	0.27	162.0	159.0	165.0	Irrigated	Faris et al., 2014
SLNS	5A	6.7	0.33	163.0	117.0	209.0	Rainfed	Peng et al., 2003
TGW	5A	5.1	0.09	163.2	150.4	176.0	Irrigated	Avni et al., 2018
GY	5A	10.0	0.50	164.0	133.0	195.0	Rainfed	Peng et al., 2003
GNPS	5A	10.4	0.40	164.0	133.0	195.0	Rainfed	Peng et al., 2003
GNSL	5A	8.7	0.39	171.0	123.0	219.0	Rainfed	Peng et al., 2003
GY	5A	4.0	0.30	193.0	188.2	197.8	Irrigated	Rehman Arif et al., 2020
SW	5A	2.1	0.09	277.7	261.8	293.6	Rainfed	Rehman Arif et al., 2020
TGW	5A	2.6	0.10	277.7	263.4	292.0	Rainfed	Rehman Arif et al., 2020
GY	5B	3.6	0.09	0.0	0.0	20.6	Rainfed	Roncallo et al., 2017
TGW	5B	4.5	0.08	12.9	0.0	61.5	Rainfed	Thanh et al., 2013
GNPS	5B	2.3	0.07	13.4	0.0	33.8	Irrigated	Rehman Arif et al., 2020
TGW	5B	4.5	0.06	25.9	0.0	86.9	Rainfed	Thanh et al., 2013
SL	5B	3.0	0.10	33.3	19.0	47.6	Irrigated	Rehman Arif et al., 2020
TGW	5B	4.6	0.06	43.2	22.0	64.5	Irrigated	Avni et al., 2018
TGW	5B	2.0	0.05	62.0	33.4	90.6	Irrigated	Rehman Arif et al., 2020
GW	5B	5.9	0.18	75.2	71.5	78.9	Rainfed	Desiderio et al., 2019
TGW	5B	5.4	0.15	75.2	70.7	79.7	Rainfed	Desiderio et al., 2019
TGW	5B	4.2	0.12	75.2	69.6	80.8	Rainfed	Desiderio et al., 2019

(*Continued*)

Table 1 (*Continued*)

Trait	Chr	LOD	R2	Peak CI (cM)	Left CI (cM)	Right_CI (cM)	Condition	Reference
TGW	5B	2.9	0.15	75.8	64.0	87.6	Rainfed	Roncallo et al., 2017
TGW	5B	8.0	0.22	85.2	82.1	88.3	Rainfed	Desiderio et al., 2019
GA	5B	8.3	0.18	85.2	81.4	89.0	Rainfed	Desiderio et al., 2019
GY	5B	4.9	0.07	101.9	86.5	117.3	Rainfed	Zaïm et al., 2020
GNPS	5B	5.4	0.06	106.0	91.4	120.6	Irrigated	Faris et al., 2014
GWPS	5B	4.6	0.07	106.0	94.4	117.6	Irrigated	Faris et al., 2014
GY	5B	4.1	0.06	116.9	103.1	130.7	Rainfed	Zaïm et al., 2020
HI	5B	4.7	0.00	117.3	96.2	138.4	Rainfed	Peleg et al., 2009
HI	5B	4.7	0.00	117.3	96.2	138.4	Irrigated	Peleg et al., 2009
GNPS	5B	4.4	0.33	132.2	127.9	136.5	Irrigated	Rehman Arif et al., 2020
HI	5B	3.3	0.11	143.2	130.2	156.2	Rainfed	Rehman Arif et al., 2020
GY	5B	2.7	0.09	143.2	127.3	159.1	Rainfed	Rehman Arif et al., 2020
GNPS	5B	2.1	0.07	143.2	122.8	163.6	Irrigated	Rehman Arif et al., 2020
GA	5B	4.1	0.10	179.5	172.8	186.2	Rainfed	Desiderio et al., 2019
SNM	5B	6.1	0.07	195.8	183.3	208.3	Rainfed	Zaïm et al., 2020
GL	5B	5.8	0.12	205.8	200.1	211.5	Rainfed	Desiderio et al., 2019
GNPS	5B	3.2	0.11	218.3	205.3	231.3	Rainfed	Rehman Arif et al., 2020
HI	5B	2.8	0.10	228.6	214.3	242.9	Irrigated	Rehman Arif et al., 2020
GNPS	5B	2.8	0.09	232.7	216.8	248.6	Irrigated	Rehman Arif et al., 2020
GNPS	5B	2.3	0.07	233.1	212.7	253.5	Irrigated	Rehman Arif et al., 2020
SW	5B	2.4	0.09	237.1	221.2	253.0	Irrigated	Rehman Arif et al., 2020
SW	5B	3.7	0.15	260.0	250.5	269.5	Irrigated	Rehman Arif et al., 2020
SW	5B	2.2	0.06	270.9	247.1	294.7	Rainfed	Rehman Arif et al., 2020
GNPS	5B	2.9	0.09	273.9	258.0	289.8	Irrigated	Rehman Arif et al., 2020
TGW	6A	8.4	0.09	5.8	0.0	7.9	Rainfed	Fatiukha et al., 2020
TGW	6A	8.4	0.09	5.8	0.0	7.9	Irrigated	Fatiukha et al., 2020
TGW	6A	2.7	0.03	8.0	0.0	34.2	Rainfed	Maccaferri et al., 2016
TGW	6A	3.8	0.04	32.2	17.0	47.5	Irrigated	Graziani et al., 2014
TGW	6A	4.3	0.33	43.8	41.8	45.8	Rainfed	Graziani et al., 2014
TW	6A	5.3	0.19	54.1	50.4	57.8	Rainfed	Desiderio et al., 2019
SLNS	6A	4.9	0.19	55.7	46.5	64.9	Rainfed	Roncallo et al., 2017
TGW	6A	3.5	0.05	60.5	38.1	82.9	Rainfed	R, 2018
TGW	6A	3.8	0.08	60.5	46.2	74.8	Irrigated	Avni et al., 2018
TGW	6A	4.4	0.06	60.5	38.9	82.1	Irrigated	Avni et al., 2018
GNSL	6A	3.1	0.14	64.0	51.5	76.5	Irrigated	Giunta et al., 2018
GY	6A	2.6	0.06	64.9	37.1	92.7	Rainfed	Roncallo et al., 2017
TGW	6A	4.1	0.10	97.2	85.4	109.0	Rainfed	Zaïm et al., 2020
GW	6A	5.6	0.20	97.4	93.9	100.9	Rainfed	Desiderio et al., 2019
TW	6A	5.4	0.19	97.4	93.8	101.0	Rainfed	Desiderio et al., 2019
GNPS	6A	3.8	0.17	105.0	94.7	115.3	Irrigated	Giunta et al., 2018
GP	6B	6.2	0.22	0.0	0.0	3.2	Rainfed	Desiderio et al., 2019
GL	6B	5.9	0.18	0.0	0.0	3.8	Rainfed	Desiderio et al., 2019
GA	6B	4.4	0.15	0.5	0.0	5.0	Rainfed	Desiderio et al., 2019

(*Continued*)

Table 1 *(Continued)*

Trait	Chr	LOD	R2	Peak CI (cM)	Left CI (cM)	Right_CI (cM)	Condition	Reference
GL	6B	8.5	0.27	0.5	0.0	3.0	Rainfed	Desiderio et al., 2019
GP	6B	5.4	0.15	1.0	0.0	5.7	Rainfed	Desiderio et al., 2019
GP	6B	4.9	0.18	1.0	0.0	4.9	Rainfed	Desiderio et al., 2019
GL	6B	5.7	0.15	1.0	0.0	5.6	Rainfed	Desiderio et al., 2019
GL	6B	6.8	0.18	2.0	0.0	5.8	Rainfed	Desiderio et al., 2019
GP	6B	5.8	0.16	2.0	0.0	6.3	Rainfed	Desiderio et al., 2019
GW	6B	3.9	0.04	2.2	2.0	2.4	Rainfed	Mangini et al., 2021
GL	6B	4.5	0.05	4.1	3.9	4.3	Rainfed	Mangini et al., 2021
GL	6B	2.6	0.03	4.1	3.9	4.3	Rainfed	Mangini et al., 2021
TGW	6B	6.1	0.05	5.1	0.0	29.0	Rainfed	Zaïm et al., 2020
GL	6B	3.7	0.11	8.2	1.9	14.5	Rainfed	Desiderio et al., 2019
GA	6B	3.9	0.12	8.6	2.7	14.5	Rainfed	Desiderio et al., 2019
SNP	6B	5.4	0.23	17.0	9.5	24.5	Irrigated	Giunta et al., 2018
SNP	6B	6.2	0.26	31.0	24.4	37.6	Irrigated	Giunta et al., 2018
GA	6B	3.5	0.12	35.1	29.4	40.8	Rainfed	Desiderio et al., 2019
TGW	6B	4.4	0.05	48.5	26.0	70.9	Irrigated	Avni et al., 2018
TGW	6B	3.0	0.12	51.2	39.3	63.1	Rainfed	Elouafi and Nachit, 2004
GA	6B	7.6	0.24	53.8	51.0	56.6	Rainfed	Desiderio et al., 2019
GA	6B	4.2	0.14	54.2	49.5	58.9	Rainfed	Desiderio et al., 2019
TGW	6B	4.1	0.06	57.0	37.5	76.5	Irrigated	Avni et al., 2018
GNPS	6B	3.6	0.16	59.0	48.4	69.6	Irrigated	Giunta et al., 2018
TGW	6B	4.4	0.15	59.8	48.1	71.5	Rainfed	Roncallo et al., 2017
TGW	6B	4.2	0.16	64.9	53.9	75.9	Rainfed	Roncallo et al., 2017
SLNS	6B	3.5	0.16	69.0	58.0	80.0	Irrigated	Giunta et al., 2018
HI	6B	6.7	0.05	69.2	47.8	90.6	Rainfed	Peleg et al., 2009
HI	6B	6.7	0.04	69.2	47.8	90.6	Irrigated	Peleg et al., 2009
TW	6B	3.0	0.09	74.2	58.3	90.1	Rainfed	Elouafi and Nachit, 2004
GNSL	6B	3.0	0.09	76.4	57.8	95.0	Rainfed	Roncallo et al., 2017
SLNS	6B	4.0	0.18	87.0	77.3	96.7	Irrigated	Giunta et al., 2018
SNP	6B	3.3	0.15	87.0	75.4	98.6	Irrigated	Giunta et al., 2018
TGW	6B	7.5	0.09	88.1	87.9	88.3	Rainfed	Mangini et al., 2021
SNP	6B	4.3	0.19	90.0	80.9	99.1	Irrigated	Giunta et al., 2018
GL	6B	7.8	0.08	94.5	89.8	99.1	Rainfed	Mangini et al., 2021
TGW	6B	12.7	0.13	95.7	87.0	104.4	Irrigated	Tzarfati et al., 2014
GNPS	6B	2.8	0.10	97.0	79.3	114.7	Rainfed	Roncallo et al., 2017
TGW	6B	3.1	0.12	97.4	82.4	112.4	Rainfed	Roncallo et al., 2017
TGW	6B	4.7	0.17	107.0	101.1	112.9	Rainfed	Blanco et al., 2012
SLNS	6B	3.8	0.08	116.0	77.0	155.0	Rainfed	Peng et al., 2003
GY	6B	2.1	0.07	121.7	101.3	142.1	Irrigated	Rehman Arif et al., 2020
GY	6B	3.0	0.09	123.0	115.3	130.7	Irrigated	Maccaferri et al., 2008
SNP	6B	4.0	0.18	125.0	115.3	134.7	Irrigated	Giunta et al., 2018
TGW	6B	2.6	0.08	144.1	126.2	162.0	Irrigated	Rehman Arif et al., 2020

(Continued)

Table 1 *(Continued)*

Trait	Chr	LOD	R2	Peak CI (cM)	Left CI (cM)	Right_CI (cM)	Condition	Reference
TGW	6B	2.7	0.09	160.6	144.7	176.5	Rainfed	Rehman Arif et al., 2020
TGW	7A	5.3	0.10	9.2	1.2	17.2	Rainfed	Zaïm et al., 2020
GWPS	7A	2.3	0.08	16.1	2.0	30.1	Irrigated	Golabadi et al., 2011
SLNS	7A	3.7	0.17	41.2	31.1	51.2	Rainfed	Patil et al., 2013
TW	7A	3.0	0.17	55.9	47.5	64.3	Rainfed	Elouafi and Nachit, 2004
SW	7A	2.0	0.08	57.8	39.9	75.7	Rainfed	Rehman Arif et al., 2020
TW	7A	4.1	0.17	62.9	51.2	74.5	Irrigated	Patil et al., 2013
SLNS	7A	5.0	0.23	63.9	56.3	71.5	Rainfed	Roncallo et al., 2017
GYPS	7A	4.7	0.17	64.4	56.6	72.3	Rainfed	Blanco et al., 2012
GNPS	7A	4.3	0.06	72.0	57.4	86.6	Irrigated	Faris et al., 2014
GWPS	7A	5.0	0.03	72.0	43.9	100.1	Irrigated	Faris et al., 2014
GNPS	7A	3.0	0.10	74.8	56.9	92.7	Rainfed	Roncallo et al., 2017
SNM	7A	6.1	0.24	74.8	67.6	82.0	Rainfed	Roncallo et al., 2017
GL	7A	5.4	0.07	76.4	59.0	93.8	Rainfed	Russo et al., 2014
TGW	7A	3.1	0.08	81.0	74.5	87.4	Irrigated	Patil et al., 2013
HI	7A	2.8	0.14	87.3	74.5	100.1	Rainfed	Roncallo et al., 2017
SW	7A	2.6	0.07	98.8	78.4	119.2	Irrigated	Rehman Arif et al., 2020
HI	7A	2.4	0.07	108.6	88.2	129.0	Rainfed	Rehman Arif et al., 2020
GP	7A	6.9	0.15	108.7	113.3	113.3	Rainfed	Desiderio et al., 2019
SNP	7A	5.3	0.36	109.0	82.0	136.0	Rainfed	Peng et al., 2003
GY	7A	7.1	0.09	110.9	98.8	123.4	Rainfed	Milner et al., 2016
TGW	7A	2.1	0.05	113.4	84.8	142.0	Irrigated	Rehman Arif et al., 2020
GY	7A	3.0	0.08	130.0	121.8	138.2	Irrigated	Maccaferri et al., 2008
SLNS	7A	7.8	0.12	132.0	125.1	138.9	Irrigated	Faris et al., 2014
TGW	7A	3.5	0.08	133.8	119.3	148.3	Rainfed	R, 2018
GY	7A	4.4	0.04	138.7	119.5	157.9	Rainfed	Zaïm et al., 2020
TGW	7A	4.0	0.05	148.1	123.8	172.4	Irrigated	Avni et al., 2018
GNP	7A	2.0	0.20	154.0	82.0	226.0	Rainfed	Peng et al., 2003
SL	7A	2.1	0.07	217.8	197.4	238.2	Rainfed	Rehman Arif et al., 2020
TGW	7A	2.9	0.10	220.8	206.5	235.1	Irrigated	Rehman Arif et al., 2020
GY	7B	3.0	0.04	0.0	0.0	16.0	Irrigated	Maccaferri et al., 2008
TGW	7B	2.1	0.06	0.1	0.0	62.8	Rainfed	Golabadi et al., 2011
GY	7B	3.9	0.12	7.0	0.0	21.4	Rainfed	Roncallo et al., 2017
SLNS	7B	5.7	0.07	10.0	0.0	21.0	Irrigated	Faris et al., 2014
TGW	7B	3.2	0.12	10.1	0.0	39.9	Rainfed	Golabadi et al., 2011
GY	7B	3.0	0.04	12.5	0.0	27.4	Irrigated	Maccaferri et al., 2008
HI	7B	14.1	0.22	14.1	11.5	16.7	Rainfed	Peleg et al., 2009
HI	7B	14.1	0.13	14.1	11.5	16.7	Irrigated	Peleg et al., 2009
TGW	7B	3.0	0.12	14.9	0.0	30.0	Rainfed	Roncallo et al., 2017
GY	7B	8.6	0.13	17.8	0.8	34.8	Rainfed	Peleg et al., 2009
GY	7B	8.6	0.08	17.8	0.8	34.8	Irrigated	Peleg et al., 2009
GNM	7B	3.8	0.04	21.1	5.2	37.1	Irrigated	Graziani et al., 2014

(Continued)

Table 1 (*Continued*)

Trait	Chr	LOD	R2	Peak CI (cM)	Left CI (cM)	Right_CI (cM)	Condition	Reference
TGW	7B	3.3	0.03	23.9	4.1	43.8	Rainfed	Graziani et al., 2014
GNM	7B	3.8	0.07	25.9	17.0	34.9	Irrigated	Graziani et al., 2014
GY	7B	3.0	0.08	38.0	29.7	46.3	Rainfed	Maccaferri et al., 2008
GY	7B	3.7	0.12	44.1	41.5	46.7	Rainfed	Mangini et al., 2021
GP	7B	5.8	0.12	55.9	50.1	61.7	Rainfed	Desiderio et al., 2019
GWPS	7B	3.1	0.10	59.4	57.9	60.8	Irrigated	Patil et al., 2013
GL	7B	20.3	0.16	63.8	59.6	68.0	Rainfed	Desiderio et al., 2019
GL	7B	13.5	0.19	63.8	60.2	67.4	Rainfed	Desiderio et al., 2019
TGW	7B	8.4	0.04	71.6	69.4	76.9	Rainfed	Fatiukha et al., 2020
TGW	7B	8.4	0.07	71.6	69.4	76.9	Rainfed	Fatiukha et al., 2020
TGW	7B	8.4	0.01	71.6	69.4	76.9	Rainfed	Fatiukha et al., 2020
TGW	7B	8.4	0.03	71.6	69.4	76.9	Irrigated	Fatiukha et al., 2020
TGW	7B	8.4	0.05	71.6	69.4	76.9	Irrigated	Fatiukha et al., 2020
TGW	7B	3.0	0.12	74.3	59.7	88.9	Rainfed	Roncallo et al., 2017
SLNS	7B	4.7	0.20	82.0	73.6	90.4	Irrigated	Giunta et al., 2018
GYPS	7B	3.3	0.13	86.2	75.5	96.9	Rainfed	Blanco et al., 2012
TGW	7B	3.4	0.13	86.2	75.8	96.6	Rainfed	Blanco et al., 2012
GYPS	7B	4.2	0.16	86.2	77.7	94.7	Rainfed	Blanco et al., 2012
SL	7B	3.9	0.18	89.9	82.0	97.9	Irrigated	Rehman Arif et al., 2020
SNP	7B	2.9	0.13	90.0	77.0	103.0	Irrigated	Giunta et al., 2018
SW	7B	3.1	0.09	92.2	76.3	108.1	Irrigated	Rehman Arif et al., 2020
GNPS	7B	4.3	0.19	94.0	84.8	103.2	Irrigated	Giunta et al., 2018
SNP	7B	4.0	0.18	100.0	90.3	109.7	Irrigated	Giunta et al., 2018
GW	7B	4.5	0.10	108.1	107.9	108.3	Rainfed	Mangini et al., 2021
GY	7B	3.9	0.12	110.6	95.6	125.6	Rainfed	Zaïm et al., 2020
GY	7B	3.0	0.07	113.0	103.5	122.5	Rainfed	Maccaferri et al., 2008
GNSL	7B	3.3	0.12	117.4	102.2	132.6	Rainfed	Roncallo et al., 2017
SL	7B	2.4	0.10	117.6	103.3	131.9	Irrigated	Rehman Arif et al., 2020
GNSL	7B	2.6	0.09	124.0	71.0	177.0	Rainfed	Peng et al., 2003
TGW	7B	4.1	0.40	140.5	136.1	144.9	Rainfed	Roncallo et al., 2017
TGW	7B	11.9	0.41	149.6	146.1	153.1	Irrigated	Nagel et al., 2014
GA	7B	26.5	0.53	149.6	146.9	152.3	Irrigated	Nagel et al., 2014
GL	7B	37.3	0.58	149.6	147.1	152.1	Irrigated	Nagel et al., 2014
GA	7B	3.4	0.11	167.8	154.8	180.8	Irrigated	Nagel et al., 2014
GNPS	7B	2.1	0.07	172.6	152.2	193.0	Rainfed	Rehman Arif et al., 2020
GA	7B	21.0	0.59	173.1	170.7	175.5	Irrigated	Nagel et al., 2014
GL	7B	32.0	0.64	173.1	170.9	175.3	Irrigated	Nagel et al., 2014
HI	7B	2.9	0.09	184.4	168.5	200.3	Irrigated	Rehman Arif et al., 2020
SLNS	7B	4.3	0.14	184.6	172.3	196.9	Rainfed	Roncallo et al., 2017
GY	7B	2.5	0.08	218.2	200.3	236.1	Irrigated	Rehman Arif et al., 2020
GY	7B	3.0	0.09	219.1	203.2	235.0	Rainfed	Rehman Arif et al., 2020
TGW	7B	2.7	0.09	275.2	259.3	291.1	Irrigated	Rehman Arif et al., 2020

Overall, out of the 590 QTL projected on the consensus map, 421 were grouped into 76 MQTL. The rest of the QTL (169) remained as single QTL since they did not overlap with any MQTL interval, the QTL overlapped with more than one MQTL due to their large CI, or because the predicted QTL peaks were not included within any MQTL. 28 and 48 MQTL were identified under irrigated and rainfed conditions, respectively.

Under rainfed conditions, the number of QTL per MQTL varied from 2 on several chromosomes to more than 20 on chromosomes 1B (YIELD_MQTL1B.1_D; 47.7 cM) and 3B (YIELD_MQTL3B.1_D; 9.8 cM), with an average of 5.89 QTL per MQTL. Under irrigation conditions, the number of QTL per MQTL varied from 2 to 8 on chromosome 3B (YIELD_MQTL3B.3_I; 206.94 cM), with an average of 3.75 QTL per MQTL. The number of MQTL per chromosome varied from 1 on chromosome 1B (under irrigated conditions) to 5 on chromosome 2 (A and B) under rainfed conditions. In addition, no MQTL were detected on chromosome 1A under irrigated conditions. The number of traits involved per MQTL ranged from 1 to 9 in the MQTL YIELD_MQTL1B.1_D (1B), which also contains the largest number of studies (7). Finally, the CI of the MQTL ranged from 0.12 to 25.96 cM with an average of 6.79 cM, which is significantly lower than the average of CI (24.3 cM) considering the original QTL. Interestingly, in 12 regions of the genome (on all chromosomes except 1A, 4A, 5A, and 6B), the MQTL detected under both water conditions overlapped.

Twelve regions of the genome were overlapping for both rainfed and irrigated conditions. A new MQTL was detected on chromosome 5A (YIELD_MQTL5A.1_D) relating to genes activated only during stress conditions. QTL for stress conditions were also reported by Soriano et al. (2021) on chromosome 5A. Chromosome 5A seems to have an important role in both yield and adaptation. This may be due to the presence of the vernalization genes *Vrn-A1*. Favorable alleles for this gene during breeding help develop a spring habit for flowering without cold requirements. This can be used as a strategy for introgression of important target traits from non-adapted pre-breeding materials combining the most favorable vernalization alleles (Soriano et al., 2021).

Arriagada et al. (2022) described an interesting MQTL on chromosome 2A YIELD_MQTL2A.2_D (map position 51.86 cM) linked to TGW. HI was co-localized with a MQTL previously described for different traits in durum wheat by Soriano et al. (2021). These authors identified a MQTL on the chromosome 2A at 50.8 cM (durumMQTL2A.3) associated with traits related to abiotic stress. Specifically, normalized difference vegetation index (NDVI) and chlorophyll content (SPAD) were identified in that genetic region, which are associated with grain yield under drought stress (Cairns et al., 2012). Considering the map position of the two MQTL, this strong MQTL for stress and for yield under rainfed conditions can be useful in durum wheat breeding programs, in which the objective is to increase grain yield under drought conditions.

5 Candidate genes for yield-related traits under drought stress conditions

QTL involved in grain-yield-related traits in durum wheat grown under different water regimes were projected onto the durum wheat consensus map (Maccaferri et al., 2015) for further comparison. Gene annotations for the most important marker-trait associations (MTAs) was performed using the high confidence genes reported for the wheat genome sequence (Svevo browser), available at https://wheat.pw.usda.gov/GG3/ jbrowse_Durum_Svevo. The marker locations were defined by flanking marker positions and CI of the MQTL. Gene model regulations were obtained through in-silico expression analysis, using RNAseq data, filtered for drought and drought combined with heat stress, allowing identification of up-regulated genes. Gene models were primarily identified by 'Chinese spring' annotated sequences and subsequently homologous genes from 'Svevo'. Gene models involved in drought stress during plant development and spike drought during early booting were analyzed using in-silico expression data from a database defined by the markers flanking the MQTL.

A total of 44 genes were detected and used to determine differentially expressed genes (DEG) up/down regulated under drought/heat conditions using the RNAseq data available at http://www.wheat-expression.com/. During the exposure to water stress conditions, the 15 most expressed genes, showing the higher expression level (tmp > 3) were associated to MQTL under both irrigated and rainfed conditions (Fig. 1).

A candidate gene identified on chromosome 2A and associated with a MQTL for harvest index and spike length was the CBL-interacting protein kinase 2-like gene. This is involved in CIPK serine-threonine protein kinase interaction with activation of kinase occurring in a calcium-dependent manner. This gene plays a positive regulatory effect in drought stress response. Wang et al. (2016) found that the over-expression of the TaCBL-CIPK2 gene confers drought tolerance in transgenic tobacco plants by regulating stomatal closure.

Another important gene detected on chromosome 2A was the endo-1,4-beta-xylanase 1-like gene, involved in the hydrolyzation of xylan backbones into shorter, soluble xylo-oligo saccharides. Xylanase is strongly expressed in tolerant barley genotypes under drought stress through mobilizing nutrients from the aleurone layer and endosperm in the developing seed (Hajibarat et al., 2022).

Among the gene models detected, different disease resistance RGA proteins were identified: specifically RGA3, RGA4, and RGA5 on chromosomes 5A, 5B, and 7B under both irrigated and rainfed condition. The RGA genes have been identified primarily in response to biotic stresses such as fungal pathogens (Huang and Gill, 2001; Césari et al., 2014) and subsequently for

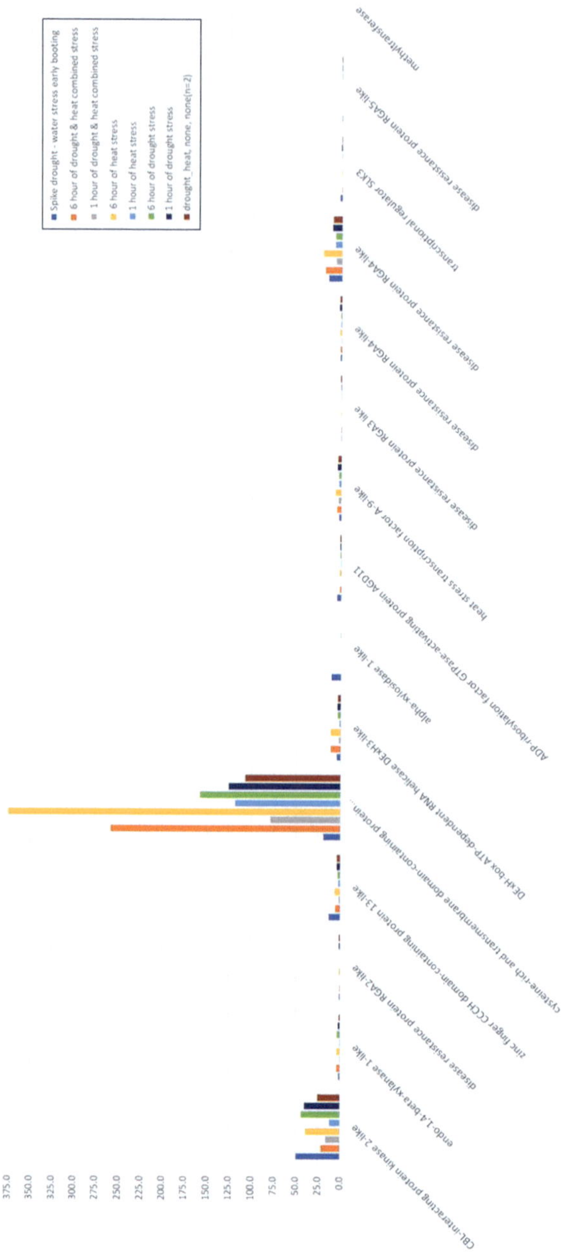

Figure 1 Expressed genes identified in MQTL under irrigated and rainfed conditions.

drought stress, due to interaction with other proteins which positively affect ABA biosynthesis in seed germination (Skubacz et al., 2016) and flag leaves (Onyemaobi et al., 2021).

On chromosome 3A, the zing finger CCCH domain protein 13-like gene was found to have a function in plant development and tolerance to abiotic stresses such as salt, drought, flooding, cold temperatures, and oxidative stress (Han et al., 2021). Another candidate is the cysteine rich and transmembrane domain-containing protein WIH1-like gene, which is involved in megasporogenesis and germ cell formation from somatic precursor cells (Lieber et al., 2011).

A *DExH-box ATP-dependent RNA helicase DExH3-like* (*DExHbox RHs*) gene, which is involved in biotic and abiotic stresses response as well as plant development, was also identified and associated to MQTL on chromosome 3B for most of the traits for grain weight and grain number per area which have been considered. Recently, the relationship between *DExHbox RHs* and temperature stress tolerance has been reported in Arabidopsis (Liu and Imai, 2018). Another gene detected on chromosome 3B was the *α-xylosidase 1-like* gene, which helps maintain the mechanical integrity of the primary cell wall in growing tissues. In *Arabidospis* mutant for *α-xylosidase* the expression of genes encoding specific abscisic acid and gibberellin enzymes has been altered (Shigeyama et al., 2016). Considering that abscisic acid is involved in plant adaptation to environmental stresses (Audran et al., 2001), an indirect correlation between the expression of the *α-xylosidase 1-like* can be assumed in response to water regimes. One additional gene identified on chromosome 3B was *ADP-ribosylation factors GTPase-activating* protein AGD11, which has a function in diverse physiological and molecular activities, including in conferring tolerance to biotic and abiotic stresses in rice and foxtail millet (Muthamilarasan et al., 2016).

Heat stress transcription factors have been detected for MQTL on chromosome 4B. These factors play a crucial role in response to high temperature, salinity, and drought because they adversely affect survival, growth, and reproduction by regulating the expression of stress-responsive genes, such as heat shock proteins (Guo et al., 2016). One gene was identified on chromosome 7A associated with a MQTL for TW, GWPS, GNPS, and TGW. This was the transcriptional regulator SLK3, which encodes a regulator of *AGAMOUS* gene and functions together with a repressor of the *AGAMOUS* gene, the *LEUNIG* gene. One experiment in *Arabidopsis* with loss-of-function mutants of *AGAMOUS* showed that repression of the gene by transcriptional regulator SLK3 induced a replacement of stamens with petals and carpels with a new flower (Franks et al., 2002).

On chromosome 7B, a *methyltransferase* involved in DNA methylation at cytosine residues and required for gene expression control and genome stability (Thomas et al., 2014), was detected, correlating to a MQTL for TGW,

GL, SLNS. This gene appeared to be expressed as a response to stress for the regulation of developmental events such as dormancy (Gianoglio et al., 2017), and in response to damaged proteins (Krzewska et al., 2021).

6 Conclusion

Drought is one of the main environmental factors that affects the grain yield and the plant architecture of durum wheat, for this reason there is the necessity to develop new cultivars well adapted to abiotic stress conditions. Yield and their components are complex traits controlled by many QTLs with small additive effects; in this context, hundreds of QTLs have been identified for yield-related traits in durum wheat. However, very few QTLs have been utilized in breeding programs. The use of MQTL analysis provides valuable information for QTL fine mapping and key genes for cloning. In this chapter, a total of 76 MQTLs distributed on all 14 chromosomes were described, and 12 regions of the durum wheat genome contain stable MQTLs that have been identified under both irrigated and rainfed conditions. Additionally, some candidate genes associated with drought stress tolerance and yield were reported within these MQTL regions. These efforts allow the identification of alleles that can be used to increase durum wheat grain yields under different water regimes and to obtain new genotypes adapted to climate change, necessary to meet the demand for food safety.

7 References

Arcade, A., Labourdette, A., Falque, M., Mangin, B., Chardon, F., Charcosset, A. and Joets, J. (2004). BioMercator: integrating genetic maps and QTL towards discovery of candidate genes. *Bioinformatics* 20(14), 2324–2326. doi: 10.1093/bioinformatics/bth230.

Arriagada, O., Marcotuli, I., Gadaleta, A. and Schwember, A. R. (2020). Molecular mapping and genomics of grain yield in durum wheat: a review. *Int. J. Mol. Sci.* 21(19), 7021. doi: 10.3390/ijms21197021.

Arriagada, O., Gadaleta, A., Marcotuli, I., Maccaferri, M., Campana, M., Reveco, S., Alfaro, C., Matus, I. and Schwember, A. R. (2022). A comprehensive meta-QTL analysis for yield-related traits of durum wheat (*Triticum turgidum* L. var. durum) grown under different water regimes. *Front. Plant Sci.* 13, 984269. doi: 10.3389/fpls.2022.984269.

Audran, C., Liotenberg, S., Gonneau, M., North, H., Frey, A., Tap-Waksman, K., Vartanian, N. and Marion-Poll, A. (2001). Localisation and expression of zeaxanthin epoxidase mRNA in Arabidopsis in response to drought stress and during seed development. *Funct. Plant Biol.* 28(12), 1161–1173. doi: 10.1071/PP00134.

Avni, R., Oren, L., Shabtay, G., Assili, S., Pozniak, C., Hale, I., Ben-David, R., Peleg, Z. and Distelfeld, A. (2018). Genome based meta-QTL analysis of grain weight in tetraploid wheat identifies rare alleles of GRF4 associated with larger grains. *Genes* 9(12), 636.

Ayed, S., Othmani, A., Bouhaouel, I. and Teixeira da Silva, J. A. (2021). Multi-Environment screening of durum wheat genotypes for drought tolerance in changing climatic events. *Agronomy* 11(5), 875. doi: 10.3390/agronomy11050875.

Blanco, A., Mangini, G., Giancaspro, A., Giove, S., Colasuonno, P. and Simeone, R. (2012). Relationships between grain protein content and grain yield components through quantitative trait locus analyses in a recombinant inbred line population derived from two elite durum wheat cultivars. *Mol. Breed.* 30, 79–92.

Cairns, J. E., Sanchez, C., Vargas, M., Ordoñez, R. and Araus, J. L. (2012). Dissecting maize productivity: ideotypes associated with grain yield under drought stress and well-watered conditions. *J. Integr. Plant Biol.* 54(12), 1007–1020. doi: 10.1111/j.1744-7909.2012.01156.x.

Césari, S., Kanzaki, H., Fujiwara, T., Bernoux, M., Chalvon, V., Kawano, Y., Shimamoto, K., Dodds, P., Terauchi, R. and Kroj, T. (2014). The NB-LRR proteins RGA4 and RGA5 interact functionally and physically to confer disease resistance. *EMBO J.* 33(17), 1941–1959. doi: 10.15252/embj.201487923.

Chardon, F., Virlon, B., Moreau, L., Falque, M., Joets, J., Decousset, L., Murigneux, A. and Charcosset, A. (2004). Genetic architecture of flowering time in maize as inferred from quantitative trait loci meta-analysis and synteny conservation with the rice genome. *Genetics* 168(4), 2169–2185. doi: 10.1534/genetics.104.032375.

Cobb, J. N., Biswas, P. S. and Platten, J. D. (2019). Back to the future: Revisiting MAS as a tool for modern plant breeding. Back to the future. *Theor. Appl. Genet.* 132(3), 647–667. doi: 10.1007/s00122-018-3266-4.

Colasuonno, P., Marcotuli, I., Blanco, A., Maccaferri, M., Condorelli, G. E., Tuberosa, R., Parada, R., de Camargo, A. C., Schwember, A. R. and Gadaleta, A. (2019). Carotenoid pigment content in durum wheat (Triticum turgidum L. var durum): an overview of quantitative trait loci and candidate genes. *Front. Plant Sci.* 10, 1347. doi: 10.3389/fpls.2019.01347.

Colasuonno, P., Marcotuli, I., Gadaleta, A. and Soriano, J. M. (2021). From genetic maps to QTL cloning: an overview for durum wheat. *Plants* 10(2), 315. doi: 10.3390/plants10020315

Dahl (2017). Global durum outlook. Available at: http://www.italmopa.com/wp-content/uploads/2017/05/144_all_1.pdf (accessed on Aug 15, 2021).

Desiderio, F., Zarei, L., Licciardello, S., Cheghamirza, K., Farshadfar, E., Virzi, N., Sciacca, F., Bagnaresi, P., Battaglia, R., Guerra, D., Palumbo, M., Cattivelli, L. and Mazzucotelli, E. (2019). Genomic regions from an Iranian landrace increase kernel size in durum wheat. *Front. Plant Sci.* 10, 448. doi: 10.3389/fpls.2019.00448.

Dettori, M., Cesaraccio, C. and Duce, P. (2017). Simulation of climate change impacts on production and phenology of durum wheat in Mediterranean environments using CERES-Wheat model. *Field Crops Res.* 206, 43–53. doi: 10.1016/j.fcr.2017.02.013.

Elouafi, I. and Nachit, M. M. (2004). A genetic linkage map of the Durum Triticum dicoccoides backcross population based on SSRs and AFLP markers, and QTL analysis for milling traits. *Theor. Appl. Genet.* 108(3), 401–413. doi: 10.1007/s00122-003-1440-8.

Faris, J. D., Zhang, Q., Chao, S., Zhang, Z. and Xu, S. S. (2014). Analysis of agronomic and domestication traits in a durum cultivated emmer wheat population using a high-density single nucleotide polymorphism-based linkage map. *Theor. Appl. Genet.* 127(11), 2333-2348. doi: 10.1007/s00122-014-2380-1.

Fatiukha, A., Filler, N., Lupo, I., Lidzbarsky, G., Klymiuk, V., Korol, A. B., Pozniak, C., Fahima, T. and Krugman, T. (2020). Grain protein content and thousand kernel weight QTLs identified in a durum wild emmer wheat mapping population tested in five environments. *Theor. Appl. Genet.* 133(1), 119–131. doi: 10.1007/s00122-019-03444-8.

Food and Agriculture Organization of the United Nations [FAO] (2018). *The Future of Food and Agriculture Alternative Pathways to 2050*. Rome: Food and Agriculture Organization of the United Nations.

Franks, R. G., Wang, C., Levin, J. Z. and Liu, Z. (2002). SEUSS, a member of a novel family of plant regulatory proteins, represses floral homeotic gene expression with LEUNIG. *Development* 129(1), 253–263. doi: 10.1242/dev.129.1.253.

Gianoglio, S., Moglia, A., Acquadro, A., Comino, C. and Portis, E. (2017). The genome-wide identification and transcriptional levels of DNA methyltransferases and demethylases in globe artichoke. *PLoS ONE* 12(7), e0181669. doi: 10.1371/journal.pone.0181669.

Giunta, F., De Vita, P., Mastrangelo, A. M., Sanna, G. and Motzo, R. (2018). Environmental and genetic variation for yield-related traits of durum wheat as affected by development. *Front. Plant Sci.* 9, 8. doi: 10.3389/fpls.2018.00008.

Goffinet, B. and Gerber, S. (2000). Quantitative trait loci: A meta-analysis. *Genetics* 155(1), 463–473. doi: 10.1093/genetics/155.1.463

Golabadi, M., Arzani, A., Mirmohammadi Maibody, S. A. M., Sayed Tabatabaei, B. E. and Mohammadi, S. A. (2011). Identification of microsatellite markers linked with yield components under drought stress at terminal growth stages in durum wheat. *Euphytica* 177(2), 207–221. doi: 10.1007/s10681-010-0242-8.

Graziani, M., Maccaferri, M., Royo, C., Salvatorelli, F. and Tuberosa, R. (2014). QTL dissection of yield components and morpho-physiological traits in a durum wheat elite population tested in contrasting thermo-pluviometric conditions. *Crop Pasture Sci.* 65(1), 80–95. doi: 10.1071/CP13349.

Guo, B., Sleper, D. A., Lu, P., Shannon, J. G., Nguyen, H. T. and Arelli, P. R. (2006). QTLs associated with resistance to soybean cyst nematode in soybean: meta-analysis of QTL locations. *Crop Sci.* 46, 202–208. doi: 10.2135/cropsci2005. 04-0036.

Guo, M., Liu, J. H., Ma, X., Luo, D. X., Gong, Z. H. and Lu, M. H. (2016). The plant heat stress transcription factors (HSFs): structure, regulation, and function in response to abiotic stresses. *Front. Plant Sci.* 7, 114. doi: 10.3389/fpls.2016.00114.

Gupta, P. K., Balyan, H. S., Sharma, S. and Kumar, R. (2020). Genetics of yield, abiotic stress tolerance and biofortification in wheat (Triticum aestivum L.). *Theor. Appl. Genet.* 133(5), 1569–1602. doi: 10.1007/s00122-020-03583-3.

Hajibarat, Z., Saidi, A., Shahbazi, M., Zeinalabedini, M., Gorji, A. M., Mirzaei, M., Salekdeh, G. H. and Ghaffari, M. R. (2022). Comparative proteome analysis of the penultimate internodes of Barley genotypes differing in stem reserve remobilization under drought stress. *Res Square* [Preprint]. doi: 10.21203/rs.3.rs-1178453/v1.

Han, G., Qiao, Z., Li, Y., Wang, C. and Wang, B. (2021). The roles of CCCH zinc-Finger proteins in plant abiotic stress tolerance. *Int. J. Mol. Sci.* 22(15), 8327. doi: 10.3390/ijms22158327.

Huang, L. and Gill, B. S. (2001). An RGA" like marker detects all known Lr21 leaf rust resistance gene family members in Aegilops tauschii and wheat. *Theor. Appl. Genet.* 103(6–7), 1007–1013. doi: 10.1007/s001220100701.

Islam, M. S., Ontoy, J. and Subudhi, P. K. (2019). Meta-analysis of quantitative trait loci associated with seedling-stage salt tolerance in rice (Oryza sativa L.). *Plants* 8(2), 33. doi: 10.3390/plants8020033.

Krzewska, M., Dubas, E., Biowska, G., Nowicka, A., Janas, A., Zieliński, K., Surówka, E., Kopeć, P., Mielczarek, P. and Żur, I. (2021). Comparative proteomic analysis provides new insights into regulation of microspore embryogenesis induction in winter triticale (Triticosecale Wittm.) after 5-azacytidine treatment. *Sci. Rep.* 11, 22215. doi: 10.1038/s41598-021-01671-y.

Kumar, A., Saripalli, G., Jan, I., Kumar, K., Sharma, P. K., Balyan, H. S. and Gupta, P. K. (2020). Meta-QTL analysis and identification of candidate genes for drought tolerance in bread wheat (Triticum aestivum L.). *Physiol. Mol. Biol. Plants* 26(8), 1713–1725. doi: 10.1007/s12298-020-00847-6.

Li, J., Zhang, Z., Liu, Y., Yao, C., Song, W., Xu, X., Zhang, M., Zhou, X., Gao, Y., Wang, Z., Sun, Z. and Zhang, Y. (2019). Effects of microsprinkling with different irrigation amount on grain yield and water use efficiency of winter wheat in the North China Plain. *Agric. Water Manag.* 224, 105736. doi: 10.1016/j.agwat.2019.105736.

Liang, W., Zhang, Z., Wen, X., Liao, Y. and Liu, Y. (2017). Effect of nonstructural carbohydrate accumulation in the stem pre-anthesis on grain filling of wheat inferior grain. *Field Crops Res.* 211, 66–76. doi: 10.1016/j.fcr.2017.06.016.

Lieber, D., Lora, J., Schrempp, S., Lenhard, M. and Laux, T. (2011). Arabidopsis WIH1 and WIH2 genes act in the transition from somatic to reproductive cell fate. *Curr. Biol.* 21(12), 1009–1017. doi: 10.1016/j.cub.2011.05.015.

Liu, Y. and Imai, R. (2018). Function of plant DExD/H-Box RNA helicases associated with ribosomal RNA biogenesis. *Front. Plant Sci.* 9, 125. doi: 10.3389/fpls.2018.00125.

Maccaferri, M., El-Feki, W., Nazemi, G., Salvi, S., Canè, M. A., Colalongo, M. C., Stefanelli, S. and Tuberosa, R. (2016). Prioritizing quantitative trait loci for root system architecture in tetraploid wheat. *J. Exp. Bot.* 67(4), 1161–1178. doi: 10.1093/jxb/erw039.

Maccaferri, M., Harris, N. S., Twardziok, S. O., Pasam, R. K., Gundlach, H., Spannagl, M., Ormanbekova, D., Lux, T., Prade, V. M., Milner, S. G., Himmelbach, A., Mascher, M., Bagnaresi, P., Faccioli, P., Cozzi, P., Lauria, M., Lazzari, B., Stella, A., Manconi, A., Gnocchi, M., Moscatelli, M., Avni, R., Deek, J., Biyiklioglu, S., Frascaroli, E., Corneti, S., Salvi, S., Sonnante, G., Desiderio, F., Marè, C., Crosatti, C., Mica, E., Özkan, H., Kilian, B., De Vita, P., Marone, D., Joukhadar, R., Mazzucotelli, E., Nigro, D., Gadaleta, A., Chao, S., Faris, J. D., Melo, A. T. O., Pumphrey, M., Pecchioni, N., Milanesi, L., Wiebe, K., Ens, J., MacLachlan, R. P., Clarke, J. M., Sharpe, A. G., Koh, C. S., Liang, K. Y. H., Taylor, G. J., Knox, R., Budak, H., Mastrangelo, A. M., Xu, S. S., Stein, N., Hale, I., Distelfeld, A., Hayden, M. J., Tuberosa, R., Walkowiak, S., Mayer, K. F. X., Ceriotti, A., Pozniak, C. J. and Cattivelli, L. (2019). Durum wheat genome highlights past domestication signatures and future improvement targets. *Nat. Genet.* 51(5), 885-895. doi: 10.1038/s41588-019-0381-3.

Maccaferri, M., Ricci, A., Salvi, S., Milner, S. G., Noli, E., Martelli, P. L., Casadio, R., Akhunov, E., Scalabrin, S., Vendramin, V., Ammar, K., Blanco, A., Desiderio, F., Distelfeld, A., Dubcovsky, J., Fahima, T., Faris, J., Korol, A., Massi, A., Mastrangelo, A. M., Morgante, M., Pozniak, C., N'Diaye, A., Xu, S. and Tuberosa, R. (2015). A high-density, SNP-based consensus map of tetraploid wheat as a bridge to integrate durum and bread wheat genomics and breeding. *Plant Biotechnol. J.* 13(5), 648–663. doi: 10.1111/pbi.12288.

Maccaferri, M., Sanguineti, M. C., Corneti, S., Ortega, J. L. A., Salem, M. B., Bort, J., DeAmbrogio, E., del Moral, L. F., Demontis, A., El-Ahmed, A., Maalouf, F., Machlab, H., Martos, V., Moragues, M., Motawaj, J., Nachit, M., Nserallah, N., Ouabbou, H., Royo, C., Slama, A. and Tuberosa, R. (2008). Quantitative trait loci for grain yield and adaptation of durum wheat (Triticum durum Desf.) across a wide range of water availability. *Genetics* 178(1), 489–511. doi: 10.1534/genetics.107.077297.

Mangini, G., Blanco, A., Nigro, D., Signorile, M. A. and Simeone, R. (2021). Candidate genes and quantitative trait loci for grain yield and seed size in durum wheat. *Plants* 10(2), 312. doi: 10.3390/plants10020312.

Marcotuli, I., Colasuonno, P., Hsieh, Y. S. Y., Fincher, G. B. and Gadaleta, A. (2020). Non-starch polysaccharides in durum wheat: a review. *Int. J. Mol. Sci.* 21(8), 2933. doi: 10.3390/ijms21082933.

Marcotuli, I., Gadaleta, A., Mangini, G., Signorile, A. M., Zacheo, S. A., Blanco, A., Simeone, R. and Colasuonno, P. (2017). Development of a high-density SNP-based linkage map and detection of QTL for b-glucans, protein content, grain yield per spike and heading time in durum wheat. *Int. J. Mol. Sci.* 18(6), 1329. doi: 10.3390/ijms18061329.

Martinez, A. K., Soriano, J. M., Tuberosa, R., Koumproglou, R., Jahrmann, T. and Salvi, S. (2016). Yield QTLome distribution correlates with gene density in maize. *Plant Sci.* 242, 300–309. doi: 10.1016/j.plantsci.2015.09.022.

Martínez-Moreno, F., Solís, I., Noguero, D., Blanco, A., Özberk, İ., Nsarellah, N., Elias, E., Mylonas, I. and Soriano, J. M. (2020). Durum wheat in the Mediterranean rim: historical evolution and genetic resources. *Genet. Resour. Crop Evol.* 67(6), 1415–1436.

Milner, S. G., Maccaferri, M., Huang, B. E., Mantovani, P., Massi, A., Frascaroli, E., Tuberosa, R. and Salvi, S. (2016). A multiparental cross population for mapping QTL for agronomic traits in durum wheat (Triticum turgidum ssp. durum). *Plant Biotechnol. J.* 14(2), 735–748. doi: 10.1111/pbi.12424.

Muthamilarasan, M., Mangu, V. R., Zandkarimi, H., Prasad, M. and Baisakh, N. (2016). Structure, organization, and evolution of ADP-ribosylation factors in rice and foxtail millet, and their expression in rice. *Sci. Rep.* 6, 24008.

Nagel, M., Navakode, S., Scheibal, V., Baum, M., Nachit, M., Röder, M. S. and Börner, A. (2014). The genetic basis of durum wheat germination and seedling growth under osmotic stress. *Biol. Plant.* 58(4), 681–688. doi: 10.1007/s10535-014-0436-3

Nehe, A., Akin, B., Sanal, T., Evlice, A. K., Ünsal, R., Dinçer, N., Demir, L., Geren, H., Sevim, I., Orhan, Ş., Yaktubay, S., Ezici, A., Guzman, C. and Morgounov, A. (2019). Genotype x environment interaction and genetic gain for grain yield and grain quality traits in Turkish spring wheat released between 1964 and 2010. *PLoS ONE* 14(7), e0219432. doi: 10.1371/journal.pone.0219432.

Onyemaobi, O., Sangma, H., Garg, G., Wallace, X., Kleven, S., Suwanchaikasem, P., Roessner, U. and Dolferus, R. (2021). Reproductive stage drought tolerance in wheat: importance of stomatal conductance and plant growth regulators. *Genes* 12(11), 1742. doi: 10.3390/genes12111742.

Patil, R. M., Tamhankar, S. A., Oak, M. D., Raut, A. L., Honrao, B. K., Rao, V. S. and Misra, S. C. (2013). Mapping of QTL for agronomic traits and kernel characters in durum wheat (Triticum durum Desf.). *Euphytica* 190(1), 117–129. doi: 10.1007/s10681-012-0785-y.

Peleg, Z. V. I., Fahima, T., Krugman, T., Abbo, S., Yakir, D. A. N., Korol, A. B. and Saranga, Y. (2009). Genomic dissection of drought resistance in durum wheat wild emmer wheat recombinant inbreed line population. *Plant Cell Environ.* 32(7), 758–779. doi: 10.1111/j.1365-3040.2009.01956.x.

Peng, J. H., Ronin, Y., Fahima, T., Röder, M. S., Li, Y., Nevo, E. and Korol, A. (2003). Domestication quantitative trait loci in Triticum dicoccoides, the progenitor of wheat. *Proc. Natl. Acad. Sci. U. S. A.* 100(5), 2489–2494. doi: 10.1073/pnas.252763199

Quraishi, U. M., Abrouk, M., Murat, F., Pont, C., Foucrier, S., Desmaizieres, G., Confolent, C., Rivière, N., Charmet, G., Paux, E., Murigneux, A., Guerreiro, L., Lafarge, S., Le Gouis, J., Feuillet, C. and Salse, J. (2011). Cross-genome map based dissection of a nitrogen use efficiency Ortho-metaQTL in bread wheat unravels concerted cereal genome evolution. *Plant J.* 65(5), 745–756. doi: 10.1111/j.1365-313X.2010.04461.x.

Rehman Arif, M. A., Attaria, F., Shokat, S., Akram, S., Waheed, M. Q., Arif, A. and Börner, A. (2020). Mapping of QTLs associated with yield and yield related traits in durum wheat (Triticum durum Desf.) under irrigated and drought conditions. *Int. J. Mol. Sci.* 21(7), 2372. doi: 10.3390/ijms21072372.

Reynolds, M. P., Mujeeb-Kazi, A. and Sawkins, M. (2005). Prospects for utilising plant-adaptive mechanisms to improve wheat and other crops in drought-and salinity-prone environments. *Ann. Appl. Biol.* 146, 239–259. doi: 10.1111/j.1744-7348.2005.040058.x.

Roncallo, P. F., Akkiraju, P. C., Cervigni, G. L. and Echenique, V. C. (2017). QTL mapping and analysis of epistatic interactions for grain yield and yieldrelated traits in Triticum turgidum L. var. durum. *Euphytica* 213(12), 277. doi: 10.1007/s10681-017-2058-2.

Russo, M. A., Ficco, D. B. M., Laido, G., Marone, D., Papa, R., Blanco, A., Gadaleta, A., De Vita, P. and Mastrangelo, A. M. (2014). A dense durum wheat T. dicoccum linkage map based on SNP markers for the study of seed morphology. *Mol. Breed.* 34(4), 1579–1597.

Shariatipour, N., Heidari, B. and Richards, C. M. (2021). Meta-analysis of QTLome for grain zinc and iron contents in wheat (Triticum aestivum L.). *Euphytica* 217(5), 86. doi: 10.1007/s10681-021-02818-8.

Sharma, J. S., Running, K. L. D., Xu, S. S., Zhang, Q., Haugrud, A. R. P., Sharma, S., McClean, P. E. and Faris, J. D. (2019). Genetic analysis of threshability and other spike traits in the evolution of cultivated emmer to fully domesticated durum wheat. *Mol. Genet. Genomics* 294(3), 757–771. doi: 10.1007/s00438-019-01544-0.

Shew, A. M., Tack, J. B., Nalley, L. L. and Chaminuka, P. (2020). Yield reduction under climate warming varies among wheat cultivars in South Africa. *Nat. Commun.* 11(1), 4408. doi: 10.1038/s41467-020-18317-8.

Shigeyama, T., Watanabe, A., Tokuchi, K., Toh, S., Sakurai, N., Shibuya, N. and Kawakami, N. (2016). a-Xylosidase plays essential roles in xyloglucan remodelling, maintenance of cell wall integrity, and seed germination in Arabidopsis thaliana. *J. Exp. Bot.* 67(19), 5615–5629. doi: 10.1093/jxb/erw321.

Skubacz, A., Daszkowska-Golec, A. and Szarejko, I. (2016). The role and regulation of ABI5 (ABA-Insensitive 5) in plant development, abiotic stress responses and phytohormone crosstalk. *Front. Plant Sci.* 7, 1884. doi: 10.3389/fpls.2016.01884.

Soriano, J. M. and Alvaro, F. (2019). Discovering consensus genomic regions in wheat for root-related traits by QTL meta-analysis. *Sci. Rep.* 9(1), 10537. doi: 10.1038/s41598-019-47038-2.

Soriano, J. M., Colasuonno, P.,Marcotuli, I. and Gadaleta, A. (2021). Meta-QTL analysis and identification of candidate genes for quality, abiotic and biotic stress in durum wheat. *Sci. Rep.* 11(1), 11877. doi: 10.1038/s41598-021-91446-2.

Soriano, J. M., Malosetti, M., Roselló, M., Sorrells, M. E. and Royo, C. (2017). Dissecting the old Mediterranean durum wheat genetic architecture for phenology, biomass and yield formation by association mapping and QTL meta-analysis. *PLoS ONE* 12(5), e0178290. doi: 10.1371/journal.pone.0178290.

Takai, T., Fukuta, Y., Shiraiwa, T. and Horie, T. (2005). Time-related mapping of quantitative trait loci controlling grain-filling in rice (Oryza sativa L.). *J. Exp. Bot.* 56(418), 2107–2118. doi: 10.1093/jxb/eri209.

Thanh, P. T., Vladutu, C. I., Kianian, S. F., Thanh, P. T., Ishii, T., Nitta, M., Nasuda, S. and Mori, N. (2013). Molecular genetic analysis of domestication traits in emmer wheat. I: Map construction and QTL analysis using an F2 pupulation. *Biotechnol. Biotechnol. Equip.* 27(2), 3627–3637. doi: 10.5504/BBEQ.2013.0008.

Thomas, M., Pingault, L., Poulet, A., Duarte, J., Throude, M., Faure, S., Pichon, J. P., Paux, E., Probst, A. V. and Tatout, C. (2014). Evolutionary history of methyltransferase 1 genes in hexaploid wheat. *BMC Genomics* 15(1), 922. doi: 10.1186/1471-2164-15-922.

Tyagi, S., Mir, R. R., Balyan, H. S. and Gupta, P. K. (2015). Interval mapping and meta-QTL analysis of grain traits in common wheat (Triticum aestivum L.). *Euphytica* 201(3), 367–380. doi: 10.1007/s10681-014-1217-y.

Tzarfati, R., Barak, V., Krugman, T., Fahima, T., Abbo, S., Saranga, Y. and Korol, A. B. (2014). Novel quantitative trait loci underlying major domestication traits in tetraploid wheat. *Mol. Breed.* 34(4), 1613–1628. doi: 10.1007/s11032-014-0182-4.

Veyrieras, J. B., Goffinet, B. and Charcosset, A. (2007). MetaQTL: A package of new computational methods for the meta-analysis of QTL mapping experiments. *BMC Bioinformatics* 8, 49. doi: 10.1186/1471-2105-8-49.

Wang, Y., Sun, T., Li, T., Wang, M., Yang, G. and He, G. (2016). A CBL-Interacting protein kinase TaCIPK2 confers drought tolerance in transgenic tobacco plants through regulating the stomatal movement. *PLoS ONE* 11(12), e0167962. doi: 10.1371/journal.pone.0167962.

Würschum, T., Leiser, W. L., Langer, S. M., Tucker, M. R. and Longin, C. F. H. (2018). Phenotypic and genetic analysis of spike and kernel characteristics in wheat reveals long-term genetic trends of grain yield components. *Theor. Appl. Genet.* 131(10), 2071–2084. doi: 10.1007/s00122-018-3133-3.

Zaïm, M., Kabbaj, H., Kehel, Z., Gorjanc, G., Filali-Maltouf, A., Belkadi, B., Nachit, M. M. and Bassi, F. M. (2020). Combining QTL analysis and genomic predictions for four durum wheat populations under drought conditions. *Front. Genet.* 11, 316. doi: 10.3389/fgene.2020.00316.

Zhang, L. Y., Liu, D. C., Guo, X. L., Yang, W. L., Sun, J. Z., Wang, D. W. and Zhang, A. (2010). Genomic distribution of quantitative trait loci for yield and yield-related traits in common wheat. *J. Integr. Plant Biol.* 52(11), 996–1007. doi: 10.1111/j.1744-7909.2010.00967.x.

Zhang, X., Shabala, S., Koutoulis, A., Shabala, L. and Zhou, M. (2017). Meta-analysis of major QTL for abiotic stress tolerance in barley and implications for barley breeding. *Planta* 245(2), 283–295. doi: 10.1007/s00425-016-2605-4.

Zheng, T., Hua, C., Li, L., Sun, Z., Yuan, M., Bai, G., Humphreys, G. and Li, T. (2021). Integration of meta-QTL discovery with omics: towards a molecular breeding platform for improving wheat resistance to Fusarium head blight. *Crop J.* 9(4), 739–749. doi: 10.1016/j.cj.2020.10.006.

Chapter 10

Developing corn hybrids with improved performance under water deficits

Elhan Ersoz, University of Illinois at Urbana-Champaign and Umbrella Genetics, USA; and Robert J. Bensen, Umbrella Genetics, USA

1 Introduction

Agriculture is the largest consumer of freshwater globally. The demand for water in agriculture is projected to increase due to population growth, urbanization, and changing dietary patterns. The Food and Agriculture Organization (FAO) estimates that currently a staggering 70% of global freshwater resources are consumed by agriculture, and the demand for water in agriculture will increase by at least 20%, putting further pressure on already strained water resources. Therefore, sustainable management of freshwater resources in agriculture is crucial to ensure food security, support livelihoods, and promote economic development while also safeguarding the environment.

http://dx.doi.org/10.19103/AS.2022.0109.14

Developing crop varieties with improved tolerance for water deficits is crucial in addressing the challenges posed by climate change and water scarcity in agriculture. As projected by the United States Department of Agriculture (USDA) in 2015 (https://www.usda.gov/media/press-releases/2015/12/02/usda-report-warns-climate-change-likely-impede-progress-global-food), climate change is expected to result in widespread droughts and increased water demand for agriculture, which will affect food security and the global food system. In response, the FAO (https://www.fao.org/state-of-food-agriculture/2020/en/) advocates for the development of water-efficient agricultural practices and crop varieties with enhanced water-deficit tolerance. Such crops will require less water, reducing the overall demand for freshwater resources, and enable sustainable crop production even in arid regions. This approach aligns with the 2019 UN Water policy brief, which recommends the adoption of climate-smart water management strategies to ensure water security and support food production in a changing climate. By utilizing water-efficient crop varieties, farmers can sustainably produce food, improve resilience to droughts, and reduce the risk of water conflicts and food insecurity (Abberton et al., 2015; Gleick, 2014; Homer-Dixon, 1994; Scheffran et al., 2012; Varshney et al., 2021; Millet et al., 2016; Zeitoun and Warner, 2006).

Limited water availability and droughts have significant economic impacts on farmers, as well as food security, especially in regions of the world where corn is a major staple crop. With the exception of a limited number of varieties developed by large seed companies like Syngenta (Agrisure Artesian®), Corteva (Aquamax®), and Bayer (Droughtgard®), modern corn hybrids are often highly susceptible to water deficits, which can significantly reduce their productivity and result in significant yield losses. Studies have shown that under severe water deficits, corn yields can be reduced by up to 50%, leading to significant economic losses for farmers (Hatfield et al., 2011; Lobell et al., 2014). The economic impacts of water deficits on corn productivity extend beyond crop losses to include higher production costs due to the need for additional irrigation, fertilizers, and other inputs to compensate for reduced soil moisture. These costs can be particularly challenging for small-scale farmers who may lack the financial resources to invest in expensive irrigation systems or purchase additional inputs. Moreover, drought-induced losses in corn productivity can have ripple effects on the wider economy, including reduced availability of corn-based products such as animal feed, ethanol, and food items, leading to higher prices for consumers. The economic impacts of drought on corn productivity highlight the urgent need for research and development of water-efficient crop varieties and sustainable agricultural practices to improve resilience to water deficits and ensure food security in a changing climate.

Corn provides critically important calories to the human diet. Water deficits during its cultivation can have major negative impacts on its productivity

(Boyer, 1982). There is a large body of published work focused on revealing how water deficits impact corn productivity (see Campos et al., 2004; Tuberosa et al., 2007; Hu and Xiong, 2014; Tardieu et al., 2018; Yang and Yan, 2021, for reviews). It is clear from this body of work that the impacts of water deficits and the plant's response systems to it are complex, and while much progress has been made there remains much more to be learned. Fortunately, the genetic and phenotypic tools supporting this research area are improving exponentially, and the application of these tools by basic and applied research efforts is creating actionable knowledge that can be used to rapidly create and select novel varieties capable of delivering improved productivity in water-deficient growing environments.

In this chapter, the authors will review the key points in corn plant development when water deficits can be most impactful on productivity, and review progress made in genotyping and phenotyping that enable selection for and generation of improved productivity in water-limited growing conditions. Also, technical approaches to experimental design and data analysis that can facilitate the discovery, validation, and implementation of research outputs at a commercial level will be described.

It should be noted that the literature is rich with attempts at the discovery of underlying genetic factors, or candidate genes related to tolerance to water deficits. A select few of these discoveries will be highlighted later in this chapter and elsewhere in this book. However, there are fewer reports of the successful implementation of these discoveries to truly develop water-deficit tolerant corn hybrids, perhaps hindered by hardships and expensive/strenuous physical demands of real-world experimental trialing under limited water availability, as well as the lack of a well-tested theoretical framework for analytical approaches that can quantitatively measure and objectively evaluate the change in crop performance.

In this chapter, we will share a product-development-centric view to discovery and discuss how working backward from a product concept toward developing water-efficient corn hybrids, in a theoretical framework that not only covers the required data types but also presents alternative approaches to the discovery of candidates/targets that would be most suitable for rapid and successful implementation through breeding. We will conclude with a presentation of a forward-looking analysis of the opportunities and challenges going forward in this field.

This chapter will limit its focus to reports where corn is used as the model plant. It also recognizes that water deficits are often experienced in tandem with additional environmental challenges to plant productivity such as high temperatures and increased sensitivity to disease pressure. Separating these impacts and focusing on the water-deficit responses can be challenging and can obfuscate the proper interpretation of causality for the impacts.

2 The importance of corn

Corn (also known as maize) is grown on more than 160 million hectares across 6 continents (Awika, 2011). At over 1.2 billion tons per year, the United States is the largest producing nation, followed by China and Brazil (http://www.worldagriculturalproduction.com/crops/corn.aspx). About 70% of total corn production occurs in these three countries. Corn provides about 15% of global human caloric intake, which is similar to the percentages provided by rice and wheat, making these three grains accountable for nearly half of humankind's dietary calories (https://www.worldatlas.com/articles/most-important-staple-foods-in-the-world.html). In Africa, more than 90% of the corn produced is directly consumed by humans. Conversely, in the major corn-growing markets its primary use has historically been for livestock feed. Recently, its uses have been diversified to include as a food sweetener and, in the last decade and a half, for ethanol production. Currently, 15 billion gallons of ethanol are produced in the United States annually, consuming nearly 45% of the domestic production of corn (https://afdc.energy.gov/data/10339).

Corn was domesticated about 10 000 years ago, likely in the uplands of Southern Mexico/Central America (Matsuoka et al., 2002). Corn has demonstrated an amazing ability to adapt to and thrive in very diverse growing environments. It is now grown in tropical, sub-tropical, and temperate climates. A number of factors influence productivity including daily high and low temperatures, disease and insect pressure, soil types and associated nutrient availability, and importantly: wide ranges of water availability. This adaptability has been enabled and accelerated by the selective practices of growers, and more recently by the efforts of breeders at public research institutions and commercial seed companies.

Genetically complex traits, such as end-of-season grain yield, and genetically simple traits, such as resistance to specific diseases and male sterility, have both been significantly improved by using 'eyes in the field' and traditional selective breeding practices (Duvick, 2005). However, grain yield losses due to limited available water continue to be a major source of unrealized potential. This is particularly true in regions that have highly variable annual or seasonally limited rainfall. The western third of the United States 'corn belt' typically experiences water deficits at flowering time and during the subsequent grain fill period (Boyer, 1982). The wide use of irrigation, when available and using lower planting densities to reduce water demands have helped to mitigate this risk factor. However, drying aquifers and increased energy costs for irrigation systems have placed additional burdens and risks on the grower. Limited available water at the time of planting is also detrimental to productivity, particularly in a second major corn-growing region, northwest China (Yuan et al., 2015). Reducing the negative impact of water deficits presents an economically and emotionally important, though challenging, target for basic and product development-oriented scientists.

3 Impacts of water deficits during the life cycle of the corn plant

To optimize the potential utility of the genetic elements that are identified as impacting tolerance to water deficits, a foundational understanding of the key developmental time points and phenotypes to be monitored is needed (Tuberosa, 2012). While water deficits occurring anytime during the life cycle of the corn plant can negatively impact productivity, these key time points are when deficits are most impactful.

When attempting to understand and improve tolerance to limited available water, it is important to start with a broad view of the entire life cycle of the corn plant as it relates to its demand for water and to then focus on these key developmental periods. The commercial corn plant in major production markets is a hybrid that has a life cycle that broadly ranges from 90 to 140 days. This diversity has enabled corn to be successfully grown across a multitude of latitudes. The life cycle can be described as consisting of two major phases (Kiesselbach, 1999, https://store.extension.iastate.edu/product/Corn-Growth-and-Development , Iowa State University, USDA):

- The vegetative (V) phase; and
- The reproductive (R) phase.

While the length in days of the V-phase varies between hybrids, the R-phase is more consistent at approximately 60 days in length. Typically, the longer the life cycle the more productive the individual plant is in terms of net biomass accumulation in both its vegetative and reproductive tissues. Importantly, plants with longer vegetative phases generally deliver higher end-of-season grain yields.

Corn requires comparatively less water than rice and more water than wheat. The water demand of the corn plant during its V- and R-phases naturally varies through its life cycle. Demand for water peaks at the start of the R-phase. At this time, the male pollen is shed from the tassel and receptive female silks emerge from the ear husk. This peak period of demand continues during an approximately 15-day period of rapid cell division in the developing kernel following fertilization. Water deficits at this time can have serious negative impacts on the coordination of pollen shed and silk emergence, known as the Anthesis Silking Interval (ASI). While pollen shed is typically less sensitive to water deficits, delays in silk emergence, particularly in response to severe deficits, lead to reduced or missed fertilization with obvious negative impacts on grain yield. Since commercial corn plants typically develop only one ear per plant, the availability of adequate levels of water during this peak water use period is critical, and the resulting damage is not remediable.

At the time of pollination, growth in the leaves, stem, and roots is minimal, but these tissues are high-demand sinks for water. This demand is particularly high in the leaves and water deficits can be the limiting factor for the net production and transfer of carbohydrates to the developing seed. Water deficits in the days immediately following pollination can lead to reduced final cell numbers in the developing kernels resulting in smaller final seed size, again negatively impacting final grain yield. Yield losses due to kernel abortion at this time are also a potential consequence of water deficits (Zinselmeier et al., 1999).The demand for water continues at a lower but significant levels during the subsequent grain fill period of 15–45 days after pollination (dap) and eventually falls significantly during the final grain dry down period (45–60 dap).

Understanding the developmentally dependent impacts of water deficits on corn productivity, as overviewed earlier, is foundational for driving actionable research results. End-of-season grain yield is the ultimate test of how well a corn plant tolerates water deficits. However, yield is an extremely complex trait with many contributing elements. To help address this complexity, researchers have often chosen to focus their work on key phenotypes that are believed to be predictive contributors to final grain yield. Not surprisingly, the list of phenotypes examined and the dissection of their genetic basis is extensive (Tuberosa et al., 2002). It includes germination rates; seedling survival or stand establishment; root growth rates; coordination of pollen shed and silk emergence (ASI); seed set and kernel abortion rates; leaf water status; chlorophyll content; evapotranspiration and photosynthetic rates; stomatal density and sensitivity to deficits; and leaf morphology, surface wax levels, and leaf area indices. Technical approaches used to identify key metabolic, proteomic, and genetic elements contributing to tolerance toward water deficits have been equally numerous. They include the application of candidate compounds or their antagonists; analysis of transcript, metabolic, and proteomic profiles; screens for variant genotypes; mapping of genotypic and phenotypically diverse populations; and the evaluation of the impact of transgenes. A number of these phenotypes and technical approaches and the progress made in examining and assigning functions to them are summarized in other chapters of this book.

Understanding the mechanisms by which corn productivity is modulated under water deficits and developing quantitative measurement techniques for assessing the responses under experimental conditions would allow scientists and breeders to apply techniques such as directional selection to develop new varieties with improved properties. The process starts with quantitatively defining breeding objectives and follows with defining product concepts. In

Section 4, we will briefly describe potential breeding objectives for improving performance under water-deficit conditions

4 Breeding objectives for improving performance under water deficits

Breeding for improved tolerance to water deficits is important for ensuring sustainable crop production under water-limited conditions. By identifying and selecting quantitatively measurable traits that can improve a plant's ability to avoid water deficits, utilize water more efficiently, or respond more effectively to water deficits; breeding efforts aim to develop crop varieties that can thrive even in regions where water availability can be limited.

Some of the key breeding objectives to improve tolerance include (Table 1):

1. **Avoidance: developing plants with deep and efficient root systems:** A deep and efficient root system can help plants access additional soil moisture from deeper soil layers. Selection for anatomical features that modify root architecture and improve transport rates and resistance to cavitation are also viable target phenotypes. Successful improvement

Table 1 Key breeding objectives to improve tolerance to water deficits

Objective	Measurements	Throughput cost
Developing plants with deep and efficient root systems	Steep root growth angles, few nodal roots and sparse lateral branching, and low architectural plasticity in response to environmental cues. Root depth, steeper root angles, reduced production of crown roots, reduced lateral root branching density, greater root hair formation, reduced cortical cell file number, increased cortical cell size, and root cortical senescence	Low
Increasing plant water use efficiency	Higher stomatal conductance, improved aquaporin function	Medium
Enhancing plant water conservation mechanisms	Lower stomatal conductance, waxiness, or hairiness of the leaves, reduced transpiration.	Medium
Increasing photosynthetic efficiency under limited water availability	Increased chlorophyll content, improved Rubisco activity, and improved activity for C4 photosynthesis components.	Medium
Enhancing yield under limited water availability by comparative evaluation	Comparative plot yield between well-watered vs. water-deficit grown populations.	High

for these traits can be thought of as an avoidance strategy: avoiding water deficits by improving access and transport systems.

2. **Water use efficiency (WUE): increasing plant WUE:** WUE is defined as creating more net biomass per unit of water used. There are a number of potential mode of action (MOA) to improve WUE. They typically are a function of improved phototranspiration rates and can be achieved by a number of strategies. One well-established approach is to improve the photosynthetic capacity of the plant, which can help it to maintain higher rates of photosynthesis even when water is limited. This can be accomplished by selecting plant varieties with higher levels of chlorophyll or improved Rubisco activity, which can result in a greater rate of carbon fixation and improved productivity. Corn is a C4 plant, and overexpressing genes involved in C4 photosynthesis, which is a more efficient form of photosynthesis, can result in improved productivity under water deficits. Stomatal factors including density and response sensitivity are targets for breeding. Also, aquaporin function and leaf wax levels can impact WUE.
3. **Optimizing responses to water deficits:** Maintenance of cellular water potential and gradients through the accumulation of osmotic compounds is an example of improving tolerance to water deficits. There are a number of osmolytes that are candidate compounds for modification (see the 'Glycine betaine' section as an example). It has also recently become clear that not all plant responses to water deficits result in the plant being more productive. See 'Serine/Threonine Protein Phosphatases' section for an example of a water-deficit response that results in a negative outcome that is reversible by genetic engineering.
4 **Enhancing yield under limited water availability by comparative evaluation:** Alternatively, one can bypass all the temporal predictive phenotypes and indicators and go straight to end-of-season grain yield per plot under well-watered and water-limited growth conditions for comparative evaluation and selection. Although predictive phenotypes and indicators can provide valuable information on the underlying factors contributing to yield differences and can help to identify specific traits and mechanisms that could be targeted for crop improvement, the ultimate measure of performance is end-of-season grain yield.

Understandably, the first three strategies of the four listed earlier are typically labor intensive, low throughput, and sensitive to subtle variations in environmental conditions during the measurement period and hence may require repeated measurements and replicates over the growing season for utility. A fundamental aspect for a better understanding and interpretation of the results is a good monitoring of the water status of the plants to be selected

(Blum and Tuberosa, 2018; Varshney et al., 2018). These strategies have been commonly used in public breeding and research programs, which have been focused on hypothesis-driven efforts and have worked with relatively small numbers of comparative entries in trials. Conversely, strategy 4 has been the preferred method for most large commercial breeding programs and product development efforts. With the creation of well-characterized diversity panels in the public sector, they too are now leveraging the power of directly co-screening populations for both hypothesis-driven phenotypes and end-of-season grain yield (Genomes to Fields, https://www.genomes2fields.org/home/).

5 Hypothesis-driven and genetic mapping studies to identify candidate genes

5.1 Hypothesis driven

Hypothesis-driven discovery and validation of genes that can affect tolerance to water deficits in corn using transgenes as a tool has been actively pursued by both academic and commercial enterprises since the mid-1990s when the tools enabling such efforts first became available. An excellent comprehensive review of a major seed company's program for creating and evaluating the performance of hypothesis-driven transgenes directed at improving agronomic performance at commercializable levels, including end-of-season grain yield in water-limited field trials was recently published (Simmons et al., 2021). In this review, it was reported that the program had evaluated 1671 unique genes, with 57% of them targeted at improving tolerance to 'abiotic stress'. Slightly over half of the genes evaluated in the program were corn genes, the next most common gene source was *Arabidopsis*, and less than 3% of the genes were of non-plant origin. The genes represented a broad swath of functionality but were enriched for transposable elements (TEs) and specific pathways and networks related to water-deficit tolerance. Overexpression, RNAi, knockout, and gene editing strategies were applied, with most using a constitutive ubiquitin promoter. Efficacy screens for the transgenes were performed in multi-location field trials in commercial-quality genetic backgrounds. A minimal improvement of 2–3 bushels per acre (equivalent to 0.13–0.202 Mg/ha) was required for advancement to more rigorous validation testing. From this large screening program, 13 transgenes (1.4%) reached a validated status for improved tolerance to water deficits. Nine of the 13 were linked to plant hormones, 5 of which were linked to ethylene (see the 'Ethylene biosynthesis' section) and 3 to abscisic acid (ABA). There are many lessons to be learned from the successes of and challenges faced by this program, which are generously shared in the review.

5.2 Trehalose 6-phosphate phosphatase

Modifications to assimilate production or assimilation processes impact grain development in corn. Insufficient assimilate availability during water deficits has been shown to limit grain development in corn (Schussler and Westgate, 1991a, b). This limit was shown to be partially reversed by the infusion of sucrose into corn stems (Zinselmeier et al., 1995a, b).

The endogenous levels of the metabolite trehalose 6-phosphate (T6P) have been shown to impact assimilation processes. Trehalose, which accumulates at low levels, functions in carbohydrate storage, stress protection, and metabolic regulation (Satoh-Nagasawa et al., 2006). Its biosynthesis occurs in two steps. First, T6P is made from UDP-glucose and glucose 6-phosphate by T6P synthase, and T6P is then converted to trehalose by T6P phosphatase (TPP). RAMOSA3, a genetic locus in maize that influences inflorescence architecture, has been shown to encode a functional TPP enzyme, further pointing to a role for trehalose metabolism in developmental signaling and morphogenesis in developing corn ears (Satoh-Nagasawa et al., 2006). Overexpression of a rice TPP gene, using a rice floral promoter (MADS6), was shown to reduce the concentration of T6P in corn and result in an increase in the concentration of sucrose in ear spikelets. Overexpression also increased both the kernel set and the harvest index of transgenic plants. In addition, field trials showed that the engineered trait improved yields from 9% to 49% under non-drought or mild drought conditions and from 31% to 123% under more severe drought conditions, compared to yields in non-transgenic controls (Nuccio et al., 2015). The reported yield improvements were attributed to enhanced seed set. To understand the mechanistic basis of the increased yield more precisely, gene expression and metabolite profiles in leaves and developing female reproductive tissue were examined (Oszvald et al., 2018).

5.3 Cold shock proteins

Alternate model systems can serve as sources of genes that positively impact plant performance when introduced as transgenes. For example, a class of proteins found broadly in bacteria known as cold shock proteins (CSPs) were evaluated as transgene candidates at Monsanto (now Bayer). CSPs are notable for their induced expression levels at low temperatures in bacteria when most other protein synthesis is severely repressed (Etchegaray and Inouye, 1999; Budkina et al., 2020). Their role in bacteria is thought to be to serve as RNA-chaperones, stabilizing or aiding in the coupling of transcription and translation of mRNAs, perhaps even stabilizing the expression of select transcripts in order to facilitate a successful passage of the organism through a stressful period, in this case, low temperatures. Homologous corn proteins with similar predicted functions have also been identified in plants (Karlson et al., 2002), predicting the existence of common response networks between bacteria and plants based on these conserved proteins. Two bacterial genes: *Escherichia coli CSPA* and *Bacillus subtilis CSPB* were introduced as transgenes in corn plants, under a constitutive 35S promoter. Improved performance of the transgenic plants in water-limited growing conditions at flowering time was demonstrated for a number of phenotypes including leaf extension rates, chlorophyll content, and photosynthetic rates. Importantly, a reproducible significant increase in end-of-season grain yield for plants experiencing moderate water deficits at flowering time, without a yield penalty in well-watered growing conditions, was reported for both CspA and CspB (Castiglioni et al., 2008). A product containing the CspB transgene, branded DroughtGard® by Bayer, is the first commercial transgenic product marketed as improving tolerance to water deficits and recently represents as much as 10% of market share in western US corn belt states (https://www.fbn.com/community/blog/market-share-yield-and-price-of-the-droughtgard-corn-trait). This putative MOA of selectively stabilizing RNA transcripts is a novel approach to improving plant performance and deserves further investigation aimed at building on the present set of knowledge concerning the CSP class of genes, potentially enabling further improvement in their yield impacts in water-limited growing conditions.

5.4 Vegetative storage protein: ZmLOX6

Many plants can temporarily store what is thought to be excess available nitrogen in their leaves by sequestering it in proteins known as vegetative storage proteins (VSPs) (Staswick, 1994). These proteins serve as mobile sources of nitrogen for the plant. ZmLOX6, which is annotated as a lipoxygenase, is a VSP found in corn (Abbaraju et al., 2022). *ZmLOX6* transcripts were shown to be most abundant in leaf tissues early in the reproductive phase, and accumulation of ZmLOX6 protein was shown to be induced as much as ten-fold in leaf tissues in response to exogenously applied nitrate. Transgenic overexpression of *ZmLOX6*, using a mesophyll-specific promoter, drove accumulation rates of this protein to 4% of the total cellular protein in leaf tissue. As grain development progressed and leaf senescence began, the protein was shown to be mobilized from the leaf tissue. In well-watered field trials, transgenic plants showed improved grain yields in the 5–10% range, as tested in a commercial hybrid background. In field trials where water was limited at flowering time, an even larger 10–20% grain yield improvement over non-transgenic controls was reported for multiple events. Thus, *ZmLOX6* overexpression appears to be an excellent strategy for improving yield stability in plants experiencing flowering time water deficits. This observation represents a novel approach to improving tolerance to water deficits by manipulating the availability of an essential nutrient at a critical point in the grain-filling phase of the corn plant.

5.5 Ethylene biosynthesis: ZmACS6

Ethylene is a plant hormone known to be associated with a number of facets of plant growth and development. Its accumulation is induced by any of a number of abiotic stresses, including water deficits. In the reproductive organs of corn, one of its impacts is to trigger cell death in and therefore abortion of developing kernels (Young et al., 1997). Hypothesizing that this response is possibly too conservative, it was investigated whether downregulating its induction might result in better grain yield outcomes. The biosynthesis of ethylene is a two-step process with methionine converted through an intermediate to ethylene. 1-Aminocyclopropane-1-carboxylate synthase synthase catalyzes the first committed step of this process. The second step is considered to be constitutive in plants. A transgenic gene-silencing construct of *ZmACS6* RNAi under a ubiquitin promoter was introduced into corn plants and its effect on plant performance in abiotic stress environments was examined (Habben et al., 2014). Field trials where water was limited during flowering time demonstrated as much as a 0.58 Mg/ha (9.3 Bu/A) increase in grain yield in transgenic hybrids created using proprietary lines.

5.6 Nuclear factor Y B

Transcription Factors (TFs) are inviting transgene candidates for modifying plant responses to water deficits for a number of reasons. Plant responses to water deficits have been shown to be complex, involving multiple genetic elements and networks. The possibility of modifying whole networks with single transgenes using TFs represents a promising opportunity.

The C-repeat-binding factor (CBF)/dehydration-responsive element-binding (DREB) proteins belong to the AP2/ethylene-responsive element-binding protein family (Gilmour et al., 2004, Maruyama et al., 2004). They have been shown to modulate the expression of genes with CBF/DRE boxes in their promoters. Increased expression of CBF/DREB proteins has been shown to confer drought tolerance in multiple plant species (Zhang et al., 2004). However, overexpression has been associated with off-types. The use of drought-responsive promoters to induce CBF/DREB expression under water-limited conditions has demonstrated a reduction in these off-types (Kasuga et al.,1999).

More than 1500 putative *Arabidopsis* TFs were evaluated as constitutively expressed transgenes in *Arabidopsis*. The screens included a gauntlet of abiotic stress treatments, including dehydration. A number of the screened TFs demonstrated enhanced tolerance to the treatments, including *AtNF-YB1*. As a follow-up to the *Arabidopsis* studies, an orthologous maize CAAT box TF, *ZmNF-YB2*, under a constitutive rice actin promoter, was transformed into corn plants and evaluated for improved tolerance to water deficits. In field trials, where water was limited at flowering time, the presence of the transgene resulted in enhanced photosynthetic capacity, chlorophyll content, stomatal conductance, reduction of leaf temperature, reduced wilting, higher photosynthetic rates, and importantly improved grain yield (Nelson et al., 2007). Confirmation of the transferability of transgene efficacy for improved performance in water deficits from *Arabidopsis* to corn for this gene supports the potential utility of using model systems for candidate gene discovery.

5.7 Glycine betaine

Genetically engineering higher levels of metabolites associated with maintaining cellular turgor through modifying the osmotic potential in cells experiencing water deficits has been a target of interest for some time. One such metabolite, glycine betaine (GB), is known to accumulate in response to water deficits in corn and has been shown to improve tolerance to water deficits when applied exogenously to corn plants (Chen et al., 2000). The *betA* gene from *E. coli*, encoding a choline dehydrogenase hypothesized to be rate limiting for GB biosynthesis, was transformed into corn plants and GB levels were reported to be only moderately raised. However, even this low level of GB increase was shown to improve tolerance to water deficits at seed germination and seedling stages, and end-of-season grain yields (Quan et al., 2004).

In light of the only modest increase in GB that resulted from the transgenic introduction of the *betA* gene, an attempt was made to identify the genetic locus that may be controlling GB levels in corn plants. An examination of naturally occurring GB levels in leaves of commercial corn lines revealed that there were two well-populated classes of lines: those that were classified as non-accumulators and those classified as accumulators that accumulated as much as 100-fold greater levels of GB. The existence of these two distinct classes presented an opportunity to identify loci associated with the observed GB level differences between the classes using a differential transcript expression screen. The screen identified a novel gene, *GB1*, with significantly higher levels of expression in the leaves of accumulating lines and showing demonstrably inducible expression levels in water-limited conditions (Castiglioni et al., 2018). In support of these observations, the *GB1* locus was mapped and shown to be tightly linked with a previously mapped locus, *Bet1* that had been associated with GB accumulators (Rhodes et al., 1993). However, sequence analysis of the *GB1* locus indicated the absence of a choline oxidation domain, suggesting its role in modulating GB levels was not a direct function of it acting as a GB biosynthetic enzyme.

Introducing *GB1* as a transgene, under control of the rice actin promoter, into low-accumulating lines resulted in accumulation levels as high as those found in accumulator lines (Castiglioni et al., 2018). This result demonstrated for the first time that large increases in GB accumulation can be engineered. However, constitutive overexpression of GB1 led to pleiotropic effects, thus limiting the ability to evaluate its impact on tolerance to water deficits, but the authors did report improved germination rates in transgenic plants at low temperatures. A more refined approach using tissue-specific or water-deficit inducible promoters for *GB1* may have utility for creating transgenic corn plants that have improved performance in water-limited growing conditions without the off-types. Non-transgenic approaches using map-based selection at the GB1 locus, and therefore higher GB accumulation states, are also a viable path forward with this promising metabolite.

5.8 Target genes: summary

We have just described a sample of a growing list of genes that have been introduced as transgenes and evaluated to varying degrees for their impacts on tolerance to water deficits in corn (see Table 2). In the examples cited, diverse technical approaches were used for their discovery and they represent multiple response MOAs. Collectively, these studies demonstrate that introducing single transgenes can deliver significantly improved tolerance to water deficits as measured by a number of phenotypes. Mapping and characterizing the impact on tolerance of genetic variants for these genes in commercial germplasm would add value to these discoveries, and if proven to be efficacious they could be integrated or selected-for in commercial products. The admittedly modest improvements in end-of-season grain yield reported to date for most of these should not discourage further screening and testing of promising candidates. Ultimately, transgenes with greater impacts, a small select number of transgenes in concert, or selection for desired alleles of single or multiple genes in breeding programs may deliver greater positive levels of tolerance.

5.9 Diversity panel-based genome-wide association studies

In the public sector, diversity panel-based screens directed more selectively at a single or small number of high-priority but less complex phenotypes than yield have also successfully identified loci responsive to water deficits. In some cases, the efficacy of these loci was subsequently confirmed for additional tolerance-associated phenotypes, thus strengthening the value of the initial discoveries. For example, diversity screens have identified loci that can overcome soil surface water deficits at germination and early seedling stages of growth. Some of these loci have also been confirmed in follow-up evaluations for additional tolerance-related phenotypes (Liu and Qin, 2021).

Table 2 Hypothesis-driven transgenic studies: annotated genes associated with tolerance to water deficits that have been validated in corn

Gene/annotation	Initial selection screen/ phenotype	Validation method in maize	Water-limited grain yield/ additional phenotypes	Reference
Oryza sativa TPP T6P phosphatase	Overexpression	Overexpression/field trial	Yes/enhanced seed set	Nuccio et al. (2015)
Bacillus subtilis CSPA and B RNA chaperone	Overexpression in *Arabidopsis thaliana*/abiotic stress screens	Overexpression/field trial	Yes/leaf growth, chlorophyll content, Ps rates, higher kernel number	Castiglioni et al. (2008)
ZmLOX6 vegetative storage protein	None reported	Overexpression/field trial	Yes	Abbaraju et al. (2022)
ZmACX6 ethylene biosynthesis	None reported	Overexpression/field trial	Yes/shorter ASI, higher kernel number	Habben et al. (2014)
ZmARGOS1, ZmARGOS8		Overexpression	Yes/	Gao et al. (2022) and Shi et al. (2015)
ZmNF-YB2 transcription factor	Overexpression in *Arabidopsis thaliana*/abiotic stress screens	Overexpression field trial	Yes/Ps capacity & rates, chlorophyll content, stomatal conductance, lower leaf temperature, less wilting	Nelson et al. (2007)
ZmGB1 glycine betaine biosynthesis	Corn inbred panel/ transcript profile	Overexpression	Not tested/cold germination	Castiglioni et al. (2018)

Table 3 Genetic diversity panel: annotated genes associated with tolerance to water deficits that have been validated in corn

Gene/annotation	Initial selection screen/ phenotype	Validation method in maize	Water-limited grain yield/ additional phenotypes	Reference
ZmNAC111 transcription factor	Diversity panel seedling survival rate	Overexpression	Yes	Mao et al. (2015)
ZmVPP1 vacuolar pyrophosphatase	Diversity panel seedling survival rate	Overexpression	Yes	Wang et al. (2016)
ZmPP2C-A10 Ser/Thr protein phosphorylase	Diversity panel seedling survival screen			Xiang et al. (2017)
ZmPP2C26 Ser/Thr protein phosphorylase	Diversity panel seedling survival screen	Overexpression knockout	Not reported/root hair length	Lu et al. (2022)
ZmGRXCC14				Ding et al. (2019)
ZmTIP1 S-acyltransferase	Diversity panel seedling survival screen	Overexpression knockout	Yes/root hair length	Wang et al. (2016); Zhang et al. (2020)
ZmSRO1d				Gao et al. (2022)
ZmDREB				Liu et al. (2022)

A number of genetic elements identified in these screens have been mapped to loci known to be water-deficit response related including *ZmNAC111* (Mao et al., 2015), *ZmVPP1* (Wang et al., 2016), *ZmPP2C-A10* (Xiang et al., 2017), *ZmPP2C26* (Lu et al., 2022), *ZmGRXCC14* (Ding et al., 2019), *ZmTIP1* (Zhang et al., 2020), *ZmSRO1d* (Gao et al., 2022), and *ZmDREB* (Liu et al., 2013) and were the subject of subsequent studies (see Table 3).

6 Transgenic approaches to improving tolerance to water deficits

Introduction of candidate genetic elements as transgenes in corn, with the aim of evaluating their ability to improve tolerance to water deficits, has served to validate a number of hypothesis-driven and genetic mapping-based discoveries. In fact, a number of various MOAs for improved tolerance have been validated using this technology, and at least one transgenic product from these efforts has been made commercially available. A few representative examples are summarized in the following sections.

6.1 NAC-type transcription factor

NAC TFs are a large family of plant-specific TFs that contain a common NAC-binding domain. They have been broadly associated with biotic and abiotic stress responses (see Nuruzzaman et al., 2013).

In a diversity genome-wide association studies (GWAS) screen, the expression level of a NAC-type TF, designated as *ZmNAC111,* was positively associated with seedling survival rates (SRs) in both moderate and severe water deficits. Overexpression of *35S:ZmNAC111* transgenes in corn was shown to confer improved tolerance to water deficits as evaluated for SR and WUE (Mao et al., 2015). In addition, transcript analysis in *35S:ZmNAC111* plants revealed that a number of genes known to be drought-responsive were induced by the transgene, including pathways responsive to ABA. Interestingly, a low expressing variant of *ZmNAC111* that appears to be a post-domestication variant is commonly found in temperate stiff-stalk and non-stiff-stalk lines, suggesting that this variant may be important to a reported increased susceptibility to water deficits in temperate germplasm. These results reinforce the importance of TFs in plant responses to water deficits through their regulation of gene response networks. They also identify a specific locus that may play a role in tolerance to water deficits and has been 'lost' in a population of foundational temperate inbred lines.

6.2 Vacuolar proton-pumping pyrophosphatases

Among the various proton-pumping enzymes, vacuolar pyrophosphatases form a distinct class of proton pumps, which are characterized by their ability to translocate protons across a membrane using energy released by hydrolysis of the phosphoanhydride bond of inorganic pyrophosphate.

A diversity screening for tolerance to water deficits at the seedling stage identified 42 candidate genes, of which *ZmVPP1* was the most significant contributor. An insertion in the *ZmVPP1* promoter was shown to increase its expression in genotypes with tolerance to water deficits (Wang et al., 2016). Transgenic overexpression of *ZmVPP1* was shown to result in enhanced plant height, fresh weight, ear weight, and grain yield in subsequent hybrid field trials exposed to water deficits beginning at late vegetative stage and through early flowering time points (Jia et al., 2020). The demonstration that the early seedling screen efficacy translated into full-season field trials where water deficits were not introduced until the late vegetative phase of the plant life cycle supports the robustness of the high-throughput seedling screen in being predictive of efficacy for yield in water-limited field trials.

In a follow-up to the transgenic *ZmNAC111* and *ZmVPP* reports, noted earlier, materials containing both transgenes were created. Co-expression of transgenic *ZmNAC111* with *ZmVPP1* demonstrated additional tolerance to water deficits over the single transgenes (Liu et al., 2013). This result indicates that coupling transgenes, or favorable alleles of candidate genes, may prove to be a feasible tactic for additively improving drought tolerance levels in plants.

6.3 Serine/threonine protein phosphatases

Studies have uncovered numerous genetic elements whose induced expression in response to water facilitates improved tolerance. However, not all deficit-induced responses are beneficial to plant performance (Blum and Tuberosa, 2018). In the case of members of two phosphatase clades, induced expression appears to suppress tolerance.

Clade A PP2C phosphatases (PP2C-A), which are conserved in most plant species, play important roles in mediating ABA signaling. Higher expressing alleles of one such phosphatase, *ZmPP2C-A10*, were shown to negatively correlate with tolerance in a diversity-based SR screen. A deletion mutant of *ZmPP2C-A10* lacks a critical endoplasmic reticulum response element required for water deficit-induced expression. Transgenic overexpression of the deletion mutant was shown to block ABA signaling (Xiang et al., 2017). Thus, it would appear that the ABA signal induction caused by the deficit treatment may, in this case, be a counterproductive response. Notably, leaf ABA accumulation at a given water-stress level in maize at an early growth stage significantly decreased during 50 years of conventional breeding as shown in Sanguineti et al. (2006). In fact, under the deficit conditions tested, it was shown to be beneficial to block the response. This may be an example of the corn plant overreacting to a perceived abiotic threat. This is an informative example that points to the fact that apparent protective response pathways need to be evaluated for their utility (Blum, 2011).

A clade B member of the maize PP2C family, *ZmPP2C26*, was shown to negatively regulate seedling survival under water deficits by dephosphorylating *ZmMAPK3* and *ZmMAPK7* in corn. Isoforms of the *ZmPP2C26* gene generated by alternate splicing were reported in different lines. The *ZmPP2C26S* variant was shown to have lost its ability to interact with MAPK7, but to have higher phosphatase activity than the *ZmPP2C26L* variant, and was more sensitive to water deficits in seedling SR screens (Lu et al., 2022). A Mu-knockout mutant, *zmpp2c26*, also demonstrated improved seedling SRs compared to wild-type lines coupled with higher phosphorylation levels. In addition, overexpression of *ZmPP2C26L* and *ZmPP2C26S* significantly decreased tolerance to water deficits in *Arabidopsis* and rice coupled with shorter root lengths and lower chlorophyll content and photosynthetic rates. Similar to the PP2C-A10 results, it appears that blocking this response pathway can lead to improved seedling tolerance to water deficits.

6.4 S-acyltransferase

S-acyltransferases are involved in influencing a number of biological processes and act through a post-translational modification that adds a 16-carbon palmitate to a cysteine residue of target proteins. A locus, annotated as *ZmTIP1*, was associated with improved tolerance to water deficits at the seedling stage in a corn diversity screen (Wang et al., 2016). Tip growth defective (TIP) genes have been shown to positively regulate root hair length in plants. Variants for *ZmTIP1* in the screened material indicated that the association with tolerance was not due to changes in the coding region, but rather in the promoter, suggesting that differences between the lines for expression were linked to their differences in tolerance. As predicted from this observation, transgenic overexpression of *ZmTIP1* was shown to positively impact root hair length (Zhang et al., 2020). Conversely, a transposon-mediated knockout mutant of zmtip1 had the opposite effect on root hairs and tolerance. Further, overexpression of *ZmTIP1* resulted in improved grain yield in subsequent field trials experiencing moderate levels of water deficits that were introduced beginning at 21 days after sowing

Modifying acetylation of proteins represents yet another MOA for improving tolerance to water deficits.

7 Hybrid breeding for improved performance under water deficits

The details of how commercial hybrid breeding programs are generally organized have been covered in many excellent reviews and are out of the scope for this review. Here, we will briefly describe the requirements of commercial breeding for WUE with an emphasis on modality:

- Forward genetics; and
- Reverse genetics.

It also looks at the tools and techniques that can be leveraged for each modality.

The key prerequisites for corn hybrid breeding are a deep understanding of corn genetics and physiology, access to advanced genetic engineering technologies, multiple inbred lines to utilize as hybrid parents, and the ability to conduct appropriately managed field trials and to evaluate phenotypic effects in different environmental conditions.

7.1 Reverse genetics approaches

Reverse genetics approaches are often applied to pre-determined individual gene targets, such as the ones described earlier in this chapter. The process typically involves the use of genetic engineering technologies to manipulate specific genes that are involved in desirable traits, in an inbred background that could then be used for hybrid production. The three canonical steps for this process are as follows:

1. **Generation of gene knockout, knockdown, or overexpression lines for the targets:** Once candidate genes have been identified, the next step is to generate corn lines in which these target gene(s) are either knocked out, knocked down, or overexpressed, if possible, to generate an allelic series for the target gene(s) of interest. This can be done using a variety of molecular manipulation techniques, such as CRISPR-Cas9 gene editing, RNA interference, or T-DNA mutagenesis.
2. **Evaluation of phenotypic effects:** The resulting T0 lines with modified genes are then evaluated for their phenotypic effects and crossed to one or more testers to create hybrids for phenotype assessment. This may involve conducting field trials under various environmental conditions and/or market segments to determine the performance of the modified lines compared to the other existing hybrid lines.
3 **Hybridization and selection:** The final step in forward genetics-based corn breeding is to hybridize the modified lines with other lines to

create new hybrid combinations. This can be done using traditional breeding techniques, such as crossing and selection, or more advanced techniques, such as marker-assisted selection in the commercial hybrid parent portfolio populations. The resulting hybrids are then evaluated for their performance, and the best-performing hybrids can be further developed and commercialized.

Notably, reverse genetics approaches often have a very narrow scope, focusing on individual genes as a representative for specific modes of action in selected genetic backgrounds, which often requires evaluation of the allele of interest in hybrids of multiple backgrounds obtained by backcross introgression

This limitation used to be a major challenge for the development of commercially competitive varieties, as the consumer demand often is for the latest and greatest varieties available at any given time. Most commercially competitive corn varieties are not suitable for transformation and clonal propagation, due to issues around low levels of totipotency under tissue culture conditions. This drawback, however, has been reportedly overcome by the discovery of morphogenic regulators *Baby boom (Bbm)* and *Wuschel2 (Wus2)* (Lowe et al., 2016). Overexpression of these two genes in any given maize inbred line leads to higher transformation frequencies, making transformation feasible in numerous previously non-transformable maize inbred lines. While we have still not attained the 'holy grail' of a truly genotype-independent transformation method for maize, this breakthrough in transformation technology opened the doors much wider than with previous methods for broader utilization of reverse genetics approaches for commercial variety development. In combination with recent advances in truly targeted genetic engineering techniques such as CRISPR-Cas9 system, sets up the stage for exciting breakthroughs for the utility of reverse genetics for varietal improvement in the near future.

7.2 Forward genetics approaches and genetic mapping

Forward breeding for improved performance under water deficits is broader in its scope for discovery and implementation, although it may require pre-development of specialized populations for this purpose. Key steps are as follows:

1. **Discovery population:** The classic approach is to develop a discovery population of test-cross hybrids that can be leveraged for genetic mapping studies to identify genetic loci that can be targeted for any given breeding objective. These populations can vary in genetic diversity from two alleles per locus to *N* alleles per locus where *N* is the number of founding parents for the population and requires the

material included to be pre-characterized to capture a wide range of genetic diversity. Once the set of inbred lines to be included in the study is determined, the breeder often starts with identifying test-cross partners. An ideal set of testers would (i) have high general combining ability, (ii) be from non-overlapping germplasm segments pools, and if possible (iii) cover the range of flowering times available in each pool. Due to the fragmentation of global elite maize germplasm into heterotic groups, the combining ability of a single tester may not be sufficient to cover the whole set of inbred lines, and multiple testers may be needed to generate hybrids for the whole collection. The resulting hybrid population(s) would then be used for phenotypic screens for any and all breeding objectives. This hybrid population is not genotyped directly on the individuals phenotyped but instead, the inbred lines of parents and testers are either fully sequenced or genotyped, as genotyping and sequencing of inbred lines are much less error-prone than hybrid genotyping or sequencing.

2. **Validation population(s):** The targets identified via genetic mapping in a discovery population should then be evaluated in a collection of biparental populations to refine the expected effect sizes per allele and rank-ordering of sampled allele per locus allelic series (best to worst allele rankings for the phenotype per target locus). Genetic mapping hits often have large confidence intervals that span many genes and lower effect sizes as they rely on standing genetic variation sampled for the experiment, and relative position of the locus within the genome – as the recombination rate is not uniform across the genome of maize, resulting with large variations in the extent of linkage disequilibrium. This process can be expedited by the application of doubled-haploidization (DH) to progeny from each cross whereby generation of stable homozygous recombinant inbred lines can be quickly generated to increase the resolution of mapping. It is advantageous to utilize some sort of connected breeding design such as NCII for the collection of bi-parental populations (e.g. Muthoni and Shimmelis, 2020) to unify the results generated through this stage, thus boosting statistical power. Following DH generation, hybrids with select testers would be generated for each family (progeny from each biparental cross) prior to field trialing and phenotypic assessment for selection, and genotyping of the parents and testers can be carried out as described in the first stage for the discovery population. Provided that the validation population parents are included in the discovery population, genotyping may not be required at this stage as it would be available from the previous stage.
3 **Implementation populations:** Following the validation step, markers linked to the targets of interest are developed and are used to screen

the broader germplasm base available for the alleles or haplotypes of interest. Note that haplotype here refers to linked allelic states across multiple target loci intended for implementation. Allelic states at various unlinked loci can be in coupling or repulsion states with equal probability. The aim for breeding is often to compile/stack all alleles constituting the ideotype identified during discovery and validation for implementation. This is the stage where large populations of individuals to select from would be advantageous as the probability of finding a line with the target ideotype at random decreases exponentially with an increasing number of individual target loci that make up the ideotype.

8 Genotyping assays and data: linking phenotype and genotype

The process of genetic mapping often requires two main data types: genotype and phenotype. Understanding which traits to measure, what instruments to use, and how many replicate measurements need to be performed to ensure low measurement errors are all criteria that should be considered prior to performing large-scale phenotyping in a field experiment (Tuberosa, 2012). Additionally, pre-developing genotyping assays that can best assess the variability and genetic diversity within the base population can constitute an expensive and time-consuming operational challenge. Fortunately, recent advances in next-generation sequencing technologies, which can generate large amounts of data at a low cost, coupled with recent computational advances in bioinformatics as well as cloud-based and on-demand availability of computer power enable rapid processing of tera-bases of sequence data, thereby alleviating some of the prior concerns around cost and accuracy of genetic mapping and marker-assisted breeding (MAB) practices.

8.1 DNA sequencing or genotyping

The utility of bi-parental QTL mapping of genetically simple traits, i.e. those traits with a small number of major contributing loci, has been demonstrated for many traits (Savli et al., 2007; Landi et al., 2010). However, using bi-parental screens for mapping QTLs for complex traits, i.e. those with many minor and no major contributing loci, such as tolerance to water deficits, has proven to be more challenging. Mapping loci with small, but significant, contributions to a trait clearly requires more power than bi-parental screens can deliver. Back-crossing to create near iso-genic lines can increase the resolution of the mapping, but it is time-consuming and resource-demanding particularly if one is pursuing multiple loci.

With complex traits, it is not uncommon for the effects detected during discovery stages to diminish or disappear as the back-cross process proceeds in later stages of breeding. Near iso-genic lines may also not deliver the desired level of precision. The advent of GWAS in the early 2000s, which are based on linkage disequilibrium of populations of genetically diverse materials, has further improved the precision of mapping trait-linked loci (Tuberosa et al., 2007, and see Wang and Qin, 2017, for review). Coupled with the availability of 1M or more single nucleotide polymorphisms (SNPs), the panels have enabled mapping phenotypes to gene-level resolution (Wu et al., 2021). Germplasm panels used in GWAS screens typically contain 100s of inbred line entries of sufficient genetic diversity to allow for precise mapping (Fu et al., 2013). The maize genetics community has cooperatively worked to create a genetically diverse nested association mapping (NAM) panel containing historically important foundational inbred lines (Yu et al., 2008), 26 of which now have high-quality genome assemblies (Hufford et al., 2021), and is estimated to capture approximately 80% of the species level diversity available in maize. Recent studies have reported that the genetic diversity captured within this population is not limited to various forms of sequence polymorphism but also revealed significant levels of non-allelic variation, better known as presence/absence polymorphisms or presence absence variants (PAVs). Investigations into PAVS revealed that only about 60% of the genes identified in any given inbred line genome are shared across a variety of the lines in this population. This led to the development of the pan-genome concept - which is a synthetic genome assembly that is generated through compilations of sequences across multiple whole genome sequenced inbred lines (Gui et al., 2023).

Further reduction of sequencing cost also enabled redirection of available funds to deeper omic screens, such as full-transcriptomic evaluation of study populations at multiple time points (see Fig. 1). It has long been hypothesized that the actual mechanism of DNA sequence polymorphisms' association with phenotype is achieved through variation in allelic expression levels of target genes. There are a number of factors that can influence gene expression, and thus the association of a genetic element for a given trait. These include transacting factors such as TFs and alternative splicing; presence or absence of TEs; and epigenetic factors such as methylation patterns and chromatin structure. The NAM panel, when coupled with SNPs and additional MaizeGDB data tracks, has enabled progress to be made by research groups for hundreds of traits (see Woodhouse et al., 2021).

Private industry-based panels that include commercially important lines have also been created and studied for various traits. The power of these panels to deliver actionable results has made forward genetic mapping-based breeding efforts cheaper and more efficient. These systematic and structured

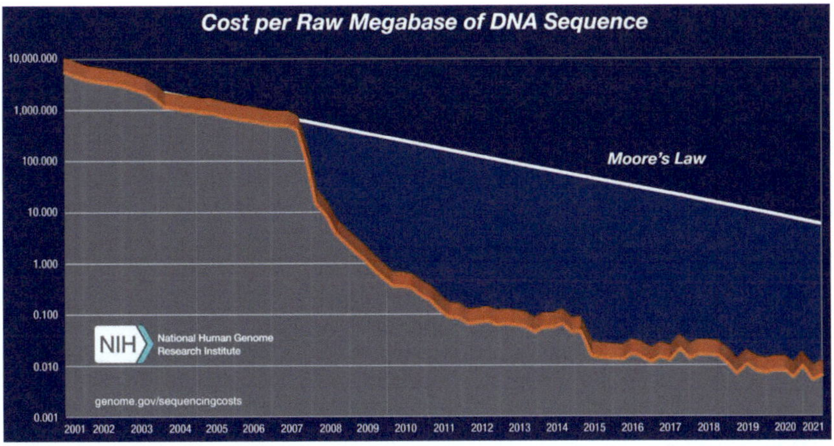

Figure 1 Cost per raw megabase of DNA sequence as reported by National Institutes for Health (NIH) as of 2021. Since 2007, the costs have dropped rapidly and dramatically beyond expectations under Moore's law and are currently under 1 ¢ (US$) per megabase of raw DNA sequence.

studies quickly became the tool of choice for many researchers to understand and improve plant responses to water deficits.

8.2 Phenotyping and phenomics

As described in Section 4, a large number of phenotypes could be associated with improved tolerance to water deficits in corn. They range from reasonably simple traits, such as germination rates and seedling survival, to complex multifaceted traits such as WUE, and even end-of-season grain yield. The trait(s) one selects to focus on have usually been hypothesis-driven: root architecture, photorespiration rates, osmotic adjustment, or sugar sensing, for example, mostly built on crop modeling studies, such as those described by Sinclair et al. in Chapter 1 of this volume.

Many of these 'surrogate' phenotypes have been used for an initial identification of linked loci and, in turn, have been evaluated for their informativeness for subsequent water-limited grain yield trials. Translating what is typically a small but real improvement reported in the initial trait screen to a significant and reproducible improvement in grain yield is challenging and many initial findings are left unproven in subsequent yield trials.

In a recent 'tour de force' phenotyping example, a genetic panel containing 368 public lines was monitored for a large number of image-based phenotypes (i-traits) over a 98-day growth period, essentially the corn plant's life cycle. Using SNPs to map genotypes in the panel, 2318 candidate genes impacting tolerance to water deficits were identified, demonstrating the power of this approach (Wu et al., 2021). A number of the candidate genes were shown to co-localize with previously identified water-deficit responsive QTLs. In addition, of the 26 910 i-traits measured in this study, 10 080 i-traits were linked to water-deficit responses, and of these, 15 were prioritized as i-traits for use in selective breeding efforts for tolerance to water deficits. Based on the known complexity of plant responses to water deficits and their near-universal impact on plant growth and development, the large number of loci identified in this study is not surprising. However, the results serve as a reminder of the daunting challenges involved in characterizing and improving tolerance to water deficits.

9 Analytical methods to facilitate breeding: genome-wide association studies, genomic selection and beyond

Throughout the last decade, there have been several innovations in breeding analytics that allow faster and more accurate selection of valuable alleles and individuals during breeding selection. During the early 2000s, GWAS approaches were established as the most efficient and successful genetic mapping approaches albeit with various drawbacks and challenges that led to various improvements in the methods and approaches leveraged for implementation of the discoveries generated.

In classical QTL mapping approaches, the implementation is quite straightforward as often the parents of the discovery population have been selected to be a high-line and a low-line for the phenotypes of interest. The implementation target is often an individual progeny that mostly resembles the low-line parent, but improved with respect to the original low-line parent at target loci by swapping the original allele with the high-line parent's allele. The process often involves the generation of additional recombinant lines at the target locus, preferably in isogenic or near-isogenic backgrounds of the low-line parent, to assess the background-specific effects of alleles of interest in the specific background of interest.

With GWAS, however, the background dependence creates complications for implementation as there is a risk that the effect sizes per alleles estimated in the discovery may not hold or even change sign during implementation when integrated into a specific genetic background. To alleviate this drawback of classical diverse line population-based GWAS result validation, family-based association methods have been proposed. Family-based association aims to balance between levels of diversity and suitability for implementation of the discoveries from genetic mapping studies, to validate and refine the effect size per allele at each locus conditional on the background.

Family-based association mapping leverages a mixed model-variance components analytical framework, where the effects attributed to individual loci are modeled as separate variance components conditional on the background for each allele at the target locus. The background is often modeled as a matrix of relative similarities between individuals that make up the population, often referred to as G matrix or genetic relationship matrix generated through pairwise comparison of individuals across a moderate number of genetic marker loci. This approach allows not only for controlling effects of ancestral relationships between the individuals due to the population differentiation (structure) but also can account for the breeding design leveraged for the study, in a hierarchical and conditional model framework, that would have otherwise been too complicated to solve with the existing statistical approaches (Zhang et al., 2020) due to issues around number of samples being much smaller than number of variables in the models ($N << p$ problem).

Further, the actual entries utilized for discovery are recombinant inbred lines and can be evaluated for their breeding values (BVs) directly and subjected to advancement and directional selection.

For this process, a standard toolbox of analytical methods involves mixed-model association mapping, also known as the animal model, followed by genomic selection-based BV calculations, such as GBLUP or RR-BLUP that utilize a similar relationship-based representation of the model parameters.

In the next few subsections, we will provide extensions to the standard mixed-effects association model that would be suitable for applications to comparative studies for water-deficit response improvement via pairwise assessment of varieties under well-watered and water-deficit treatment conditions.

9.1 Standard mixed-effects model

Standard mixed model association is often denoted as

$$y = X\beta + Zu + \varepsilon$$

where

y is a vector of observed phenotypic values
X is a design matrix for the fixed effects β
β is a vector of fixed effects

Z is a design matrix for the random effects u
u is a vector of random effects with a normal distribution ~N(0, G)
ε is a vector of residuals with a normal distribution ~N(0, R)

G is the variance-covariance matrix for the random effects
R is the variance-covariance matrix for the residuals
Conditional modeling of allelic effects based on genetic backgrounds

To model the effects of individual alleles at a genetic locus conditional on the genetic background, we can include both fixed and random effects for the locus of interest. For example, if we are interested in modeling the effect of allele A at locus L, and we have information on the genotypes of individuals at other loci, we can write the mixed effects model as:

$$y = X\beta + ZLAL + Zu + \varepsilon$$

where

X, β, Z, u, and ε are as defined earlier

ZLA is a design matrix for the random effects of the locus of interest L, with one row for each individual and one column for each allele at L (in this example, we only consider the effects of allele A at L)

ZLAL is a vector of random effects for each individual corresponding to their genotype at locus L and allele A (e.g. 0 for homozygous reference, 1 for heterozygous, and 2 for homozygous alternate), with a normal distribution $\sim N(0, \sigma^2_A)$

σ^2_A is the variance of the random effect for allele A at locus L.

In matrix notation, ZLAL can be constructed as a binary matrix where each row represents an individual and each column represents a specific genotype at L and allele A. For example, if we have 100 individuals and 3 genotypes at L (AA, AB, and BB), ZLAL would be a 100 × 3 matrix with 1's in the appropriate columns and 0's elsewhere.

Note that the specific design of ZLAL will depend on the genetic model being used (e.g. additive, dominant, and recessive) and the number of alleles being considered at the locus. The variance-covariance matrix G will also need to be constructed based on the genetic relatedness between individuals.

9.2 Multiple treatment intensities or developmental stage targets as fixed effects

Specifically for mapping loci responsive to water deficits, the earlier model can be modified to include treatment as a fixed effect by including it in the design matrix X. For example, if we assume that the treatment effects are binary and is denoted by treatment variable T that takes a value of 0 for control, and value of 1 for treatment we can then include this variable as a fixed effect in the previous model.

The new model would be:

$$y = X\beta + ZLAL + Zu + \varepsilon$$

where:

X is now a design matrix that includes a column for the treatment variable T, with a 0 or 1 for each individual depending on their treatment status.

β is now a vector of fixed effects that includes an intercept term and a coefficient for the treatment variable *T*.

In matrix notation, we can write *X* as:

$X = [1\ T]$

where the first column is a vector of 1's for the intercept term, and the second column is the treatment variable *T*.

The new model that includes a fixed effects term due to treatment would then be:

$y = \beta_0 + \beta_1 T + ZLAL + Zu + \varepsilon$

where:

β_0 is the intercept term

β_1 is the coefficient for the treatment variable *T*

The effect of the treatment variable on the response variable can be estimated by the coefficient β1. The effects of individual alleles at the genetic locus can still be estimated by the random effects ZLAL.

Multiple treatment intensities or developmental stage targets as random effects

One can further imagine trialing conditions where various levels and timing of water deficits have been experimentally tested, either by modulating the timing of the deficits or modulating the intensity of deficits at a given developmental time point.

If the treatment variable has multiple levels it would best be treated as a random variable. This approach would require the use of a mixed-effects model that includes both fixed and random effects for the treatment variable. This can be achieved by introducing a new design matrix for the random effects of the treatment variable, and including it in the overall model as follows:

$y = X\beta + ZT\gamma T + ZLAL + Zu + \varepsilon$

where

y, *X*, β, ZLAL, Zu, and ε are as defined in the previous model

ZT is a design matrix for the random effects of the treatment variable, with one row for each individual and one column for each treatment level (e.g. 0 for control, 1 for treatment 1, 2 for treatment 2, and so on).

γT is a vector of random effects for each treatment level, with a normal distribution ~N(0, D)

D is the variance-covariance matrix for the random effects of the treatment variable

In matrix notation, ZT can be constructed as a binary matrix where each row represents an individual and each column represents a specific treatment level. For example, if we have 100 individuals and 3 treatment levels (control, treatment 1, and treatment 2), ZT would be a 100 × 3 matrix with 1's in the appropriate columns and 0's elsewhere.

Note that the specific design of ZT will depend on the number of treatment levels being considered. The variance–covariance matrix D will also need to be constructed based on the assumptions made about the random effects of the treatment variable.

The new model allows for both fixed and random effects of the treatment variable, which can be estimated separately by the fixed effects β and the random effects γT, respectively. The effects of individual alleles at the genetic locus can still be estimated by the random effects ZLAL.

9.3 Estimation of breeding values and line selection

BVs can be calculated based on the mixed effects models by using the estimated random effects for each individual. The BV represents the additive genetic value of an individual for the trait of interest and is estimated by summing the effects of all alleles at the genetic locus that are included in the model.

To calculate BVs for each individual, we first need to estimate the random effects of the genetic locus and the treatment variable using the mixed effects model. The random effects for each individual can be obtained by multiplying the corresponding row of the design matrix ZLAL and ZT (if the treatment is treated as a random variable) by their respective estimated random effects. For example, the BV for individual *i* can be estimated as follows:

Breeding value (BV)*i* = Σ*j* (ZLAL)*ij* × (AL)j + Σ*k* (ZT)*ik* × (γ*T*)*k*

where:

(ZLAL)*ij* is the *j*th element of the *i*th row of the design matrix ZLAL, representing the presence (1) or absence (0) of the *j*th allele for individual *i*

(AL)*j* is the estimated random effect of the *j*th allele, obtained from the mixed-effects model

(ZT)*ik* is the *k*th element of the *i*th row of the design matrix ZT, representing the presence (1) or absence (0) of the *k*th treatment level for individual *i* (if the treatment is treated as a random variable)

(γT)*k* is the estimated random effect of the *k*th treatment level, obtained from the mixed-effects model

The BV for each individual represents the sum of the estimated effects of all alleles and treatment levels for that individual, weighted by their respective random effects.

Extending the single locus models to ideotype models:

The model described earlier can easily be extended to consider multiple targets that constitute the desired ideotype. Suppose we have 10 independently segregating target loci that we would like to consider for BV calculations. We can extend the previous mixed effects model to include all 10 loci by adding the appropriate random effects for each locus. The model would be:

$$y = X\beta + Z_1L_1AL_1 + Z_2L_2AL_2 + \ldots + Z_{10}L_{10}AL_{10} + ZT\gamma T + Zu + \varepsilon$$

where:

y, X, β, ZT, γT, Zu, and ε are as defined in the previous models

$Z_1L_1, Z_2L_2, \ldots, Z_{10}L_{10}$ are design matrices for the random effects of each locus, with one row for each individual and one column for each allele at each locus (e.g. under additive effects assumption, 0 for homozygous reference, 1 for heterozygous, and 2 for homozygous alternate).

$AL_1, AL_2, \ldots, AL_{10}$ are vectors of random effects for each allele at each locus, with a normal distribution ~N(0, G)

G is the variance-covariance matrix for the random effects of all alleles at all loci

In matrix notation, the design matrices $Z_1L_1, Z_2L_2, \ldots, Z_{10}L_{10}$ can be constructed as binary matrices where each row represents an individual and each column represents a specific allele at a specific locus.

The BV for each individual can then be estimated as:

Breeding value $(BV)i = \Sigma j\ (Z1L1)ij \times (AL1)j + \Sigma k\ (Z2L2)ik \times (AL2)k + \ldots + \Sigma p\ (Z10L10)ip \times (AL10)p + \Sigma q\ (ZT)iq \times (\gamma T)q$

where:

$(Z1L1)ij$ is the jth element of the ith row of the design matrix Z1L1, representing the presence (1), absence (0), or heterozygous state (0.5) of the jth allele at the first locus for individual i

$(AL1)j$ is the estimated random effect of the jth allele at the first locus, obtained from the mixed effects model

$(Z2L2)ik$ is the kth element of the ith row of the design matrix Z2L2, representing the presence (1), absence (0), or heterozygous state (0.5) of the kth allele at the second locus for individual i, and so on for the remaining loci

$(AL2)k$, ..., $(AL10)p$ are the estimated random effects of the alleles at the remaining loci

$(ZT)iq$ is the qth element of the ith row of the design matrix ZT, representing the presence (1) or absence (0) of the qth treatment level for individual i (if the treatment is treated as a random variable)

$(\gamma T)q$ is the estimated random effect of the qth treatment level, obtained from the mixed effects model

The BV for each individual represents the sum of the estimated effects of all alleles and treatment levels for that individual, weighted by their respective random effects. Note that the variance-covariance matrix G will need to be constructed based on the assumptions made about the random effects of all alleles at all loci.

Note that despite the apparent complexity of these proposed models they neither include interactions between target loci being evaluated, nor additional models of allelic interactions per locus, other than codominance. Obviously, such models would become increasingly complex that would result in reduction of statistical power and increased uncertainty in the conclusions. Therefore, it is prudent for researchers to use their best judgment as to how much complexity their experimental design can afford prior to engaging in modeling effect sizes and selecting individuals based on their experiments' design parameters.

10 Beyond association mapping and genomic selection: next-generation modeling approaches with machine learning

Provided that data from such a well-defined experiment exists, one would be prudent to develop a predictive model based on allelic states at the target loci of interest. Previous studies have reported that the possibility to generalize such predictive models is limited due to massive amounts of stratification and differentiation within modern maize breeding populations.

However, the data generated from such a structured experiment can be leveraged to train a neural network model to either predict BVs for previously untested individuals, or for prediction of performance phenotypes of these individuals, based solely on marker data input.

In the example case mentioned earlier for BV calculation, the input to the neural network would be the genotypes of the individuals at the 10 target loci, and the output would be the predicted BVs. It would be plausible to model this with a standard feed-forward neural network architecture, with multiple hidden layers and an output layer consisting of a single node.

For the loss function, mean squared error (MSE) is a common choice for regression problems. The NN model can be trained using stochastic gradient descent or another optimization algorithm that minimizes the MSE loss.

To design the neural network architecture, we can start with a simple architecture and gradually increase the complexity until we achieve satisfactory performance on a known-truth test data set. A reasonable starting point would be to use a single hidden layer with 10-20 nodes, and then add more hidden layers or nodes as needed. We can experiment with different activation functions (e.g. ReLU, sigmoid, the hyperbolic tangent function) and regularization techniques (e.g. dropout, L1/L2 regularization) to improve the generalization performance of the neural network models.

Neural network models have additional benefits in comparison to the statistical association models presented earlier. Particularly, a fully connected network of nodes in the hidden layer would enable the evaluation of potential interactions between every locus incorporated into the experiment. In this scenario, the inputs would be the genotypes at each locus per individual, and the output would be the phenotypes.

It is worth noting that neural network models are black box models, and the interpretation of the model coefficients is not straightforward as in mixed models. It is recommended to carefully validate the neural network model and compare its performance with the mixed model solutions for known-truth test cases to ensure that it is providing useful and accurate predictions.

Specific to the multiple treatments scenario example presented earlier, other loss functions for gradient descent other than MSE may be considered.

For instance, it is possible to use a Siamese neural network with a Siamese loss function to predict BVs for a pair of experiments where the sole difference is the treatment conditions. Another name for a Siamese neural network is a 'twin neural network'. This type of neural network gets its name from the Siamese twins, who are identical twins that are physically connected to each other. Similarly, in a Siamese neural network, two or more identical subnetworks are connected to each other and share the same parameters, similar to how the bodies of the Siamese twins are connected. A Siamese neural network is defined by two or more identical subnetworks that share the same input parameters and is commonly used for learning similarity metrics or distance metrics between pairs of inputs. In the context of BV prediction, the two inputs to the Siamese network would be the genotypes of two individuals, and the output would be a similarity score or distance measure between the two individuals' BVs.

To use a Siamese neural network for BV prediction, you can train the network to predict the pairwise similarity or distance between individuals based on their genotypes, and then use this similarity/distance measure to estimate BVs. One common approach to this is to define a Siamese loss function that penalizes the difference between the predicted similarity/distance and the true similarity/distance. The Siamese loss function can be defined as follows:

$L = (1/2N) \times$ Sum[(y_true - y_pred)^2]

where N is the number of pairwise comparisons in the training set, y_true is the true similarity/distance between two individuals, and y_pred is the predicted similarity/distance between the same two individuals based on their genotypes.

The convergence criteria for this would imply minimum difference between responses under WW and WD conditions, ensuring there is no major directionality change in the designation of alleles as favorable or unfavorable, but only minor variation in the slope of the effect size performance attributed to allelic variants at a locus that changes proportionally under different treatment conditions.

It is worth noting that using sophisticated NN architectures such as Siamese neural network with a Siamese loss function may require more training data and computational resources compared to a standard neural network with MSE loss. Additionally, – typical attribute of NN models – the interpretation of the model coefficients may be less straightforward with a Siamese neural network. Therefore, it is important to carefully consider options and validate and compare the performance of any given model with other models to ensure that it is providing accurate and useful predictions.

11 Conclusion

Improving the productivity of corn where water is limiting is the stated long-term objective of many basic and applied research efforts. Although final vegetative biomass can be an important product in some growing regions, the major focus for improving corn is to increase its end-of-season grain yield. Grain yield can be reduced by water deficits at any time during the life cycle of the corn plant. However, there are key time points that deserve to be focused upon due to their potentially large impacts on productivity. They are germination and early seedling establishment, floral development and pollination, early seed development, and stay green. Improving plant performance during these critical time points presents the largest opportunity to positively impact final plant productivity.

Hypothesis-driven research efforts have delivered an improved understanding of how corn plants respond to water deficits. Three MOAs have been broadly studied: (1) avoiding water deficits through better access to water; (2) improving WUE, and (3) reducing the impacts of water deficits by favorably modifying cellular responses to water-limited conditions. Hypothesis-driven research continues to increase our knowledge of how corn plants respond to water deficits. However, going forward, there is tremendous promise in the power of diversity screens and high-resolution genetic mapping studies not only to build on our understanding of known response networks but also to discover novel networks and their enabling genetic elements. This power is further enhanced when coupled with higher quality and throughput phenotyping, genotyping and analytical tools. These tools have combined to bring our understanding to a level that enables targeted gene modification and rapid MAB.

These successful applications have admittedly, delivered only modest improvements in performance. To date, no single genetic element has been engineered and proven to be the 'silver bullet' to deliver large impacts. Likewise, the MAB products even though they may involve selection for multiple loci also have only delivered moderate improvements. The complexity of the sensing and response system to water deficits remains challenging to characterize. To overcome these barriers, identifying and dissecting response networks and their controlling elements such as TFs have been frequently targeted. However, there still remain large gaps in our understanding of the complexity, hierarchy, and temporal response patterns to water deficits. Which of the myriad of MOAs that corn plants use to tolerate water deficits will offer the best avenue forward, perhaps in combinations, toward more impactful improved performance? One thing we know for sure: the responses of corn plants to water deficits are complex and optimizing their responses will be equally if not more complex.

12 Acknowledgements

The authors wish to thank the editor and publisher for the opportunity to share our perspectives on the current status of academic and commercial efforts to better understand how water deficits impact corn productivity and what can be done to ameliorate those impacts. We also wish to thank our fellow research scientists and organizations for their dedication to the goal of improving crop productivity and thus, helping to feed the people of the world.

13 References

Abbaraju, H. K. R., Gupta, R., Appenzeller, L. M., Fallis, L. P., Hazebroek, J., Zhu, G., Bourett, T. M., Howard, R. J., Weers, B., Lafitte, R. H., Hakimi, S. M., Schussler, J. R., Loussaert, D. F., Habben, J. E. and Dhugga, K. S. (2022). A vegetative storage protein improves drought tolerance in maize. *Plant Biotechnol. J.* 20(2), 374-389.

Abberton, M., Batley, J., Bentley, A., Bryant, J., Cai, H., Cockram, J. and Yano, M. (2015). Global agricultural intensification during climate change: a role for genomics. *Plant Biotechnol. J.* 14(4), 1095–1098.

Awika, J. (2011). *Major Cereal Grains Production and Use around the World. Advances in Cereal Science: Implications to Food Processing and Health Promotion ACS Symposium Series*, American Chemical Society, Washington, DC.

Blum, A. (2011). Drought resistance and its improvement. Plant breeding for water-limited environments. In *Springer Link Textbook*, 53–152, Springer, New York, ISBN: 978-1-4419-7491-4.

Blum, A. and Tuberosa, R. (2018). Dehydration survival of crop plants and its measurement. *J. Exp. Bot.* 69(5), 975–981.

Boyer, J. S. (1982). Plant productivity and environment. *Science* 218(4571), 443–448.

Budkina, K., Zlobin, N., Kononova, S., Ovchinnikov, L. P. and Babakov, A. V. (2020). Cold shock domain proteins: structure and interaction with nucleic acids. *Biochem. Mosc.* 85(Suppl. 1), 1–19. https://doi.org/10.1134/S0006297920140011.

Campos, H., Cooper, M., Habben, J. E., Edmeades, G. O. and Schussler, J. R. (2004). Improving drought tolerance in maize: a view from industry. *Field Crops Res.* 90(1), 19–34.

Castiglioni, P., Warner, D., Bensen, R. J., Anstrom, D. C., Harrison, J., Stoecker, M., Abad, M., Kumar, G., Salvador, S., D'Ordine, R., Navarro, S., Back, S., Fernandes, M., Targolli,

J., Dasgupta, S., Bonin, C., Luethy, M. H. and Heard, J. E. (2008). Bacterial RNA chaperones confer abiotic stress tolerance in plants and improved grain yield in maize under water-limited conditions. *Plant Physiol.* 147(2), 446–455.

Castiglioni, P., Bell, E., Lund, A., Rosenberg, A., Galligan, M., Hinchey, B., Bauer, S., Nelson, D. and Bensen, R. (2018). Identification of *GB1*, a gene whose constitutive overexpression increases glycinebetaine content in maize and soybean. *Plant Direct* 2, 1–7.

Chen, W. P., Li, P. H. and Chen, T. H. H. (2000). Glycinebetaine increases chilling tolerance and reduces chilling-induced lipid peroxidation in Zea mays L. *Plant Cell & Environ.* 23(6), 609–618. https://doi.org/ 10.1046/j.1365-3040.2000.00570.x.

Ding, S., He, F., Tang, W., Du, H. and Wang, H. (2019). Identification of maize CC-type glutaredoxins that are associated with response to drought stress. *Genes* 10(8), 610.

Duvick, D. N. (2005). The contribution of breeding to yield advances in maize (Zea mays L.). *Adv. Agron.* 86, 83–145.

Etchegaray, J. P. and Inouye, M. (1999). CspA, CspB and CspG, major cold shock proteins of Escherichia coli, are induced at low temperature and under conditions that completely block protein synthesis. *J. Bacteriol.* 181(6), 1827–1830.

Fu, J., Cheng, Y., Linghu, J., Yang, X., Kang, L., Zhang, Z., Zhang, J., He, C., Du, X., Peng, Z., Wang, B., Zhai, L., Dai, C., Xu, J., Wang, W., Li, X., Zheng, J., Chen, L., Luo, L., Liu, J., Qian, X., Yan, J., Wang, J. and Wang, G. (2013). RNA sequencing reveals the complex regulatory network in the maize kernel. *Nat. Commun.* 4, 2832. https://doi.org/10.1038/ncomms3832.

Gao, H., Cui, J., Liu, S., Wang, S., Lian, Y., Bai, Y., Zhu, T., Wu, H., Wang, Y., Yang, S., Li, X., Zhuang, J., Chen, L., Gong, Z. and Qin, F. (2022). Natural variations of ZmSRO1d modulate the trade-off between drought resistance and yield by affecting ZmRBOHC-mediated stomatal ROS production in maize. *Mol. Plant* 15(10), 1558–1574.

Gilmour, S. J., Fowler, S. G. and Thomashow, M. F. (2004). Arabidopsis transcriptional activators CFB1, CFB2, and CFB3 have matching functional activities. *Plant Mol. Biol.* 54(5), 767–781.

Gleick, P. H. (2014). Water, drought, climate change, and conflict in Syria. *Weather Clim. Soc.* 6(3), 331–340. https://doi.org/10.1175/WCAS-D-13-00059.1.

Gui, S., Martinez-Rivas, F. J., Wen, W., Meng, M., Yan, J., Usadel, B. and Fernie, A. R. (2023). Going broad and deep: sequencing-driven insights into plant physiology, evolution, and crop domestication. *Plant J.* 113(3), 446–459. https://doi.org/10.1111/tpj.16070.

Habben, J. E., Bao, X., Bate, N. J., DeBruin, J. L., Dolan, D., Hasegawa, D., Helentjaris, T. G., Lafitte, R. H., Lovan, N., Mo, H., Reimann, K. and Schussler, J. R. (2014). Transgenic alteration of ethylene biosynthesis increases grain yield in maize under field drought-stress conditions. *Plant Biotechnol. J.* 12(6), 685–693.

Hatfield, J. L., Boote, K. J., Kimball, B. A., Ziska, L. H., Izaurralde, R. C., Ort, D., Thomson, A. M. and Wolfe, D. (2011). Climate impacts on agriculture: implications for crop production. *Agron. J.* 103(2), 351–370. https://doi.org/10.2134/agronj2010.0303.

Homer-Dixon, T. F. (1994). Environmental scarcities and violent conflict: evidence from cases. *Int. Sec.* 19(1), 5–40. https://doi.org/10.2307/2539147.

Hu, H. and Xiong, L. (2014). Genetic engineering and breeding of drought-resistant crops. *Annu. Rev. Plant Biol.* 65, 715–741.

Hufford, M., Seetharam, A., Woodhouse, M., Chougule, K., Ou, S., Liu, J., et al. (2021). De novo assembly, annotation, and comparative analysis of 26 diverse maize genomes. bioRxiv. Available at: http://biorxiv.org/lookup/ doi/10.1101/2021.01.14.426684.

Jia, T., Jing-jing, L., Wang, L., Cao, Y., Ma, J., Wang, H., Zhang, D. and Li, H. (2020). Evaluation of drought tolerance in ZmVPP1-overexpressing transgenic inbred maize lines and their hybrids. *J. Integr. Agric.* 19(9), 2177–2187.

Karlson, D., Nakaminami, K., Toyomasu, T. and Imai, R. (2002). A cold-regulated nucleic acid-binding protein of winter wheat shares a domain with bacterial cold shock proteins. *J. Biol. Chem.* 277(38), 35248–35256.

Kasuga, M., Liu, Q., Miura, S., Yamaguchi-Shinozaki, K. and Shinozaki, K. (1999). Improving plant drought, salt, and freezing tolerance by gene transfer of a single stress-inducible transcription factor. *Nat. Biotechnol.* 17(3), 287–291.

Kiesselbach, T. (1999). *The Structure and Reproduction of Corn. 50th Anniversary Edition*, Cold Spring Harbor Laboratory Press, ISBM, 0-87969-556-0.

Landi, P., Giuliani, S., Salvi, S., Ferri, M., Tuberosa, R. and Sanguineti, M. C. (2010). Characterization of root-yield-1.06, a major constitutive QTL for root and agronomic traits in maize across water regimes. *J. Exp. Bot.* 61(13), 3553–3562.

Liu, S., Wang, X., Wang, H., Xin, H., Yang, X., Yan, J., Li, J., Tran, L. S., Shinozaki, K., Yamaguchi-Shinozaki, K. and Qin, F. (2013). Genome-wide analysis of ZmDREB genes and their association with natural variation in drought tolerance at seedling stage of Zea mays L. *PLoS Genet.* 9(9), e1003790.

Liu, S. and Qin, F. (2021). Genetic dissection of maize drought tolerance for trait improvement. *Mol. Breed.* 41(2), 8. https://doi.org/10.1007/s11032-020-01194-w.

Liu, S., Liu, X., Zhang, X., Chang, S., Ma, C. and Qin, F. (2022). Co-expression of ZmVPP1 with ZmNAC111 confers robust drought resistance in maize. *Genes* 14(1), 8. https://doi.org/10.3390/genes14010008.

Lobell, D. B., Roberts, M. J., Schenker, W., Braun, N., Little, B. B., Rejesus, R. M. and Hammer, G. L. (2014). Greater Sensitivity to drought accompanies maize increase in the U.S. midwest. *Science* 344(6183), 516–519. https://doi.org/10.1126/science.1251423.

Lowe, K., Wu, E., Wang, N., Hoerster, G., Hastings, C., Cho, M. J., Scelonge, C., Lenderts, B., Chamberlin, M., Cushatt, J., Wang, L., Ryan, L., Khan, T., Chow-Yiu, J., Hua, W., Yu, M., Banh, J., Bao, Z., Brink, K., Igo, E., Rudrappa, B., Shamseer, P. M., Bruce, W., Newman, L., Shen, B., Zheng, P., Bidney, D., Falco, C., Register, J., Zhao, Z. Y., Xu, D., Jones, T. and Gordon-Kamm, W. (2016). Morphogenic regulators *Baby boom* and *Wuschel* improve monocot transformation. *Plant Cell* 28(9), 1998–2015.

Lu, F., Li, W., Peng, Y., Cao, Y., Qu, J., Sun, F., Yang, Q., Lu, Y., Zhang, X., Zheng, L., Fu, F. and Yu, H. (2022). ZmPP2C26 alternative splicing variants negatively regulate drought tolerance in maize. *Front. Plant Sci.* 13, 851531.

Mao, H., Wang, H., Liu, S., Li, Z., Yang, X., Yan, J., Li, J., Tran, L. S. and Qin, F. (2015). A Transposable element in a NAC gene is associated with drought tolerance in maize seedlings. *Nat. Commun.* 6, 8326.

Maruyama, K., Sakuma, Y., Kasuga, M., Ito, Y., Seki, M., Goda, H., Shimada, Y., Yoshida, S., Shinozaki, K. and Yamaguchi-Shinozaki, K. (2004). Identification of cold-inducible downstream genes of the Arabidopsis DREB1A/CBF3 transcriptional factor using two microarray systems. *Plant J.* 38(6), 982–993.

Matsuoka, Y., Vigouroux, Y., Goodman, M. M., Sanchez, G., Buckler, E. and Doebley, J. (2002). A single domestication for maize shown by multilocus microsatellite genotyping. *Proc. Natl Acad. Sci. USA* 99(9), 6080–6084. https://doi.org/10.1 073/pnas.052125199.

Millet, E. J., Welcker, C., Kruijer, W., Negro, S., Coupel-Ledru, A., Nicolas, S. D., Laborde, J., Bauland, C., Praud, S., Ranc, N., Presterl, T., Tuberosa, R., Bedo, Z., Draye, X., Usadel,

B., Charcosset, A., Van Eeuwijk, F. and Tardieu, F. (2016). Genome-wide analysis of yield in Europe: allelic effects vary with drought and heat scenarios. *Plant Physiol.* 172(2), 749–764.

Muthoni, J. and Shimmelis, H. (2020). Heat and drought stress and their implications on potato production under dry African tropics. *Aust. J. Crop Sci.* 14(9), 1405–1414. https://doi.org/10.21475/ajcs.20.14.09.p2402.

Nelson, D. E., Repetti, P. P., Adams, T. R., Creelman, R. A., Wu, J., Warner, D. C., Anstrom, D. C., Bensen, R. J., Castiglioni, P. P., Donnarummo, M. G., Hinchey, B. S., Kumimoto, R. W., Maszle, D. R., Canales, R. D., Krolikowski, K. A., Dotson, S. B., Gutterson, N., Ratcliffe, O. J. and Heard, J. E. (2007). Plant nuclear factor YB transcription factor subunits confer drought tolerance and lead to improved corn yields on water-limited conditions. *Proc. Natl Acad. Sci. U. S. A.* 104(42), 16450–16455.

Nuccio, M. L., Wu, J., Mowers, R., Zhou, H. P., Meghji, M., Primavesi, L. F., Paul, M. J., Chen, X., Gao, Y., Haque, E., Basu, S. S. and Lagrimini, L. M. (2015). Expression of trehalose-6- phosphate phosphatase in maize ears improves yield in well-watered and drought conditions. *Nat. Biotechnol.* 33(8), 862–869.

Nuruzzaman, M., Sharoni, A. M. and Kikuchi, S. (2013). Roles of NAC transcription factors in the regulation of biotic and abiotic stress responses in plants. *Front. Microbiol.* 4, 248.

Oszvald, M., Primavesi, L. F., Griffiths, C. A., Cohn, J., Basu, S. S., Nuccio, M. L. and Paul, M. J. (2018). Trehalose 6-phosphate regulates photosynthesis and assimilate partitioning in reproductive tissue. *Plant Physiol.* 176(4), 2623–2638.

Quan, R., Shang, M., Zhang, H., Zhao, Y. and Zhang, J. (2004). Engineering of enhanced glycine betaine synthesis improves drought tolerance in maize. *Plant Biotechnol. J.* 2(6), 477–486.

Rhodes, D., Yang, W., Samaras, Y., Wood, K., Bonham, C., Rhodes, J. and Burr, B. (1993). Map locations of genes conferring glycinebetaine and trigonelline accumulation in maize. *Plant Physiol.* 102(S-160i) (abstract No. 917). https://doi.org/10.1111/j.1467-7652.2004.00093.x. PMID: 17147620.

Salvi, S., Sponza, G., Morgante, M., Tomes, D., Niu, X., Fengler, K. A., Tuberosa, R., Ananiev, E. V., Svitashev, S., Bruggemann, E., Li, B., Hainey, C. F., Radovic, S., Zaina, G., Rafalski, J. A., Tingey, S. V., Miao, G. H., Phillips, R. L. and Tuberosa, R. (2007). Conserved noncoding genomic sequences associated with a flowering-time quantitative trait locus in maize. *Proc. Natl Acad. Sci. U. S. A.* 104(27), 11376–11381.

Satoh-Nagasawa, N., Nagasawa, N., Malcomber, S., Sakai, H. and Jackson, D. (2006). A trehalose metabolic enzyme controls inflorescence architecture in maize. *Nature* 441(7090), 227–230. https://doi.org/10.1038/nature04725. PMID: 16688177.

Scheffran, J., Brzoska, M., Kominek, J., Link, P. M. and Schilling, J. (2012). Climate change and violent conduct. *Science* 336(6083), 869–871. https://doi.org/10.1126/science.1221339.

Schussler, J. R. and Westgate, M. E. (1991a). Maize kernel set at low water potential: II. Sensitivity to reduced assimilates at pollination. *Crop Sci.* 31(5), 1196–1203.

Schussler, J. R. and Westgate, M. E. (1991b). Maize kernel set at low water potential: I. Sensitivity to reduced assimilates during early kernel growth. *Crop Sci.* 31(5), 1189–1195.

Shi, J., Habben, J. E., Archibald, R. L., Drummond, B. J., Chamberlin, M. A., Williams, R. W., Lafitte, H. R. and Weers, B. P. (2015). Overexpression of ARGOS genes modifies plant sensitivity to ethylene, leading to improved drought tolerance in both

Arabidopsis and maize. *Plant Physiol.* 169(1), 266–282. https://doi.org/10.1104/pp.15.00780.

Simmons, C. R., Lafitte, H. R., Reimann, K. S., Brugière, N., Roesler, K., Albertsen, M. C., Greene, T. W. and Habben, J. E. (2021). Successes and insights of an industry biotech program to enhance maize agronomic traits. *Plant Sci.* 307, 110899. https://doi.org/10.1016/j.plantsci.2021.110899.

Staswick, P. E. (1994). Storage proteins of vegetative plant tissues. *Annu. Rev. Plant Physiol. Plant Mol. Biol.* 45(1), 303–322.

Tardieu, F., Simonneau, T. and Muller, B. (2018). The physiological basis of drought tolerance in crop plants: A scenario-dependent probabilistic approach. *Annu. Rev. Plant Biol.* 69, 733–759.

Tuberosa, R. (2012). Phenotyping for drought tolerance of crops in the genomics era. *Front. Physiol.* 3, 347–347.

Tuberosa, R., Salvi, S., Sanguineti, M. C., Landi, P., Maccaferri, M. and Conti, S. (2002). Mapping QTLs regulating morpho-physiological traits and yield: case studies, shortcomings and perspectives in drought-stressed maize. *Ann. Bot.* 89(7) (Spec No), 941–963.

Tuberosa, R., Salvi, S., Giuliani, S., Sanguineti, M. C., Bellotti, M., Conti, S. and Landi, P. (2007). Genome-wide approaches to investigate and improve maize response to drought. *Crop Sci.* 47(S3), S-120.

Varshney, R. K., Barmukh, R., Roorkiwal, M., Qi, Y., Kholova, J., Tuberosa, R., Reynolds, M. P., Tardieu, F. and Siddique, K. H. M. (2021). Breeding custom-designed crops for improved drought adaptation. *Adv. Genet.* 2(3), e202100017. https://doi.org/10.1002/ggn2.202100017.

Varshney, R. K., Tuberosa, R. and Tardieu, F. (2018). Progress in understanding drought tolerance: from alleles to cropping systems. *J. Exp. Bot.* 69(13), 3175–3179.

Wang, H. and Qin, F. (2017). Genome-wide association study reveals natural variations contributing to drought resistance in crops. *Front. Plant Sci.* 8, 1110. https://doi.org/10.3389/fpls.2017.01110.

Wang, X., Wang, H., Liu, S., Ferjani, A., Li, J., Yan, J., Yang, X. and Qin, F. (2016). Genetic variation in ZmVPP1 contributes to drought tolerance in maize seedlings. *Nat. Genet.* 48(10), 1233–1241.

Woodhouse, M. R., Cannon, E. K., Portwood, J. L., Harper, L. C., Gardiner, J. M., Schaeffer, M. L. and Andorf, C. M. (2021). A pan-genomic approach to genomic databases using maize as a model system. *BMC Plant Biol.* 21(1), 385. https://doi.org/10.1186/s12870-021-03173-5.

Wu, X., Feng, H., Wu, D., Yan, S., Zhang, P., Wang, W., Zhang, J., Ye, J., Dai, G., Fan, Y., Li, W., Song, B., Geng, Z., Yang, W., Chen, G., Qin, F., Terzaghi, W., Stitzer, M., Li, L., Xiong, L., Yan, J., Buckler, E., Yang, W. and Dai, M. (2021). Using high-throughput multiple optical phenotyping to decipher the genetic architecture of maize drought tolerance. *Genome Biol.* 22(1), 185. https://doi.org/10.1186/s13059-021-02377-0.

Xiang, Y., Sun, X., Gao, S., Qin, F. and Dai, M. (2017). Deletion of an endoplasmic reticulum stress response element in a ZmPP2C-A gene facilitates drought tolerance of maize seedlings. *Mol. Plant* 10(3), 456–469.

Yang, N. and Yan, J. (2021). New genomic approaches for enhancing maize genetic improvement. *Curr. Opin. Plant Biol.* 60, 101977.

Young, T. E., Gallie, D. R. and DeMason, D. A. (1997). Ethylene-mediated programmed cell death during maize endosperm development of wild-type and shrunken2 genotypes. *Plant Physiol.* 115(2), 737–751.

Yu, J., Holland, J. B., McMullen, M. D. and Buckler, E. S. (2008). Genetic design and statistical power of nested association mapping in maize. *Genetics* 178(1), 539–551.

Yuan, Z., Yan, D., Yang, Z., Yin, J. and Yuan, Y. (2015). Temporal and spatial variability of drought in Huang-Huai-Hai River Basin, China. *Theor. Appl. Climatol.* 122(3–4), 755–769.

Zeitoun, M. and Warner, J. (2006). Hydro-hegemony: a framework for analysis of trans-boundary water conflicts. *Water Policy* 8(5), 435–460. https://doi.org/10.2166/wp.2006.054.

Zhang, J. Z., Creelman, R. A. and Zhu, J. K. (2004). From laboratory to field. Using information from Arabidopsis to engineer salt, cold, and drought tolerance in crops. *Plant Physiol.* 135(2), 615–621.

Zhang, X., Mi, Y., Mao, H., Liu, S., Chen, L. and Qin, F. (2020). Genetic variation in ZmTIP1 contributes to root hair elongation and drought tolerance in maize. *Plant Biotechnol. J.* 18(5), 1271–1283.

Zinselmeier, C., Lauer, M. J. and Boyer, J. S. (1995a). Reversing drought-induced losses in grain yield: sucrose maintains embryo growth in maize. *Crop Sci.* 35(5), 1390–1400.

Zinselmeier, C., Westgate, M. E., Schussler, J. R. and Jones, R. J. (1995b). Low water potential disrupts carbohydrate metabolism in maize (Zea mays L.) ovaries. *Plant Physiol.* 107(2), 385–391.

Zinselmeier, C., Jeong, B. R. and Boyer, J. S. (1999). Starch and the control of kernel potentials. *Plant Physiol.* 121(1), 25–36. Available at: http://www.jstor.org/stable/4278913.

Index